Lecture Notes in Computer Science 11532

Commenced Publication in 1973
Founding and Former Series Editors:
Gerhard Goos, Juris Hartmanis, and Jan van Leeuwen

Editorial Board Members

René van Bevern · Gregory Kucherov (Eds.)

Computer Science – Theory and Applications

14th International Computer Science Symposium in Russia, CSR 2019
Novosibirsk, Russia, July 1–5, 2019
Proceedings

 Springer

Editors
René van Bevern ⓘ
Novosibirsk State University
Novosibirsk, Russia

Gregory Kucherov ⓘ
Laboratoire d'Informatique Gaspard-Monge
Marne-la-Vallée, France

ISSN 0302-9743   ISSN 1611-3349   (electronic)
Lecture Notes in Computer Science
ISBN 978-3-030-19954-8   ISBN 978-3-030-19955-5   (eBook)
https://doi.org/10.1007/978-3-030-19955-5

LNCS Sublibrary: SL1 – Theoretical Computer Science and General Issues

This Springer imprint is published by the registered company Springer Nature Switzerland AG
The registered company address is: Gewerbestrasse 11, 6330 Cham, Switzerland

Preface

The 14th International Computer Science Symposium in Russia (CSR 2019) was held during July 1–5, 2019, in Novosibirsk, Russian Federation. It was organized by Novosibirsk State University, Ershov Institute of Informatics Systems, and Sobolev Institute of Mathematics. This was the 14th edition of the annual series of meetings; previous editions were held in St. Petersburg (2006), Ekaterinburg (2007), Moscow (2008), Novosibirsk (2009), Kazan (2010), St. Petersburg (2011), Nizhny Novgorod (2012), Ekaterinburg (2013), Moscow (2014), Listvyanka (2015), St. Petersburg (2016), Kazan (2017), and Moscow (2018). The symposium covers a broad range of topics in theoretical computer science, ranging from fundamental to application-related.

This year, CSR was organized as a part of the Computer Science Summer in Russia and was held in parallel with the A.P. Ershov Informatics Conference (PSI conference series). The distinguished CSR keynote lecture was given by Andrew Yao. The lecture was delivered at a joint CSR-PSI distinguished lecture session, along with the lecture of Moshe Vardi, a keynote speaker of PSI, attended by CSR participants as well.

The seven other CSR invited plenary speakers were Michael Fellows (University of Bergen, Norway), Eric Fusy (École Polytechnique, France), Giuseppe Italiano (LUISS University, Italy), Meena Mahajan (Institute of Mathematical Sciences, India), Petros Petrosyan (Yerevan State University, Armenia), David Woodruff (Carnegie Mellon University, USA), and Dmitry Zhuk (Moscow State University, Russia). The Program Committee included 24 scientists and was chaired by Gregory Kucherov (CNRS and University of Paris-Est Marne-la-Vallée).

This volume contains the accepted papers as well as abstracts of invited lectures. We received 71 submissions in total. Each paper was reviewed by at least three Program Committee (PC) members. As a result, the PC selected 31 papers for presentation at the symposium and publication in these proceedings. The reviewing process was run using the EasyChair conference system.

The PC also selected two papers to receive the Best Paper Award and the Best Student Paper Award. These awards were sponsored by Yandex and Springer. The winners were:

- Best Paper Award: Ludmila Glinskih and Dmitry Itsykson, "On Tseitin Formulas, Read-Once Branching Programs and Treewidth"
- Best Student Paper Award: Jan-Hendrik Lorenz, "On the Complexity of Restarting"

We would like to thank many people and organizations that contributed to the organization of CSR 2019. In particular, our thanks go to:

- All invited speakers for accepting to give a talk at the conference
- The PC members, who graciously gave their time and energy
- All reviewers and additional reviewers for their expertise
- The members of the local Organizing Committee: Denis Ponomaryov (Ershov Institute of Informatics Systems), Oxana Tsidulko (Sobolev Institute

of Mathematics, Novosibirsk, Russia), Pavel Emelyanov (Novosibirsk State University), and Anastasia Karpenko (Novosibirsk State University).
- Our sponsors: Springer, the European Association for Theoretical Computer Science, Yandex, and UMA.TECH.

April 2019

René van Bevern
Gregory Kucherov

Organization

Program Committee

Maxim Babenko	Yandex and HSE Moscow, Russia
Petra Berenbrink	University of Hamburg, Germany
René van Bevern	Novosibirsk State University, Russia
Olaf Beyersdorff	Friedrich Schiller University Jena, Germany
Manuel Bodirsky	TU Dresden, Germany
Vladimir Braverman	Johns Hopkins University, USA
Holger Dell	Saarland University and Cluster of Excellence, MMCI, Germany
Michael Elkin	Ben-Gurion University of the Negev, Israel
Pierre Fraigniaud	CNRS and University of Paris Diderot, France
Anna Frid	Aix Marseille Université, France
Pawel Gawrychowski	University of Wroclaw, Poland
Dora Giammarresi	University of Rome Tor Vergata, Italy
Elena Grigorescu	Purdue University, USA
Gregory Kucherov	CNRS - LIGM, France
Christophe Paul	CNRS - LIRMM, France
Valentin Polishchuk	Linköping University, Sweden
Artem Pyatkin	Novosibirsk State University, Sobolev Institute of Mathematics, Russia
Alexander Rabinovich	Tel Aviv University, Israel
Kunihiko Sadakane	The University of Tokyo, Japan
Arseny Shur	Ural Federal University, Russia
Jacobo Torán	Universität Ulm, Germany
Igor Walukiewicz	CNRS - LaBRI, France
Sergey Yekhanin	Microsoft, USA

Additional Reviewers

Akhmedov, Maxim	Bshouty, Nader	Fernau, Henning
Arseneva, Elena	Cai, Leizhen	Fernández, Maribel
Baburin, Alexey	Carl, Merlin	Filtser, Arnold
Bampis, Evripidis	Chandrasekaran, Karthik	Ganian, Robert
Berthe, Valerie	Chlamtac, Eden	Gańczorz, Michał
Biermeier, Felix	Cohen, Liron	Gharibian, Sevag
Bodor, Bertalan	Cseh, Ágnes	Hirahara, Shuichi
Bodwin, Greg	Das, Bireswar	Hosseinpour, Hamed
Bonichon, Nicolas	Dershowitz, Nachum	Huang, Chien-Chung
Brunet, Paul	Dudek, Bartlomiej	Husfeldt, Thore

Abstract of Invited Talks

Two Heresies

Michael R. Fellows

Department of Informatics, University of Bergen, Bergen, Norway
michael.fellows@uib.no

The advent of parameterization has allowed two important heresies concerning how we model and analyze computational complexity to be articulated. The first concerns the handling of real numbers. The second concerns decision problems and how these interface with scientific methodology. The talk will survey the basics of parameterized complexity and how these heresies naturally follow, and some recent results that illustrate them.

Schnyder Woods and the Combinatorics of Triangulations in Genus 0 and 1

Eric Fusy

École Polytechnique, France
fusy@lix.polytechnique.fr

Schnyder woods are combinatorial structures on planar triangulations (maximal planar graphs embedded on the sphere) that can be formulated as certain partitions of the edges into three spanning trees. These structures have originally been introduced by Schnyder in 1989 to give a new planarity criterion for graphs in terms of the order dimension of their incidence posets. Since then, they have found many algorithmic applications, for instance in graph drawing, succinct encoding of meshes, efficient routing in planar networks, geometric spanners, random generation of planar graphs, etc.

I will present some of these applications, with an emphasis on bijective enumeration, giving in particular combinatorial proofs of the beautiful formulas $s_n = \frac{6(2n)!(2n+2)!}{n!(n+1)!(n+2)!(n+3)!}$ for the total number of Schnyder woods on triangulations with $n+3$ vertices and $t_n = \frac{2(4n+1)!}{(n+1)!(3n+2)!}$ for the number of planar triangulations with $n+3$ vertices. I will also explain how the bijective approach can be extended to count triangulations on the torus.

Based on joint work with Dominique Poulalhon, Gilles Schaeffer, Olivier Bernardi, and Benjamin Lévêque.

2-Connectivity in Directed Graphs

Giuseppe F. Italiano

LUISS University, Rome, Italy

We survey some recent theoretical and experimental results on 2-edge and 2-vertex connectivity in directed graphs. Despite being complete analogs of the corresponding notions on undirected graphs, in digraphs 2-connectivity has a much richer and more complicated structure. For undirected graphs it has been known for over 40 years how to compute all bridges, articulation points, 2-edge- and 2-vertex-connected components in linear time, by simply using depth first search. In the case of digraphs, however, the very same problems have been much more challenging and have been tackled only recently.

QBF Proof Complexity: An Overview

Meena Mahajan

The Institute of Mathematical Sciences, HBNI, Chennai 600113, India
meena@imsc.res.in

How do we prove that a false QBF is indeed false? A proof of which size is needed? The special case when all quantifiers are existential is the well-studied setting of propositional proof complexity. Expectedly, universal quantifiers change the game significantly. Several proof systems have been designed in the last couple of decades to handle QBFs. Lower bound paradigms from propositional proof complexity cannot always be extended—in most cases feasible interpolation and consequent transfer of circuit lower bounds works, but obtaining lower bounds on size by providing lower bounds on width fails dramatically in one of the simplest QBF proof systems. A central lower bound paradigm for QBFs (with no analogue in the propositional world) involves viewing the QBF as a 2-player game and considering the winning strategies of the player who can win. Lower bounds follow from the hardness of computing these strategies in restricted models. Thus circuit lower bounds and QBF proof lower bounds are intimately intertwined. Lower bounds also follow from a semantic cost measure on winning strategies.

This talk will provide a broad overview of some of these developments. A brief introduction to the topic can be found in [1].

Keywords: Proof complexity · Quantified Boolean formulas · Resolution · Lower bounds

Reference

1. Mahajan, M.: Lower bound techniques for QBF proof systems. In: STACS, LIPIcs, vol. 96, pp. 2:1–2:8. Schloss Dagstuhl - Leibniz-Zentrum fuer Informatik (2018)

Interval Edge-Colorings of Graphs: Variations and Generalizations

Petros A. Petrosyan[1,2]

[1]Department of Informatics and Applied Mathematics, Yerevan State University, 0025, Armenia
petros_petrosyan@ysu.am
[2]Department of Applied Mathematics and Informatics,
Russian-Armenian University, 0051, Armenia
http://ysu.am/persons/en/Petros-Petrosyan

Abstract. An edge-coloring of a graph G with colors $1,\ldots,t$ is called an *interval t-coloring* if all colors are used and the colors of edges incident to each vertex of G are distinct and form an interval of integers. The concept of interval edge-coloring of graphs was introduced by Asratian and Kamalian more than 30 years ago and was motivated by the problems in scheduling theory. In the last 10 years different types of variations and generalizations of interval edge-colorings were studied. In this talk we will give a survey of the topic and present a recent progress in the study of interval edge-colorings and their various variations and generalizations.

Introduction

We use [33] for terminology and notation not defined here. We consider graphs that are finite, undirected, and have no loops or multiple edges and multigraphs that may contain multiple edges but no loops. Let $V(G)$ and $E(G)$ denote the sets of vertices and edges of a multigraph G, respectively. The degree of a vertex $v \in V(G)$ is denoted by $d_G(v)$ and the maximum degree of G by $\Delta(G)$.

A proper edge-coloring of a multigraph G is a mapping $\alpha : E(G) \to \mathbf{N}$ such that $\alpha(e) \neq \alpha(e')$ for every pair of adjacent edges e and e' in G. If α is a proper edge-coloring of G and $v \in V(G)$, then the *spectrum of a vertex v*, denoted by $S(v, \alpha)$, is the set of colors appearing on edges incident to v. If α is a proper edge-coloring of a multigraph G and $v \in V(G)$, then the smallest and largest colors of the spectrum $S(v, \alpha)$ are denoted by $\underline{S}(v, \alpha)$ and $\overline{S}(v, \alpha)$, respectively.

An *interval t-coloring* of a multigraph G is a proper edge-coloring α of G with colors $1, \ldots, t$ such that all colors are used and for each $v \in V(G)$, the set $S(v, \alpha)$ is an interval of integers. The notion of interval colorings was introduced by Asratian and Kamalian [5] in 1987 and was motivated by the problem of finding compact school timetables, that is, timetables such that the lectures of each teacher and each class are scheduled at consecutive periods. This problem corresponds to the problem of finding an interval edge-coloring of a bipartite multigraph. There are graphs that do not have interval colorings (e.g. odd cycles, complete graphs of an odd order). For bipartite graphs, it is known that all subcubic graphs have interval colorings [15]. However, for every positive integer $\Delta \geq 11$, there exists a bipartite graph with maximum degree Δ that has no interval coloring [27]. Generally, it is an *NP*-complete problem to determine whether a bipartite graph has an interval coloring [31]. There are many papers devoted to this topic, in particular, surveys on the topic can be found in some books (see, for example, [4, 17, 22]).

In this talk we will discuss a recent progress in the study of interval edge-colorings and their various variations and generalizations.

Variations and Generalizations on Interval Edge-Colorings

One of the partial cases of an interval edge-coloring is a *sequential edge-coloring* which was considered in the early 1980s by Asratian [1] and Caro, Chonheim [9]. A *sequential t-coloring* of a multigraph G is a proper edge-coloring α of G with colors $1, \ldots, t$ such that all colors are used and for each $v \in V(G)$, $S(v, \alpha) = \{1, 2, \ldots, d_G(v)\}$. Sequential edge-colorings correspond to the problems of constructing a compact school timetable, when all groups and (or) teachers begin at the same time. This type of edge-coloring is also related to sum edge-colorings of graphs.

Let $G = (V, E)$ be a multigraph and $R \subseteq V$. An *interval (R, t)-coloring* of a multigraph G is a proper edge-coloring α of G with colors $1, \ldots, t$ such that all colors are used and for each $v \in R$, the set $S(v, \alpha)$ is an interval of integers. This type of interval colorings was the main subject of study in the doctoral thesis of Kamalian [18]. In the case of bipartite graphs with bipartition (X, Y), interval (R, t)-colorings with $R = X$ or $R = Y$ are also called *one-sided interval t-colorings*. Some new results on one-sided interval colorings of bipartite graphs were published in the last 5 years by Kamalian [20], Casselgren and Toft [11]. This type of coloring is of particular interest for graphs that do not have interval colorings.

In [26] it was given a generalization of interval edge-colorings of graphs, where the authors suggested the concept of *interval (t, k)-colorings* of graphs and studied the problem of the existence such colorings for some classes of graphs. Let k be a non-negative integer. An *interval (t, k)-coloring* of a graph G is a proper edge-coloring α of G with colors $1, \ldots, t$ such that all colors are used and the colors of edges incident to each vertex v satisfy the following condition:

$$d_G(v) - 1 \leq \overline{S}(v, \alpha) - \underline{S}(v, \alpha) \leq d_G(v) + k - 1.$$

The case $k = 1$ was also considered by some authors under the name "near-interval colorings". In particular, Casselgren and Toft [10] proved that some classes of bipartite graphs admit near-interval colorings. On the other hand, in [26] it was shown that there are bipartite graphs having no near-interval colorings. Recently, Petrosyan [25] proved that all graphs with maximum degree at most four admit near-interval colorings. On the other hand, he constructed bipartite graphs with maximum degree at least 18 that have no near-interval colorings.

Another important generalizations of interval edge-colorings are *cyclic interval colorings* which were introduced by de Werra and Solot [32] in 1991. A proper edge-coloring of a graph G with colors $1, 2, \ldots, t$ is called a *cyclic interval t-coloring* if for each vertex v of G the edges incident to v are colored by consecutive colors, under the condition that color 1 is considered as consecutive to color t. Cyclic interval colorings correspond to the problems of constructing a production schedule when the production cycle is repeated, tasks that are being processed at the beginning and at the end of the production cycle can be considered as non-preemptive. Generally, it is an *NP*-complete problem to determine whether a bipartite graph has a cyclic interval coloring [23]. There are many papers devoted to this topic [2, 10, 12, 19, 24, 29].

An interesting variant of interval colorings is *improper interval colorings* which was suggested by Hudák, Kardoš, Madaras and Vrbjarová [16]. An edge-coloring of a graph G with colors $1, \ldots, t$ is called an *improper interval t-coloring* if for each vertex v of G the edges incident to v are colored by consecutive colors. Since each graph has such a coloring, so it is natural to study the maximum number of colors used in improper interval colorings of a graph. In [16], they derived some bounds on the parameter and determined the exact value of this parameter for some classes of graphs.

Finally, we would like to mention that by some authors were introduced and studied measures of closeness for a graph to be interval colorable. A first attempt to introduce such a measure was done by Giaro, Kubale and Małafiejski [13] in 1999. The *deficiency* of a graph is the minimum number of pendant edges whose attachment to the graph makes the resulting graph interval colorable. The authors obtained many interesting results. In particular, they showed that there are graphs whose deficiency approaches the number of vertices [14]. The last years many papers were published on this topic [6–8, 21, 28]. In [30], Petrosyan and Sargsyan suggested another measure of closeness for a graph to be interval colorable. The *resistance* of a graph is the minimum number of edges that should be removed from a given graph to obtain an interval colorable graph. The connection between the deficiency and the resistance of a graph should be useful for study interval colorings of graphs. Recently, Asratian, Casselgren and Petrosyan [3] introduced a new measure of closeness for a graph to be cyclically interval colorable. The *cyclic deficiency* is an analogue of deficiency for cyclic interval colorings. In particular, they proved that the difference between the deficiency and cyclic deficiency can be arbitrarily large.

References

1. Asratian, A.S.: Investigation of some mathematical model of scheduling theory. Doctoral Thesis, Moscow (1980)
2. Asratian, A.S., Casselgren, C.J., Petrosyan, P.A.: Some results on cyclic interval edge colorings of graphs. J. Graph Theory **87**, 239–252 (2018)
3. Asratian, A.S., Casselgren, C.J., Petrosyan, P.A.: Cyclic deficiency of graphs. Discrete Appl. Math. (2019, to appear)
4. Asratian, A.S., Denley, T.M.J., Haggkvist, R.: Bipartite Graphs and their Applications. Cambridge University Press, Cambridge (1998)
5. Asratian, A.S., Kamalian, R.R.: Interval colorings of edges of a multigraph. Appl. Math. **5**, 25–34 (1987). (in Russian)

6. Borowiecka-Olszewska, M., Drgas-Burchardt, E.: The deficiency of all generalized Hertz graphs and minimal consecutively non-colourable graphs in this class. Discrete Math. **339**, 1892–1908 (2016)
7. Borowiecka-Olszewska, M., Drgas-Burchardt, E., Hałuszczak, M.: On the structure and deficiency of k-trees with bounded degree. Discrete Appl. Math. **201**, 24–37 (2016)
8. Bouchard, M., Hertz, A., Desaulniers, G.: Lower bounds and a tabu search algorithm for the minimum deficiency problem. J. Comb. Optim. **17**, 168–191 (2009)
9. Caro, Y., Chonheim, J.: Generalized 1-factorization of trees. Discrete Math. **33**, 319–321 (1981)
10. Casselgren, C.J., Toft, B.: On interval edge colorings of biregular bipartite graphs with small vertex degrees. J. Graph Theory **80**, 83–97 (2015)
11. Casselgren, C.J., Toft, B.: One-sided interval edge-colorings of bipartite graphs. Discrete Math. **339**, 2628–2639 (2016)
12. Casselgren, C.J., Petrosyan, P.A., Toft, B.: On interval and cyclic interval edge colorings of (3,5)-biregular graphs. Discrete Math. **340**, 2678–2687 (2017)
13. Giaro, K., Kubale, M., Małafiejski, M.: On the deficiency of bipartite graphs. Discrete Appl. Math. **94**, 193–203 (1999)
14. Giaro, K., Kubale, M., Małafiejski, M.: Consecutive colorings of the edges of general graphs. Discrete Math. **236**, 131–143 (2001)
15. Hansen, H.M.: Scheduling with minimum waiting periods. MSc Thesis, Odense University, Odense, Denmark (1992). (in Danish)
16. Hudák, P., Kardoš, F., Madaras, T., Vrbjarová, M.: On improper interval edge colourings. Czechoslovak Math. J. **66**, 1119–1128 (2016)
17. Jensen, T.R., Toft, B.: Graph Coloring Problems. Wiley Interscience (1995)
18. Kamalian, R.R.: Interval edge colorings of graphs. Doctoral Thesis, Novosibirsk (1990)
19. Kamalian, R.R.: On cyclically-interval edge colorings of trees. Bul. Acad. Ştiinç Repub. Mold. Mat. **1**(68), 50–58 (2012)
20. Kamalian, R.R.: On one-sided interval edge colorings of biregular bipartite graphs. Algebra and Discrete Math. **192**, 193–199 (2015)
21. Khachatrian, H.H.: Deficiency of outerplanar graphs. Phys. Math. Sci. **51**(1), 3–9 (2017). Proceedings of the Yerevan State University
22. Kubale, M.: Graph Colorings. American Mathematical Society (2004)
23. Kubale, M., Nadolski, A.: Chromatic scheduling in a cyclic open shop. Eur. J. Oper. Res. **164**, 585–591 (2005)
24. Nadolski, A.: Compact cyclic edge-colorings of graphs. Discrete Math. **308**, 2407–2417 (2008)
25. Petrosyan, P.A.: Near-interval edge-colorings of graphs. In: Alekseev, V.B., Romanov, D.S., Danilov, B.R. (eds.) Proceedings of the X International Conference "Discrete Models in Theory of Control Systems", pp. 22–24. Moscow State University, Moscow (2018)
26. Petrosyan, P.A., Arakelyan, H.Z., Baghdasaryan, V.M.: A generalization of interval edge-colorings of graphs. Discrete Appl. Math. **158**, 1827–1837 (2010)
27. Petrosyan, P.A., Khachatrian, H.H.: Interval non-edge-colorable bipartite graphs and multigraphs. J. Graph Theory **76**, 200–216 (2014)
28. Petrosyan, P.A., Khachatrian, H.H.: Further results on the deficiency of graphs. Discrete Appl. Math. **226**, 117–126 (2017)
29. Petrosyan, P.A., Mkhitaryan, S.T.: Interval cyclic edge-colorings of graphs. Discrete Math. **339**, 1848–1860 (2016)
30. Petrosyan, P.A., Sargsyan, H.E.: On resistance of graphs. Discrete Appl. Math. **159**, 1889–1900 (2011)

31. Sevast'janov, S.V.: Interval colorability of the edges of a bipartite graph. Metody Diskret. Analiza **50**, 61–72 (1990). (in Russian)
32. de Werra, D., Solot, Ph.: Compact cylindrical chromatic scheduling. SIAM J. Discrete Math. **4**(4), 528–534 (1991)
33. West, D.B.: Introduction to Graph Theory. Prentice-Hall, New Jersey (2001)

Sketching as a Tool for Numerical Linear Algebra

David P. Woodruff

Computer Science Department, Carnegie Mellon University, Pittsburgh, USA
dwoodruf@cs.cmu.edu

In this tutorial, I'll give an overview of near optimal algorithms for regression, low rank approximation, and a variety of other problems. The results are based on the sketch and solve paradigm, which is a tool for quickly compressing a problem to a smaller version of itself, for which one can then run a slow algorithm on the smaller problem. These lead to the fastest known algorithms for fundamental machine learning and numerical linear algebra problems, which run in time proportional to the number of non-zero entries of the input. Time-permitting I'll discuss extensions to tensors, NP-hard variants of low rank approximation, and robust variants of the above problems.

Fintech and Its Scientific Drivers

Andrew Chi-Chih Yao

Institute for Interdisciplinary Information Sciences,
Tsinghua University, Beijing, China
andrewcyao@tsinghua.edu.cn

Abstract. FinTech can be seen as the meeting of minds between economics and computer science in the digital age. Among its major intellectual foundations are reliable distributed computing and cryptography from the side of computer science, and efficient mechanism design for financial activities from the side of economics. In this talk we discuss some recent work in auction and blockchain from this perspective. For example, is it true that more revenue can always be extracted from an auction where the bidders are more willing to pay than otherwise? Can more revenue be extracted when the bidders are more risk-tolerant than otherwise? We also present some new results on blockchain fees. These results help shed light on some structural questions in economics whose answers are non-obvious.

Keywords: Fintech · Distributed computing · Blockchain fees

A Proof of CSP Dichotomy Conjecture[1]

Dmitriy Zhuk

Lomonosov Moscow State University
zhuk@intsys.msu.ru
http://intsys.msu.ru/staff/zhuk/

Abstract. We prove the CSP Dichotomy conjecture, and therefore characterize the complexity of the Constraint Satisfaction Problem for all constraint languages on a finite set.

Keywords: Constraint satisfaction problem · Computational complexity · CSP Dichotomy conjecture

Many combinatorial problems (graph coloring, solving systems of equations, and so on) can be expressed as constraint satisfaction problems, where the *Constraint Satisfaction Problem (CSP)* is a problem of deciding whether there is an assignment to a set of variables subject to some specified constraints. This class of problems is known to be NP-complete in general, but certain restrictions on the constraint language (set of allowed constraints) can ensure tractability.

In 1998 Feder and Vardi [2] conjectured that for any constraint language CSP is either solvable in polynomial time, or NP-complete. Later this conjecture (known as CSP Dichotomy conjecture) was formulated in the following form: CSP over a constraint language Γ can be solved in polynomial time if Γ admits a weak near-unanimity polymorphism, and it is NP-complete otherwise. The hardness result has been known since 2001, but the other part remained open until 2017 when two different polynomial algorithms for the weak near-unanimity case were suggested [1, 3].

We present one of the two algorithms, which proves the CSP Dichotomy Conjecture.

References

1. Bulatov, A.A.: A dichotomy theorem for nonuniform CSPs. In: 2017 IEEE 58th Annual Symposium on Foundations of Computer Science (FOCS), pp. 319–330. IEEE (2017)
2. Feder, T., Vardi, M.Y.: The computational structure of monotone monadic SNP and constraint satisfaction: a study through datalog and group theory. SIAM J. Comput. **28**(1), 57–104 (1998)
3. Zhuk, D.: A proof of CSP dichotomy conjecture. In: 2017 IEEE 58th Annual Symposium on Foundations of Computer Science (FOCS), pp. 331–342. IEEE (2017)

---

[1] Supported by RFBR, grant 19-01-00200.

Contents

Approximability and Inapproximability for Maximum k-Edge-Colored
Clustering Problem   1
    Yousef M. Alhamdan and Alexander Kononov

The Non-hardness of Approximating Circuit Size   13
    Eric Allender, Rahul Ilango, and Neekon Vafa

Reconstructing a Convex Polygon from Its ω-cloud   25
    Elena Arseneva, Prosenjit Bose, Jean-Lou De Carufel,
    and Sander Verdonschot

A Space-Efficient Parameterized Algorithm for the Hamiltonian Cycle
Problem by Dynamic Algebraization   38
    Mahdi Belbasi and Martin Fürer

Quantum Algorithm for Distribution-Free Junta Testing   50
    Aleksandrs Belovs

On Induced Online Ramsey Number of Paths, Cycles,
and Trees   60
    Václav Blažej, Pavel Dvořák, and Tomáš Valla

Approximations of Schatten Norms via Taylor Expansions   70
    Vladimir Braverman

Nearly Linear Time Isomorphism Algorithms for Some Nonabelian
Group Classes   80
    Bireswar Das and Shivdutt Sharma

Belga B-Trees   93
    Erik D. Demaine, John Iacono, Grigorios Koumoutsos,
    and Stefan Langerman

Eventually Dendric Shifts   106
    Francesco Dolce and Dominique Perrin

On Decidability of Regular Languages Theories   119
    Sergey Dudakov and Boris Karlov

Minimizing Branching Vertices in Distance-Preserving Subgraphs   131
    Kshitij Gajjar and Jaikumar Radhakrishnan

On Tseitin Formulas, Read-Once Branching Programs and Treewidth   143
  Ludmila Glinskih and Dmitry Itsykson

Matched Instances of Quantum Satisfiability (QSat) – Product State
Solutions of Restrictions   156
  Andreas Goerdt

Notes on Resolution over Linear Equations   168
  Svyatoslav Gryaznov

Undecidable Word Problem in Subshift Automorphism Groups   180
  Pierre Guillon, Emmanuel Jeandel, Jarkko Kari, and Pascal Vanier

Parameterized Complexity of Conflict-Free Set Cover   191
  Ashwin Jacob, Diptapriyo Majumdar, and Venkatesh Raman

Forward Looking Huffman Coding   203
  Shmuel Tomi Klein, Shoham Saadia, and Dana Shapira

Computational Complexity of Real Powering and Improved Solving
Linear Differential Equations   215
  Ivan Koswara, Svetlana Selivanova, and Martin Ziegler

On the Quantum and Classical Complexity of Solving Subtraction Games . . .   228
  Dmitry Kravchenko, Kamil Khadiev, and Danil Serov

Derandomization for Sliding Window Algorithms with Strict Correctness   237
  Moses Ganardi, Danny Hucke, and Markus Lohrey

On the Complexity of Restarting   250
  Jan-Hendrik Lorenz

On the Complexity of MIXED DOMINATING SET   262
  *Jayakrishnan Madathil, Fahad Panolan, Abhishek Sahu,
  and Saket Saurabh*

Uniform CSP Parameterized by Solution Size is in W[1]   275
  Ruhollah Majdoddin

On the Parameterized Complexity of Edge-Linked Paths   286
  Neeldhara Misra, Fahad Panolan, and Saket Saurabh

The Parameterized Complexity of Dominating Set and Friends Revisited
for Structured Graphs   299
  Neeldhara Misra and Piyush Rathi

Transition Property for Cube-Free Words   311
  Elena A. Petrova and Arseny M. Shur

A Polynomial Time Delta-Decomposition Algorithm for Positive DNFs 325
   Denis Ponomaryov

Unpopularity Factor in the Marriage and Roommates Problems 337
   Suthee Ruangwises and Toshiya Itoh

AND Protocols Using only Uniform Shuffles 349
   Suthee Ruangwises and Toshiya Itoh

Sybil-Resilient Conductance-Based Community Growth 359
   Ouri Poupko, Gal Shahaf, Ehud Shapiro, and Nimrod Talmon

Author Index 373

Approximability and Inapproximability for Maximum k-Edge-Colored Clustering Problem

Yousef M. Alhamdan[1]($\boxtimes$)(iD) and Alexander Kononov[1,2]($\boxtimes$)(iD)

[1] Mathematics and Mechanics Department, Novosibirsk State University,
Novosibirsk, Russia
yousify32@gmail.com
[2] Laboratory "Mathematical Models of Decision Making",
Sobolev Institute of Mathematics, Novosibirsk, Russia
alvenko@math.nsc.ru

Abstract. We consider the Max k-Edge-Colored Clustering problem (abbreviated as MAX-k-EC). We are given an edge-colored graph with k colors. Each edge of the graph has a positive weight. We seek to color the vertices of the graph so as to maximize the total weight of the edges which have the same color at their extremities. The problem was introduced by Angel et al. [2]. We give a polynomial-time algorithm for MAX-k-EC with an approximation factor $\frac{49}{144} \approx 0.34$, which significantly improves the best previously known factor $\frac{7}{23} \approx 0.304$, obtained by Ageev and Kononov [1]. We also present an upper bound of $\frac{241}{248} \approx 0.972$ on the inapproximability of MAX-k-EC. This is the first inapproximability result for this problem.

Keywords: Clustering · Edge-colored graph · Randomized rounding

1 Introduction

In this paper, we study the Max k-Edge-Colored Clustering problem introduced by Angel et al. [2]. Given an undirected graph $G = (V, E)$ with colors of edges $c : E \rightarrow \{1, \ldots, k\}$ and weights $w : E \rightarrow Q^+$. Our goal is to color vertices of G so as to maximize the total weight of edges which have the same color at their extremities. Denote the set of colors by $\mathcal{C}$, i.e., $\mathcal{C} = \{1, \ldots, k\}$. Given a vertex coloring, we say that an edge of G is *stable* if both its extremities have the same color as the color of the edge[1]. Thus, we are interested to assign a color from $\mathcal{C}$ to each vertex of G so as to maximize the total weight of stable edges.

---

[1] Note that, "stable" is defined as "matched" in Angel et al. [2] and Ageev and Kononov [1]. But, we follow Cai and Leung [5] definition since it makes sense.

This work was appeared as part of the first author's MSc thesis.

© Springer Nature Switzerland AG 2019
R. van Bevern and G. Kucherov (Eds.): CSR 2019, LNCS 11532, pp. 1–12, 2019.
https://doi.org/10.1007/978-3-030-19955-5_1

As observed by Cai and Leung [5], the MAX-k-EC problem can be considered as the optimization counterpart of the Vertex-Monochromatic Subgraph problem or the Alternating Path Removal problem. In the Vertex-Monochromatic Subgraph problem it is required to find the largest subgraph where every vertex has one color for its incident edges or, what is the same, remove the smallest number of edges to destroy all the alternating paths, as required in the Alternating Path Removal problem. Although a huge number of papers are devoted to monochromatic subgraphs and alternating paths [4,9], Angel et al. [2] were the first to consider the optimization problem concerning these concepts.

The MAX-k-EC problem is also a natural generalization of Maximum Weight Matching Problem. Indeed, if each edge has its own color the problem coincides with the edge packing problem which is equivalent to finding a maximum weight matching. Our problem can also be considered as an extension of the centralized version of the information-sharing model introduced by Kleinberg and Ligett [10]. In their model, the edges are not colored. Two adjacent vertices share information only if they are colored with the same color but some pairs of vertices are forbidden to color in one color. As Kleinberg and Ligett mention, it is interesting to expand their model by considering different categories of information. In the MAX-k-EC problem every edge-color corresponds to a different information category and two adjacent vertices share information if their color is the same as the color of the edge that connects them.

On the other hand, MAX-k-EC is a special case of the combinatorial allocation problem [6]. We associate each color with a player and each vertex with an item, where items have to be allocated to competing players by a central authority, with the goal of maximizing the total utility provided to the players. Every player has utility functions derived from the different subsets of items. In the most general setting, utility functions are arbitrary functions satisfying the monotonicity property. Feige and Vondrak [6] consider subadditive, fractional subadditive and submodular functions and present constant factor approximation algorithms. It is easy to check that in our problem the function is supermodular. For the general supermodular function no polynomial time algorithm can find an allocation within a factor $n^{(1-\varepsilon)}$ from optimum for any $\varepsilon > 0$ unless NP = ZPP.

Angel et al. [2,3] presented a polynomial-time algorithm for the MAX-k-EC problem on edge-bicoloured graphs by a reduction to the maximum independent set problem on bipartite graphs. Moreover, they showed the strongly NP-hardness of the problem for edge-tricoloured bipartite graphs. Recently, Cai and Leung [5] expanded the last result and proved that MAX-k-EC is NP-hard in the strong sense even on edge-tricoloured planar bipartite graphs of maximum degree four. Angel et al. [2,3] obtained an LP-based approximation algorithm for the MAX-k-EC problem and showed that it finds a solution such that the weight of stable edges is at least $\frac{1}{e^2}$ from the optimum. Later, Ageev and Kononov [1] gave a better analysis of the same algorithm of [2,3] and improved the ratio from $\frac{1}{e^2} = 0.135$ to 0.25. They also presented a $\frac{7}{23}$-approximation algorithm based on the same LP introduced in [2,3]. Cai and Leung [5] derived two FPT algorithms for the MAX-k-EC problem when they consider the numbers of stable edges and unstable edges, respectively, as a fixed parameter.

1.1 Our Contributions

In Sect. 2 we present a new algorithm for the MAX-k-EC problem with approximation ratio $\frac{49}{144} \approx 0.34$. As in [2,3], we formulate the problem as an integer linear program and develop a randomized rounding scheme for the linear programming relaxation. We design a two-phase scheme, where the first phase selects a set of desired edges randomly and independently for each color, while the second phase colors vertices, taking into account the selection of the edges made in the first phase.

In Sect. 3 we provide an inapproximability result showing that there is no ρ-approximation algorithm for the MAX-k-EC problem for constant $\rho > \frac{241}{248}$ unless $P = NP$. Our proof uses an L-reduction from the well-known MAX-E3-SAT problem.

2 Two-Phase Randomized Approximation Algorithm

Consider two sets of variables z_e, $e \in E$ and x_{vi}, $v \in V$, $i \in C$, where $z_e = 1$ if both endpoints of e are colored with the same color as e and $z_e = 0$ otherwise and $x_{vi} = 1$ if v is colored with color i and $x_{vi} = 0$ otherwise. Angel et al. [2,3] introduced the following integer linear program (ILP) for MAX-k-EC.

$$\text{maximize } \sum_{e \in E} w_e z_e \tag{1}$$

$$\text{subject to } \sum_{i \in C} x_{vi} = 1, \qquad\qquad \forall v \in V \tag{2}$$

$$z_e \leq \min\{x_{vc(e)}, x_{uc(e)}\} \qquad \forall e = [v, u] \in E \tag{3}$$

$$x_{vi}, z_e \in \{0, 1\}, \qquad\qquad \forall v \in V, i \in C, e \in E \tag{4}$$

The first set of constraints ensures that each vertex is colored in exactly one color, and the second ensures that an edge e is stable if its color is the same as the color of both its extremities.

The LP-relaxation (LP) of (1)–(4) is obtained by replacing the constraints $x_{vi} \in \{0, 1\}$ and $z_e \in \{0, 1\}$ by $x_{vi} \geq 0$ and $z_e \geq 0$, respectively. The following two-phase algorithm was considered in [1,2]. In the first phase, it starts with solving LP and then works in k iterations, by considering each color i, $1 \leq i \leq k$, independently from the others. For each color, the algorithm picks r at random in $(0, 1)$ and chooses all edges of this color with $z_e^* \geq r$. When an edge is chosen, this means that both its endpoints get the color of this edge. Since in general a vertex can be adjacent to differently colored edges, it may get more than one colors. In the second phase, the algorithm chooses randomly one of these colors. Denote by $\lambda_v(l, c)$ the probability with which the algorithm chooses the color c

if l colors were assigned to v at the first phase of the algorithm. We present the algorithm below.

Algorithm 1. Algorithm 2-PHASE

1: **Phase I**:
2: Solve LP and let z_e^* be the values of variables z_e.
3: **for** each color $c \in C$ **do**
4:     Let r be a random value in $[\mathbf{0,1}]$.
5:     Choose the c-colored edges e with $z_e^* \geq r$ and give color c to both of e's endpoints.
6: **end for**
7: **Phase II**:
8: **for** each vertex $v \in V$ **do**
9:     Let vertex v got l colors.
10:     assign randomly one of l colors to v, each with the probability $\lambda_v(l, c)$.
11: **end for**

Remind that in the second phase the algorithm from [1] picks each color with equal probability, i.e. $\lambda_v(l, c) = \frac{1}{l}$ for all colors assigned to v at the first phase of the algorithm. In our algorithm we change this property. Let (x^*, z^*) be an optimal solution of the LP. Following [1] we say that an edge e is *big* if $z_e^* > \frac{1}{2}$; otherwise an edge e is *small*. We say that a vertex v is *heavy* if it is extrimity of at least one big edge; otherwise we say that vertex v is *light*. We say that a color i is a *heavy* color for a vertex v if v is incident to an i-colored big edge, otherwise a color i is a *light* color for a vertex v. We note that each vertex has at most one heavy color. If the vertex v got two colors: a heavy color i and a light color j then we set $\lambda_v(2, i) = \frac{1}{3}$ and $\lambda_v(2, j) = \frac{2}{3}$ else we set $\lambda_v(l, c) = \frac{1}{l}$ for all l colors assigned to v at the first phase of the algorithm. We call this algorithm *2-Phase with Heavy and Light Vertices* (2-PHLV). This simple idea leads to an improved approximation guarantee for MAX-k-EC, even though it contradicts the intuition; since in **Phase II** we increase the probability of picking light colors. Although at first glance we made minor changes to the algorithm from [1], we will have to refine the analysis of new algorithm because the previous one strongly relied on the specifics of the old algorithm.

2.1   Analysis

In this subsection, we give a worst-case analysis of algorithm 2-PHLV. Let X_{vi} denote the event where vertex v gets color i after **Phase I** of the algorithm. Since the first phase of the algorithm 2-PHLV coincides with the first phases of the algorithms RR and RR2, presented in [2] and [1], respectively, the following simple statements are valid.

Lemma 1. *[2] For any edge $e \in E$, the probability that e is chosen in* **Phase** *I is z_e^*.*

Lemma 2. *[2] For every vertex $v \in V$ and for all $i \in C$ we have:*

$$Pr[X_{vi}] = \max\{z_e^* : e = [v, u] \in E \ \& \ c(e) = i\}.$$

Lemma 3. *[2] For every vertex $v \in V$, $\sum_{i \in C} Pr[X_{vi}] \leq 1$.*

Remind that the vertex v can get several colors after **Phase I**. However, in general this number will be small. Let Y_{vi} denote the event where vertex v is colored with i after **Phase II** of the algorithm. The following lemmas give a lower bound for the probability that color i was assigned to vertex v in **Phase II**. We consider three possible cases: a heavy vertex and a heavy color, a heavy vertex and a light color, a light vertex and a light color.

Lemma 4. *Assume that a heavy vertex v gets a heavy color q in **Phase I** of Algorithm 2-PHLV, then $Pr[Y_{vq}|X_{vq}] \geq \frac{2}{3}$.*

Proof. The probability that a vertex v is colored with a color q in **Phase II** depends on how many colors a vertex v received in **Phase I**. Without loss of generality, assume that the edges with colors $1, \ldots, t$ and q are incident to the vertex v. By the law of total probability we have

$$Pr[Y_{vq}|X_{vq}] \geq \prod_{i=1}^{t}(1 - Pr[X_{vi}]) + \frac{1}{3}\sum_{i=1}^{t} Pr[X_{vi}]\prod_{j \neq i}(1 - Pr[X_{vj}]) \quad (5)$$

The first term is the probability that a vertex v is colored with color q and no additional color is chosen. The second term is the probability that the vertex v is colored with color q under the condition that it received one additional color in **Phase I**. Since the color q is heavy, we have $\lambda_v(2, q) = \frac{1}{3}$ and the vertex v will be colored in color q with probability $1/3$. We drop all the remaining terms of the formula because they are equal to zero in the worst case.

To simplify computations we set $X_i = Pr[X_{vi}]$ and consider the right-hand-size of (5) as a function f_{vq} of variables $X_1, X_2, ..., X_t$. We have

$$f_{vq} = \prod_{i=1}^{t}(1 - X_i) + \frac{1}{3}\sum_{i=1}^{t} X_i \prod_{j \neq i}(1 - X_j).$$

Taking into account that color q is heavy, from (2) we have $\sum_{i=1}^{t} X_i \leq \frac{1}{2}$. In order to obtain a lower bound for $Pr[Y_{vq}|X_{vq}]$, consider the minimum value of f_{vq} over all choises of X_i subject to the constraint $\sum_{i=1}^{t} X_i \leq \frac{1}{2}$. Putting the first two variables out of the summation and the product, we get

$$f_{vq} = (1 - X_1)(1 - X_2)\prod_{i=3}^{t}(1 - X_i) + \frac{1}{3}(X_1(1 - X_2) + X_2(1 - X_1))\prod_{i=3}^{t}(1 - X_i)$$

$$+ \frac{1}{3}(1 - X_1)(1 - X_2)\sum_{i=3}^{t} X_i \prod_{j \geq 3, j \neq i}(1 - X_j) \quad (6)$$

Then from (6) it follows that

$$
f_{vq} = (1 - \frac{2}{3}X_1 - \frac{2}{3}X_2)\prod_{i=3}^{t}(1 - X_i) + \frac{1}{3}(1 - X_1 - X_2)\sum_{i=3}^{t}X_i\prod_{j\geq3,j\neq i}(1 - X_j)
$$

$$
+ \frac{1}{3}X_1X_2(\prod_{i=3}^{t}(1 - X_i) + \sum_{i=3}^{t}X_i\prod_{j\geq3,j\neq i}(1 - X_j)) \quad (7)
$$

Consider f_{vq} as a function of two variables X_1 and X_2. Assume that $X_1 + X_2 = \gamma$, where $\gamma \leq \frac{1}{2}$ is a constant. Let $X_1 \geq X_2 > 0$. If we increase X_1 and decrease X_2 by δ, $0 < \delta \leq X_2$, then the first two terms of (7) do not change and the last term decreases and therefore the function f_{vq} decreases as well. It follows that the minimum of f_{vq} is attained at $X_1 = \gamma$ and $X_2 = 0$. By repeating this argument we get that the minimum of f_{vq} is attained when $X_1 = \frac{1}{2}$ and $X_i = 0$, $i = 2, \ldots, t$. Finally, we get $Pr[Y_{vq}|X_{vq}] \geq f_{vq} \geq \frac{2}{3}$.

Lemma 5. *Assume that a heavy vertex v gets a light color q in **Phase I** of Algorithm 2-PHLV, then $Pr[Y_{vq}|X_{vq}] \geq \frac{5}{8}$.*

Proof. The probability that a vertex v is colored with a color q in **Phase II** depends on what other colors were also chosen for it in **Phase I**. Without loss of generality, assume that the edges with colors $1, \ldots, t$ and q are incident to the vertex v and let color 1 be the heavy color. By the law of total probability we have

$$
Pr[Y_{vq}|X_{vq}] \geq \prod_{i=1}^{t}(1 - Pr[X_{vi}]) + \frac{2}{3}Pr[X_{v1}]\prod_{i=2}^{t}(1 - Pr[X_{vi}])
$$

$$
+ \frac{1}{2}(1 - Pr[X_{v1}])\sum_{i=2}^{t}Pr[X_{vi}]\prod_{j\geq2,j\neq i}(1 - Pr[X_{vj}])
$$

$$
+ \frac{1}{3}\sum_{i=1}^{t}\sum_{j=i+1}^{t}Pr[X_{vi}]Pr[X_{vj}]\prod_{\hat{i}=1,\hat{i}\neq i,\hat{i}\neq j}(1 - Pr[X_{v\hat{i}}])
$$

The first term is the probability that a vertex v is colored with color q and no additional color is chosen. The second term is the probability that the vertex v is colored with color q under the condition that it receives an additional heavy color in **Phase I**. Since the vertex q is light, we have $\lambda_v(2, q) = \frac{2}{3}$ and the vertex v will be colored in color q with probability 2/3. The third term is the probability that the vertex v is colored with color q under the condition that it receives exactly one additional light color in **Phase I**. In this case, the vertex v will be colored in color q with probability 1/2. The last term is the probability that the vertex v is colored with color q under the condition that for v two more colors were chosen. We drop all the remaining terms of the formula because they are equal to zero in the worst case.

By setting $A = \prod_{i=3}^{t}(1 - Pr[X_{vi}])$, $B = \sum_{i=3}^{t}Pr[X_{vi}]\prod_{j\geq3,j\neq i}(1 - Pr[X_{vj}])$ and

$C = \sum_{i=3}^{t} \sum_{j=i+1}^{t} Pr[X_{vi}]Pr[X_{vj}] \prod_{i \geq 3, i \neq i, i \neq j}(1 - Pr[X_{vi}])$ we can rewrite this expression as

$$Pr[Y_{vq}|X_{vq}] \geq (1 - Pr[X_{v1}])(1 - Pr[X_{v2}])A + \frac{2}{3}Pr[X_{v1}](1 - Pr[X_{v2}])A$$

$$+ \frac{1}{2}(1 - Pr[X_{v1}])Pr[X_{v2}]A + \frac{1}{2}(1 - Pr[X_{v1}])(1 - Pr[X_{v2}])B + \frac{1}{3}Pr[X_{v1}]Pr[X_{v2}]A$$

$$+ \frac{1}{3}(Pr[X_{v1}](1 - Pr[X_{v2}]) + (1 - Pr[X_{v1}])Pr[X_{v2}])B + \frac{1}{3}(1 - Pr[X_{v1}])(1 - Pr[X_{v2}])C.$$

Discarding the last term and setting $X_i = Pr[X_{vi}]$ we get

$$Pr[Y_{vq}|X_{vq}] \geq f_{vq} \doteq (1 - X_1)(1 - X_2)A + \frac{2}{3}X_1(1 - X_2)A + \frac{1}{2}(1 - X_1)X_2A$$

$$+ \frac{1}{2}(1 - X_1)(1 - X_2)B + \frac{1}{3}X_1X_2A + \frac{1}{3}(X_1(1 - X_2) + (1 - X_1)X_2)B.$$

In order to obtain a lower bound for $Pr[Y_{vq}|X_{vq}]$, we first show that the minimum of f_{vq} is attained when $X_1 = \frac{1}{2}$. After multiplying the terms with each other, we get

$$f_{vq} = (1 - \frac{1}{3}X_1 - \frac{1}{2}X_2 + \frac{1}{6}X_1X_2)A + (\frac{1}{2} - \frac{1}{6}X_1 - \frac{1}{6}X_2 - \frac{1}{6}X_1X_2)B$$

$$= (1 - \frac{1}{3}X_1 - \frac{1}{3}X_2)A + (\frac{1}{2} - \frac{1}{6}X_1 - \frac{1}{6}X_2)B - \frac{1}{6}(X_1X_2B + X_2(1 - X_1)A).$$

Let us consider f_{vq} as a function of two variables X_1 and X_2. Assume that $X_1 + X_2 = \gamma$. Since color 1 is heavy then $\frac{1}{2} \leq X_1 \leq \gamma$ and $X_2 \leq \frac{1}{2}$. If we decrease X_1 and increase X_2 by δ, $0 < \delta \leq X_1 - \frac{1}{2}$, then the expression $X_1X_2B + X_2(1 - X_1)A$ increases. It follows that f_{vq} reaches a minimum when $X_1 = \frac{1}{2}$.

Now, substitute X_1 by $\frac{1}{2}$. Thus, we obtain

$$f_{vq} \geq (\frac{5}{6} - \frac{5}{12}X_2)\prod_{i=3}^{t}(1 - X_i) + (\frac{5}{12} - \frac{1}{4}X_2)\sum_{i=3}^{t}X_i \prod_{j \geq 3, j \neq i}(1 - X_j). \quad (8)$$

Rewrite the right-hand side of (8) as

$$f_{vq} = (\frac{5}{6} - \frac{5}{12}X_2)(1 - X_3)\prod_{i=4}^{t}(1 - X_i) + (\frac{5}{12} - \frac{1}{4}X_2)X_3\prod_{i=4}^{t}(1 - X_i)$$

$$+ (\frac{5}{12} - \frac{1}{4}X_2)(1 - X_3)\sum_{i=4}^{t}X_i \prod_{j=4, j \neq i}^{t}(1 - X_j) \quad (9)$$

Discarding the last term in (9) we get

$$f_{vq} \geq \frac{5}{6}(1 - \frac{1}{2}X_2 - \frac{1}{2}X_3 + \frac{1}{5}X_2X_3)\prod_{i=4}^{t}(1 - X_i).$$

Taking into account that $X_2 \leq \frac{1}{2}$ and $X_3 \leq \frac{1}{2}$, we finally obtain $f_{vq} \geq \frac{5}{8}$.

The following result directly follows from Lemma 4(c) in [1].

Lemma 6. *Assume that a light vertex v gets a color q in* **Phase I** *of Algorithm 2-PHLV, then $Pr[Y_{vq}|X_{vq}] \geq \frac{7}{12}$.*

Suppose that $e = (u, v)$ has a color c and it is chosen in the first phase of Algorithm 2-PHLV and let $Pr[e$ is stable] denote the probability that both vertices v and u get the color c. We want to prove that

$$Pr[e \text{ is stable}] \geq Pr[Y_{uc}]Pr[Y_{vc}]. \tag{10}$$

Similar results were proven for the algorithms RR and RR2, presented in [2] and [1], respectively. Algorithm 2-PHLV differs from algorithms RR and RR2 only by the choice of probabilities $\lambda_v(l, c)$ in the second phase of the algorithm. Unfortunately, the previous proofs of (10) are not suitable for our algorithm. Here, we introduce a sufficient condition for probabilities $\lambda_v(l, c)$ such that (10) holds. The Algorithm 2-PHASE specifies values of $\lambda_v(l, c)$ only if the edge of color c incident to the vertex v was chosen in the first phase of the algorithm. Without loss of generality, we assume that $\lambda_v(l, c) = 0$ otherwise.

Lemma 7. *Let $e = (u, v)$ has a color c and it is chosen in the first phase of Algorithm 2-PHASE. If the sequences $\lambda_v(1, c), \ldots, \lambda_v(k, c)$ and $\lambda_u(1, c), \ldots, \lambda_u(k, c)$ are non-increasing then $Pr[e$ is stable] $\geq Pr[Y_{uc}|X_{uc}]Pr[Y_{vc}|X_{vc}]$.*

Due to space limitation, the proof of the lemma is removed in this version.

Theorem 1. *The expected approximation ratio of Algorithm 1 is bounded by $\frac{49}{144}$.*

Proof. Let OPT denote the sum of the weights of the stable edges in an optimal solution. Since z^* is an optimal solution of the LP, we have $OPT \leq \sum_{e \in E} w_e z_e^*$.

Consider an edge $e \in E$ is chosen in **Phase I** of Algorithm 2-PHLV. This occurs with probability z_e^* by Lemma 1. Suppose an edge $e = [u, v]$ has a color c, then by Lemma 7, the probability that the both endpoints of e are colored with c at least $Pr[Y_{vc}|X_{vc}]Pr[Y_{uc}|X_{uc}]$. Lemmata 4–6 imply that the expected contribution of the edge e is at least $\frac{49 w_e z_e^*}{144}$.

The expected weight of the stable edges in a solution obtained by Algorithm 2-PHLV is

$$W = \sum_{e \in E} w_e Pr[e \text{ is stable}] \geq \frac{49}{144} \sum_{e \in E} w_e z_e^* \geq \frac{49}{144} OPT.$$

3   Inapproximability

To establish MAXSNP-hardness of MAX-k-EC we give an L-reduction from the well-known Maximum 3-Satisfiability problem (MAX-E3-SAT), which cannot be approximated within $\frac{7}{8} + \epsilon$ for $\epsilon > 0$ unless **P = NP** [7]. MAX-E3-SAT: We are given a set of Boolean variables $x_1, x_2, \ldots, x_n$ and a collection of disjunctive clauses $C_1, C_2, \ldots, C_m$. Every clause has exactly three literals. Find an assignment of Boolean values to the variables which satisfies as many clauses as possible.

Definition 1 (L-reducibility [11]**)**
Given two optimization problems Π_1 and Π_2, we say we have an L-reduction $(\Pi_1 \leq_L \Pi_2)$ with parameters α and β from Π_1 to Π_2, if for some $\alpha, \beta > 0$

- *For each instance I of Π_1 we can compute in polynomial time an instance I' of Π_2;*
- *$OPT(I') \leq \alpha\, OPT(I)$;*
- *Given a solution of value B' to I', we can compute in polynomial time a solution of value B to I such that $|OPT(I) - B| \leq \beta |OPT(I') - B'|$.*

For any given instance I of MAX-E3-SAT we construct a particular instance I' of Max-k-EC, i.e., we define a particular edge-colored graph G. Up to renaming of variables we have four different types of clauses: $(x_i \vee x_j \vee x_k)$, $(x_i \vee x_j \vee \bar{x}_k)$, $(x_i \vee \bar{x}_j \vee \bar{x}_k)$, and $(\bar{x}_i \vee \bar{x}_j \vee \bar{x}_k)$. We start the construction of G from the gadgets given in Fig. 1, where the color assigned to each edge is indicated. For the rest of the proof we assume that $\mathcal{C} = R(ed), B(lue), G(reen)$.

For each variable x_i of the MAX-E3-SAT x_i we create a vertex v_i. We construct a gadget for every clause C_i, $i = 1, \ldots, m$. The gadget has seven vertices and nine edges. Six edges join vertices of the gadget and three edges join three vertices of the gadget with three vertices corresponding to variables of C_i.

First type: Assume that all literals are positive i.e., when the clause is $(x_i \vee x_j \vee x_k)$, then we construct the gadgets in Fig. 1(a).

Second type: Assume that two literals are positive and one literal is negative. That is the clause is of the form $(x_i \vee x_j \vee \bar{x}_k)$, then we construct the gadgets in Fig. 1(b).

Third type: Assume that two literals are negative and one literal is positive. That is the clause is of the form $(x_i \vee \bar{x}_j \vee \bar{x}_k)$, then we construct the gadgets in Fig. 1(c).

Forth type: Assume that all literals are negative i.e., when the clause is $(\bar{x}_i \vee \bar{x}_j \vee \bar{x}_k)$, then we construct the gadgets in Fig. 1(d).

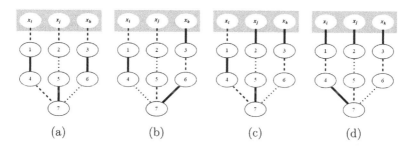

(a)   (b)   (c)   (d)

Fig. 1. Gadgets for four types of clauses. Let green, red and blue edges be dashed, thick, and dotted edges, respectively. (a) $(x_i \vee x_j \vee x_k)$. (b) $(x_i \vee x_j \vee \bar{x}_k)$. (c) $(x_i \vee \bar{x}_j \vee \bar{x}_k)$. (d) $(\bar{x}_i \vee \bar{x}_j \vee \bar{x}_k)$.

We note that the constructed graph does not contain any odd cycle and so it is bipartite. Also, for every vertex of the constructed graph the edges that are incident to this vertex are colored with at most three different colors, i.e. the chromatic degree of the graph is equal to three. The following lemma follows directly from the construction of the graph G.

Lemma 8. *The maximum contribution that any gadget can have is exactly 4 and is obtained when at least one of the vertices corresponding to variable has the same color as the edge that connects it with the rest of the gadget. Otherwise the maximum contribution that can be achieved is 3.*

Theorem 2. *MAX-E3-SAT $\leq_L$ MAX-3-EC for $\alpha = \frac{31}{7}$, $\beta = 1$.*

Proof. First we claim that for an instance I of the MAX-E3-SAT problem there is a truth-assignment that satisfies at least B clauses if and only if there is a vertex coloring with $3m + B$ stable edges. To prove the only-if direction, consider a truth assignment that satisfies at least B clauses. In the graph G, color green all the vertices that correspond to variables that are true and red all the vertices that correspond to false variables. In this way, for each satisfied clause the corresponding gadget in the optimum coloring will have pay-off four. We color all unsatisfied clauses in such a way that each unsatisfied clause has three stable edges. Hence, the total pay-off will be at least $3m + B$. For the opposite direction, suppose that the graph G has a coloring with $3m + B$ stable edges. Since each one of the gadgets has either 3 or 4 stable edges, there must exist at least B gadgets with 4 stable edges. Let us assign the value true to the variables with green corresponding vertices and the value false to the rest of the variables. Notice now that each one of the gadgets with four stable edges corresponds to a satisfied clause. Since such gadgets are at least B, there are at least B clauses that are satisfied and the if direction holds too. Thus, we obtain that $OPT(I') = 3m + OPT(I)$.

Johnson [8] presents the simple algorithm for the maximum satisfiability problem that satisfies at least $\frac{7}{8}$ of the clauses of a MAX-E3-SAT instance, so that $OPT(I) \geq \frac{7}{8}m$.

$$OPT(I') = 3m + OPT(I) \leq 3(\frac{8}{7}OPT(I)) + OPT(I) = \frac{31}{7}OPT(I)$$

Suppose that we have a solution of value $B' = 3m + B$ to the instance I'. Assigning a value of true to variables with green corresponding vertices, and false for the remaining variables we obtain a solution of value B to the instance I. It follows that $|OPT(I) - B| = |OPT(I') - 3m - B| \leq |OPT(I') - B'|$ and we have an L-reduction with parameters $\alpha = \frac{31}{7}$ and $\beta = 1$.

Since L-reductions preserve approximability within constant factors, an L-reduction from a MAX-E3-SAT problem to MAX-k-EC implies that the latter problem is MAXSNP-hard. In particular, MAXSNP-hard problems do not admit a PTAS.

Corollary 1
MAX-k-EC cannot be approximated better than $\frac{241}{248}$ unless $P = NP$, even though the given graph is bipartite and its edges are colored with at most three different colors.

Proof. If there is an L-reduction with parameters α and β from maximization problem Π_1 to maximization problem Π_2, and there is an ρ-approximation algorithm for Π_2, then there is an $(1 - \alpha\beta(1 - \rho))$-approximation algorithm for Π_1, see Theorem 16.5 in [11]. Given a ρ'-approximation algorithm for MAX-k-EC, we then have a $\rho = (1 - \frac{31}{7}(1 - \rho'))$-approximation algorithm for the MAX-E3-SAT problem. Hastad [7] proved that no ρ-approximation algorithm exists for MAX-E3-SAT for constant $\rho > \frac{7}{8}$ unless P $=$ NP. By simple calculations, we obtain that no ρ-approximation algorithm exists for MAX-E3-SAT for constant $\rho > \frac{241}{248}$ unless P $=$ NP.

4   Conclusion

For the MAX-k-EC problem we have presented a new approximation algorithm with approximation ratio $\frac{49}{144}$, improving the previous known ratio $\frac{7}{23}$. We also have obtained the first upper bound of $\frac{241}{248}$ for the inapproximability of the MAX-k-EC problem, even for the case when the given graph is bipartite and its edges are colored with at most three different colors. Our approximation algorithm is based on a randomized rounding of a natural LP-relaxation. First, we notice that our algorithm can be derandomized using the method of conditional expectations. Second, note that Theorem 1 provides a lower bound for the integrality gap of the integer linear program (1)–(4). We propose an overview of the results on (in)approximability of the MAX-k-EC problem in Fig. 2. Therefore a number of problems remained open for further research.

- Is there a better approximation factor than $\frac{49}{144}$?
- It is interesting to determine what is the maximum possible integrality gap of the integer linear program (ILP) (1)–(4). It is given by Ageev and Kononov [1] that the upper bound of ILP is $\frac{2}{3}$ by giving a trivial example of triangle with unit edge weights and a unique color for each edge.
- We note that all known approximation algorithms for the MAX-k-EC problem are based on the ILP (1)–(4). It is interesting to design an algorithm using other lower bounding scheme.
- Is there a better upper bound of the inapproximability of the MAX-k-EC problem?

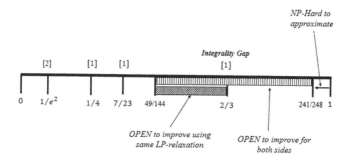

Fig. 2. Overview of results on (in)approximability the MAX-k-EC problem

References

1. Ageev, A., Kononov, A.: Improved approximations for the max k-colored clustering problem. In: Bampis, E., Svensson, O. (eds.) WAOA 2014. LNCS, vol. 8952, pp. 1–10. Springer, Cham (2015). https://doi.org/10.1007/978-3-319-18263-6_1
2. Angel, E., Bampis, E., Kononov, A., Paparas, D., Pountourakis, E., Zissimopoulos, V.: Clustering on k-edge-colored graphs. In: Chatterjee, K., Sgall, J. (eds.) MFCS 2013. LNCS, vol. 8087, pp. 50–61. Springer, Heidelberg (2013). https://doi.org/10.1007/978-3-642-40313-2_7
3. Angel, E., Bampis, E., Kononov, A., Paparas, D., Pountourakis, E., Zissimopoulos, V.: Clustering on k-edge-colored graphs. Discrete Appl. Math. **211**, 15–22 (2016)
4. Bang-Jensen, J., Gutin, G.: Alternating cycles and paths in edge-coloured multi-graphs: a servey. Discrete Math. **165–166**, 39–60 (1997)
5. Cai, L., Leung, O.-Y.: Alternating path and coloured clustering (2018). arXiv:1807.10531
6. Feige, U., Vondrak, J.: Approximation algorithm for allocation problem improving the factor of $1 - \frac{1}{e}$. In: 47th Annual IEEE Symposium on Foundations of Computer Science, pp. 667–676 (2006)
7. Hastad, J.: Some optimal inapproximability results. J. ACM **48**, 798–859 (2001)
8. Johnson, D.S.: Approximation algorithm for combinatorial problems. J. Comput. Syst. Sci. **9**(3), 256–278 (1974)
9. Kano, M., Li, X.: Monochromatic and heterochromatic subgraphs in edge-colored graphs - a survey. Graphs Comb. **24**, 237–263 (2008)
10. Kleinberg, J.M., Ligett, K.: Information-sharing and privacy in social networks. Games Econ. Behav. **82**, 702–716 (2013)
11. Wiliiamson, D.P., Shmoys, D.B.: The Design of Approximation Algorithms. Cambridge University Press, New York (2011)

The Non-hardness of Approximating Circuit Size

Eric Allender[1(✉)], Rahul Ilango[1], and Neekon Vafa[2]

[1] Rutgers University, Piscataway, NJ, USA
allender@cs.rutgers.edu, rahul.ilango@rutgers.edu
[2] Harvard University, Cambridge, MA, USA
nvafa@college.harvard.edu

Abstract. The Minimum Circuit Size Problem (MCSP) has been the focus of intense study recently; MCSP is hard for SZK under rather powerful reductions [4], and is provably not hard under "local" reductions computable in TIME($n^{0.49}$) [22]. The question of whether MCSP is NP-hard (or indeed, hard even for small subclasses of P) under some of the more familiar notions of reducibility (such as many-one or Turing reductions computable in polynomial time or in AC^0) is closely related to many of the longstanding open questions in complexity theory [7,8,16–18,20,22].

All prior hardness results for MCSP hold also for computing somewhat weak approximations to the circuit complexity of a function [3,4,9,16,21,27]. (Subsequent to our work, a new hardness result has been announced [19] that relies on more exact size computations.) Some of these results were proved by exploiting a connection to a notion of time-bounded Kolmogorov complexity (KT) and the corresponding decision problem (MKTP). More recently, a new approach for proving improved hardness results for MKTP was developed [5,7], but this approach establishes only hardness of extremely good approximations of the form $1 + o(1)$, and these improved hardness results are not yet known to hold for MCSP. In particular, it is known that MKTP is hard for the complexity class DET under nonuniform $\leq_m^{\mathsf{AC}^0}$ reductions, implying MKTP is not in AC$^0[p]$ for any prime p [7]. It was still open if similar circuit lower bounds hold for MCSP. (But see [13,19].) One possible avenue for proving a similar hardness result for MCSP would be to improve the hardness of approximation for MKTP beyond $1 + o(1)$ to $\omega(1)$, as KT-complexity and circuit size are polynomially-related. In this paper, we show that this approach cannot succeed.

More specifically, we prove that PARITY does not reduce to the problem of computing superlinear approximations to KT-complexity or circuit size via AC^0-Turing reductions that make $O(1)$ queries. This is significant, since approximating any set in P/poly AC^0-reduces to just *one* query of a much worse approximation of circuit size or KT-complexity [24].

Supported by NSF grants CCF-1514164 and CCF-1559855. This work was done [in part] while author Eric Allender was visiting the Simons Institute for the Theory of Computing.

© Springer Nature Switzerland AG 2019
R. van Bevern and G. Kucherov (Eds.): CSR 2019, LNCS 11532, pp. 13–24, 2019.
https://doi.org/10.1007/978-3-030-19955-5_2

For weaker approximations, we also prove non-hardness under more powerful reductions. Our non-hardness results are unconditional, in contrast to conditional results presented in [7] (for more powerful reductions, but for much worse approximations). This highlights obstacles that would have to be overcome by any proof that MKTP or MCSP is hard for NP under AC^0 reductions. It may also be a step toward confirming a conjecture of Murray and Williams, that MCSP is not NP-complete under logtime-uniform $\leq_m^{AC^0}$ reductions [22].

Keywords: NP-completeness · Minimum Circuit Size Problem · Reductions · Time-bounded Kolmogorov complexity

1   Introduction

The Minimum Circuit Size Problem (MCSP) is the problem of determining whether a (given) Boolean function f (represented as a bitstring of length 2^k for some k) has a circuit of size at most a (given) threshold θ. Although the complexity of MCSP has been studied for more than half a century (see [21, 28] for more on the history of the problem), recent interest in MCSP traces back to the work of Kabanets and Cai [21], who connected the problem to questions involving the natural proofs framework of Razborov and Rudich [26].

Since then, there has been a flurry of research on MCSP [3–8, 15–18, 20, 22, 24], but still the exact complexity of MCSP remains unknown. MCSP is in NP, but it remains an important open question whether MCSP is NP-complete.

MCSP is Likely Not in P. There is good evidence for believing MCSP $\notin$ P. If MCSP is in P, then there are no cryptographically-secure one-way functions [21]. Furthermore, [4] shows MCSP is hard for SZK under BPP-Turing reductions, so if MCSP $\in$ P then SZK $\subseteq$ BPP, which seems unlikely.

Showing MCSP is NP-Hard Would Be Difficult. Murray and Williams [22] have shown that if MCSP is NP-hard under polynomial-time many-one reductions, then EXP $\neq$ ZPP, which is a likely separation but one that escapes current techniques. Results from [4, 18, 22] also give various likely (but difficult to show) consequences for MCSP being hard under more restrictive forms of reduction. We note that it has been suggested that MCSP might well be complete for NP [20]. In this regard, it may also be relevant to note that $MCSP^{QBF}$ is complete for PSPACE under ZPP-Turing reductions [3].

The Hardness of Both MCSP and Approximating MCSP Have Important Consequences for Complexity Theory. We have already mentioned that if MCSP is NP-hard under polynomial-time reductions, then EXP $\neq$ ZPP [22]. In a recent development, Hirahara [15] shows that if a certain approximation to MCSP is NP-hard, then NP $\neq$ BPP implies that NP is difficult to compute even on average. In another recent development, [25] and [23] show that even seemingly meager $n^{1+\epsilon}$ circuit lower bounds on certain approximations to MCSP imply results such as NP $\not\subseteq$ P/poly.

MCSP is Not Hard for NP in Limited Settings. Murray and Williams [22] show MCSP is not NP-hard under a certain type of "local" reductions computable in $\mathsf{TIME}(n^{0.49})$. This is significant, since many well-known NP-complete problems are complete under local reductions computable in even logarithmic time. (A list of such problems is given in [22].)

Many Hardness Results for MCSP Also Hold for Approximate Versions of MCSP. In various settings, the power of MCSP to distinguish between circuits of size θ and $\theta + 1$ is not fully used. Rather, in [3,4,9,20,24,27], the reduction succeeds assuming only that reliable answers are given to queries on instances of the form (T, θ), where either the truth table T requires circuits of size $\geq \theta = |T|/2$ or T can be computed by circuits of size $\leq |T|^\delta$, for some $\delta > 0$.

This is an appropriate time to call attention to one such reduction to approximations to MCSP. Corollary 6 of [24] shows that, for every $\delta > 0$, for every solution S to MCSP$[n^\delta, n/2]$, for every set $A \in \mathsf{P/poly}$, there is a $c > 1$ and a set A' that differs from A on at most $(1/2 - 1/n^c)2^n$ of the strings of each length n, such that $A' \leq_{tt}^{\mathsf{AC}^0} S$ via a reduction[1] that makes only *one query*. (That is, $A' \leq_{1-tt}^{\mathsf{AC}^0} S$.) Stated another way, any set in $\mathsf{P/poly}$ can be "approximated" with just one query to a weak approximation of MCSP. (Changing the solution S will yield a different set A'.)

There is No Known Many-One Hardness Result for MCSP, But One is Known for a Related Problem. MKTP, the minimum time-bounded Kolmogorov complexity problem, is loosely the "program version" of MCSP. It is known [7] that MKTP is hard for DET under (non-uniform) NC^0 many-one reductions; it is conjectured that the same is true for MCSP. Time-bounded Kolmogorov complexity is polynomially-related to circuit complexity [3], so one natural way to extend the hardness result of [7] from MKTP to MCSP would be to stretch the very small gap given in the reduction of DET to MKTP.

1.1 Our Contributions, and Related Prior Work

We address the following questions based on prior work:

1. Can the non-hardness result of Murray and Williams [22] be extended to more powerful reductions? Both [22] and [8] conjecture that MCSP is not NP-complete under uniform AC^0 reductions.
2. Can the conditional theorem of [7], establishing the non-NP-hardness of very weak approximations to MCSP under cryptographic assumptions, be improved, to show non-NP-hardness of MCSP for stronger approximations?
3. The worst-case to average case reduction given by [15] is conditional on the NP-hardness of a certain approximation to MCSP. Can we say anything about the NP-hardness of this problem in, say, the context of limited reductions?
4. Finally, can the result of [7], showing that MKTP is hard for DET under $\leq_m^{\mathsf{AC}^0}$ reductions, be extended, to hold for MCSP as well, by increasing the gap?

---

[1] Although Corollary 6 of [24] does not mention the number of queries, inspection of the proof shows that only one query is performed.

Our results give the following replies to these questions

1. For superlinear approximations to MCSP, one can, in fact, give much stronger non-hardness results than [22], showing non-hardness even under non-uniform AC^0 many-one reductions and even limited types of AC^0 Turing reductions. To our knowledge, this is the first known non-hardness result for any variant of MCSP under non-uniform AC^0 reductions. While AC^0 reductions are provably less powerful than polynomial time reductions, most natural examples of NP-complete problem are easily seen to be complete under AC^0 (and even NC^0!) reductions [10].

2. [7] shows that, if cryptographically-secure one-way functions exist, then $\epsilon(n)$-GapMCSP is not hard for NP under P/poly-Turing reductions[2] for some $\epsilon(n) = n^{o(1)}$. Our result gives a trade-off, where we reduce the gap dramatically but also weaken the type of reduction. In particular, our results imply that if one-way functions exist, then $\epsilon(n)$-GapMCSP is NP-intermediate under $\leq_m^{AC^0}$ and $\leq_{k-tt}^{AC^0}$ reductions, where $\epsilon(n) = o(n)$.

3. We show that the approximation to MCSP considered by [15] is actually *not* NP-hard under AC^0 reductions.

4. Our work rules out one natural way to extend the MKTP hardness results to MCSP. One might have hoped that the reduction given by [7] could be extended to a larger gap and hence apply to MCSP (since MKTP and MCSP are polynomially related [3]). However, we show that this is impossible.

Our main theorem is an impossibility result in the setting of $\epsilon(\theta)$-GapMCSP, which is the promise version of MCSP with a multiplicative $\epsilon(\theta)$ gap where θ is the threshold.

Theorem 1. PARITY $\not\leq_m^{AC^0} \epsilon(\theta)$-GapMCSP *where* $\epsilon(\theta) = o(\theta)$.

We note that this is not the first work to describe non-hardness of approximation under AC^0 reductions. Arora [11] is credited by [1], with showing that no AC^0 reduction f can have the property that $x \in$ PARITY implies $f(x)$ has a very large clique, and $x \notin$ PARITY implies $f(x)$ has only very small cliques. Our work differs from that of [11] in several respects. Arora shows that AC^0 reductions cannot prove very *strong* hardness of approximations for a problem where strong inapproximability results are already known. We show that AC^0 reductions cannot establish even very *weak* inapproximability results for MCSP. Also, our techniques allow us to move beyond $\leq_m^{AC^0}$ reductions, to consider AC^0-Turing reducibility.

All of the theorems that we state in terms of MCSP hold also for MKTP, with identical proofs. For the sake of readability, we present the theorems and proofs only in terms of MCSP. Note that some proofs are omitted, due to space limitations.

---

[2] The problem ϵ-GapMCSP is defined somewhat differently in [7] than here. See Sect. 2. Thus the form of $\epsilon(n)$ looks different here than in [7].

2   Preliminaries

We use $\backslash$ to denote set difference, and for $n \in \mathbb{N}$, $[n]$ denotes the set $\{1, \ldots, n\}$.

2.1   Defining MCSP

For any binary string T of length 2^k, we define $\mathrm{CC}(T)$ to be the size of the smallest circuit (using only NOT gates and AND and OR gates of fan-in 2) that computes the function given by truth table T written in lexicographic order, where, for concreteness, circuit size is defined to be the number of AND and OR gates, although our arguments work for other reasonable notions of circuit size.

Throughout the paper, we use various approximate notions of the minimum circuit size problem, given as follows:

Definition 1 (Gap MCSP). *For any function $\epsilon : \mathbb{N} \to \mathbb{N}$, let $\epsilon(n)$-GapMCSP be the promise problem (Y, N) where*

$$Y := \{(T, \theta) \mid \mathrm{CC}(T) < \epsilon(\theta)\}, \quad and \quad N := \{(T, \theta) \mid \mathrm{CC}(T) > \theta\},$$

where θ is written in binary.

Note that this definition differs in minor ways from the way that ϵ-GapMCSP was defined in [7]. The definition presented here allows for finer distinctions than the definition that was used in [7].

Our results for non-hardness under $\leq_T^{\mathsf{AC}^0}$ reductions are best stated in terms of a restricted version of ϵ-GapMCSP, where the thresholds are fixed, for inputs of a given size: This variant of MCSP has been studied previously in [16, 22]; the analogous problem defined in terms of KT-complexity is denoted R_{KT} in [3].

Definition 2 (Parameterized Gap MCSP). *For any functions $\ell, g : \mathbb{N} \to \mathbb{N}$ such that $\ell(n) \leq g(n)$, we define the language $\mathsf{MCSP}[\ell, g]$ to be the promise problem (Y, N) where*

$$Y := \{T \mid \mathrm{CC}(T) < \ell(|T|)\}, \quad and \quad N := \{T \mid \mathrm{CC}(T) > g(|T|)\}.$$

2.2   Complexity Classes and Reductions

We assume the reader is familiar with basic complexity classes such as P and NP. As we work extensively with non-uniform NC^0 and AC^0, we refer to the text by Vollmer [29] for background on these circuit classes. Throughout this paper, unless otherwise explicitly mentioned, we refer to the non-uniform versions of these circuit classes.

Let $\mathcal{C}$ be a class of circuits. For any languages A and B, we write $A \leq_m^{\mathcal{C}} B$ if there is a function f computed by a circuit family $\{C_n\} \in \mathcal{C}$ such that $f(x) \in B \iff x \in A$. We write $A \leq_T^{\mathcal{C}} B$ if there is a circuit family in $\mathcal{C}$ computing A with B-oracle gates. In particular, since we are primarily concerned with $\mathcal{C} = \mathsf{AC}^0$, we denote this as $A \leq_T^{\mathsf{AC}^0} B$. We write $A \leq_{tt}^{\mathsf{AC}^0} B$ if there is an

AC^0 circuit family computing A with B-oracle gates, where there is no directed path from any oracle gate to another, i.e. if the reduction is non-adaptive. If, furthermore, the non-adaptive reduction has the property that each of the oracle circuits contains at most k oracle gates, then we write $A \leq^{\mathsf{AC}^0}_{k-tt} B$.

Let $Y \subseteq \{0,1\}^*$ and $N \subseteq \{0,1\}^*$ be disjoint. Then $\Pi = (Y,N)$ is a *promise problem*. A language L is a *solution* to a promise problem $\Pi = (Y,N)$ if $Y \subseteq L$ and $N \cap L = \emptyset$. For two promise problems Π_1 and Π_2, some type of reducibility r (many-one, truth table, or Turing), and a circuit class $\mathcal{C}$, we say $\Pi_1 \leq^{\mathcal{C}}_r \Pi_2$ if there is a *single* family of oracle circuits $\{C_n\}$ in $\mathcal{C}$ such that for every solution S_2 of Π_2, there is a solution S_1 of Π_1 such that C_n computes an r-reduction from S_1 to S_2.

2.3   Boolean Strings and Functions

For an $x \in \{0,1\}^n$ and a set of indices $B \subseteq [n]$, we let x^B denote the Boolean string obtained by flipping the ith bit of x for each $i \in B$.

A *partial string* (or *restriction*) is an element of $\{0,1,?\}^*$. Define the *size* of a partial string p to be the number of bits in which it is $\{0,1\}$-valued. We say a partial string $p \in \{0,1,?\}^n$ *agrees* with a binary string $x \in \{0,1\}^n$ if they agree on all $\{0,1\}$-valued bits. If $x \in \{0,1\}^n$ is a binary string and $B \subseteq [n]$, then $x|_B$ denotes the partial string given by replacing the jth bit of x with ? for each $j \in [n] \setminus B$. We say a partial string p_1 *extends* a partial string p_2 if p_1 is equal to p_2 on all bits where p_2 is $\{0,1\}$-valued.

A *partial Boolean function* on n variables is a function $f : I \rightarrow \{0,1\}$ where $I \subseteq \{0,1\}^n$. For a promise problem $\Pi = (Y,N)$ and $n \in \mathbb{N}$, we let $\Pi|_n$ be the partial Boolean function that decides membership in Y on instances of length n which satisfy the promise. (In particular, $\Pi|_n : I := (Y \cup N) \cap \{0,1\}^n \rightarrow \{0,1\}$.)

We will make use of two well-studied complexity measures on Boolean functions: block sensitivity and certificate complexity. We refer the reader to a detailed survey by Hatami, Kulkarni, and Pankratov [14] for background on these notions. For completeness, we provide the definitions of the two measures that we need. In our context, we will use these measures on partial Boolean functions. Let $I \subseteq \{0,1\}^n$ and let $f : I \rightarrow \{0,1\}$ be a partial Boolean function. For an input $x \in I$, define the *block sensitivity of f at x*, denoted $bs(f,x)$, to be the maximum number of non-empty, disjoint sets $B_1, \ldots, B_k$ such that $x^{B_i} \in I$ and $f(x) \neq f(x^{B_i})$ for all i. (Here, by "$f(y) \neq f(z)$" we require that f is defined at both y and z.) Define the 0-*block sensitivity of f* be $bs_0(f) := \max_{x : f(x)=0} bs(f,x)$. For an input $x \in I$, define the *certificate complexity of f at x*, denoted $c(f,x)$, to be the size of the smallest set $B \subseteq [n]$ such that $f(y) = f(x)$ for all $y \in I$ that agree with $x|_B$. Define the 0-*certificate complexity of f* to be $c_0(f) := \max_{x : f(x)=0} c(f,x)$.

3   Non-Hardness Under NC^0 Reductions

In this section, we prove our main lemmas, showing that problems that are NC^0-reducible to ϵ-GapMCSP have sublinear 0-certificate complexity and also have bounded 0-block sensitivity. Whenever we will have occasion to use these lemmas, it will be in situations when we are able to assume that the NC^0 reduction is computing a function f satisfying the condition that there is a bound $\gamma(n) > 0$ such that, for all n, there is a $\theta \geq \gamma(n)$ such that, for all x of length n, $f(x)$ is of the form $(T(x), \theta)$. (In particular, the threshold θ is the same for all inputs of length n.) We will call such an NC^0 reduction a γ-honest reduction.

Lemma 1. *Let $\epsilon(\theta) = o(\theta)$, and let $\Pi = (Y, N)$ be a promise problem, where $\Pi \leq_m^{NC^0} \epsilon$-GapMCSP via a γ-honest reduction f computed by an NC^0 circuit family C_n of depth $\leq d$, where $\gamma(n) \geq \log\log n$. Let $k \geq 1$. Then there is an n_0 (that depends only on ϵ, k and d) such that for all $n \geq n_0$, if $N|_n \neq \emptyset$, then $c_0(\Pi|_n) \leq n/k$.*

Proof. Let $p = 2^d$, let $p' = \binom{2pk+1}{p}$, and let K be a constant that is specified later (and which depends only on k and d). Since $\epsilon(\theta) = o(\theta)$, we can pick a constant s_0 such that $\binom{p'}{2}\epsilon(s) + K < s$ for all $s \geq s_0$.

Pick $n_0 \geq 2^{2^{s_0}}$, and let $n \geq n_0$.

For contradiction, suppose $c_0(\Pi|_n) > n/k$. Let $x \in N \cap \{0, 1\}^n$ be a 0-valued instance with $c_0(\Pi|_n, x) > n/k$. Then, for all $S \subseteq [n]$ with $|S| \leq n/k$, there is an x_S such that x_S agrees with $x|_S$ and such that $\Pi|_n(x_S) = 1$. (That is, $x_S \in Y$.)

Let (T, θ) be the truth table produced by C_n on input x. Since $x \in N$ and C_n is a reduction, we know that any circuit computing T has size at least θ.

For each $S \subseteq [n]$ with size at most n/k, let T_S be the truth table produced by C_n on input x_S. Since $x_S \in Y$, we know that T_S has a circuit D_S of size at most $\epsilon(\theta)$.

We aim to build a "small" circuit computing T, which would contradict that T has high complexity. Recall that $p = 2^d$, and that $p' = \binom{2pk+1}{p}$.

Claim. There exists sets $S_1, \ldots S_{p'} \subseteq [n]$ such that

- $|S_i| \leq \frac{n}{2k}$ for all i, and
- for any set $P \subseteq [n]$ with $|P| \leq p$, we have that $P \subseteq S_i$ for some i.

Proof. (Proof of Claim) Pick sets $V_1, \ldots, V_{2pk+1} \subseteq [n]$ of size at most $\frac{n}{2pk}$ whose union is $[n]$. Let $\mathcal{V} = \{V_1, \ldots, V_{2pk+1}\}$. Now let each of $S_1, \ldots, S_{\binom{2pk+1}{p}}$ be the union of some p sets chosen from $\mathcal{V}$. Each S_i has size at most $p\frac{n}{2pk} = \frac{n}{2k}$. Let $P \subseteq [n]$ be an arbitrary set of size p. Since $\bigcup_{V \in \mathcal{V}} V = [n]$, every element e of P lies within some $V \in \mathcal{V}$. Then P is contained in the union of some p sets from $\mathcal{V}$, so $P \subseteq S_i$ for some i.

For each $i \neq j \in [p']$, let $S_{i,j} = S_{j,i} = S_i \cup S_j$. Note that $|S_{i,j}| \leq n/k$.

Our circuit C for computing T works as follows. On input r, for each $i \in [p']$, see if $D_{S_{i,1}}(r) = \cdots = D_{S_{i,p'}}(r)$. If so, then output $D_{S_{i,1}}(r)$. The size of this circuit is at most $\binom{p'}{2}\epsilon(\theta) + K$ (for some fixed constant K) since each of the $\binom{p'}{2}$ $D_{S_{i,j}}$ circuits has size at most $\epsilon(\theta)$ and the other "unanimity" condition is a

Boolean function on $\binom{p'}{2}$ variables (of in fact linear size) and so can be computed with circuit of some size $K = O(p')^2$ (that depends only on k and d).

Now, we argue that C on input r correctly computes the rth bit of T. Let $r \in [m]$ be arbitrary. For convenience, on an input $y \in \{0,1\}^n$ let $C_n^r(y)$ denote the rth output of $C_n(x)$. Recall the rth bit of T is defined to be $C_n^r(x)$. We must show two things. First, that there exists an i such that $D_{S_{i,1}}(r) = \cdots = D_{S_{i,p'}}(r)$ and second, that if for some i we have that $D_{S_{i,1}}(r) = \cdots = D_{S_{i,p'}}(r)$, then $D_{S_{i,1}}(r) = C_n^r(x)$.

Since C_n has depth d, the rth output of C_n can depend on at most 2^d input wires $W \subseteq [m]$. Hence, on any input y such that $y|_W = x|_W$, we have that $C_n^r(y) = C_n^r(x)$. Since $p = 2^d$, by the claim, there exists some S_{i*} such that $W \subseteq S_{i*}$. Therefore, for all j we have that $x_{S_{i*,j}}|_W = x|_W$, so $D_{S_{i*,j}}(r) \overset{\text{def}}{=} C_n^r(x_{S_{i*,j}}) = C_n^r(x)$.

This implies both things we must show. First, we know that $D_{S_{i*,1}}(r) = \cdots = D_{S_{i*,p'}}(r)$ since they each equal $C_n^r(x)$. Second, if for some i, we have that $D_{S_{i,1}}(r) = \cdots = D_{S_{i,p'}}(r)$, then we also have that $D_{S_{i,1}}(r) = D_{S_{i,i*}}(r) = C_n^r(x)$.

Thus we have that T can be computed by a circuit of size at most $\binom{p'}{2}\epsilon(\theta) + K$, which is less than θ, since $\theta \geq \log \log n \geq s_0$. This contradicts that $\mathrm{CC}(T) > \theta$.

Next, we note that one can improve the bounds given by Lemma 1 assuming a larger gap.

Lemma 2. Let $\epsilon(\theta) < \theta^\alpha$, and let $\Pi = (Y, N)$ be a promise problem, where $\Pi \leq_m^{NC^0} \epsilon$-GapMCSP via a γ-honest reduction f computed by an NC^0 circuit family C_n of depth $\leq d$, where $\gamma(n) \geq n^\beta$. Then for all δ such that $\delta_0 = \beta(1 - \alpha)/2^{d+1} > \delta > 0$ there is an n_0 such that for all $n \geq n_0$, if $N|_n \neq \emptyset$, then $c_0(\Pi|_n) \leq n^{1-\delta}$.

Proof. Similar to the proof of Lemma 1. $\qquad\blacksquare$

Next, we present a variant of Lemma 2, but restricted to the parameterized version of MCSP. This variant is useful in extending our non-hardness results to $\leq_T^{AC^0}$ reductions that make $n^{o(1)}$ queries.

Lemma 3. Let $\Pi = (Y, N)$ be a promise problem. If $\Pi \leq_m^{NC^0} \mathrm{MCSP}[\ell, g]$ with $\ell(m) = o(g(m)/m^\delta)$ for some $\delta > 0$, then $c_0(\Pi|_n) \leq n^\epsilon$ for some $\epsilon < 1$ for all but finitely many n where $N|_n \neq \emptyset$, where ϵ depends only on the depth of the NC^0 circuit family and δ.

Proof. Again, the proof is similar to the proof of Lemma 1. $\qquad\blacksquare$

We also can show that problems NC^0-reducible to ϵ-GapMCSP have bounded 0-block sensitivity. The proof is omitted.

Lemma 4. Let $\epsilon(\theta) = o(\theta)$, and let $\Pi = (Y, N)$ be a promise problem, where $\Pi \leq_m^{NC^0} \epsilon$-GapMCSP via a γ-honest reduction f computed by an NC^0 circuit family C_n of depth $\leq d$, where $\gamma(n) \geq \log \log n$. Then there is an n_0 (that depends only on ϵ and d) such that for all $n \geq n_0$, if $N|_n \neq \emptyset$, then $bs_0(\Pi|_n) < s$, where s is a constant that depends only on d.

4   Non-hardness Under Many-One AC^0 Reductions

To extend our non-hardness results to AC^0 we make use of a version of a theorem given in [1] that was first proved by [2,12] that says randomly restricting a family of AC^0 circuits yields a family of NC^0 circuits with high probability.

Lemma 5 (Lemma 7 in [1]). *Let C_n be a family of n-input (multi-output) AC^0 circuits. Then there exists an $a > 0$ such that for all $n \in \mathbb{N}$ there exists a restriction of C_n to $\Omega(n^{1/a})$ input variables that transforms C_n into a (multi-output) NC^0 circuit.*

Theorem 2. PARITY $\not\leq_m^{\mathsf{AC}^0}$ ϵ-GapMCSP *where $\epsilon(n) = o(n)$.*

Proof. Suppose not. Then there is a family of AC^0 circuits C_n that many-one reduces PARITY to ϵ-GapMCSP. By Lemma 5, there is an a such that we can transform each C_n into an NC^0 circuit D_m on $m = \Omega(n^{1/a})$ variables, computing a reduction f from either PARITY or ¬PARITY (depending on the parity of the restriction) to ϵ-GapMCSP. For each input x of length n, $f(x)$ is of the form $(T(x), \theta(x))$. Since there are only $O(\log n)$ output gates in the $\theta(x)$ field, and each output gate depends on only $O(1)$ input variables, all of the output gates for $\theta(x)$ can be fixed by setting only $O(\log n)$ input variables. Furthermore, we claim that there is some setting of these $O(\log n)$ input variables, such that the resulting value of θ is greater than $\log n / \log \log n$. If this were not the case, then the $\leq_m^{\mathsf{AC}^0}$ reduction of PARITY (or ¬PARITY) on $m = \Omega(n^{1/a})$ variables to ϵ-GapMCSP has the property that $\theta(x)$ is always less than $\log n / \log \log n$. But, as in the proof of Theorem 1.3 of [22], instances of MCSP where θ is $O(\log n / \log \log n)$ can be solved with a depth-3 AC^0 circuit of polynomial size. Thus this would give rise to AC^0 circuits for PARITY, contradicting the well-known circuit lower bounds of [2,12].

Thus we can set $O(\log n)$ additional variables, and obtain circuits that reduce PARITY (or ¬PARITY) on $m' = m - O(\log n) = \Omega(n^{1/(a+1)})$ variables to ϵ-GapMCSP, where furthermore this reduction satisfies the hypotheses of Lemmas 4 and 1. But this contradicts the fact that both PARITY and ¬PARITY on m' variables have 0-certificate complexity and 0-block-sensitivity m'.

5   Non-hardness Under Limited Turing AC^0 Reductions

With some work, we can extend our non-hardness results beyond many-one reductions to some limited Turing reductions.

In our proofs that deal with AC^0-Turing reductions, we will need to replace some oracle gates with "equivalent" hardware – where this hardware will provide answers that are consistent with *some* solution to the promise problem ϵ-GapMCSP, but might not be consistent with the particular solution that is provided as an oracle. In order to ensure that this doesn't cause any problems, we introduce the notion of a "sturdy" AC^0-Turing reduction:

Definition 3. *Let $\Pi_1 = (Y_1, N_1)$ and $\Pi_2 = (Y_2, N_2)$ be promise problems. A family $\{C_n\}$ of AC^0-oracle circuits is a sturdy $\leq^{AC^0}_T$ reduction from Π_1 to Π_2 if, for every pair of solutions S, S' to Π_2, every oracle gate G in C_n, and every $x \in Y_1 \cup N_1$, there is a solution S'' such that $C_n^S(x) = C_n^{S''}(x) = C_n^S[G \to S'](x)$, where the notation $C_n^S[G \to S']$ refers to the circuit C_n with oracle S, but where the oracle gate G answers queries according to the solution S' instead of S.*

Lemma 6. *Let Π be any promise problem. If $\Pi \leq^{AC^0}_{tt} \epsilon(n)$-GapMCSP via a reduction of depth d, then $\Pi \leq^{AC^0}_{tt} \epsilon(n)$-GapMCSP via a sturdy reduction of depth $5d$ with the same number of oracle gates. If $\Pi \leq^{AC^0}_T \epsilon(n)$-GapMCSP via a reduction of depth d, then $\Pi \leq^{AC^0}_T \epsilon(n)$-GapMCSP via a sturdy reduction of depth $5d$ with the same number of oracle gates.*

Proof. Due to space limitations, we provide a sketch only. We modify C_n, so that each oracle query is checked against queries that were asked "earlier" in the computation, and the computation uses only the oracle answer from the first time a query was asked. Since each query is given an answer that is consistent with *some* solution, the new circuit gives the same answers as a new solution (which we denote as S''). Since C_n is a reduction, we get the same answer when using S or S''.

Theorem 3. *Let $k \geq 1$, and let $\epsilon(n) = o(n)$. Then $\mathsf{PARITY} \not\leq^{AC^0}_{k\text{-}tt} \epsilon$-GapMCSP.*

With a larger gap, we can rule out nonadaptive reductions that use $n^{o(1)}$ queries.

Theorem 4. *Let $\epsilon(n) < n^\alpha$ for some $1 > \alpha > 0$. Then for any circuit family $\{C_n\}$ computing an $\leq^{AC^0}_{tt}$ reduction of PARITY to ϵ-GapMCSP, there is a $\delta > 0$ such that, for all large n, $\{C_n\}$ makes at least n^δ queries.*

If we consider the parameterized version of MCSP, rather than ϵ-GapMCSP, we obtain non-hardness even under $\leq^{AC^0}_T$ reductions.

Theorem 5. *Let $\ell(m) = o(g(m)/m^\delta)$ for some $1 > \delta > 0$. Then for any circuit family $\{C_n\}$ computing an $\leq^{AC^0}_T$ reduction of PARITY to $\mathsf{MCSP}[\ell, g]$, there is an $\epsilon > 0$ such that, for all large n, $\{C_n\}$ makes at least n^ϵ queries.*

6   Open Questions

There remain several open questions. The true complexity of MCSP remains a mystery. We have made progress in understanding the hardness of an approximation to MCSP, but how far can Theorem 2 be extended? Can we reduce the gap in the theorem to some constant factor approximations? Does the impossibility result hold when AC^0 is replaced with, say, $AC^0[2]$ many-one reductions? Does the DET-hardness of MKTP [7] also hold for MCSP, given that we have ruled out any large gap reduction?

Acknowledgments. Much of this work was done in the 2018 DIMACS REU, organized by Lazaros Gallos, Parker Hund, and many others. We thank Michael Saks, Shuichi Hirahara, Avishay Tal, and John Hitchcock for helpful discussions.

References

1. Agrawal, M., Allender, E., Rudich, S.: Reductions in circuit complexity: an isomorphism theorem and a gap theorem. J. Comput. Syst. Sci. **57**(2), 127–143 (1998)
2. Ajtai, M.: Σ_1^1-formulae on finite structures. Ann. Pure Appl. Log. **24**, 1–48 (1983)
3. Allender, E., Buhrman, H., Koucký, M., van Melkebeek, D., Ronneburger, D.: Power from random strings. SIAM J. Comput. **35**(6), 1467–1493 (2006)
4. Allender, E., Das, B.: Zero knowledge and circuit minimization. Inf. Comput. **256**, 2–8 (2017)
5. Allender, E., Grochow, J.A., van Melkebeek, D., Moore, C., Morgan, A.: Minimum circuit size, graph isomorphism, and related problems. SIAM J. Comput. **47**(4), 1339–1372 (2018)
6. Allender, E., Hellerstein, L., McCabe, P., Pitassi, T., Saks, M.: Minimizing disjunctive normal form formulas and AC^0 circuits given a truth table. SIAM J. Comput. **38**(1), 63–84 (2008)
7. Allender, E., Hirahara, S.: New insights on the (non)-hardness of circuit minimization and related problems. In: Proceedings of 42nd International Symposium on Mathematical Foundations of Computer Science (MFCS 2017) (2017)
8. Allender, E., Holden, D., Kabanets, V.: The minimum oracle circuit size problem. Comput. Complex. **26**(2), 469–496 (2017)
9. Allender, E., Koucký, M., Ronneburger, D., Roy, S.: The pervasive reach of resource-bounded Kolmogorov complexity in computational complexity theory. J. Comput. Syst. Sci. **77**(1), 14–40 (2011)
10. Allender, E., Loui, M.C., Regan, K.W.: Reducibility and completeness. In: Atallah, M.J., Blanton, M. (eds.) Algorithms and Theory of Computation Handbook, pp. 23–23. Chapman & Hall/CRC, New York (2010)
11. Arora, S.: AC^0-reductions cannot prove the PCP theorem (1995, unpublished Manuscript)
12. Furst, M., Saxe, J.B., Sipser, M.: Parity, circuits, and the polynomial-time hierarchy. Math. Syst. Theory **17**(1), 13–27 (1984)
13. Golovnev, A., Ilango, R., Impagliazzo, R., Kabanets, V., Kolokolova, A., Tal, A.: $AC^0[p]$ lower bounds against MCSP via the coin problem. Technical report TR19-018, Electronic Colloquium on Computational Complexity (ECCC) (2019). To appear in ICALP 2019
14. Hatami, P., Kulkarni, R., Pankratov, D.: Variations on the sensitivity conjecture. Theory Comput. Grad. Surv. **4**, 1–27 (2011)
15. Hirahara, S.: Non-black-box worst-case to average-case reductions within NP. In: 59th IEEE Symposium on Foundations of Computer Science (FOCS), pp. 247–258 (2018)
16. Hirahara, S., Santhanam, R.: On the average-case complexity of MCSP and its variants. In: Proceedings of 32nd Conference on Computational Complexity (CCC). LIPIcs-Leibniz International Proceedings in Informatics, vol. 79. Schloss Dagstuhl-Leibniz-Zentrum fuer Informatik (2017)

17. Hirahara, S., Watanabe, O.: Limits of minimum circuit size problem as oracle. In: Proceedings of 31st Conference on Computational Complexity (CCC). LIPIcs-Leibniz International Proceedings in Informatics, vol. 50. Schloss Dagstuhl-Leibniz-Zentrum fuer Informatik (2016)

18. Hitchcock, J., Pavan, A.: On the NP-completeness of the minimum circuit size problem. In: FSTTCS (2015)

19. Ilango, R.: $AC^0[p]$ lower bounds and NP-hardness for variants of MCSP. Technical report TR19-021, Electronic Colloquium on Computational Complexity (ECCC) (2019)

20. Impagliazzo, R., Kabanets, V., Volkovich, I.: The power of natural properties as oracles. In: LIPIcs-Leibniz International Proceedings in Informatics, vol. 102. Schloss Dagstuhl-Leibniz-Zentrum fuer Informatik (2018)

21. Kabanets, V., Cai, J.Y.: Circuit minimization problem. In: Proceedings of 32nd ACM Symposium on Theory of Computing (STOC), New York, NY, USA, pp. 73–79 (2000)

22. Murray, C.D., Williams, R.R.: On the (non) NP-hardness of computing circuit complexity. Theory Comput. **13**(1), 1–22 (2017)

23. Oliveira, I., Pich, J., Santhanam, R.: Hardness magnification near state-of-the-art lower bounds. In: Electronic Colloquium on Computational Complexity 158 (2018)

24. Oliveira, I., Santhanam, R.: Conspiracies between learning algorithms, circuit lower bounds and pseudorandomness. In: Proceedings of 32nd Conference on Computational Complexity (CCC), vol. 79, pp. 18:1–18:49. Schloss Dagstuhl-Leibniz-Zentrum fuer Informatik (2017)

25. Oliveira, I.C., Santhanam, R.: Hardness magnification for natural problems. In: Symposium on Foundations of Computer Science (FOCS), pp. 65–76 (2018)

26. Razborov, A., Rudich, S.: Natural proofs. In: Proceedings of 26th ACM Symposium on Theory of Computing (STOC), New York, NY, USA, pp. 204–213 (1994)

27. Rudow, M.: Discrete logarithm and minimum circuit size. Inf. Process. Lett. **128**, 1–4 (2017)

28. Trakhtenbrot, B.: A survey of Russian approaches to perebor (brute-force searches) algorithms. IEEE Ann. Hist. Comput. **6**(4), 384–400 (1984)

29. Vollmer, H.: Introduction to Circuit Complexity: A Uniform Approach. Springer, Heidelberg (2013). https://doi.org/10.1007/978-3-662-03927-4

Reconstructing a Convex Polygon
from Its ω-cloud

Elena Arseneva[1(✉)], Prosenjit Bose[2], Jean-Lou De Carufel[3],
and Sander Verdonschot[2]

[1] St. Petersburg State University, Saint-Petersburg, Russia
`ea.arseneva@gmail.com`
[2] Carleton University, Ottawa, Canada
`jit@scs.carleton.ca`, `sander@cg.scs.carleton.ca`
[3] University of Ottawa, Ottawa, Canada
`jdecaruf@uottawa.ca`

Abstract. An ω-wedge is the closed set of points contained between
two rays that are emanating from a single point (the apex), and are
separated by an angle $\omega < \pi$. Given a convex polygon P, we place the
ω-wedge such that P is inside the wedge and both rays are tangent to P.
The set of apex positions of all such placements of the ω-wedge is called
the ω-*cloud* of P.

We investigate reconstructing a polygon P from its ω-cloud. Previous work on reconstructing P from probes with the ω-wedge required
knowledge of the points of tangency between P and the two rays of the
ω-wedge in addition to the location of the apex. Here we consider the
setting where the *maximal* ω-cloud alone is given. We give two conditions
under which it uniquely defines P: (i) when $\omega < \pi$ is fixed/given, or (ii)
when what is known is that $\omega < \pi/2$. We show that if neither of these
two conditions hold, then P may not be unique. We show that, when the
uniqueness conditions hold, the polygon P can be reconstructed in $O(n)$
time with $O(1)$ working space in addition to the input, where n is the
number of arcs in the input ω-cloud.

1 Introduction

"Geometric probing considers problems of determining a geometric structure or
some aspect of that structure from the results of a mathematical or physical measuring device, a probe." [16, Page 1] Many probing tools have been studied in
the literature such as finger probes [7], hyperplane (or line) probes [8,12], diameter probes [15], x-ray probes [9,11], histogram (or parallel x-ray) probes [13],
half-plane probes [17] and composite probes [6,12] to name a few. For example,
diameter probes measure the width of the polygon at a certain direction; such
measurements in all directions yield the diameter function of the polygon.

E. A. was partially supported by F.R.S.-FNRS, and by SNF Early PostDoc Mobility
project P2TIP2-168563. P. B., J. C., and S. V. were partially supported by NSERC.

R. van Bevern and G. Kucherov (Eds.): CSR 2019, LNCS 11532, pp. 25–37, 2019.
https://doi.org/10.1007/978-3-030-19955-5_3

A geometric probing problem can be considered as a *reconstruction* problem. Can one reconstruct an object given a set of probes? For diameter probes this is not the case: there exists a class of (curved) *orbiform* shapes such as the celebrated *Reuleaux triangle* that have the same diameter function as the circle. What is more surprising is that this is not the case even for polygonal shapes: there are uncountable families of polygons with the same diameter function, where the polygons need not be regular, nor does their diameter function need to be constant [15]. In this paper we show, that an alternative probing device called an ω-*wedge* yields a function that is free from these drawbacks.

The ω-wedge, first studied by Bose et al. [4], is the closed set of points contained between two rays that are emanating from a single point, the *apex* of the wedge. The angle ω formed by the two rays is such that $0 < \omega < \pi$. A single probe of a convex n-gon P is *valid* when P is inside the wedge and both rays of the wedge are tangent to P, see Fig. 1a.[1] A valid probe returns the coordinates of the apex and of the two points of contact between the wedge and the polygon. Using this tool, a convex n-gon can be reconstructed using between $2n - 3$ and $2n + 5$ probes, depending on the value of ω and the number of *narrow vertices* (vertices whose internal angle is at most ω) in P [4]. As an ω-wedge rotates around P, the locus of positions of the apex of the ω-wedge describes a curve called an ω-*cloud*, see Fig. 1c. The ω-cloud finds applications in diverse geometric algorithms [1, 2, 14].

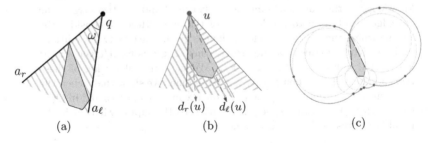

Fig. 1. A convex polygon P (filled with gray color), and (a) A minimal ω-wedge; (b) Narrow vertex u of P, wedges $W_\ell(u)$ and $W_r(u)$ (shaded with rising and falling tiling pattern, respectively) and their directions $d_\ell(u)$ and $d_r(r)$ (dashed lines); (c) The ω-cloud Ω of P: the arcs (bold curved lines), pivots (filled disk marks), and all the supporting circles (light thin curved lines).

The ω-cloud can be seen as a generalization of the diameter function introduced by Rao and Goldberg [15]. In their paper, a diameter probe consists of two parallel calipers turning around a convex object P in the plane. The function that returns the distance between these calipers as they turn around P

---

[1] In [4] probing with an ω-wedge is defined for a wider class of convex objects. Since here we focus on the objects being convex polygons, we restrict our definition accordingly.

is called a *diameter function*. Rao and Goldberg [15] show that two different convex polygons can have the same diameter function, and the function need not be a constant. This implies that recovering the exact shape and orientation of a convex n-gon given only its diameter function is not always possible and additional information is required. An ω-wedge can be seen as two non-parallel calipers turning around a convex object P in the plane. As we prove in this paper, two different convex polygons cannot have the same (non-constant) ω-cloud function, as opposed to the case of Rao and Goldberg's tool. This clearly shows the advantage of the ω-cloud against the latter tool.

Related to our probing method is also the method of Fleischer and Wang [10]. In their method, a convex n-gon P is placed inside a circle of radius 1. A camera that sees only the silhouette of P can be placed anywhere on the circle. The angle with which the camera sees P together with the position of the camera is a probe of P. Let α be the largest angle of P. They prove that if no two camera tangents overlap, then $\left\lceil \frac{3\pi}{\pi-\alpha} \right\rceil$ probes are necessary and sufficient. Otherwise, approximately $\left\lceil \frac{4\pi}{\pi-\alpha} \right\rceil$ probes are sufficient. In our method, the apex of the ω-wedge that turns around P can be seen as a camera. Thus our method is a variant of theirs because instead of fixing the circle on which the camera can move, we fix the angle from which P can be seen by the camera.

In this paper, we prove that no two convex polygons have the same ω-cloud (see Theorem 2 in Sect. 2). We further consider a harder, but more natural problem of reconstructing a polygon from its *maximal* ω-cloud, which is a variant of an ω-cloud where the consecutive arcs with the same supporting circle are merged in one arc (see Sect. 2.1). We give two conditions under which the maximal ω-cloud uniquely determines P: (i) if ω with $\omega < \pi$ is fixed/given, or (ii) without fixing ω, but with a guarantee that $\omega < \pi/2$. We show that if neither of the conditions (i), (ii) hold, then P may not be unique. Finally, we show that, when conditions (i) or (ii) hold, the polygon P can be reconstructed in $O(n)$ time and $O(1)$ working space in addition to the input, where n is the number of arcs in the input ω-cloud (see Sect. 3). Due to space constraints, some of the proofs are omitted. They can be found in the full version of this paper [3].

2   Properties of the ω-cloud

Let P be an n-vertex convex polygon in $\mathbb{R}^2$. For any vertex v of P, let $\alpha(v)$ be the internal angle of P at v. Let ω be a fixed angle with $0 < \omega < \pi$. Consider an ω-*wedge* W; recall that it is the set of points contained between two rays a_ℓ and a_r emanating from the same point q (the *apex* of W), such that the angle between the two rays is exactly ω. See Fig. 1a. We call the ray a_ℓ (resp., a_r) that bounds W from the left (resp., right) as seen from q, the *left* (resp., *right*) *arm* of W. We say that an ω-wedge W is *minimal* for P if P is contained in W and the arms of W are tangent to P. The *direction* of W is given by the bisector ray of the two arms of W. For each direction, there is a unique minimal ω-wedge.

Definition 1 (ω-cloud [5]). *The ω-cloud of P is the locus of the apices of all minimal ω-wedges for P.*

Let Ω denote the ω-cloud of P. We define an *arc* Γ of Ω as a maximal connected portion of Ω such that for every point of Γ the corresponding minimal ω-wedge is combinatorially the same, i.e., its left and right arm touch the same pair of vertices of P. Due to the inscribed angle theorem, an arc of the ω-cloud is a circular arc, and Ω consists of a circular sequence of circular arcs, where each pair of consecutive arcs shares an endpoint, see also Bose et al. [5] for more details. We refer to such a sequence as a *circular arc sequence*. The *supporting circle* of a circular arc is the circle containing the arc. The points on Ω that are the intersection of two consecutive arcs of Ω are called *pivots*. Note that if $\omega \geq \pi/2$, two consecutive arcs of the ω-cloud can lie on the same supporting circle. In this case we call the pivot separating them a *hidden pivot*. For example, points b, d, f in Fig. 3a are hidden pivots for the polygon $abcdef$, when $\omega = 5/6$. The total number of pivots (including hidden ones) is between n and $2n$ [5].

A vertex v of P is called *narrow* if the angle $\alpha(v)$ is at most ω. Note that a pivot of Ω coincides with a vertex of P if and only if that vertex is narrow. In this case, we also call such pivot *narrow*. If $\alpha(v) < \omega$, we call v (the vertex or the pivot) *strictly narrow*. The portion of Ω between two points $s, t \in \Omega$, unless explicitly stated otherwise, is the portion of Ω one encounters when traversing Ω from s to t clockwise, excluding s and t. We denote this portion of Ω by Ω_{st}. The *angular measure* of an arc Γ of Ω is the angle spanned by Γ, measured from the center of its supporting circle; the angular measure of a *portion* of an arc is defined similarly. For two points s, t on Ω, the *total angular measure* of Ω from s to t, denoted as $D_\Omega(s, t)$, is the sum of the angular measures of all arcs in Ω_{st} (including the at most two non-complete arcs of Ω).

Each point x in the interior of an arc corresponds to a unique minimal ω-wedge $W(x)$ with direction $d(x)$. Let u be a pivot of Ω. If u is not strictly narrow, u also corresponds to a unique minimal ω-wedge $W(u)$ with direction $d(u)$. Otherwise, u corresponds to a closed interval of directions $[d_\ell(u), d_r(u)]$, where the angle between $d_\ell(u)$ and $d_r(u)$ equals $\omega - \alpha(u)$. See Fig. 1b. Let $W_\ell(u)$ and $W_r(u)$ denote the minimal ω-wedges with apex at u and directions respectively $d_\ell(u)$ and $d_r(u)$. For points x on Ω that are not strictly narrow pivots, we define $d_r(x)$ and $d_\ell(x)$ both to be equal to $d(x)$, and $W_\ell(x)$ and $W_r(x)$ equal to $W(x)$.

We now give a crucial property of the ω-cloud.

Lemma 1. *Let s and t be two points on Ω such that Ω_{st} contains no narrow pivots. Then the angle between $d_r(s)$ and $d_\ell(t)$ is $D_\Omega(s, t)/2$.*

Proof. Suppose first that Ω_{st} is a single arc, see Fig. 2a. The angle α between $d_r(s)$ and $d_\ell(t)$ equals the angle β between the left arms of the two minimal ω-wedges corresponding to these directions, which in turn equals angle β'. By the inscribed angle theorem, β' is half the central angle spanning the arc Ω_{st}, which, by definition of the angular measure, equals $D_\Omega(s, t)/2$.

Now suppose that Ω_{st} consists of several arcs, which are either arcs of Ω or (at most two) connected portions of those arcs. By assumption, none of the pivots separating these arcs are narrow (although s and t may be, see the definition of Ω_{st}). Consider the change of direction of the minimal ω-wedge as its apex moves from s to t. After traversing an arc with endpoints u and v, by the above observation, the direction changes by exactly $D_\Omega(u,v)/2$. Since none of the pivots on Ω_{st} are narrow, there is no change in direction as the ω-wedge passes through each pivot. Therefore the total change in direction for the ω-wedge is the sum of the changes induced by the traversed arcs, that is, $D_\Omega(s,t)/2$.   □

The following are two implications of Lemma 1, and a simple fact that derives from the definition of the ω-cloud.

Corollary 1. *The sum of the angular measures of all arcs of Ω is $2(2\pi - \sum_{v \in S}(\omega - \alpha(v)))$, where S is the set of all narrow vertices of P. In particular, P has no strictly narrow vertices if and only if the sum of angular measures of the arcs of Ω is 4π.*

Corollary 2. *For any arc Γ of Ω, the angular measure of Γ is at most $2(\pi - \omega)$.*

Lemma 2. *Let u be a non-narrow pivot of Ω. Let x be the second point of intersection between the supporting circles of the two arcs of Ω adjacent to u (the first point of intersection is u itself). The minimal ω-wedge with the apex at u touches x with one of its arms.*

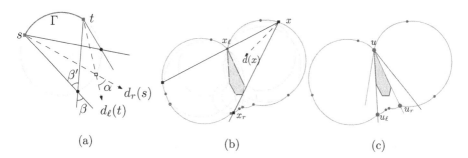

(a)   (b)   (c)

Fig. 2. (a) Illustration for the proof of Lemma 1, the case when Ω_{st} is a single arc. Vertices of P are marked as filled disks. (b) Point x in the interior of an arc of Ω, wedge $W(x)$ with its direction $d(x)$, points x_ℓ and x_r. (c) Narrow pivot u, wedges $W_\ell(u)$ and $W_r(u)$, points u_ℓ and u_r.

Characterization of narrow pivots. Let x be a point on Ω. It corresponds to two minimal ω-wedges $W_\ell(x)$, $W_r(x)$ (which coincide if x is not a narrow pivot). Consider the open ray of the right arm of $W_\ell(x)$. A small neighborhood of its origin is in the interior of the region bounded by Ω, thus the ray intersects Ω at least once. Among the points of this intersection, let x_ℓ be the one closest to x. Define the point x_r analogously for the left arm of $W_r(x)$. See Fig. 2b, c.

Lemma 3. *(a) Neither $\Omega_{x_\ell x}$ nor Ω_{xx_r} contains a narrow pivot. (b) If x is a narrow pivot, then $D_\Omega(x_\ell, x) = D_\Omega(x, x_r) = 2(\pi - \omega)$. (c) If x is not narrow, then either $D_\Omega(x, x_r) = 2(\pi - \omega)$, or x_r is the first narrow pivot following x in the clockwise direction. A symmetric statement holds for x_ℓ.*

The above lemma characterizes narrow pivots in terms of the position of the corresponding minimal ω-wedges. Now we characterize them in terms of the adjacent arcs and their measures, the information that can be used by the reconstruction algorithm.

Lemma 4. *Let u be a pivot of Ω, and let v and w be the points on Ω such that $D_\Omega(v, u) = D_\Omega(u, w) = 2(\pi - \omega)$.*

(a) If pivot u is narrow, then the supporting circles of all the arcs of Ω_{vw} pass through u.

(b) Pivot u is narrow, if at least one of the following conditions is satisfied: (i) Ω_{vu} consists of a single arc; (ii) there is an arc Γ of Ω_{vu} that is not incident to u, such that the supporting circle of Γ contains u.
A symmetric statement holds for Ω_{uw}.

Observation 1. *If u is a narrow pivot followed by a non-narrow pivot, then the condition (ii) of item (b) of Lemma 4 for Ω_{uw} is satisfied, and the supporting circle of the arc following w does not pass through u.*

Characterization of hidden pivots. Recall that a pivot is called hidden if its incident arcs have the same supporting circle.

Lemma 5. *Let u be a hidden pivot of Ω, let Γ_ℓ and Γ_r be the two arcs of Ω incident to u, and let v and w be the other endpoints of Γ_ℓ and Γ_r, respectively. Then v, u, and w are all narrow and each of Γ_ℓ, Γ_r has angular measure $2(\pi - \omega)$.*

Proof. Since u is a hidden pivot, Γ_ℓ and Γ_r are supported by the same circle C. Consider the minimal ω-wedge as its apex traverses Γ_ℓ. Its arms are touching two vertices of P, both lying on C. Since u is a hidden pivot, the vertex touched by the left arm of the wedge is u (otherwise, there would be no possibility to switch from Γ_ℓ to Γ_r at point u). Let a be the vertex of P touched by the right arm of the wedge. When the apex of the wedge reaches u, the wedge becomes $W_\ell(u)$, and its right arm is passing through a and u. If polygon P had a vertex between a and u, that vertex must lie outside W_ℓ, which is impossible. Thus a is actually v. Therefore, $D_\Omega(v, u) = D_\Omega(a, u)$, which by Lemma 3b equals $2(\pi - \omega)$. A symmetric argument for u, v, and Γ_r completes the proof.   □

We proceed with the main result of this section.

Theorem 2. *For a circular arc sequence Ω, there is at most one convex polygon P and one angle ω with $0 < \omega < \pi$ such that Ω is the ω-cloud of P.*

2.1 Maximal ω-cloud

Given the ω-cloud Ω of P, let the *maximal ω-cloud* of P, denoted Ω^*, be the result of merging all the pairs of consecutive arcs in Ω that have same supporting circle; equivalently Ω^* is the result of removing all the hidden pivots from Ω.

The maximal ω-cloud is a natural modification of the ω-cloud for the reconstruction task, see Sect. 3; it reflects the situation when as an input we are given a locus of the apices of all the minimal ω-wedges without any additional information, i.e., without the coordinates of the hidden pivots. The following two statements are corollaries of Lemma 5.

Lemma 6. *The maximal ω-cloud of P is a circle C if and only if $k = \pi/(\pi-\omega)$ is an integer and P is a regular k-gon inscribed in C.*

Proof. Suppose the maximal ω-cloud Ω^* of P is a circle C. Since all arcs of Ω^* of P are supported by the circle C, all the pivots of Ω are hidden. Applying Lemma 5 to each pivot, we obtain that the pivots of Ω are exactly the vertices of P, and each arc has measure $2(\pi-\omega)$. Therefore the number of vertices of P is $2\pi/(2(\pi-\omega))$, and the claim follows. The other direction of the claim directly follows from the definition of an ω-cloud. $\qquad\square$

Lemma 7. *An arc of Ω^* is greater than $2(\pi-\omega)$ if and only if this arc is a concatenation of at least two co-circular arcs of Ω separated by hidden pivots, each of measure $2(\pi-\omega)$.*

Lemmas 6 and 7 together with Theorem 2 imply a uniqueness result for Ω^*:

Theorem 3. *For an angle ω and a circular arc sequence Ω^* of at least two arcs, there is at most one convex polygon P such that Ω^* is the maximal ω-cloud of P.*

Another uniqueness result, useful for our reconstruction algorithm, follows from Theorem 2 since for $\omega < \pi/2$ the ω-cloud and the maximal ω-cloud coincide.

Corollary 3 (of Theorem 2). *For a circular arc sequence Ω, there is at most one convex polygon P and one angle $\omega < \pi/2$ such that Ω is the maximal ω-cloud of P.*

For the above statement it is necessary that $\omega < \pi/2$: indeed, Fig. 3a shows a construction that proves the following.

Proposition 1. *There is a circular arc sequence that is the maximal ω-cloud of P and the maximal ω'-cloud of P' for distinct angles ω, ω' and polygons P, P'.*

3 Reconstructing P from Its Maximal ω-cloud

In this section we let Ω^* be a circular arc sequence. Our goal is to reconstruct the convex polygon P for which Ω^* is the maximal ω-cloud for some angle ω, or determine that no such polygon exists.

Note that if Ω^* is a single arc, i.e., it is a circle C, then P is not unique. By Lemma 6, it is a regular $\pi/(\pi - \omega)$-gon inscribed in C. The position of its vertices on C is impossible to identify given only Ω^* and ω. Therefore we assume that Ω^* has at least two arcs.

First, in Sect. 3.1, we consider ω to be given, and such that $0 < \omega < \pi$; in this case the maximal ω-cloud of P may differ from its ω-cloud. Afterwards, in Sect. 3.2, we consider the setting where ω is not known, but it is known that $0 < \omega < \pi/2$; in this case the two variations of the ω-cloud coincide, and we use Ω instead of Ω^* to denote the input.

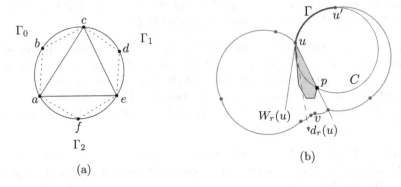

(a)

(b)

Fig. 3. (a) A circular arc sequence of 3 arcs, that is the maximal $2/3\pi$−cloud for triangle ace and the maximal $5/6$−cloud for hexagon $abcdef$. (b) Illustration for Lemma 8.

3.1   An ω-aware Reconstruction Algorithm

Here we assume that ω is given, and that $0 < \omega < \pi$. The main difficulty in the reconstruction task is caused by strictly narrow vertices, as the turn of the minimal ω-wedge at those vertices is not reflected in the ω-cloud, see Corollary 1. In our reconstruction algorithm, we first find all the non-hidden strictly narrow pivots of Ω^*, and then treat the connected portions of Ω^* between those pivots separately. Lemma 8 gives a procedure to process such a portion.

For two points u and v on Ω^*, let P_{uv} be the union of the edges and vertices of P touched by the arms of the minimal ω-wedge as its apex traverses Ω^*_{uv}. Since Ω^*_{uv} is connected, when the apex of the wedge traverses Ω^*_{uv}, for each of its two arms, the point of contact with P traverses a connected portion of P; thus P_{uv} consists of at most two connected portions of P. Note that it is possible that one of the portions is a single vertex.

Lemma 8. *Given a portion Ω^*_{uv} of Ω^* that does not contain any strictly narrow pivots, and the direction $d_r(u)$ of the rightmost minimal ω-wedge $W_r(u)$ of u, the portion P_{uv} of P that corresponds to Ω^*_{uv} can be reconstructed in time linear in the number of arcs in Ω^*_{uv}. This requires $O(1)$ working space in addition to the input.*

Proof. Let Γ be the arc of Ω_{uv}^* incident to u, let u' be the other endpoint of Γ, and let C be the supporting circle of Γ. See Fig. 3b. By knowing the value of ω and the direction $d_r(u)$, we determine the wedge $W_r(u)$. The intersection between the wedge $W_r(u)$ and the circle C determines the two vertices of P touched by the minimal ω-wedge as its apex traverses arc Γ. In Fig. 3b, these two points are u and p. The direction of the leftmost minimal ω-wedge at u', $W_\ell(u')$, is $d_r(u) + D_\Omega(u, u')/2$ due to Lemma 1. If u' is inside Ω_{uv}^*, then u' is not a strictly narrow vertex, and thus there is a unique minimal ω-wedge $W(u')$ at u', $W(u') = W_\ell(u')$. Therefore, for each arc of Ω_{uv}^* we find the pair of vertices of P that induces that arc. Moreover, by visiting the pivots of Ω_{uv}^* one by one, we find the vertices of each of the two chains of P_{uv} ordered clockwise. To avoid double-reporting vertices of P, we keep the startpoints of the two chains, and whenever one chain reaches the startpoint of the other one, we stop reporting the points of the former chain.

This procedure visits the pivots of Ω_{uv}^* one by one, and performs $O(1)$ operations at each pivot, namely, finding the intersection between a given wedge and a given circle. No additional information needs to be stored.   □

Reconstruction Algorithm. As an input, we are given an angle ω, $0 < \omega < \pi$, and a circular arc sequence Ω^* which is not a single circle. We now describe an algorithm to check if Ω^* is the ω-cloud of some convex polygon P, and to return P if this is the case. It consists of two passes through Ω^*, which are detailed below. During the first pass we compute a list S of all strictly narrow vertices of P that are not hidden pivots. With each such vertex u, we store the supporting lines of the two edges of P incident to u. In the second pass we use this list to reconstruct P.

First pass. We iterate through the pivots of Ω^* (recall that these are not hidden by the definition of Ω^*). For the currently processed pivot u, we maintain the point v on Ω^* such that $D_\Omega(v, u) = 2(\pi - \omega)$. If pivot u is narrow, we jump to the point on Ω^* at the distance $2(\pi - \omega)$ from u. Moreover, if u is strictly narrow, we add u to the list S. If u is not narrow, we process the next pivot of Ω^*. We now give the details.

Let Γ be the arc of Ω^* incident to u and following it in clockwise direction. Let Γ_r be the arc following Γ, and C_r be the supporting circle of Γ_r. We consider several cases depending on the angular measure $|\Gamma|$ of Γ:

(a) $|\Gamma| < 2(\pi - \omega)$. See Fig. 4.
   (i) Circle C_r passes through u, see Fig. 4a. Then u is narrow by Lemma 4b. By tracing Ω^*, find the point w on it such that $D_\Omega(u, w) = 2(\pi - \omega)$. In case u is strictly narrow (i.e., $\angle vuw < \omega$), add u to the list S with the lines through vu and uw (these are the intended lines due to Lemma 3). Set v to u and u to w (regardless the later condition).
   (ii) Circle C_r does not pass through u, see Fig. 4b. Then u is not narrow by Lemma 4a. Set u to be the other endpoint of Γ, and update v accordingly.

(b) $|\Gamma| = 2(\pi - \omega)$. Then u is narrow by Lemma 4b. Let w be the other endpoint of Γ. Update S, v, and u as in item a(i).

(c) $|\Gamma| = 2t(\pi - \omega)$ for some integer $t > 1$. Then Γ is in fact multiple arcs separated by hidden pivots, see Lemma 5 and Corollary 2. Let p be the other endpoint of Γ. Let w and w' be the points on Γ such that $D_\Omega(u, w) = 2(\pi - \omega)$ and $D_\Omega(w', p) = 2(\pi - \omega)$. Update S as in item a(i). Set u to p, and v to w'.

(d) Otherwise, stop and report that Ω^* is not the maximal ω-cloud of any polygon.

Before starting the above procedure, we need to find the starting positions v and u. We choose v arbitrarily and traverse Ω^* until we reach the corresponding position of u (i.e., $D_\Omega(v, u) = 2(\pi - \omega)$). Since we already have traversed the portion of Ω^* between v and u, in order to perform exactly one pass through Ω^*, we will have to finish the procedure as soon as the pointer u has reached the initial position of v (not the initial position of u). However, this is still enough for creating the complete list S. Indeed, by Lemma 3b, there is at most one narrow pivot between v and u. There is no such narrow pivot if and only if the left arm of the minimal ω-wedge $W_r(v)$ and the right arm of $W_\ell(u)$ coincide, and in this case there is nothing to add to S. If there is such a narrow pivot w, then neither u nor v is narrow. By Lemma 3c, both the left arm of wedge $W(v)$ and the right arm of $W(u)$ pass through w. Thus w can be found as their intersection.

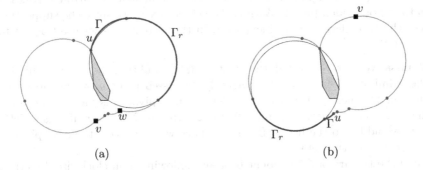

(a)   (b)

Fig. 4. First pass of the ω-aware algorithm: pivot u is (a) strictly narrow, (b) non-narrow

Second pass. In case list S is empty, we apply the procedure of Lemma 8 to the whole Ω^*. In particular, as both the start and the endpoint, we take the point x with which we completed the first pass of the algorithm; the point x' such that $D_\Omega(x', x) = 2(\pi - \omega)$ is already known from the first pass. Then $d_r(x) = d(x)$ is the direction of the minimal ω-wedge with the apex at x and the right arm passing through x'.

Suppose now that list S contains k vertices. They subdivide Ω^* into k connected portions that are free from strictly narrow non-hidden pivots. Each portion is treated as follows:

- If it is a single maximal arc of measure $2t(\pi-\omega)$, we simply separate it by $t-1$ equidistant points, and those points are exactly the vertices of the portion of P corresponding to the considered component of Ω^*, see Lemma 5.
- Otherwise it is a portion free from any strictly narrow pivots. We process this portion by the procedure of Lemma 8.

Correctness and complexity of this algorithm are analyzed in the following.

Theorem 4. *Given an angle ω such that $0 < \omega < \pi$, and a circular arc sequence Ω^* of n arcs with $n > 1$, there is an algorithm to check if Ω^* is the maximal ω-cloud of some convex polygon P, and to return P if this is the case. The algorithm works in $O(n)$ time, making two passes through the input, and it uses $O(1)$ working space. The constants in big-Oh depend only on ω.*

Proof. Consider the first pass of the algorithm. Cases a(i), b, and c are exactly the cases where u is a narrow pivot, see Lemmas 4 and 5. The case d corresponds to an impossible situation by Lemma 5 and Corollary 2. In case a(i), we update u to be w, thus we are not processing the pivots between u and w; Lemma 3 guarantees we do not miss any narrow pivot. Correctness of the second pass is implied by Lemmas 5 and 3a.

Each pivot of Ω^* is visited exactly once during the first pass (it is either checked as the pivot u, traversed in the case a(i), or skipped in the case c). The second pass as well visits each pivot once. The time spent during each such visit is $O(1+t)$, where t is the number of hidden pivots on the maximal arc adjacent to the visited pivot. The storage required by the algorithm is the storage required for the list S, i.e., it is $O(k)$.

Observe finally that by Lemma 3, for any pair u, v of narrow pivots of Ω^*, $D_\Omega(u, v) \geq 2(\pi - \omega)$. By Corollary 1 the total angular measure of the arcs of Ω^* is at most 4π. Thus the total number of narrow pivots of Ω^* (including the hidden pivots) is at most $\lfloor 2\pi/(\pi - \omega) \rfloor$, which is a constant if ω is fixed. $\square$

3.2   An ω-oblivious Reconstruction Algorithm

Assume that $\omega < \pi/2$. Thus there are no hidden pivots in the ω-cloud of P, and therefore the input sequence equals the (not maximal) ω-cloud Ω. We now give the reconstruction algorithm, which is the main result of this section.

Theorem 5. *Given a circular arc sequence Ω, there is an algorithm that finds the convex polygon P such that Ω is the ω-cloud of P for some angle ω with $0 < \omega < \pi/2$, if such a polygon exists. Otherwise, it reports that such a polygon does not exist. The algorithm works in $O(n)$ time, making two passes through the input, and it uses $O(1)$ working space.*

Proof. If Ω is a single arc, i.e., it is a circle, we return the negative answer, as a circle cannot be an ω-cloud for $\omega < \pi/2$ by Lemma 6. Below we assume that Ω has at least two arcs.

If the total angular measure of Ω is 4π, then no pivot is strictly narrow due to Corollary 1. Starting from any pivot u of Ω and applying Lemma 2, we find the placement of one of the arms of the minimal ω-wedge with the apex u. Since no pivot is strictly narrow, Lemma 3c determines whether it is the right or the left arm. This gives us the angle and the placement of the wedge. We can perform the procedure of Lemma 8 to reconstruct the polygon.

If the total angular measure of Ω is less than 4π, there must be a strictly narrow pivot. We perform a pass through Ω to find a maximal sequence of arcs whose supporting circles contain the same pivot u, and such that the sequence starts at u. If we found such sequence of at least two arcs, the distance between u and the endpoint of this sequence is $2(\pi - \omega)$ due to Lemma 4 and Observation 1. This determines the value of ω. We run the ω-aware reconstruction procedure of Sect. 3.1. If there is no such sequence of at least two arcs, then by Observation 1 all pivots must be narrow, but since $\omega < \pi/2$, the polygon P in that case must be a line segment. Whether Ω is an ω-cloud of a line segment for some ω can be checked in one pass through Ω.

We can combine finding the above maximal sequence with counting the total angular measure of Ω in one pass; this pass requires $O(1)$ working storage. The claim then follows from Theorem 4. $\square$

References

1. Abellanas, M., Bajuelos, A., Hurtado, F., Matos, I.: Coverage restricted to an angle. Oper. Res. Lett. **39**(4), 241–245 (2011)
2. Aloupis, G., et al.: Highway hull revisited. Comput. Geom. **43**(2), 115–130 (2010)
3. Arseneva, E., Bose, P., De Carufel, J.L., Verdonschot, S.: Reconstructing a convex polygon from its ω-cloud. ArXiv e-prints (2019). arXiv:1801.02162
4. Bose, P., Carufel, J.D., Shaikhet, A., Smid, M.: Probing convex polygons with a wedge. Comput. Geom. **58**, 34–59 (2016)
5. Bose, P., Mora, M., Seara, C., Sethia, S.: On computing enclosing isosceles triangles and related problems. Int. J. Comput. Geom. Appl. **21**(01), 25–45 (2011)
6. Bruckstein, A., Lindenbaum, M.: Reconstruction of polygonal sets by constrained and unconstrained double probing. Ann. Math. Artif. Intell. **4**, 345–361 (1991)
7. Cole, R., Yap, C.: Shape from probing. J. Algorithms **8**(1), 19–38 (1987)
8. Dobkin, D., Edelsbrunner, H., Yap, C.K.: Probing convex polytopes. In: Proceedings of STOC 1986, pp. 424–432. ACM (1986)
9. Edelsbrunner, H., Skiena, S.: Probing convex polygons with x-rays. SIAM J. Comput. **17**(5), 870–882 (1988)
10. Fleischer, R., Wang, Y.: On the camera placement problem. In: Dong, Y., Du, D.-Z., Ibarra, O. (eds.) ISAAC 2009. LNCS, vol. 5878, pp. 255–264. Springer, Heidelberg (2009). https://doi.org/10.1007/978-3-642-10631-6_27
11. Gardner, R.J.: X-rays of polygons. Discrete Comput. Geom. **7**, 281–293 (1992)
12. Li, S.: Reconstruction of polygons from projections. Inf. Process. Lett. **28**(5), 235–240 (1988)

13. Meijer, H., Skiena, S.: Reconstructing polygons from x-rays. Geometriae Dedicata **61**, 191–204 (1996)
14. Moslehi, Z., Bagheri, A.: Separating bichromatic point sets by minimal triangles with a fixed angle. Int. J. Found. Comput. Sci. **28**(04), 309–320 (2017)
15. Rao, A., Goldberg, K.: Shape from diameter: recognizing polygonal parts with a parallel-jaw gripper. Int. J. Robotic Res. **13**(1), 16–37 (1994)
16. Skiena, S.: Problems in geometric probing. Algorithmica **4**(4), 599–605 (1989)
17. Skiena, S.: Probing convex polygons with half-planes. J. Algorithms **12**(3), 359–374 (1991)

A Space-Efficient Parameterized Algorithm for the Hamiltonian Cycle Problem by Dynamic Algebraization

Mahdi Belbasi and Martin Fürer[✉]

Department of Computer Science and Engineering, Pennsylvania State University,
University Park, PA 16802, USA
{mub329,furer}@cse.psu.edu

Abstract. An NP-hard graph problem may be intractable for general graphs but it could be efficiently solvable using dynamic programming for graphs with bounded treewidth. Employing dynamic programming on a tree decomposition usually uses exponential space. In 2010, Lokshtanov and Nederlof introduced an elegant framework to avoid exponential space by algebraization. Later, Fürer and Yu modified the framework in a way that even works when the underlying set is dynamic, thus applying it to tree decompositions.

In this work, we design space-efficient algorithms to count the number of Hamiltonian cycles and furthermore solve the Traveling Salesman problem, using polynomial space while the time complexity is only slightly increased. This might be inevitable since we are reducing the space usage from an exponential amount (in dynamic programming solutions) to polynomial. We give an algorithm to count the number of Hamiltonian cycles in time $\mathcal{O}((4k)^d n M(n \log n))$ using $\mathcal{O}(kdn \log n)$ space, where $M(r)$ is the time complexity to multiply two integers, each of which being represented by at most r bits. Then, we solve the more general Traveling Salesman problem in time $\mathcal{O}((4k)^d poly(n))$ using space $\mathcal{O}(Wkdn \log n)$, where k and d are the width and the depth of the given tree decomposition and W is the sum of weights. Furthermore, this algorithm counts the number of Hamiltonian Cycles.

1 Introduction

Dynamic programming (DP) is largely used to avoid recomputing subproblems. It may decrease the time complexity, but it uses auxiliary space to store the intermediate values. This auxiliary space may go up to exponential in the value of the parameter. This means both the running time and the space complexity are exponential for some algorithms solving those NP-complete problems. Space complexity is a crucial aspect of algorithm design, because we typically run out of space before running out of time. To fix this issue, Lokshtanov and Nederlof [13] introduced a framework which works on a static underlying set. The problems they considered were Subset Sum, Knapsack, Traveling Salesman

© Springer Nature Switzerland AG 2019
R. van Bevern and G. Kucherov (Eds.): CSR 2019, LNCS 11532, pp. 38–49, 2019.
https://doi.org/10.1007/978-3-030-19955-5_4

(in time $\mathcal{O}(2^n k)$ using polynomial space), Weighted Steiner Tree, and Weighted Set Cover. They use DFTs, zeta transforms and Möbius transforms [16,17], taking advantage of the fact that working on zeta (or discrete Fourier) transformed values is significantly easier since the subset convolution operation converts to a pointwise multiplication operation. In all their settings, the input is a set or a graph which means the underlying set for the subproblems is static. Fürer and Yu [9] changed this approach modifying a dynamic programming algorithm applied to a tree decomposition (instead of the graph itself). The resulting algorithm uses only polynomial space and the running time does not increase drastically. By working with tree decompositions, they obtain parameterized algorithms which are exponential in the tree-depth and linear in the number of vertices. If the tree decomposition has a bounded width, then the algorithm is both fast and space-efficient. In this setting, the underlying set is not static anymore, because they are working with different bags of nodes. They show that using algebraization helps to save space even if the underlying set is dynamic. They consider perfect matchings in their paper. In recent years, there have been several results in this field where algebraic tools are used to save space when DP algorithms are applied to NP-hard problems. In 2018, Pilipczuk and Wrochna [15] applied a similar approach to solve the Minimum Dominating Set problem. Although they have not directly used these algebraic tools but it is a similar approach (in time $\mathcal{O}(3^d poly(n))$ using $poly(n)$ space).

We have to mention that there is no general method to automatically transform dynamic programming solutions to polynomial space solutions while increasing the tree-width parameter to the tree-depth in the running time.

One of the interesting NP-hard problems in graph theory is Hamiltonian Cycle. It seems harder than many other graph problems. We are given a graph and we want to find a cycle visiting each vertex exactly once. The naive deterministic algorithm for the Hamiltonian Cycle problem and the more general Traveling Salesman problem runs in time $\mathcal{O}(n!)$ using polynomial space. Later, deterministic DP and inclusion-exclusion algorithms for these two problems running in time $\mathcal{O}^*(2^n)$[1] using exponential space were given in [1,10,12]. The existence of a deterministic algorithm for Hamiltonian cycle running in time $\mathcal{O}((2 - \epsilon)^n)$, for a fixed $\epsilon > 0$ is still an open problem. There are some randomized algorithms which run in time $\mathcal{O}((2 - \epsilon)^n)$, for a fixed $\epsilon > 0$ like the one given in [3]. Although, there is no improvement in deterministic running time, there are some results on Parameterized algorithms. In 2011, Cygan et al. [8] designed a Parameterized algorithm for the Hamiltonian Cycle problem, which runs in time $4^k |V|^{\mathcal{O}(1)}$, where k is the treewidth. In 2015, Bodlaender et al. [4] introduced two deterministic single exponential time algorithms for Hamiltonian Cycle: One based on pathwidth running in time $\tilde{\mathcal{O}}(6^p p^{\mathcal{O}(1)} n^2)$[2] and the other is based on treewidth running in time $\tilde{\mathcal{O}}(15^k k^{\mathcal{O}(1)} n^2)$, where p and k are the pathwidth and the treewidth respectively. The authors also solve the Traveling Salesman problem in time $\mathcal{O}(n(2 + 2^\omega)^p p^{\mathcal{O}(1)})$ if a path decomposition of width p of G is given,

---

[1] $\mathcal{O}^*$ notation hides the polynomial factors of the expression.

[2] $\tilde{\mathcal{O}}$ notation hides the logarithmic factors of the expression.

and in time $\mathcal{O}(n(7 + 2^{\omega+1})^k k^{\mathcal{O}(1)})$, where ω denotes the matrix multiplication exponent. One of the best known upper bounds for ω is 2.3727 [18]. They do not consider the space complexity of their algorithm and as far as we checked it uses exponential space.

Recently, Curticapean et al. [6] showed that there is no positive ϵ such that the problem of counting the number of Hamiltonian cycles can be solved in $\mathcal{O}^*((6-\epsilon)^p)$ time assuming SETH. Here p is the width of the given path decomposition of the graph. They show this tight lower bound via matrix rank.

2 Preliminaries

In this section we review notations that we use later.

2.1 Tree Decomposition

In this work, we use standard definitions for tree decompositions, nice tree decompositions, and treewidth. We denote tree decompositions by $\mathcal{T} = (V_{\mathcal{T}}, E_{\mathcal{T}})$. We use k for the treewidth, and B_x for the bag of node x. Different from [7], we allow the leaves in a nice tree decomposition to have non-empty bags. It has been shown that any given tree decomposition can be converted to a nice tree decomposition with the same width in polynomial time [11].

Definition 1. *The depth of a tree decomposition is the maximum number of distinct vertices in the union of all bags on a path from the root to the leaf. We use d to denote the depth of a given tree decomposition (Although, this is not the standard definition of the depth of a tree decomposition, it is equivalent to the standard definition and is more applicable in our case).*

We have to mention that this is different from the depth of a tree. This is the depth of a tree decomposition as defined above.

Definition 2. *The tree-depth of a graph G, is the minimum over the depths of all tree decompositions of G. We use $d(G)$ (simply d) to denote the tree-depth of a graph.*

Following is the relationship between these parameters in a given graph G:

Lemma 1. *(see [14, Corollary 2.5] and [5])*
    For any connected graph G, $d(G) \geq k + 1 \geq \frac{d(G)}{\log_2 |V(G)|}$.

2.2 Algebraic Tools to Save Space

When we use dynamic programming to solve a graph problem on a tree decomposition, it usually uses exponential space. Lokshtanov and Nederlof converted some algorithms using subset convolution or union product into transformed version in order to reduce the space complexity. Later, Fürer and Yu [9] also used this approach in a dynamic setting, based on tree decompositions to solve the

Perfect Matching problem. In this work, we introduce algorithms to count the number of Hamiltonian cycles and also solve the Traveling Salesman problem. First, let us recall some definitions.

Let $\mathcal{R}[2^{\mathcal{U}}]$ be the set of all functions from the power set of the universe $\mathcal{U}$ to the ring $\mathcal{R}$. The operator $\oplus$ is the pointwise addition and the operator $\odot$ is the pointwise multiplication.

Definition 3. *A* relaxation *of a function* $f \in \mathcal{R}[2^{\mathcal{U}}]$ *is a sequence of functions* $\{f^i : f^i \in \mathcal{R}[2^{\mathcal{U}}], 0 \leq i \leq |\mathcal{U}|\}$, *where* $\forall\, 0 \leq i \leq |\mathcal{U}|$ *and* $\forall X \subseteq \mathcal{U}$, $f^i[X]$ *is defined as:*

$$f^i[X] = \begin{cases} 0 & \text{if } i < |X|, \\ f[X] & \text{if } i = |X|, \\ \text{arbitrary value} & \text{if } i > |X|. \end{cases} \tag{1}$$

Definition 4. *The* zeta transform *of a function* $f \in \mathcal{R}[2^{\mathcal{U}}]$ *is defined as:*

$$\zeta f[X] = \sum_{Y \subseteq X} f[Y]. \tag{2}$$

Definition 5. *The* Möbius transform *of a function* $f \in \mathcal{R}[2^{\mathcal{U}}]$ *is defined as:*

$$\mu f[X] = \sum_{Y \subseteq X} (-1)^{|X \setminus Y|} f[Y]. \tag{3}$$

Lemma 2. *The Möbius transform is the inverse of the zeta transform and vice versa, i.e.*

$$\mu(\zeta f[X]) = \zeta(\mu f[X]) = f[X]. \tag{4}$$

See [16,17] for the proof.

Definition 6. *Given* $f, g \in \mathcal{R}[2^{\mathcal{U}}]$ *and* $X \in 2^{\mathcal{U}}$, *the* Subset Convolution *of* f *and* g *denoted* $(f *_R g)$ *is defined as:*

$$(f *_R g)[X] = \sum_{X_1 \subseteq X} f(X_1) g(X \setminus X_1). \tag{5}$$

Definition 7. *Given* $f, g \in \mathcal{R}[2^{\mathcal{U}}]$ *and* $X \in 2^{\mathcal{U}}$, *the* Union Product *of* f *and* g *denoted* $(f *_u g)$ *is defined as:*

$$(f *_u g)[X] = \sum_{X_1 \cup X_2 = X} f(X_1) g(X_2). \tag{6}$$

Theorem 1. *([2]) Applying the zeta transform to a union product operation, results in the pointwise multiplication of zeta transforms of the outputs, i.e., given* $f, g \in \mathcal{R}[2^{\mathcal{U}}]$ *and* $X \in 2^{\mathcal{U}}$,

$$\zeta(f *_u g)[X] = (\zeta f) \odot (\zeta g)[X]. \tag{7}$$

All of the previous works, which either used DFT or the zeta transform on a given tree decomposition (such as [9], and [13]), have one common central property that the recursion in the join nodes, can be represented by a formula using a union product operation, which is a complicated operation compared to pointwise multiplication. That is why taking the zeta transform of a formula having union products makes the computation easier. As noted earlier in Theorem 1, taking the zeta transform of such a term, will result in a term having pointwise multiplication instead of union product. Instead of working on the original computation (bottom up on a nice tree decomposition), we are working on a mirrored computation on the same nice tree decomposition where the zeta-transformed values in a node are computed directly from the zeta-transformed values at the children. After doing a computation over the zeta transformed values, we can apply the Möbius transform on the outcome to get the main result (based on Theorem 2). While the direct computation keeps track of exponentially many intermediate values, the computation over the zeta transformed values partitions into exponentially many branches, and they can be executed one after another. Later, we show that this approach improves the space complexity using only polynomial space instead of exponential space.

3   Counting the Number of Hamiltonian Cycles

We are given a connected graph $G = (V, E)$ and a nice tree decomposition τ of G of width w. If H is a Hamiltonian cycle, then the induced subgraph of H on τ_x (called $H[V_x]$, where τ_x is the subtree rooted at node x, and V_x is the union of all bags in τ_x) is a set of disjoint paths with endpoints in B_x (see Fig. 1).

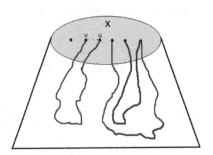

Fig. 1. $H[V_x]$ is a set of paths with endpoints in B_x

Definition 8. *A* pseudo-edge *is a pair of endpoints of a path of length ≥ 2 with endpoints in B_x and interior vertices in $V_x \setminus B_x$. We use the $\langle , \rangle$ notation for the pseudo-edges. E.g., in Fig. 1, $p = \langle u, v \rangle$ is a pseudo-edge (it does not imply that there is an edge between u and v, it just says that there is a path of length at least two in $H[V_x]$ where u and v are its endpoints). The $\langle , \rangle$ notation is a symmetrical notation since our paths are undirected, i.e., $\langle u, v \rangle = \langle v, u \rangle$. Each path is associated with a pseudo-edge.*

Lemma 3. *The degree of all vertices in $H[V_x]$ is at most 2.*

Proof. $H[V_x]$ is a subgraph of the cycle H.   □

Let X be the union of pseudo-edges (in B_x). Let $[B_x]^2$ be the set of two-element subsets of B_x, and let $X \subseteq [B_x]^2$. Let S_X be the union of vertices involved in X. Then, S_X is a subset of B_x. Let F_x be the vertices of $V_x \setminus B_x$ which are not present in the bag of the parent of x. Define $f_x[X]$ to be the number of sets of disjoint paths (except in their endpoints where they can share a vertex) whose pseudo-edge set is X (remember S_X is the union of vertices involved in X) visiting vertices of F_x exactly once (they can also visit vertices which are not in F_x, but we require them to visit at least vertices in F_x since they are not present in the proper ancestors of x). We require the root to have empty bag, thus $f_r[\emptyset]$ gives us the number of possible Hamiltonian cycles in G, where r is the root of τ. Now, we show how to compute the values of f_x for all types of nodes.

3.1 Computing $f_x[X]$

We first show how $f_x[X]$ could be computed by dynamic programming using exponential space. We introduce the recursive formula for any kind of nodes. In the next section, we will show how to compute $\zeta f_x[X]$ directly.

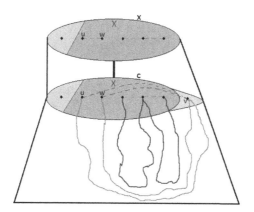

Fig. 2. Forget node x forgetting vertex v with the child c.

– **Leaf node:** Assume x is a leaf node.

$$f_x[X] = \begin{cases} 1 & \text{if } X = \emptyset, \\ 0 & \text{otherwise.} \end{cases} \tag{8}$$

Since x is a leaf node, there is no path through $V_x \setminus B_x$, so for all non-empty sets X of pseudo-edges, $f_x[X]$ is zero, and for $X = \emptyset$, there is only one set of paths, which is empty.

- **Forget node:** Assume x is a forget node (forgetting vertex v) with a child c, where $B_x = B_c \setminus \{v\}$. Any pseudo-edge $\langle u, w \rangle \in X \subseteq [B_x]^2$ can define a path starting from u, going to v possibly through $V_c \setminus B_c$ and then going to w possibly through $V_c \setminus B_c$. Here, either or both pieces of the path (from u to v, and/or from v to w) can consist of single edges (Fig. 2).

$$f_x[X] = \sum_{\langle u,w \rangle \in X} \sum_{Q \subseteq \{\langle u,v \rangle, \langle v,w \rangle\}} d_Q \, f_c[X \setminus \{\langle u,w \rangle\} \cup Q], \qquad (9)$$

where $d_Q = \begin{cases} 1 & \text{if } \langle u,v \rangle \in Q \text{ or } \{u,v\} \in E, \text{ and } \langle v,w \rangle \in Q \text{ or } \{v,w\} \in E \\ 0 & \text{otherwise.} \end{cases}$

- **Introduce vertex node:** Assume x is an introduce vertex node (introducing vertex v) with a child c, where $B_x = B_c \cup \{v\}$. The vertex v cannot be an endpoint of a pseudo-edge because v has no neighbors in V_c.

$$f_x[X] = \begin{cases} f_c[X] & \text{if } v \notin S_X, \\ 0 & \text{otherwise.} \end{cases} \qquad (10)$$

- **Join node:** Assume x is a join node with two children c_1 and c_2, where $B_x = B_{c_1} = B_{c_2}$. For any given $X \subseteq [B_x]^2$, X can be partitioned into two sets X' and $X \setminus X'$ and each of them can be the set of pseudo-edges for one of the children (Fig. 3).

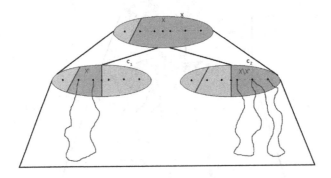

Fig. 3. Join node x with two children c_1 and c_2.

The number of such paths associated with X through V_x is equal to the sum of the products of the number of corresponding paths associated with X' and $X \setminus X'$ through V_{c_1} and V_{c_1} respectively.

$$f_x[X] = \sum_{X' \subseteq X} f_{c_1}[X'] f_{c_2}[X \setminus X'] = (f_{c_1} *_R f_{c_2})[X]. \qquad (11)$$

Here we get subset convolution and we have to convert it to union product to be able to use zeta transform. We do this conversion in the next subsection.

3.2   Computing $\zeta F_x[X]$

In this subsection, first we compute the relaxations of $f_x[X]$ for all kinds of nodes, and then we apply the zeta transform to the relaxations. In the following section let $\{f_x^i\}_{0 \le i \le \frac{k+1}{2}}$ be a relaxation of f_x.

- **Leaf node:** Assume x is a leaf node. Since $f_x[\emptyset] = 1$ and for any $X \ne \emptyset$: $f_x[X] = 0$, we can choose $f_x^i[X] = f_x[X]$ for all i and X. Then

$$(\zeta f_x^i)[X] = 1, \text{ for all } i \text{ and } X. \tag{12}$$

- **Forget node:** Assume x is a forget node (forgetting vertex v) with a child c, where $B_x = B_c \setminus \{v\}$. Thus,

$$f_x^i[X] = \sum_{\langle u,w \rangle \in X} \sum_{Q \subseteq \{\langle u,v \rangle, \langle v,w \rangle\}} d_Q \, f_c^{i'(Q)}[X \setminus \{\langle u,w \rangle\} \cup Q], \tag{13}$$

where $i'(Q) = i - 1 + |Q|$, i.e., $i'(Q)$ is the number of pseudo-edges in $X \setminus \{\langle u,w \rangle\} \cup Q$. Now we apply the zeta transform:

$$(\zeta f_x^i)[X] = \sum_{Y \subseteq X} f_x^i[Y]$$

$$= \sum_{Y \subseteq X} \sum_{\langle u,w \rangle \in Y} \sum_{Q \subseteq \{\langle u,v \rangle, \langle v,w \rangle\}} d_Q \, f_c^{i'(Q)}[Y \setminus \{\langle u,w \rangle\} \cup Q]$$

$$= \sum_{\langle u,w \rangle \in X} \sum_{Q \subseteq \{\langle u,v \rangle, \langle v,w \rangle\}} \sum_{\{\langle u,w \rangle\} \subseteq Y \subseteq X} d_Q \, f_c^{i'(Q)}[Y \setminus \{\langle u,w \rangle\} \cup Q]$$

$$= \sum_{\langle u,w \rangle \in X} \sum_{Q \subseteq \{\langle u,v \rangle, \langle v,w \rangle\}} d_Q \sum_{Y \subseteq (X \setminus \{\langle u,w \rangle\})} f_c^{i'(Q)}[Y \cup Q] \tag{14}$$

We now express $e_Q = \sum_{Y \subseteq (X \setminus \{\langle u,w \rangle\})} f_c^{i'(Q)}[Y \cup Q]$ by ζ-transforms, depending on the size of Q. We use the abbreviation $X' = X \setminus \{\langle u,w \rangle\}$.
If $Q = \emptyset$, then $e_Q = \sum_{Y \subseteq X'} f_c^{i-1}[Y] = \zeta f^{i-1}[X']$.
If $Q = \{\langle u,v \rangle\}$ or $Q = \{\langle v,w \rangle\}$, then $e_Q = \sum_{Y \subseteq X'} f_c^i[Y \cup Q] = \zeta f_c^i[X' \cup Q] - \zeta f_c^i[X']$.
If $Q = \{\langle u,v \rangle, \langle v,w \rangle\}$, then

$$e_Q = \sum_{Y \subseteq X'} f_c^{i+1}[Y \cup Q] \tag{15}$$

$$= \zeta f_c^{i+1}[X' \cup Q] - \zeta f_c^{i+1}[X' \cup \{\langle u,v \rangle\}]$$
$$- \zeta f_c^{i+1}[X' \cup \{\langle v,w \rangle\}] + \zeta f_c^{i+1}[X'].$$

With these sums computed, we can now express $(\zeta f_x^i)[X]$ more concisely.

$$(\zeta f_x^i)[X] = \sum_{\langle u,w \rangle \in X} ((d_\emptyset - d_{\{\langle u,v \rangle\}} - d_{\{\langle v,w \rangle\}} + d_{\{\langle u,v \rangle, \langle v,w \rangle\}}) \zeta f_c^{i-1}[X']$$

$$+ d_{\{\langle u,v \rangle\}} \zeta f_c^i[X' \cup \{\langle u,v \rangle\}] + d_{\{\langle v,w \rangle\}} \zeta f_c^i[X' \cup \{\langle v,w \rangle\}]$$

$$+ d_{\{\langle u,v \rangle, \langle v,w \rangle\}} \zeta f_c^{i+1}[X' \cup \{\langle u,v \rangle, \langle v,w \rangle\}]). \tag{16}$$

Note that $d_\emptyset = 1$ if $\{\langle u, v\rangle, \langle v, w\rangle\} \subseteq E$, $d_{\{\langle u,v\rangle\}} = 1$ if $\langle v, w\rangle \in E$, $d_{\{\langle v,w\rangle\}} = 1$ if $\langle u, v\rangle \in E$, and $d_{\{\langle u,v\rangle, \langle v,w\rangle\}}$ is always 1, while otherwise $d_Q = 0$. This implies that for every $\langle u, w\rangle$ in the previous equation at least one of the 4 coefficients is 0. Therefore, in each forget node, we have at most a $3|X|$ fold branching.

- **Introduce vertex node:** Assume x is an introduce vertex node (introducing vertex v) with a child c, where $B_x = B_c \cup \{v\}$. Let X_v be the set of pseudo-edges having v as one their endpoints. Therefore,

$$f_x^i[X] = \begin{cases} f_c^i[X] & \text{if } v \notin S_X, \\ 0 & \text{otherwise.} \end{cases} \tag{17}$$

$$(\zeta f_x^i)[X] = \sum_{Y \subseteq X} f_x^i[Y] = \sum_{Y \subseteq (X \setminus X_v)} f_c^i[Y] = (\zeta f_c^i)[X \setminus X_v]. \tag{18}$$

- **Join node:** Assume x is a join node with two children c_1 and c_2, where $B_x = B_{c_1} = B_{c_2}$. To compute f_x on a join node, we can use Eq. 11. In order to convert the subset convolution operation to pointwise multiplication, first we need to convert the subset convolution to a union product, and then we are able to use Theorem 1. To convert subset convolution to union product we introduce a relaxation of f_x. Let $f_x^i[X] = \sum_{j=0}^i (f_{c_1}^j *_u f_{c_2}^{i-j})[X]$ be a relaxation of f_x.

$$(\zeta f_x^i)[X] = \sum_{j=0}^i (\zeta f_{c_1}^j)[X] \cdot (\zeta f_{c_2}^{i-j})[X], \text{ for } 0 \le i \le k, \tag{19}$$

where k is the treewidth of τ.

To summarize, we present the following algorithm (Algorithm 1) for the Hamiltonian Cycle problem where a tree decomposition τ is given.

Algorithm 1. Counting the total number of the possible Hamiltonian cycles in a graph given by a nice tree decomposition

---

Input   : A nice tree decomposition τ with root r
return   : $(\zeta f)(r, \emptyset, 0)$ $//\zeta f_r^0[\emptyset]$.
procedure $(\zeta f)(x, X, i)$: $//(\zeta f)(x, X, i)$ is the representation of $(\zeta f_x^i)[X]$
    if x is a leaf node:
        return 1.
    if x is a forget node: //forgetting v
        return the value according to Equation 16.
    if x is an introduce vertex node: //introducing v
        return $(\zeta f)(c, X \setminus X_v, i)$.
    if x is a join node:
        return $\sum_{j=0}^i (\zeta f)(c_1, X, j) \cdot (\zeta f)(c_2, X, i - j)$.
end procedure

Theorem 2. *Given a graph $G = (V, E)$ and a tree decomposition τ of G, we can compute the total number of the Hamiltonian cycles of G in time $\mathcal{O}((4k)^d n M(n \log n))$ and in space $\mathcal{O}(kdn \log n)$ by using Algorithm 1 where k and d are the width and the depth of τ respectively, and $M(n)$ is the time complexity to multiply two numbers which can be encoded by at most n bits.*

Sketch of Proof. The correctness proof has been explained in 3.1 and 3.2. Running time and space complexity analysis can be found in the appendix but here is the summarized version: Branching happens only in forget nodes ($\leq d$ forget nodes from a path from the root to a leaf), where we branch into $\leq k$ branches. According to the formula for forget node: $\{u, v\}$ and $\{v, w\}$ each can be an edge or a pseudo-edge (four possibilities). There are at most $n!$ Hamiltonian cycles, so the value of f_x can be presented by at most $n \log n$ bits. It takes $M(n \log n)$ time to multiply numbers with this length. On the other hand, there are n vertices in G. All together, the running time is $\mathcal{O}((4k)^d n M(n \log n))$.

For space complexity, we only keep results for one strand at a time. There are $\leq d$ vertices on each path from the root to a leaf. On each path, we keep track of vertices in a bag (at most $k + 1$), and the number of disjoint paths can be presented by $n \log n$ bits. Therefore, the space complexity is $\mathcal{O}(kdn \log n)$. $\square$

4   The Traveling Salesman Problem

In this section, we discuss the Traveling Salesman problem (TSP).

Definition 9. *Traveling Salesman. Let $G = (V, E)$ be an undirected graph with weighted edges (nonnegative integer). In the traveling Salesman problem we are asked to find a cycle (if there is any) that visits all of the vertices exactly once with minimum cost.*

In order to solve this problem, we work with the ring of polynomials $\mathbb{Z}[x]$, where x is a variable. Our algorithm computes the polynomial $P_x(y)[X] = \sum_{i=0}^{W} a_i^x[X] y^i$, where $a_i^x[X]$ is the number of solutions in τ_x associated with X with cost i, and W is the sum of the weights of all edges.

4.1   Computing the Hamiltonian Cycles of All Costs

To find the answer to TSP, we have to find the first nonzero coefficient of $P_r(x)$, where r is the root of the given nice tree decomposition. As we did for the Hamiltonian cycle problem, we show how compute this polynomial recursively for all kinds of nodes.

- **Leaf node:** Assume x is a leaf node.

$$P_x(y)[X] = \begin{cases} 1 & \text{if } X = \emptyset, \\ 0 & \text{otherwise.} \end{cases} \tag{20}$$

- **Forget node:** Assume x is a forget node (forgetting vertex v) with a child c, where $B_x = B_c \setminus \{v\}$.

$$P_x(y)[X] = \sum_{\langle u,w \rangle \in X} \sum_{Q \subseteq \{\langle u,v \rangle, \langle v,w \rangle\}} d_Q(P_c(y))[X \setminus \{\langle u,w \rangle\} \cup Q], \quad (21)$$

where d_Q is defined in a similar way as in Eq. 9.

- **Introduce vertex node:** Assume x is an introduce vertex node (introducing vertex v) with a child c, where $B_x = B_c \cup \{v\}$.

$$P_x(y)[X] = \begin{cases} P_c(y)[X] & \text{if } v \notin S_X, \\ 0 & \text{otherwise.} \end{cases} \quad (22)$$

- **Join node:** Assume x is a join node with two children c_1 and c_2, where $B_x = B_{c_1} = B_{c_2}$.

$$f_x[X] = \sum_{X' \subseteq X} P_{c_1}(y)[X'] P_{c_2}(y)[X \setminus X']. \quad (23)$$

We skip the zeta transform part because it is similar to the Hamiltonian Cycle case.

Theorem 3. *Given a graph $G = (V, E)$ and a tree decomposition τ of G, we can solve the Traveling Salesman problem for G in time $\mathcal{O}((4k)^d \cdot poly(n))$ and in space $\mathcal{O}(\mathcal{W}kdn \log n)$ where $\mathcal{W}$ is the sum of the weights, k and d are the width and the depth of the tree decomposition τ respectively.*

The proof is omitted due to space constraints.

4.2   Conclusion

In this work, we solved Hamiltonian Cycle with polynomial space complexity and Traveling Salesman with pseudo-polynomial space complexity, while the running time is polynomial in size of the given graph and exponential in the tree-depth. Our algorithms for both problems rely on modifying a DP approach such that instead of storing all possible intermediate values, we keep track of zeta transformed values, which was first introduced in [13] for static underlying sets, and later in [9] for dynamic underlying sets.

References

1. Bellman, R.: Dynamic programming treatment of the travelling salesman problem. J. ACM (JACM) **9**(1), 61–63 (1962)
2. Björklund, A., Husfeldt, T., Kaski, P., Koivisto, M.: Fourier meets Möbius: fast subset convolution. In: Proceedings of the Thirty-Ninth Annual ACM Symposium on Theory of Computing, pp. 67–74. ACM (2007)

3. Björklund, A., Kaski, P., Koutis, I.: Directed Hamiltonicity and out-branchings via generalized Laplacians. In: 44th International Colloquium on Automata, Languages, and Programming, ICALP 2017, Warsaw, Poland, 10–14 July 2017, pp. 91:1–91:14 (2017)
4. Bodlaender, H.L., Cygan, M., Kratsch, S., Nederlof, J.: Deterministic single exponential time algorithms for connectivity problems parameterized by treewidth. Inf. Comput. **243**, 86–111 (2015)
5. Bodlaender, H.L., Gilbert, J.R., Hafsteinsson, H., Kloks, T.: Approximating treewidth, pathwidth, frontsize, and shortest elimination tree. J. Algorithms **18**(2), 238–255 (1995)
6. Curticapean, R., Lindzey, N., Nederlof, J.: A tight lower bound for counting Hamiltonian cycles via matrix rank. In: Proceedings of the Twenty-Ninth Annual ACM-SIAM Symposium on Discrete Algorithms, Society for Industrial and Applied Mathematics, pp. 1080–1099 (2018)
7. Cygan, M., et al.: Parameterized Algorithms. Springer, Heidelberg (2015). https://doi.org/10.1007/978-3-319-21275-3
8. Cygan, M., Nederlof, J., Pilipczuk, M., Pilipczuk, M., van Rooij, J.M.M., Wojtaszczyk, J.O.: Solving connectivity problems parameterized by treewidth in single exponential time. In: IEEE 52nd Annual Symposium on Foundations of Computer Science (FOCS), pp. 150–159. IEEE (2011)
9. Fürer, M., Huiwen, Y.: Space saving by dynamic algebraization based on tree-depth. Theory Comput. Syst. **61**(2), 283–304 (2017)
10. Karp, R.M.: Dynamic programming meets the principle of inclusion and exclusion. Oper. Res. Lett. **1**(2), 49–51 (1982)
11. Kneis, J., Mölle, D., Richter, S., Rossmanith, P.: A bound on the pathwidth of sparse graphs with applications to exact algorithms. SIAM J. Discrete Math. **23**(1), 407–427 (2009)
12. Kohn, S., Gottlieb, A., Kohn, M.: A generating function approach to the traveling salesman problem. In: Proceedings of the 1977 Annual Conference, pp. 294–300. ACM (1977)
13. Lokshtanov, D., Nederlof, J.: Saving space by algebraization. In: Proceedings of the Forty-Second ACM Symposium on Theory of Computing, pp. 321–330. ACM (2010)
14. Nešetřil, J., De Mendez, P.O.: Tree-depth, subgraph coloring and homomorphism bounds. Eur. J. Comb. **27**(6), 1022–1041 (2006)
15. Pilipczuk, M., Wrochna, M.: On space efficiency of algorithms working on structural decompositions of graphs. ACM Trans. Comput. Theory (TOCT) **9**(4), 18:1–18:36 (2018)
16. Rota, G.-C.: On the foundations of combinatorial theory, I. Theory of Möbius functions, Zeitschrift für Wahrscheinlichkeitstheorie und verwandte Gebiete **2**(4), 340–368 (1964)
17. Stanley, R.P.: Enumerative Combinatorics. Vol. 1, with a foreword by Gian-Carlo Rota. Corrected reprint of the 1986 original, Cambridge Studies in Advanced Mathematics, vol. 49 (1997)
18. Williams, V.V.: Multiplying matrices faster than Coppersmith-Winograd. In: Proceedings of the Forty-Fourth Annual ACM Symposium on Theory of Computing, STOC 2012, pp. 887–898. ACM, New York (2012)

Quantum Algorithm for Distribution-Free Junta Testing

Aleksandrs Belovs[✉]

Faculty of Computing, University of Latvia, Raiņa bulvāris 19, Riga 1050, Latvia
aleksandrs.belovs@lu.lv

Abstract. Inspired by a recent classical distribution-free junta tester by Chen, Liu, Serverdio, Sheng, and Xie (STOC'18), we construct a quantum tester for the same problem with complexity $O(k/\varepsilon)$, which constitutes a quadratic improvement.

This result was obtained independently from the $\widetilde{O}(k/\varepsilon)$ algorithm for this problem by Bshouty.

Keywords: Property testing · Juntas · Fourier transform

1 Introduction

The steadily growing size of data calls for algorithms that work extremely fast: with linear or, preferably, sub-linear complexity. To achieve this performance, certain assumptions must be made, usually in some sort of approximation guarantees.

Consider the problem of testing whether an object f has some property $\mathcal{P}$. Typically, this task is hardest on instances f that are close to the border: the ones that do not possess the property $\mathcal{P}$ but are extremely close to doing so. But often such meticulousness is not needed, as one can tolerate false positives that are close to true positives. What one usually wants is to filter out instances that are substantially far from having the property. The framework of property testing [9,12] does exactly this: distinguishes the objects having the property $\mathcal{P}$ from the ones that are ε-far from $\mathcal{P}$. This is an active area of research both for classical (randomised) and quantum testers. See [11] for a survey on quantum property testing.

Junta Testing. Let us focus on the case of testing juntas, which is the topic of this paper. A *k-junta* is a Boolean function $h\colon \{0,1\}^n \to \{0,1\}$ that only depends on k out of its n input variables. In other words, h is a k-junta iff there exists a subset $\{a_1, a_2, \ldots, a_k\} \subseteq [n]$ such that $h(x_1, \ldots, x_n) = \hat{h}(x_{a_1}, x_{a_2}, \ldots, x_{a_k})$ for some function $\hat{h}\colon \{0,1\}^k \to \{0,1\}$. A function $g\colon \{0,1\}^n \to \{0,1\}$ is ε-*far* from a k-junta iff g differs from any k-junta in at least $\varepsilon 2^n$ points of the hypercube

This research is partly supported by the ERDF grant number 1.1.1.2/VIAA/1/16/113.

R. van Bevern and G. Kucherov (Eds.): CSR 2019, LNCS 11532, pp. 50–59, 2019.
https://doi.org/10.1007/978-3-030-19955-5_5

$\{0,1\}^n$. A *junta tester*, given access to a Boolean function f and parameters k and ε, has to distinguish between the cases when f is a k-junta and when f is ε-far from any k-junta. It is usually assumed that n is much larger than k, and the goal is to construct a tester whose complexity does not depend on n, and is as optimal in terms of k and ε as possible.

Junta testing has interesting history with developments in classical and quantum algorithms coming hand in hand. In 2002, the problem was considered by Fischer *et al.* [8], and an algorithm with query complexity $O((k\log k)^2/\varepsilon)$ was constructed. In 2007, a quantum algorithm was constructed by Atıcı and Servedio [2], which uses $O(k/\varepsilon)$ quantum examples. But in 2009 classical algorithms caught up with an $O(k/\varepsilon + k\log k)$-query algorithm by Blais [3]. This is optimal due the lower bound by Sağlam [13] (see also [7]). In 2015, however, quantum complexity of the problem was improved to $O(\sqrt{k/\varepsilon}\log k)$ queries by Ambainis *et al.* [1]. This was shown to be almost optimal by Bun *et al.* [5]. Thus, query complexity of junta testing is well-understood both classically and quantumly.

Distribution-Free Testing. The distribution-free property testing model was introduced by Goldreich *et al.* in [9]. It is similar to the usual model of property testing, but the distance to the property $\mathcal{P}$ is measured with respect to some unknown distribution $\mathcal{D}$ over the hypercube $\{0,1\}^n$. That is, g is ε-far from a k-junta with respect to $\mathcal{D}$, iff $\Pr_{x\sim\mathcal{D}}[g(x) \neq h(x)] \geq \varepsilon$ for any k-junta h. The tester, in addition to oracle queries to f, can sample from $\mathcal{D}$. The tester should work for any distribution $\mathcal{D}$, and the complexity measure is the worst-case sum of the number of queries to f and samples from $\mathcal{D}$. Thus, distribution-free property testing is at least as hard as usual property testing.

The motivation behind distribution-free property testing model is similar to that of the PAC learning model [14]. The uniform distribution might not be the right one to measure the distance, and it might be hard to get to know what the relevant probability distribution is. Thus, we would like to make as few assumptions on the distribution as possible. But the algorithm must have some access to the distribution in order to solve the problem, and sampling is one of the weakest modes of access.

Results. Recently, Liu *et al.* [10] constructed a distribution-free randomised junta tester with complexity $\widetilde{O}(k^2)/\varepsilon$. Our main result is a distribution-free *quantum* junta tester with a (slightly better than) quadratic improvement in k.

Theorem 1. *There exists an algorithm that, given quantum membership oracle access to a Boolean function $f\colon \{0,1\}^n \to \{0,1\}$, classical sample access to a probability distribution $\mathcal{D}$ on $\{0,1\}^n$, and parameters k and ε, performs the following task. If f is a k-junta, it accepts with probability 1. If f is ε-far from any k-junta with respect to $\mathcal{D}$, it rejects with probability at least $1/2$. The algorithm uses $O(k/\varepsilon)$ queries to f and samples from $\mathcal{D}$.*

Up to our knowledge, this is the first distribution-free quantum property tester. We allow quantum membership queries to f. Additionally, we allow classical (not quantum!) sampling from $\mathcal{D}$. If one allows quantum example oracle access to $\mathcal{D}$, it is possible to get quadratic improvement in terms of ε as well, see Sect. 5. We give additional justification to our model in Sect. 4, where we show that quantum example oracle is of little use to solve this problem.

Our algorithm is inspired by the algorithm by Liu et $al.$, and it not only has smaller complexity, but is also conceptually simpler that the classical algorithm. Similarly to the tester by Atıcı and Servedio, the improvement stems from the fact that quantum algorithms can efficiently Fourier sample. Actually, the only quantum subroutine we use is the ability to Fourier sample the input function f restricted to arbitrary hypercube of $\{0,1\}^n$.

Related Results. During the review phase of this paper, classical complexity of the problem was improved to $\tilde{O}(k/\varepsilon)$ queries by Bshouty [4], which is optimal up to logarithmic factors.

2   Preliminaries

We use notation $[n] = \{1, 2, \ldots, n\}$. For $x \in \{0,1\}^n$ and $T \subseteq [n]$, x^T stands for the string x with the bits in T flipped. We write x^i instead of $x^{\{i\}}$ for $i \in [n]$.

The following ways of accessing the input function $f: \{0,1\}^n \to \{0,1\}$ are used in the paper. A *classical example* is a pair $(x, f(x))$, where x is drawn from some probability distribution (or uniformly if no distribution is specified). A *classical membership oracle* is a black-box that on a query x returns the value of $f(x)$. A *quantum example* is a quantum state of the form $\sum_{x \in \{0,1\}^n} \sqrt{\mathcal{D}_x}|x\rangle|f(x)\rangle$ for some probability distribution $\mathcal{D}$. Again, if no distribution is specified, we assume the uniform one. A *quantum example oracle* is a quantum subroutine that performs the transformation $|0\rangle \mapsto \sum_{x \in \{0,1\}^n} \sqrt{\mathcal{D}_x}|x\rangle|f(x)\rangle$. Quantum example oracles are more powerful than quantum examples: for instance, one can use quantum amplitude amplification with them. Finally, a *quantum membership oracle* is a quantum subroutine performing the transformation $|x\rangle \mapsto (-1)^{f(x)}|x\rangle$ for all $x \in \{0,1\}^n$. It is the same as the usual quantum input oracle.

In order to distinguish quantum example and membership oracles, we use terms quantum *sample* and *query*, respectively, for their execution. Classically, we use terms "sample" and "example" as synonymous. For each quantum oracle, we also assume access to its inverse.

The main technical tool we use is Fourier sampling. The *Fourier transform* (also known as the Walsh-Hadamard transform) of a Boolean function f is the linear mapping

$$\frac{1}{\sqrt{2^n}} \sum_{x \in \{0,1\}^n} (-1)^{f(x)}|x\rangle \xrightarrow{H^{\otimes n}} \sum_{S \subseteq [n]} \hat{f}(S)|S\rangle, \tag{1}$$

where $\hat{f}(z)$ are the *Fourier coefficients* of f. Here the subset S on the right-hand side can be identified with its characteristic string.

The Fourier transformation can be efficiently implemented on a quantum computer: it only requires n Hadamard gates $H = \frac{1}{\sqrt{2}} \begin{pmatrix} 1 & 1 \\ 1 & -1 \end{pmatrix}$. If the state on the right-hand side of (1) is measured, the outcome S is obtained with probability $\hat{f}(S)^2$. This is known as *Fourier sampling*. The sum of $\hat{f}(S)^2$ equals 1 since $H^{\otimes n}$ is a unitary transformation.

We only use the following two properties of the Fourier transform. First, if $\hat{f}(S) \neq 0$, then f depends on all variables in S. Second,

$$\hat{f}(\emptyset) = \frac{1}{2^n} \sum_{x \in \{0,1\}^n} (-1)^{f(x)}. \qquad (2)$$

3   The Algorithm

In this section we describe our algorithm and prove Theorem 1. Before we proceed with this task, let us introduce some notation. Everywhere in this section we assume that an input function $f \colon \{0,1\}^n \to \{0,1\}$ is fixed.

A variable $i \in [n]$ is called *relevant* if there exists an input $x \in \{0,1\}^n$ such that $f(x) \neq f(x^i)$. Note that a function f is a k-junta if and only if it has at most k relevant variables.

A *cube* $B = (x,y)$ is a subcube of the hypercube $\{0,1\}^n$ and is specified by its two opposite vertices x and y. That is, B is the set of all bit-strings in $\{0,1\}^n$ that agree to either x or y in each position. Let $I(B)$ be the set of variables where x and y disagree. A cube $B = (x,y)$ is called *relevant* if $f(x) \neq f(y)$. Note that if a cube B is relevant, then at least one variable in $I(B)$ is relevant.

A formal description of the tester is given in Algorithm 1, and it depends on two subroutines: GenerateCube and FourierSample given in Algorithms 2 and 3, respectively. The subroutine GenerateCube(S) finds a cube B such that $I(B)$ does not intersect S. The subroutine FourierSample(B) performs Fourier sampling from the function f restricted to the subcube B.

The algorithm maintains a subset of variables $S \subseteq [n]$ and a collection of cubes $\mathcal{B}$. The following invariants are maintained throughout the algorithm:

- All variables in S and all cubes in $\mathcal{B}$ are relevant.
- The set S and all $I(B)$, as B ranges over $\mathcal{B}$, are pairwise disjoint.   (3)

Claim 2. *The invariants in (3) are maintained throughout the algorithm.*

Proof. Clearly, the invariants are satisfied at the beginning of the algorithm when both S and $\mathcal{B}$ are empty. Consider all the steps of Algorithm 1 where S or $\mathcal{B}$ change. In 2(a), B is a relevant cube and $I(B)$ does not intersect S by the requirement on the subroutine GenerateCube. In 2(c), the set T returned by FourierSample has non-zero Fourier coefficient in f restricted to B. This means that all the variables in T are relevant. Also $T \subseteq I(B)$, hence, the new S does

---

Algorithm 1. Quantum Distribution-Free Junta Testing

---

1. Let $S \leftarrow \emptyset$, $\mathcal{B} \leftarrow \emptyset$
2. Repeat while $|S| + |\mathcal{B}| \leq k$, but no more than $18k$ times:
   (a) If $\mathcal{B}$ is empty, $B \leftarrow$ GenerateCube(S). If B is not fail, let $\mathcal{B} \leftarrow \{B\}$. Continue with the next iteration of the loop.
   (b) Otherwise, let $B = (x, y)$ be any cube in $\mathcal{B}$.
   (c) Let $T \leftarrow$ FourierSample(B). If $T \neq \emptyset$, remove B from $\mathcal{B}$, let $S \leftarrow S \cup T$, and continue with the next iteration of the loop.
   (d) Otherwise, generate a uniformly random subset $T \subseteq I(B)$. Let $z = x^T$ and $t = y^T$.
   (e) If $f(z) = f(t) = f(y)$, then remove B from $\mathcal{B}$, add the cubes (x, z) and (x, t) to $\mathcal{B}$, and continue with the next iteration of the loop.
   (f) If $f(z) = f(t) = f(x)$, then remove B from $\mathcal{B}$, add the cubes (z, y) and (t, y) to $\mathcal{B}$, and continue with the next iteration of the loop.
3. If $|S| + |\mathcal{B}| > k$, reject. Otherwise, accept.

Algorithm 2. GenerateCube(S) subroutine, classical version

---

1. Repeat $2/\varepsilon$ times:
   (a) Sample x from $\mathcal{D}$. Let T be a uniformly random subset of $[n] \setminus S$.
   (b) If $f(x) \neq f(x^T)$ return the cube (x, x^T).
2. Return 'fail'.

Algorithm 3. FourierSample($B = (x, y)$)

---

1. Prepare the state $|x\rangle = |x_1\rangle |x_2\rangle \cdots |x_n\rangle$ on n qubits.
2. Apply the Hadamard operator H to the qubits in $I(B)$, and get the state $\frac{1}{\sqrt{|B|}} \sum_{z \in B} |z\rangle$.
3. Apply the quantum membership oracle and obtain the state $\frac{1}{\sqrt{|B|}} \sum_{z \in B} (-1)^{f(z)} |z\rangle$.
4. Apply the Hadamard operator H to the qubits in $I(B)$ and measure them. Return the set T formed by the bits where the measurement outcome is 1.

---

not intersect any of the remaining cubes in $\mathcal{B}$. In 2(e), we have $f(z) \neq f(x)$ and $f(t) \neq f(x)$, hence the two new cubes are relevant. Also note that $t = x^{I(B) \setminus T}$. Hence, $I(x, z)$ and $I(x, t)$ are disjoint and do not intersect S nor the remaining cubes in $\mathcal{B}$. The case 2(f) is similar.

Concerning the terminating condition in Steps 2 and 3 of Algorithm 1, we have the following result.

Proposition 3. *If S and $\mathcal{B}$ satisfy the conditions in (3) and $|S| + |\mathcal{B}| > k$, then f is not a k-junta.*

Proof. Each cube $B \in \mathcal{B}$ contains a relevant variable. By the second condition in (3), all these variables are distinct and different from the variables in S.

Thus, together with the variables in S, f has more than k relevant variables. This means that f is not a k-junta.

Concerning the GenerateCube subroutine, we have the following lemma:

Lemma 4 ([10], Lemma 3.2. arXiv version). *Assume $f\colon \{0,1\}^n \to \{0,1\}$ is ε-far from any k-junta with respect to $\mathcal{D}$, and S is a subset of $[n]$ of size at most k. Then,*

$$\Pr_{x \sim \mathcal{D},\, T \subseteq [n] \setminus S} \left[f(x) \neq f(x^T) \right] \geq \varepsilon/2,$$

where x is sampled from $\mathcal{D}$, and T is a uniformly random subset of $[n] \setminus S$.

Corollary 5. *In the assumptions of Lemma 4, probability the GenerateCube(S) subroutine returns 'fail' is at most $1/2$.*

Proof. By Lemma 4, the probability that the subroutine returns 'fail' is at most $(1 - \varepsilon/2)^{2/\varepsilon} < 1/\mathrm{e} < 1/2$.

The number of iterations of the loop in Step 2 of Algorithm 1 is based on the following lemma.

Lemma 6. *Assume f is ε-far from a k-junta relative to $\mathcal{D}$, and consider one iteration of the loop in Step 2 of Algorithm 1. Let S and $\mathcal{B}$ be the values of these variables before the iteration, and S' and $\mathcal{B}'$ after the iteration. If $|S| \leq k$, then, with probability at least $1/3$, we have $2|S'| + |\mathcal{B}'| \geq 2|S| + |\mathcal{B}| + 1$.*

Proof. Assume first $\mathcal{B} = \emptyset$. Then by Corollary 5, with probability at least $1/2$, the GenerateCube subroutine does not fail and the size of $\mathcal{B}$ becomes 1, which gives $2|S'| + |\mathcal{B}'| = 2|S| + |\mathcal{B}| + 1$.

Now consider the case when $\mathcal{B}$ is not empty, and let B be the cube selected on step 2(b). Denote by $f|_B$ the function f restricted to the inputs in the subcube B. There are two cases: $f|_B$ is $1/3$-far from a constant function, or $f|_B$ is $1/3$-close to a constant function relative to the uniform distribution on B.

In the first case, by (2), the absolute value of the Fourier coefficient of $\emptyset$ is at most $2/3 - 1/3 = 1/3$. Hence, with probability $8/9$, the set T in 2(c) is non-empty. In this case, the size of $\mathcal{B}$ is reduced by 1, but the size of S grows by at least 1, which gives $2|S'| + |\mathcal{B}'| \geq 2|S| + |\mathcal{B}| + 1$.

In the second case, when f is $1/3$-close to a constant function, the probability that $f(z) = f(t)$ is at least $1/3$. In this case, the size of S does not change, but the size of $\mathcal{B}$ grows by 1, which gives $2|S'| + |\mathcal{B}'| = 2|S| + |\mathcal{B}| + 1$.

Proof (of Theorem 1). Now we can prove the theorem. If f is a k-junta, Algorithm 1 always accepts due to Proposition 3. Let us prove that if f is ε-far from a k-junta relative to $\mathcal{D}$, then the algorithm rejects with probability at least $1/2$.

Assume for a moment there is no upper bound of $18k$ on the number of iterations of the loop in Step 2 of the algorithm. Let $P(i)$ denote the number of iterations of the loop after which it holds that $2|S| + |\mathcal{B}| \geq i$ or $|S| + |\mathcal{B}| > k$.

By Lemma 6, we have that $\mathbb{E}[P(i+1) - P(i)] \leq 3$. As $P(0) = 0$, by linearity of expectation, we have that $\mathbb{E}[P(3k)] \leq 9k$. By Markov's inequality,

$$\Pr[P(3k) \geq 18k] \leq 1/2. \tag{4}$$

Note that $2|S| + |\mathcal{B}| \geq 3k$ implies $|S| + |\mathcal{B}| > k$. Thus, (4) means that Algorithm 1 rejects with probability at least $1/2$.

The complexity of the algorithm is $O(k/\varepsilon)$, as it performs $O(k)$ iterations of the loop, and each iteration costs at most $O(1/\varepsilon)$.

4   Uselessness of Quantum Examples

One interesting feature of the quantum junta tester by Atıcı and Servedio [2] is that it only uses quantum examples and not quantum or classical membership queries. This constitutes an exponential improvement since exponentially many classical examples are required to solve the problem (see Lemma 8 below). Also, this is still the best known quantum algorithm that only uses quantum examples, as the algorithm by Ambainis *et al.* [1] uses quantum membership queries.

Our algorithm in Sect. 3 uses quantum membership queries, and a natural question arises whether it is possible to attain similar complexity using only quantum examples. We show that this is impossible: contrary to the uniform case, in the distribution-free case, exponentially many quantum examples are required. As it is unclear whether quantum examples should come from the uniform distribution or from $\mathcal{D}$, we show that neither of them works.

Theorem 7. *Assume a quantum algorithm has quantum example oracle access to a Boolean function f with respect to both uniform probability distribution and $\mathcal{D}$. Then, $2^{\Omega(k)}$ executions of these oracles are required to distinguish whether f is a k-junta or $\Omega(1)$-far away from any k-junta with respect to $\mathcal{D}$.*

Recall that quantum example oracles with respect to the uniform distribution and $\mathcal{D}$ are quantum subroutines that perform the following transformations

$$|0\rangle \mapsto \frac{1}{\sqrt{2^n}} \sum_{x \in \{0,1\}^n} |x\rangle |f(x)\rangle, \qquad \text{and} \qquad |0\rangle \mapsto \sum_{x \in \{0,1\}^n} \sqrt{\mathcal{D}_x} |x\rangle |f(x)\rangle,$$

respectively. In the remaining part of this section, we sketch the proof of Theorem 7.

We start with proving that classical examples do not help to solve the problem even in the uniform case. Interestingly, the following result does not appear explicitly in prior publications, however, it is tacitly assumed in a number of papers. We add a simple proof of this lemma for completeness.

Lemma 8. *In the uniform model, $\Omega(2^{k/2})$ classical examples are required to distinguish whether a given Boolean function $f \colon \{0,1\}^n \to \{0,1\}$ is a k-junta or is $\Omega(1)$-far from any $(n-1)$-junta.*

Proof. Consider the following two probability distributions on Boolean functions $f \colon \{0,1\}^n \to \{0,1\}$. $\mathcal{H}$ is the uniform probability distribution on the functions $h(x_1, \ldots, x_n)$ that only depend on the variables $x_1, \ldots, x_k$. $\mathcal{G}$ is the uniform probability distribution on all Boolean functions f. Each function in the support of $\mathcal{H}$ is a k-junta, while a function $g \sim \mathcal{G}$ is $\Omega(1)$-far from any $(n-1)$-junta with high probability [7, Lemma 3.1]. Hence, the tester should be able to distinguish these two probability distribution with bounded error.

Let $x^{(1)}, \ldots, x^{(t)}$ be the examples obtained by the tester. If $t = o(2^{k/2})$, then, with high probability, every two inputs in this sequence disagree on their first k bits. But if this is the case, then the values $f(x_1), \ldots, f(x_t)$ form a uniformly random string in $\{0,1\}^t$ regardless whether f is sampled from $\mathcal{H}$ or from $\mathcal{G}$. Hence, the tester cannot distinguish these two probability distributions with bounded error.

Let $\mathcal{H}$ denote the set of functions that only depend on $x_1, \ldots, x_k$ and $\mathcal{G}$ denote the set of functions that are $\Omega(1)$-far from any $(n-1)$-junta. Lemma 8 shows that exponentially many classical examples are needed to distinguish these two classes. However, due to the algorithm by Atıcı and Servedio, it is possible to efficiently distinguish them using quantum examples. We proceed by modifying the classes $\mathcal{H}$ and $\mathcal{G}$ twice. First, we make the uniform quantum example oracle useless, and then the quantum example oracle with respect to $\mathcal{D}$.

We start with uniform quantum examples. The idea is to restrict the probability distribution $\mathcal{D}$ to a small subcube so that uniform examples do not give much information on the inputs in $\mathcal{D}$. Let $x_{[k]}$ denote the substring $(x_1, \ldots, x_k) \in \{0,1\}^k$ of x. Define $\mathcal{H}'$ and $\mathcal{G}'$ as classes of Boolean functions $f' \colon \{0,1\}^{n+k} \to \{0,1\}$ on the variables $(x,y) = (x_1, \ldots, x_n, y_1, \ldots, y_k)$ defined in the following way. If f comes from $\mathcal{H}$ (respectively, $\mathcal{G}$), then the corresponding function f' in $\mathcal{H}'$ (respectively, $\mathcal{G}'$) is defined by

$$f'(x_1, \ldots, x_n, y_1, \ldots, y_k) = \begin{cases} f(x_1, \ldots, x_n), & \text{if } y = x_{[k]}; \\ 0, & \text{otherwise.} \end{cases}$$

Let $\mathcal{D}'$ be the uniform probability distribution on the strings $(x,y) \in \{0,1\}^{n+k}$ satisfying $y = x_{[k]}$. Any function in $\mathcal{H}'$ is a $2k$-junta, and any function from $\mathcal{G}'$ is $\Omega(1)$-far from any $(n-1)$-junta with respect to $\mathcal{D}'$, and, hence, $\Omega(1)$-far from any $2k$-junta if $n > 2k$.

Uniform quantum examples are useless here. Indeed, the uniform quantum example is $2^{-\Omega(k)}$-close to the quantum example corresponding to the all-0 function:

$$|0\rangle \mapsto \frac{1}{\sqrt{2^{n+k}}} \sum_{x \in \{0,1\}^n} \sum_{y \in \{0,1\}^k} |x\rangle |y\rangle |0\rangle. \tag{5}$$

Hence, we can replace the uniform quantum example oracle corresponding to a function f' in $\mathcal{H}' \cup \mathcal{G}'$ with (5) and it will not significantly affect the output of the algorithm unless it makes $2^{\Omega(k)}$ samples.

Our next goal is to get rid of the quantum example oracle relative to $\mathcal{D}$. The idea is to scatter the distribution $\mathcal{D}$ so that it is impossible to make use of interference, and the quantum example oracle becomes essentially equivalent to the classical example oracle. Let $\mathcal{H}''$ and $\mathcal{G}''$ be classes of Boolean functions $f'' \colon \{0,1\}^{2n+k} \to \{0,1\}$ obtained from $\mathcal{H}'$ and $\mathcal{G}'$, respectively, by extending them with n irrelevant variables $z_1, \ldots, z_n$. For $\pi \colon \{0,1\}^n \to \{0,1\}^n$ a permutation, we define a distribution $\mathcal{D}''_\pi$ as follows. The distribution $\mathcal{D}''_\pi$ is over strings (x, y, z), where x is a uniformly random string from $\{0,1\}^n$, $y = x_{[k]}$ and $z = \pi(x)$. Clearly, all the functions in $\mathcal{H}''$ are $(2k)$-juntas and the functions in $\mathcal{H}''$ are $\Omega(1)$-far from any $(n-1)$-junta with respect to $\mathcal{D}''_\pi$. The uniform quantum samples are still not useful by the same argument as above.

Let us now prove that quantum samples from $\mathcal{D}''_\pi$ are also useless in this situation. Consider the quantum sampler from $\mathcal{D}''_\pi$. It acts as follows

$$|0\rangle \mapsto \frac{1}{\sqrt{2^n}} \sum_{x \in \{0,1\}^n} |x\rangle |x_{[k]}\rangle |\pi(x)\rangle |f(x)\rangle, \tag{6}$$

where f is from $\mathcal{H}$ or $\mathcal{G}$. But note that this state can be obtained in one query to the quantum oracle

$$|z\rangle |0\rangle \mapsto |z\rangle |\pi^{-1}(z)\rangle |f(\pi^{-1}(z))\rangle, \tag{7}$$

where we denoted $z = \pi(x)$ and permuted the registers. Hence, the complexity of the tester using the oracle in (6) is at least the complexity of the tester using the oracle in (7), which is the standard quantum oracle corresponding to a function

$$z \mapsto \left(\pi^{-1}(z), f(\pi^{-1}(z))\right).$$

Note that the problem of testing juntas using the oracle from (7) is symmetric with respect to the permutations of z because the permutation π can be arbitrary. By [6], a quantum algorithm for a symmetric function can obtain at most a cubic improvement compared to a randomised algorithm with access to the oracle. However, for a random π, access to this oracle is equivalent to uniform classical examples of the function f. Hence, the algorithm requires $2^{\Omega(k)}$ samples by Lemma 8.

5   Discussion

We have constructed a quantum algorithm with complexity $O(k/\varepsilon)$, which gives a quadratic improvement in terms of k when compared to [10]. In the algorithm, we assume we only have classical access to $\mathcal{D}$. It is also possible to have a quantum sampler from $\mathcal{D}$, that is, an oracle of the form

$$|0\rangle \mapsto \sum_{x \in \{0,1\}^n} \sqrt{\mathcal{D}_x} |x\rangle |\psi_x\rangle,$$

where $\mathcal{D}_x$ is the probability of x in $\mathcal{D}$, and ψ_x are some arbitrary unknown normalized quantum states. (Note that because of the unknown ψ_x, this is a weaker model of access than the quantum samplers we defined in Sect. 2.) In this case, it is possible to apply quantum amplitude amplification to the GenerateCube subroutine. The complexity of the subroutine becomes $O(1/\sqrt{\varepsilon})$, and the complexity of the whole algorithm becomes $O(k/\sqrt{\varepsilon})$.

So far, it is unclear whether the complexity of our quantum algorithm. In principle, it is not excluded that there exists a quantum algorithm with complexity $\tilde{O}(\sqrt{k/\varepsilon})$.

Acknowledgements. I am thankful to Srinivasan Arunachalam and Ronald de Wolf for helpful discussions about this problem.

References

1. Ambainis, A., Belovs, A., Regev, O., de Wolf, R.: Efficient quantum algorithms for (Gapped) group testing and junta testing. In: Proceedings of 27th ACM-SIAM SODA, pp. 903–922 (2016)
2. Atıcı, A., Servedio, R.A.: Quantum algorithms for learning and testing juntas. Quantum Inf. Process. **6**(5), 323–348 (2007)
3. Blais, E.: Testing juntas nearly optimally. In: Proceedings of 41st ACM STOC, pp. 151–158 (2009)
4. Bshouty, N.H.: Almost optimal distribution-free junta testing (2019)
5. Bun, M., Kothari, R., Thaler, J.: The polynomial method strikes back: tight quantum query bounds via dual polynomials. In: Proceedings of 50th ACM STOC, pp. 297–310 (2018)
6. Chailloux, A.: A note on the quantum query complexity of permutation symmetric functions. In: Proceedings of 10th ACM ITCS. LIPIcs, vol. 124, pp. 19:1–19:7. Dagstuhl (2019)
7. Chockler, H., Gutfreund, D.: A lower bound for testing juntas. Inf. Process. Lett. **90**(6), 301–305 (2004)
8. Fischer, E., Kindler, G., Ron, D., Safra, S., Samorodnitsky, A.: Testing juntas. J. Comput. Syst. Sci. **68**(4), 753–787 (2004)
9. Goldreich, O., Goldwasser, S., Ron, D.: Property testing and its connection to learning and approximation. J. ACM **45**(4), 653–750 (1998)
10. Liu, Z., Chen, X., Servedio, R.A., Sheng, Y., Xie, J.: Distribution-free junta testing. In: Proceedings of 50th ACM STOC, pp. 749–759 (2018)
11. Montanaro, A., de Wolf, R.: A survey of quantum property testing. Theory Comput. Grad. Surv. **7**, 1–81 (2016)
12. Rubinfeld, R., Sudan, M.: Robust characterizations of polynomials with applications to program testing. SIAM J. Comput. **25**(2), 252–271 (1996)
13. Sağlam, M.: Near log-convexity of measured heat in (discrete) time and consequences. In: Proceedings of 59th IEEE FOCS, pp. 967–978 (2018)
14. Valiant, L.G.: A theory of the learnable. Commun. ACM **27**(11), 1134–1142 (1984)

On Induced Online Ramsey Number
of Paths, Cycles, and Trees

Václav Blažej[1]([⊠]), Pavel Dvořák[2], and Tomáš Valla[1]

[1] Faculty of Information Technology,
Czech Technical University in Prague, Prague, Czech Republic
vaclav.blazej@fit.cvut.cz
[2] Faculty of Mathematics and Physics, Charles University, Prague, Czech Republic

Abstract. An online Ramsey game is a game between Builder and Painter, alternating in turns. They are given a fixed graph H and a an infinite set of independent vertices G. In each round Builder draws a new edge in G and Painter colors it either red or blue. Builder wins if after some finite round there is a monochromatic copy of the graph H, otherwise Painter wins. The online Ramsey number $\widetilde{r}(H)$ is the minimum number of rounds such that Builder can force a monochromatic copy of H in G. This is an analogy to the size-Ramsey number $\overline{r}(H)$ defined as the minimum number such that there exists graph G with $\overline{r}(H)$ edges where for any edge two-coloring G contains a monochromatic copy of H. In this extended abstract, we introduce the concept of induced online Ramsey numbers: the induced online Ramsey number $\widetilde{r}_{ind}(H)$ is the minimum number of rounds Builder can force an induced monochromatic copy of H in G. We prove asymptotically tight bounds on the induced online Ramsey numbers of paths, cycles and two families of trees. Moreover, we provide a result analogous to Conlon [On-line Ramsey Numbers, SIAM J. Discr. Math. 2009], showing that there is an infinite family of trees $T_1, T_2, \ldots, |T_i| < |T_{i+1}|$ for $i \geq 1$, such that

$$\lim_{i \to \infty} \frac{\widetilde{r}(T_i)}{\overline{r}(T_i)} = 0.$$

1 Introduction

For a graph H, the Ramsey number $r(H)$ is the smallest integer n such that in any two-coloring of edges of the complete graph K_n, there is a monochromatic copy of H. The size-Ramsey number $\overline{r}(H)$, introduced by Erdős, Faudree, Rousseau, and Schelp [7], is the smallest integer m such that there exists a graph G with m edges such that for any two-coloring of the edges of G one will always find a monochromatic copy of H.

V. Blažej and T. Valla acknowledge the support of the OP VVV MEYS funded project CZ.02.1.01/0.0/0.0/16_019/0000765 "Research Center for Informatics". P. Dvořák was supported by the project GAUK 1514217.

R. van Bevern and G. Kucherov (Eds.): CSR 2019, LNCS 11532, pp. 60–69, 2019.
https://doi.org/10.1007/978-3-030-19955-5_6

There are many interesting variants of the usual Ramsey function. One important concept is the *induced Ramsey number* $r_{ind}(H)$, which is the smallest integer n for which there is a graph G on n vertices such that every edge two-coloring of G contains an induced monochromatic copy of H. Erdős [8] conjectured the existence of a constant c such that every graph H with n vertices satisfies $r_{ind}(H) \leq 2^{cn}$, which would be best possible. In 2012, Conlon, Fox and Sudakov [5] proved that there is a constant c such that every graph H with n vertices satisfies $r_{ind}(H) \leq 2^{cn \log n}$. The proof uses a construction of explicit pseudorandom graphs, as opposed to random graph construction techniques used by previous attempts. For more on the topic see the excellent review by Conlon, Fox, and Sudakov [6].

The induced size-Ramsey number $\bar{r}_{ind}(H)$ is an analog of the size-Ramsey number: we define $\bar{r}_{ind}(H)$ as the smallest integer m such that there exists a graph G with m edges such that for any two-coloring of the edges of G there is always a monochromatic copy of H. In 1983, Beck [1], using probabilistic methods, proved the surprising fact that $\tilde{r}(P_n) \leq cn$, where P_n is a path of length n and c is an absolute constant. An even more surprising result came by Haxell, Kohayakawa, and Łuczak [10], who studied the induced size-Ramsey number of cycles showing that $\bar{r}_{ind}(C_n) = O(n)$. However, the proof uses random graph techniques and regularity lemma and does not provide any reasonably small multiplicative constant.

In this extended abstract, we study the online variant of size Ramsey number which was introduced independently by Beck [3] and Kurek and Ruciński [11]. The best way to define it is in term of a game between two players, Builder and Painter. An infinite set of vertices is given, in each round Builder draws a new edge and immediately it is colored by Painter in either red or blue. The goal of Builder is to force Painter to obtain a monochromatic copy of a fixed graph H (called *target graph*). The minimum number of edges which Builder must draw in order to obtain such monochromatic copy of H, assuming optimal strategy of Painter, is known as the online Ramsey number $\tilde{r}(H)$. The graph G, which is being built by Builder, is called *background graph*. The online Ramsey number is guaranteed to exist because Builder can simply create a big complete graph $K_{r(H)}$, which by Ramsey theorem trivially contains a monochromatic copy of H.

The winning condition for Builder is to obtain a copy of the target graph H. However, there are more different notions of "being a copy". This leads us to the following two definitions.

- The *online Ramsey number* $\tilde{r}(H)$ is the minimum number of rounds of the Builder-Painter game Builder has a strategy to obtain a monochromatic subgraph H.
- The *(strongly) induced online Ramsey number* $\tilde{r}_{ind}(H)$ is the minimum number of rounds of the Builder-Painter game such that Builder has a strategy to obtain a monochromatic induced subgraph H in G.

If there is no strategy of Builder to obtain the copy of H, we define the respective number as ∞.

Note that for any graph H we have $\tilde{r}(H) \leq \tilde{r}_{ind}(H)$. Also, note that the induced online Ramsey numbers provide lower bounds on the induced size-Ramsey numbers.

In 2008 Grytczuk, Kierstead and Prałat [9] studied the online Ramsey number of paths, obtaining $\tilde{r}(P_n) \leq 4n - 3$, where P_n is a path with n edges, providing an interesting counterpart to the result of Beck [1]. Also, the result by Haxell, Kohayakawa, and Łuczak [10] on induced size-Ramsey number of cycles naturally bounds the online version as well, but with no reasonable multiplicative constant.

In this extended abstract, we study the induced online Ramsey number of paths, cycles, and trees. The summary of the results for paths and cycles is as follows.

Theorem 1. *Let P_n denote the path of length n and let C_n denote a cycle with n vertices. Then*

- $\tilde{r}_{ind}(P_n) \leq 28n - 27$,
- $\tilde{r}_{ind}(C_n) \leq 367n - 27$ *for even n,*
- $\tilde{r}_{ind}(C_n) \leq 735n - 27$ *for odd n.*

A *spider* $\sigma_{k,\ell}$ is a union of k paths of length ℓ sharing exactly one common endpoint. We further show that $\tilde{r}_{ind}(\sigma_{k,\ell}) = \Theta(k^2\ell)$ and $\tilde{r}(\sigma_{k,\ell}) = \Theta(k^2\ell)$.

Although we know that $\tilde{r}(H) \leq \overline{r}(H)$, it is a challenging task to identify classes of graphs for which there is an asymptotic gap between both numbers. For complete graphs, Chvátal observed (see [7]) that $\overline{r}(K_t) = \binom{r(K_t)}{2}$. The basic question, attributed to Rödl (see [11]), is to show $\lim_{t\to\infty} \tilde{r}(K_t)/\overline{r}(K_t)$, or put differently, to show that $\tilde{r}(K_t) = o(\binom{r(K_t)}{2})$. This conjecture remains open, but in 2009 Conlon [4] showed there exists $c > 1$ such that for infinitely many t,

$$\tilde{r}(K_t) \leq c^{-t}\binom{r(K_t)}{2}.$$

In this extended abstract we contribute to this topic by showing that there is an infinite family of trees $T_1, T_2, \ldots$, with $|T_i| < |T_{i+1}|$ for $i \geq 1$, such that

$$\lim_{i\to\infty} \frac{\tilde{r}(T_i)}{\overline{r}(T_i)} = 0,$$

thus exhibiting the desired asymptotic gap. In fact, we prove a stronger statement, exhibiting the asymptotic gap even for the induced online Ramsey number.

2 Induced Paths

In this section we present an upper bound on the induced online Ramsey number of paths.

Theorem 2. *Let P_n be a path of length n. Then $\tilde{r}_{ind}(P_n) \leq 28n - 27$.*

Proof. First we build the set I of $2(7n-7)-1$ isolated edges, then at least $7n-7$ have the same color, we say this color is *abundant* in I.

Let R^0 and B^0 be the initial paths of lengths 0. In s-th step we have a red induced path $R^s = (r_0, \{r_0, r_1\}, r_1, \ldots, r_r)$ of length r and a blue induced path $B^s = (b_0, \{b_0, b_1\}, b_1, \ldots, b_b)$ of length b. We denote the concatenation of paths A and B by $A \cup B$. The removal of vertices and incident edges is denoted by $A \setminus \{v\}$. We define the potential of s-th step $p^s = 3a + 4o$ where a is the length of the path in color which is abundant in I and o is the length of path in the other color. Further, we show that we are able to maintain the invariant that there are no edges between the R^s and B^s and that $p^{s+1} > p^s$.

Assume without loss of generality that the blue edges are abundant in I. Let $g = \{x, y\}$ be an unused blue edge from the set I. One step of Builder is as follows. Builder creates an edge $e = \{r_r, b_b\}$. If Painter colored e red then Builder creates an edge $f = \{b_b, x\}$, however if e is blue then Builder creates $f = \{r_r, x\}$.

Depending on how the e and f edges were colored we end up with four different scenarios. These different cases are also depicted in Fig. 1.

$$(B^{s+1}, R^{s+1}) = \begin{cases} (B^s \cup (e, r_r, f, x, g, y), R^s \setminus \{r_r, r_{r-1}\}) & \text{if } e \text{ and } f \text{ are blue} \\ (B^s \setminus \{b_b\}, R^s \cup (f, x)) & \text{if } e \text{ is blue and } f \text{ is red} \\ (B^s \cup (f, x, g, y), R^s \setminus \{r_r\}) & \text{if } e \text{ is red and } f \text{ is blue} \\ (B^s \setminus \{b_b, b_{b-1}\}, R^s \cup (e, b_b, f, x)) & \text{if } e \text{ and } f \text{ are red} \end{cases}$$

$$p^{s+1} = \begin{cases} 3(|B^s| + 3) + 4(|R^s| - 2) = p^s + 1 & \text{if } e \text{ and } f \text{ are blue} \\ 3(|B^s| - 1) + 4(|R^s| + 1) = p^s + 1 & \text{if } e \text{ is blue and } f \text{ is red} \\ 3(|B^s| + 2) + 4(|R^s| - 1) = p^s + 2 & \text{if } e \text{ is red and } f \text{ is blue} \\ 3(|B^s| - 2) + 4(|R^s| + 2) = p^s + 2 & \text{if } e \text{ and } f \text{ are red} \end{cases}$$

We obtain a pair of paths B^{s+1}, R^{s+1} such that $p^{s+1} > p^s$ and invariant holds.

The maximum potential for which Builder did not win yet is $p^s = 7n - 7$. Therefore there are no more than $7n - 6$ steps to finish one monochromatic induced path of length n. To create the initial set I Builder creates $2(7n-7)-1$ isolated edges. In each step, Builder creates two edges. The total number of edges created by Builder is no more than $2(7n - 6) + 2(7n - 7) - 1 = 28n - 27$.   □

Note that the initial edges each span 2 vertices and in each step only the first edge can lead to a new vertex. This gives us bound on the number of vertices used in creating an induced path P_n to be at most $2(2(7n-7)-1) + 7n - 6 = 35n - 36$.

3   Cycles and Induced Cycles

In this section, we present a constructive upper bound on the online Ramsey number of cycles $\tilde{r}(C_n)$ and induced cycles $\tilde{r}_{ind}(C_n)$.

Theorem 3. *Let C_n be a cycle on n vertices, where n is even. Then, $\tilde{r}_{ind}(C_n) \leq 367n - 27$.*

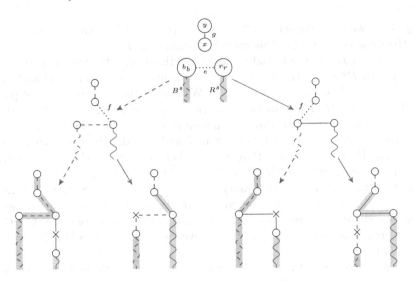

Fig. 1. One step in creating an induced monochromatic P_n (Color figure online)

Proof. First, Builder obtains disjoint paths $\rho_1, \rho_2, \ldots, \rho_9$ of length $4n/3 - 1$ and one path ρ_{10} of length $n - 2$. Instead of using Theorem 2 to create these paths separately it is more efficient to create a P_{13n} using at most $28(13n) - 27$ edges and define paths $\rho_1, \rho_2, \ldots, \rho_{10}$ as an induced subgraph of P_{13n}. Let the P_{13n} be without loss of generality red. Let $\rho_{i,j}$ denote the j-th vertex of ρ_i.

Builder will create a red C_n using $\rho_1, \rho_2, \ldots, \rho_{10}$ or three blue paths of length $n/2$ starting in u and ending in either $\rho_{10,1}$ or $\rho_{10,n-1}$. These three paths starting in the same vertex and two of them sharing a common endpoint will form a blue C_n. Each blue path will go through a separate triple of paths from $\rho_1, \rho_2, \ldots, \rho_9$ and alternate between them with each added vertex.

Let us run the following procedure three times – once for each $k \in \{1, 2, 3\}$. Let $p = \rho_{3k-2}$, $q = \rho_{3k-1}$ and $r = \rho_{3k}$. Let us define cyclic order of these paths to be p, q, r, p which defines a natural successor for each path. Builder does the following three steps, which are also depicted in Fig. 2.

1. Create edges $\{u, p_1\}$ and $\{u, p_{n-1}\}$. If both of these edges are red Builder wins immediately. If that is not the case then at least one edge $\{u, v_1\}$ where $v_1 \in \{p_1, p_{n-1}\}$ is blue.
2. Now for i from 1 to $n/2 - 1$ we do as follows:
   - Let $j := 2\lfloor i/3 \rfloor$. Let $t \in \{p, q, r\}$ such that $v_i \in t$ and set s to be the successor of t.
   - We create edges $\{v_i, s_{j+1}\}$ and $\{v_i, s_{j+n-1}\}$. If both are red Builder wins, otherwise take an edge $\{v_i, v_{i+1}\}$ where $v_{i+1} \in \{s_{j+1}, s_{j+n-1}\}$ is blue.
3. Finish the path $(u, v_1, v_2, \ldots, v_{n/2-1})$ by creating edges $\{v_{n/2-1}, \rho_{10,1}\}$ and $\{v_{n/2-1}, \rho_{10,n-1}\}$. Again if both edges are red, Builder wins immediately. Otherwise, Builder creates a blue path from u to $\rho_{10,1}$ or to $\rho_{10,n-1}$.

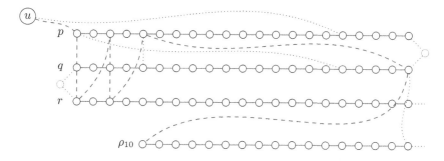

Fig. 2. Creation of $\rho_{n/2}$ for $n = 18$.

If the final circle is red then it is induced because the initial path is induced and we neither create edges connecting two vertices of ρ_k to itself, nor edges connecting v_i to any vertices between endpoints of the cycle. If the blue cycle is created it is induced because we use only odd vertices on $\rho_1, \rho_2, \ldots, \rho_9$ for creating the three blue paths and no edges are created between vertices which are further than 1 apart on these blue paths.

Note that the length of paths $\rho_1, \rho_2, \ldots, \rho_9$ is sufficient because they need to be at least $2\left\lfloor \frac{n/2-1}{3} \right\rfloor + (n-2) \leq \frac{4n-8}{3} \leq 4n/3 - 1$.

By Theorem 2 we can create the initial induced P_{13n} in $28(13n) - 27$ rounds. There are at most $3n$ additional edges, hence $\tilde{r}(C_n) \leq 367n - 27$.   □

Theorem 4. *Let C_n be a cycle on n vertices, where n is odd. Then $\tilde{r}_{ind}(C_n) \leq \tilde{r}_{ind}(C_{2n}) + n \leq 735n - 27$.*

Proof. First, we create a monochromatic cycle C_{2n}. Assume without loss of generality that this cycle is blue. Let $c_0, c_1, \ldots, c_{2n-1}$ denote vertices on the C_{2n} in the natural order and let c_i for any $i \geq 2n$ denote vertex c_j, $j = i \bmod 2n$. We join two vertices which lie $n-1$ apart on the even cycle by creating an edge $\{c_0, c_{n-1}\}$. If the edge is blue it forms a blue C_n with part of the blue even cycle (see Fig. 3). If the edge is red we can continue and create an edge $\{c_{n-1}, c_{2(n-1)}\}$ and use the same argument. This procedure can be repeated n times finishing with the edge $\{c_{(n-1)(n-1)}, c_{n(n-1)}\}$ where $c_{n(n-1)} = c_0$ because $n-1$ is even.

Let E be all the new red edges we just created, i.e., $E = \{\{c_i, c_{i+n-1}\} \mid i \in J\}$ where $J = \{j(n-1) \mid j \in \{0, 1, \ldots, n-1\}\}$. Since $\gcd(n-1, 2n) = 2$ it follows that the edges of E complete a cycle $C'_n = (\{c_0, c_2, \ldots, c_{2n-2}\}, E)$ (see Fig. 3).

Since the C_{2n} is induced then it follows trivially that the target C_n will be induced as well.

We used Theorem 3 to create an even cycle C_{2n}. Then we added n edges to form the C'. This gives us an upper bound for induced odd cycles $\tilde{r}_{ind}(C_n) \leq \tilde{r}_{ind}(C_{2n}) + n \leq 735n - 27$.   □

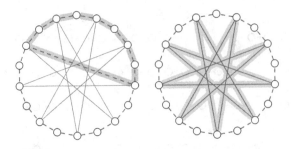

Fig. 3. Final step of building C_9 (Color figure online)

Non-induced Cycles

Although the induced cycle strategies are asymptotically tight we can get better constants for the non-induced cycles. For even cycles, we can use the non-induced path strategy to create the initial $P_{17n/2}$ in $4(17n/2) - 3$ rounds. Then we add $3n/2$ edges in the similar fashion as for the induced cycles however we can squeeze them more tightly as depicted in the Fig. 4.

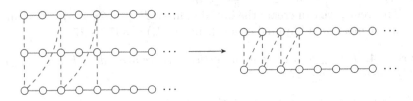

Fig. 4. More efficient construction for even non-induced cycles.

Using this method the paths $\rho_1, \rho_2, \ldots, \rho_6$ need only $5n/4$ vertices each, therefore the initial path $P_{17n/2}$ is sufficient. This gives us $\tilde{r}(C_n) \leq 71n/2 - 3$ for even n which directly translates to odd cycles and gives us $\tilde{r}(C_n) \leq \tilde{r}(C_{2n}) + n \leq 72n - 3$ for odd n.

4 Tight Bounds for a Family of Trees

We first prove a general lower bound for the online Ramsey number of graphs. It will be used to show the tightness of bounds in this section.

Lemma 1. The $\tilde{r}(H)$ is at least $VC(H)(\Delta(H) - 1)/2 + |E(H)|$ where $VC(H)$ is the vertex cover and $\Delta(H)$ is the highest vertex degree in H and $|E(H)|$ is the number of edges.

Proof. Let $deg_b(v)$ be the number of blue edges incident to the vertex v. Let us define the Painter's strategy against the target graph H as:

1. if both incident vertices have $deg_b < \Delta(H) - 1$ then color the edge blue,
2. otherwise color the edge red.

It is clear that Builder cannot create H in blue color because the blue graph can contain only vertices with degree at most $\Delta(H) - 1$. To obtain a red edge it has to have at least one incident vertex with high blue degree. The minimal number of vertices with high blue degree which are required to complete H is the vertex cover of H, therefore, Builder has to create at least $VC(H)(\Delta(H) - 1)/2$ blue edges. Then Builder has to create at least $|E(H)|$ edges to complete the target graph in red color. □

Let us define a *spider* $\sigma_{k,\ell}$ for $k \geq 3$ and $\ell \geq 2$ as a union of k paths of length ℓ that share exactly one common endpoint. Let a *center* of $\sigma_{k,\ell}$ denote the only vertex with degree equal to k.

In the following theorem we obtain an upper bound on $\widetilde{r}(\sigma_{k,\ell})$ that asymptotically matches the lower bound from Lemma 1.

Theorem 5. $\widetilde{r}_{ind}(\sigma_{k,\ell}) = \Theta(k^2\ell)$.

Proof. We describe Builder's strategy for obtaining an induced monochromatic $\sigma_{k,\ell}$. We start by creating an induced monochromatic path of length $k^2(2\ell + 1)$ which is without loss of generality blue. This path contains k^2 copies of $P_{2\ell}$ as an induced subgraph. Let $P_{i,j}$ denote the j-th vertex on path P_i. Let $\mathbb{P}^1, \mathbb{P}^2, \ldots, \mathbb{P}^k$ be k sets where each contains k disjoint induced paths. Let u be a previously unused vertex. Now for each $\mathbb{P}^j$ we do the following procedure:

1. Let $\{P^1, P^2, \ldots, P^k\} = \mathbb{P}^j$.
2. Create edges $\{\{u, w\} \mid w \in \{P_1^1, P_1^2, \ldots, P_1^k\}\}$. If there are k blue edges there is a $\sigma_{k,\ell}$ with the center in u. If that is not the case there is at least one red edge $e^1 = \{u, v^1\}$ where $v^1 \in \{P_1^1, P_1^2, \ldots, P_1^k\}$.
3. For i from 2 to ℓ we do as follows.
   - For $v^{i-1} \in P^z$ create edges $\{\{v^{i-1}, w\} \mid w \in \{P_i^1, P_i^2, \ldots, P_i^k\} - P_i^z\}$. If all of these edges are blue we have a $\sigma_{k,\ell}$ with the center in v^{i-1}, otherwise there is a red edge $\{v^{i-1}, v^i\}$ where $v^i \in \{P_i^1, P_i^2, \ldots, P_i^k\}$.
4. We obtained a red induced path $L^j = (u, \{u, v^1\}, \ldots, v^\ell)$.

If all iterations end up in obtaining a path L^j (see Fig. 5) we have k induced paths of length ℓ which all start in u and together they form a $\sigma_{k,\ell}$ with the center in u.

We built a path $P_{k^2(2\ell+1)}$ using Theorem 2 using at most $28(k^2(2\ell + 1)) - 27$ edges. During iterations, we created at most $k\ell(k - 1)$ edges. Therefore we either got a blue $\sigma_{k,\ell}$ during the process or a red $\sigma_{k,\ell}$ after using no more than $\widetilde{r}_{ind}(\sigma_{k,\ell}) \leq 57k^2\ell + 28k^2 - k\ell - 27 = O(k^2\ell)$ rounds.

The lower bound of Lemma 1 gives us $\Omega(k^2\ell)$ therefore the $\widetilde{r}_{ind}(\sigma_{k,\ell}) = \Theta(k^2\ell)$. □

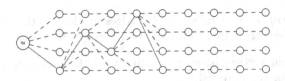

Fig. 5. Building one red leg of a spider $\sigma_{4,5}$. (Color figure online)

We can get the bound on non-induced spiders in a similar way, however, we can use several tricks to get a bound which is not far from the lower bound.

Theorem 6. $\widetilde{r}(\sigma_{k,\ell}) \leq k^2\ell + 15k\ell + 2k - 12 = O(k^2\ell)$.

Proof. We create a path $P_{4k\ell}$ using strategy by Grytczuk et al. [9] in $4(4k\ell) - 3$ rounds and split it into $2k$ paths of length 2ℓ. We follow the same strategy as in the induced case, however, we work over the same set of paths in all iterations and we exclude those vertices which are already used by some path. Choosing $2k$ paths guarantees that we have big enough set even for the last iteration. We create $2k$ edges from u and then we use $k\ell(k-1)$ to create the red paths. We either get a blue $\sigma_{k,\ell}$ in the process or a red $\sigma_{k,\ell}$ after using no more than $k^2\ell + 15k\ell + 2k - 12$ rounds.   □

5   Family of Induced Trees with an Asymptotic Gap

In 2009 Conlon [4] showed that the online Ramsey number and the size-Ramsey number differ asymptotically for an infinite number of cliques. In this section, we present a family of trees which exhibit the same property, i.e., their induced online Ramsey number and size-Ramsey number differ asymptotically.

Definition 1. *Let the centipede $S_{k,\ell}$ be a tree consisting of a path P_ℓ of length ℓ where each of its vertices is center of star S_k, i.e., a thorn-regular caterpillar.*

Note that $S_{k,\ell}$ has $(k+1)(\ell+1)$ vertices and its maximum degree is $k+2$. We will show that $S_{k,\ell}$ exhibits small induced online Ramsey number.

Theorem 7. $\widetilde{r}_{ind}(S_{k,\ell}) \leq 426k\ell - 442k + 308\ell - 295 = O(k\ell)$.

The proof of Theorem 7 was omitted from this extended abstract.

The proof uses the potential method to bound the number of created edges. We proceed in steps where each step either makes the centipede longer or we get a vertex which has k incident edges in both colors (so-called colorful star). When the potential reaches a certain threshold we either get sufficiently long centipede or enough of the colorful stars to run the induced path strategy, which enforces an induced monochromatic centipede.

Due to Beck [2] we have a lower bound for trees T which is $\overline{r}(T) \geq \beta(T)/4$ where $\beta(T)$ is defined as

$$\beta(T) = |T_0|\Delta(T_0) + |T_1|\Delta(T_1),$$

where T_0 and T_1 are partitions of the unique bipartitioning of the tree T. The β for our family of trees is $\beta(S_{k,\ell}) \approx (\ell/2 + k\ell/2)(k+2) = \Theta(k^2\ell)$, which gives us the lower bound on size-Ramsey number $\overline{r}(S_{k,\ell}) = \Omega(k^2\ell)$.

Since by Theorem 7 we have $\widetilde{r}(S_{k,\ell}) \leq \widetilde{r}_{ind}(S_{k,\ell}) = O(k\ell)$ the online Ramsey number for $S_{k,\ell}$ is asymptotically smaller than its size-Ramsey number.

Corollary 1. *There is an infinite sequence of trees* $T_1, T_2, \ldots$ *such that* $|T_i| < |T_{i+1}|$ *for each* $i \geq 1$ *and*

$$\lim_{i \to \infty} \frac{\widetilde{r}(T_i)}{\overline{r}(T_i)} = 0.$$

Acknowledgments. We would like to thank our colleagues Jiří Fiala, Pavel Veselý and Jana Syrovátková for fruitful discussions.

References

1. Beck, J.: On size Ramsey number of paths, trees, and circuits. I. J. Graph Theory **7**(1), 115–129 (1983)
2. Beck, J.: On size Ramsey number of paths, trees and circuits. II. In: Nešetřil, J., Rödl, V. (eds.) Mathematics of Ramsey Theory, vol. 5, pp. 34–45. Springer, Heidelberg (1990). https://doi.org/10.1007/978-3-642-72905-8_4
3. Beck, J.: Achievement games and the probabilistic method. Comb. Paul Erdős Eighty **1**, 51–78 (1993)
4. Conlon, D.: On-line Ramsey numbers. SIAM J. Discrete Math. **23**, 1954–1963 (2009)
5. Conlon, D., Fox, J., Sudakov, B.: On two problems in graph Ramsey theory. Combinatorica **32**(5), 513–535 (2012)
6. Conlon, D., Fox, J., Sudakov, B.: Recent developments in graph Ramsey theory. In: Surveys in Combinatorics (2015)
7. Erdős, P., Faudree, R.J., Rousseau, C.C., Schelp, R.H.: The size Ramsey number. Periodica Mathematica Hungarica **9**, 145–161 (1978)
8. Erdős, P.: Problems and results on finite and infinite graphs. In: Recent Advances in Graph Theory, Proceedings of Second Czechoslovak Symposium, Prague, pp. 183–192 (1975)
9. Grytczuk, J., Kierstead, H.A., Prałat, P.: On-line Ramsey numbers for paths and stars. Discrete Math. Theor. Comput. Sci. **10**, 3 (2008)
10. Haxell, P.E., Kohayakawa, Y., Łuczak, T.: The induced size-Ramsey number of cycles. Comb. Probab. Comput. **4**(03), 217–239 (1995)
11. Kurek, A., Rucinski, A.: Two variants of the size Ramsey number. Discuss. Math. Graph Theory **25**, 141–149 (2005)

Approximations of Schatten Norms via Taylor Expansions

Vladimir Braverman$^{(\boxtimes)}$

Johns Hopkins University, Baltimore, USA
vova@cs.jhu.edu

Abstract. In many applications of data science and machine learning data is represented by large matrices. Fast and accurate analysis of such matrices is a challenging task that is of paramount importance for the aforementioned applications. Randomized numerical linear algebra (RNLA) is an popular area of research that often provides such fast and accurate algorithmic methods for massive matrix computations. Many critical problems in RNLA boil down to approximating spectral functions and one of the most fundamental examples of such spectral functions is Schatten p norm. The p-th Schatten norm for matrix $A \in R^{n_1 \times n_2}$ is defined as the l_p norm of a vector comprised of singular values of matrix A, i.e.,

$$\|A\|_p = \left(\sum_{i=1}^{\min(n_1, n_2)} |\sigma_i(A)|^p \right)^{1/p},$$

where $\sigma_i(A)$ is the i-th singular value of A.

In this paper we consider symmetric, positive semidefinite (SPSD) matrix A and present an algorithm for computing the p-Schatten norm $\|A\|_p$. Our methods are simple and easy to implement and can be extended to general matrices. Our algorithms improve, for a range of parameters, recent results of Musco, Netrapalli, Sidford, Ubaru and Woodruff (ITCS 2018), e.g., for $p > 2$ and sufficiently small values of ϵ.

Keywords: Randomized numerical linear algebra · Schatten norms · Algorithms

1 Introduction

In many applications of data science and machine learning data is represented by large matrices. Fast and accurate analysis of such matrices is a challenging task that is of paramount importance for the aforementioned applications. Randomized numerical linear algebra (RNLA) is an popular area of research that often provides such fast and accurate algorithmic methods for massive matrix computations. Many critical problems in RNLA boil down to approximating spectral

This research was supported in part by NSF CAREER grant 1652257, ONR Award N00014-18-1-2364, and the Lifelong Learning Machines program from DARPA/MTO.

R. van Bevern and G. Kucherov (Eds.): CSR 2019, LNCS 11532, pp. 70–79, 2019.
https://doi.org/10.1007/978-3-030-19955-5_7

functions and one of the most fundamental examples of such spectral functions is Schatten p norm. The p-th Schatten norm for matrix $A \in R^{n_1 \times n_2}$ is defined as the l_p norm of a vector comprised of singular values of matrix A, i.e.,

$$\|A\|_p = \left(\sum_{i=1}^{\min(n_1, n_2)} |\sigma_i(A)|^p \right)^{1/p},$$

where $\sigma_i(A)$ is the i-th singular value of A.

1.1 Our Results

In this paper we consider symmetric, positive semidefinite (SPSD) matrix A and present an algorithm for computing $\|A\|_p$. Our method can be seen as a combination of Taylor expansion of function $(1 - x)^p$ combined with the well-known Hutchinson estimator [3] for trace estimation defined in Eq. (9). The key observation is that for the first m terms in the expansion, for sufficiently large m, the corresponding matrix polynomial is SPSD and thus the Hutchinson method is applicable. It is well known that for SPSD matrix A, $\|A\|_p^p = \mathbf{tr}\, A^p$ and thus, our algorithms provide multiplicative approximations for $\|A\|_p^p$. Thus we obtain:

Theorem 1. *Let $A \in R^{n \times n}$ be an SPSD matrix and let y be the output of Algorithm 2 on input A, ϵ, δ. Then, with probability at least $1 - \delta$, we have:*

$$\left| \|A\|_p^p - y \right| \le \epsilon \|A\|_p^p.$$

If we use Algorithm 1 to compute the Hutchinson estimator H_t and Lemma 1 to compute a 6-approximation of the largest eigenvalue $\lambda_1(A)$ then the running time of Algorithm 2 is

$$O\left(\left[\frac{\gamma(p)}{\epsilon^{2+1/p}} n^{1/p} nnz(A) + n \log n \right] \log \frac{1}{\delta} \right),$$

where

$$\gamma(p) = \frac{c(p)}{p}^{1/p} (1 + 6^p)^2, \tag{1}$$

and $c(p)$ is given in (16).

Our methods can be extended to general matrices. Indeed, to compute $\|B\|_p$ for matrix B, one can apply our methods to SPSD matrix $B^T B$ and note that $\|B\|_p = \|B^T B\|_{p/2}^{1/2}$. In Appendix A we describe, for completeness, a trivial extension of Hutchinson estimator to the matrix $B^T B$. Thus, we obtain:

Corollary 1. *Let $B \in R^{l \times n}$ be a matrix with $l \ge n$. There exists an algorithm that, on input B, ϵ, δ, outputs y such that, with probability at least $1 - \delta$, we have:*

$$\left| \|B\|_p^p - y \right| \le \epsilon \|B\|_p^p.$$

The running time of the algorithm is

$$O\left(\left[\frac{\gamma(p/2)}{\epsilon^{2+2/p}} n^{2/p} nnz(B) + n \log n \right] \log \frac{1}{\delta} \right).$$

1.2 Related Work

Musco, Netrapalli, Sidford, Ubaru and Woodruff [5] (see also the full version in [4]) provided a general approach for approximating spectral functions that works for general matrices. Below we state the results from [5] for Schatten norms.

Theorem 2. *([4, 5], Theorem 32, Schatten p-norms, $p \geq 2$, for sparse matrices) For any $p \geq 2$, and $A \in R^{n \times n}$ there is an algorithm returning $X \in (1 \pm \epsilon) \|A\|_p^p$ with high probability in time[1]*

$$\tilde{O}\left(\frac{p}{\epsilon^3} \sqrt{nnz(A)nd_s(A)} \right)$$

or

$$\tilde{O}\left(\frac{p}{\epsilon^3} \left[nnz(A)n^{\frac{1}{p+1}} + n^{1+\frac{2}{p+1}} \right] \right),$$

where $d_s(A)$ denotes the maximum number of non-zero entries in any row of A.

Theorem 3. *([4, 5], Theorem 33, Schatten p-norms, $p \leq 2$, for sparse matrices) For any $p \in (0, 2]$, and $A \in R^{n \times n}$ there is an algorithm returning $X \in (1 \pm \epsilon) \|A\|_p^p$ with high probability in time*

$$\tilde{O}\left(\frac{1}{p^3 \epsilon^{\max\{3, 1/p\}}} \left[nnz(A)n^{\frac{1/p-1/2}{1/p+1/2}} \sqrt{\gamma_s} + \sqrt{nnz(A)}n^{\frac{4/p-1}{2/p+1}} \right] \right)$$

or

$$\tilde{O}\left(\frac{1}{p^3 \epsilon^{\max\{3, 1/p\}}} \left[nnz(A)n^{\frac{1}{p+1}} + n^{1+\frac{2}{p+1}} \right] \right),$$

where $\gamma_s = \frac{d_s(A)n}{nnz(A)} \geq 1$, and $d_s(A)$ denotes the maximum number of non-zero entries in any row of A.

Our result in Corollary 1 improves the bounds in [4,5] for a range of parameters, e.g., when $d_s(A) = \Theta(n)$, $p > 2$ and $\epsilon = o\left(n^{\frac{1}{p-2}\left(\frac{1}{p(p+1)} - 2 \right)} \right)$.

Han, Malioutov, Avron, and Shin [2] use general Chebyshev polynomials to approximate a wide class of spectral functions. Ubaru, Chen, and Saad [7], use Lanczos algorithm to estimate analytic spectral functions on symmetric positive definite matrices. The methods in [2,7] work for invertible matrices and the running time depends on condition number κ. In contrast, our algorithm does not depend on κ and works for arbitrary matrices.

Our work has been inspired by the approach of Boutsidis, Drineas, Kambadur, Kontopoulou, and Zouzias [1] that uses Taylor expansion for $\log(1+x)$ to approximate the log determinant. The coefficients of this expansion have the same sign and thus the Hutchinson estimator [3] can be applied to each partial sum of this expansion. This is not the case for the Taylor expansion of $(1-x)^p$. The key idea of our approach is to find m such that a constant minus the partial sum of the first m terms is positive. Thus, we can apply the Hutchinson estimator to the corresponding matrix polynomial.

---

[1] Here $\tilde{O}$ is used to hide polylogarithmic in n factors.

1.3 Roadmap

In Sect. 1.4 we introduce necessary notations. Section 3 provides a Hutchinson estimator for a special matrix polynomial that will be used in our algorithms. Section 4 describes our main algorithm for approximating $\|A\|_p^p$. Section 2 contains necessary technical claims. In Appendix A we describe, for completeness, a trivial extension of Hutchinson estimator to handle general matrices.

1.4 Notations

We use the following symbols: $[x]$ be the integer part of x,

$$\binom{p}{k} = \frac{p(p-1)\ldots(p-k+1)}{k!}, k \geq 2, \quad \binom{p}{1} = p. \tag{2}$$

We use $\log x$ to denote the natural logarithm and dedicate lower case Latin letters for constants and real variables and upper case Latin letters for matrices. Consider the Taylor expansion

$$(1-x)^p = 1 + \sum_{k=1}^{\infty} \binom{p}{k}(-1)^k x^k, \quad -1 < x < 1. \tag{3}$$

Denote

$$h(x) = \sum_{k=1}^{\infty}(-1)^{k-1}\binom{p}{k}x^k. \tag{4}$$

It follows from (3) that for $|x| < 1$:

$$1 - (1-x)^p = h(x). \tag{5}$$

Denote

$$h_m(x) = \sum_{k=m+1}^{\infty}(-1)^{k-1}\binom{p}{k}x^k, \tag{6}$$

and

$$\tilde{h}_m(x) = \sum_{k=1}^{m}(-1)^{k-1}\binom{p}{k}x^k. \tag{7}$$

According to (5),

$$1 - (1-x)^p = h(x) = \tilde{h}_m(x) + h_m(x). \tag{8}$$

Denote by H_t be the Hutchinson estimator [3] for the trace:

$$H_t(X) = \frac{1}{t}\sum_{i=1}^{t} g_i^T X g_i, \tag{9}$$

where $g_1,\ldots,g_t \in R^n$ are i.i.d. vectors whose entries are independent Rademacher variables. Note that $H_t(I_n) = n$. We will be using the following well-known results.

Theorem 4. *(Roosta-Khorasani and Ascher [6], Theorem 1) Let $Y \in R^{n \times n}$ be an SPSD matrix. Let $t > \frac{8}{\epsilon^2} \log \frac{1}{\delta}$. Then, with probability at least $1 - \delta$:*

$$|H_t(Y) - \operatorname{tr} Y| \le \epsilon \operatorname{tr} Y. \tag{10}$$

Lemma 1. *(Boutsidis, Drineas, Kambadur, Kontopoulou, and Zouzias [1], Lemma 3) Let A be a symmetric positive semidefinite matrix. There exists an algorithm that runs in time $O((n + nnz(A)) \log n \log \frac{1}{\delta})$ and outputs α such that, with probability at least $1 - \delta$:*

$$\frac{1}{6} \lambda_1(A) \le \alpha \le \lambda_1(A).$$

2   Technical Lemmas

Lemma 2. *There exists a constant $c(p)$, (defined in (16)), that depends only on p, such that*

$$\left| \binom{p}{k} \right| \le c(p)(k+1)^{-(p+1)},$$

for any $k > [p] + 1$.

Proof. Denote

$$c_1(p) = \left\{ \prod_{i=1}^{[p]} \left| \frac{p-i+1}{i} \right|, \quad p > 1, 1, \quad 0 \le p \le 1. \right. \tag{11}$$

If $i > [p] + 1$ then $\left| \frac{p-i+1}{i} \right| = 1 - \frac{p+1}{i}$. Thus, for $k > [p] + 1$:

$$\left| \binom{p}{k} \right| = c_1(p) \frac{p - [p]}{[p] + 1} \prod_{i=[p]+2}^{k} \left(1 - \frac{p+1}{i} \right). \tag{12}$$

Since $0 < 1 - x < e^{-x}$ for $0 < x < 1$ and $\frac{p-[p]}{[p]+1} < 1$, we have:

$$\left| \binom{p}{k} \right| \le c_1(p) \exp \left(-(1+p) \sum_{i=[p]+2}^{k} \frac{1}{i} \right). \tag{13}$$

Further, it is known from Langrange formula that $\frac{1}{i} \ge \ln(i+1) - \ln i$ and thus

$$\sum_{i=[p]+2}^{k} \frac{1}{i} \ge \sum_{i=[p]+2}^{k} (\ln(i+1) - \ln i) = \ln(k+1) - \ln([p]+2). \tag{14}$$

From here and (13) we obtain

$$\left| \binom{p}{k} \right| \le c_1(p) \exp \left(-(1+p)(\ln(k+1) - \ln([p]+2)) \right)$$
$$= c_1(p)(2 + [p])^{p+1}(k+1)^{-(p+1)}. \tag{15}$$

Denote

$$c(p) = c_1(p)(2 + [p])^{p+1},\tag{16}$$

where $c_1(p)$ is defined in (11). The lemma follows.

Lemma 3. *For $m > [p] + 1$ and $|x| \le 1$:*

$$|h_m(x)| \le \frac{c(p)}{p}|x|^{m+1}(m+1)^{-p}.$$

Proof. We have:

$$|h_m(x)| \le \sum_{k=m+1}^{\infty} \left|\binom{p}{k}\right| |x|^k \le |x|^{m+1} \sum_{k=m+1}^{\infty} \left|\binom{p}{k}\right| \le |x|^{m+1} c(p) \sum_{k=m+1}^{\infty} (k+1)^{-(p+1)},$$

where the first inequality follows from the definition of $h_m(x)$ in (6), the second inequality follows since $|x| \le 1$ and the third inequality follows from Lemma 2. Let us estimate the last sum[2]. For any $k \le x \le k+1$ we have $(k+1)^{-(p+1)} \le x^{-(p+1)}$. Integrating over $[k, k+1]$ we obtain

$$(k+1)^{-(p+1)} \le \int_k^{k+1} x^{-(p+1)} dx.$$

Thus

$$\sum_{k=m+1}^{\infty} (k+1)^{-(p+1)} \le \sum_{k=m+1}^{\infty} \int_k^{k+1} x^{-(p+1)} dx = \int_{m+1}^{\infty} x^{-(p+1)} dx = \frac{1}{p}(m+1)^{-p}.$$

The lemma follows from here and the above bound.

Lemma 4. *For any $x \in [0, 1]$ and $m > [p] + 1$:*

$$1 + \frac{c(p)}{p}(m+1)^{-p} - \tilde{h}_m(x) \ge (1 - x)^p \ge 0.$$

Proof. We have $1 - \tilde{h}_m(x) = [1 - h(x)] + h_m(x) = (1 - x)^p + h_m(x)$, and, by Lemma 3, $|h_m(x)| \le \frac{c(p)}{p}(m+1)^{-p}$. So $1 - \tilde{h}_m(x) \ge (1-x)^p - \frac{c(p)}{p}(m+1)^{-p}$. The lemma follows.

3 Hutchinson Estimator for $\tilde{h}_m(X)$

The following simple algorithm computes $H_t(\tilde{h}_m(X))$.

Theorem 5. *Let $X \in R^{n \times n}$ be a matrix and let $t \ge 1$ be an integer. Then the output y of Algorithm 1 is $H_t(\tilde{h}_m(X))$ and the running time of the algorithm is $O(t \cdot nnz(X))$.*

---

[2] This and some other bounds in this paper are well known and we include their proofs for completeness.

Algorithm 1

1: **Input:** Matrix $X \in R^{n \times n}, t \geq 1$, $\triangleright$ $\tilde{h}_m$ is defined in (7)
2: **Output:** $H_t(\tilde{h}_m(X))$.
3: **Initialization:**
4: Let $g_1, g_2, \ldots, g_t \in \{-1, +1\}^n$ be i.i.d. random vectors
5:        whose entries are independent Rademacher variables.
6: **for** $i = 1, 2, \ldots, t$ **do**
7:        $v_1 = X g_i, \quad u_1 = g_i^T v_1$
8:        $a_1 = p$
9:        $S_i^1 = a_1 u_1$
10:       **for** $k = 2, \ldots, m$ **do**
11:              $v_k = X v_{k-1}$
12:              $u_k = g_i^T v_k$
13:              $a_k = a_{k-1} \frac{p-(k-1)}{k}$
14:              $S_i^k = S_i^{k-1} + (-1)^{k-1} a_k u_k$
15:       **end for**
16: **end for**
17: Return $y = \frac{1}{t} \sum_{i=1}^t S_i$.

Proof. For $k = 1, \ldots, m$ denote $a_k = \binom{p}{k}$ and note that $a_k = a_{k-1} \frac{p-(k-1)}{k}$ and $\tilde{h}_m(X) = \sum_{k=1}^m (-1)^{k-1} a_k X^k$, by (7). Fix $i \in \{1, \ldots, t\}$ and note that after the k-th iteration we have

$$u_k = g_i^T X^k g_i, \quad S_i^k = g_i^T \left(\sum_{j=1}^k (-1)^{j-1} a_j X^j \right) g_i.$$

Thus, $S_i^m = g_i^T(\tilde{h}_m(X)) g_i$. Denote $Y = \tilde{h}_m(X)$; thus the algorithm outputs the estimator $y = \frac{1}{t} \sum_{i=1}^t g_i^T Y g_i = H_t(\tilde{h}_m(X))$ and the theorem follows. The bound on the running time follows from direct computations.

4   Main Algorithm

In this section we present our main algorithm and prove it's correctness.

Algorithm 2

1: **Input:** SPSD Matrix $A \in R^{n \times n}$, $p \geq 1, \epsilon \in (0, 1), \delta \in (0, 1)$.
2: **Output:** y such that, w.p. at least $1 - \delta$, $(1 - \epsilon)\|A\|_p^p \leq y \leq (1 + \epsilon)\|A\|_p^p$.
3: **Initialization:**
4: $t > \frac{8}{\epsilon^2} \log \frac{2}{\delta}, m > \max\{7 \left(\frac{4c(p)}{p} \frac{n}{\epsilon} \right)^{1/p}, \lceil p \rceil + 1\}$ $\triangleright$ $c(p)$ is given in (16)
5: Compute α such that $\lambda_1(A) \leq \alpha \leq 6\lambda_1(A)$. $\triangleright$ Use the algorithm from Lemma 1.
6: Return $y = \alpha^p \left[n - H_t(\tilde{h}_m(I_n - \alpha^{-1}A)) \right]$. $\triangleright$ Use Algorithm 1. $\tilde{h}_m$ is a function
       given by (7).

Proof (of Theorem 1). We have for $x \in [0,1]$, $x^p = [1-(1-x)]^p = 1-h(1-x)$. Let B be a SPSD matrix with $\lambda_1(B) \leq 1$. Denote $C = I_n - B$. Since $\lambda_1(B) \in [0,1]$ we have that C is also a SPSD matrix and

$$B^p = I_n - h(C). \tag{17}$$

Denote $\beta_m = \frac{c(p)}{p}(m+1)^{-p}$. Since $m > [p] + 1$ then, according to Lemma 4 the matrix $(1+\beta_m)I_n - \tilde{h}_m(C)$ is SPSD. So, we may apply estimator H_t to this matrix. We also have

$$B^p = I_n - h(C) = I_n - \tilde{h}_m(C) - h_m(C) = [(\beta_m+1)I_n - \tilde{h}_m(C)] - [h_m(C) + \beta_m I_n].$$

Because $H_t(I_n) = n$,

$$\left| \mathbf{tr}\ B^p - \left[n - H_t(\tilde{h}_m(C))\right] \right| = \left| \mathbf{tr}\ B^p - H_t\left[(1+\beta_m)I_n - \tilde{h}_m(C)\right] + n\beta_m \right|$$

$$\leq \left| \mathbf{tr}\left[(\beta_m+1)I_n - \tilde{h}_m(C)\right] - H_t\left[(\beta_m+1)I_n - \tilde{h}_m(C)\right] \right| + |\mathbf{tr}(h_m(C))| + 2n\beta_m = \Delta_1 + \Delta_2 + 2n\beta_m.$$

Below we will bound each Δ separately.

Bounding Δ_1

According to Theorem 4 for $t > \frac{8}{\epsilon^2}\log(1/\delta)$ we have with probability at least $1 - \delta$:

$$\Delta_1 \leq \epsilon\mathbf{tr}\left[(\beta_m+1)I_n - \tilde{h}_m(C)\right] = \epsilon n\beta_m + \epsilon\mathbf{tr}\left[I_n - \tilde{h}_m(C)\right]$$

$$= \epsilon n\beta_m + \epsilon\mathbf{tr}\left[I_n - h(C)\right] + \epsilon\mathbf{tr}h_m(C) = \epsilon n\beta_m + \epsilon\mathbf{tr}\ B^p + \epsilon\mathbf{tr}\ h_m(C).$$

Bounding Δ_2

Denoting $\lambda_i = \lambda_i(B)$ we have, using Lemma 3:

$$|\mathbf{tr}\ h_m(C)| = |\sum_{i=1}^{n} h_m(1 - \lambda_i)| \leq \frac{c(p)}{p}\sum_{i=1}^{n}(1 - \lambda_i)^{m+1}(m+1)^{-p} \leq \frac{c(p)}{p}n(m+1)^{-p} = n\beta_m.$$

The final bound for B

So,

$$\Delta_1 + \Delta_2 + 2n\beta_m \leq (2\epsilon + 2)n\beta_m + \epsilon\mathbf{tr}\ B^p \leq 4n\beta_m + \epsilon\mathbf{tr}\ B^p.$$

Thus, if $n\beta_m = \frac{4c(p)}{p}n(m+1)^{-p} < \epsilon$, i.e., $m > \left(\frac{4c(p)}{p}\frac{n}{\epsilon}\right)^{1/p}$ then

$$\left| \mathbf{tr}\ B^p - \left[n - H_t(\tilde{h}_m(C))\right] \right| \leq \epsilon + \epsilon\mathbf{tr}\ B^p,$$

with probability at least $1 - \delta$.

Back to matrix A

Let α be the constant from Lemma 1 and $B = A/\alpha$. Then $\operatorname{tr} B^p = \alpha^{-p}\operatorname{tr} A^p$ and $C = I_n - B = I_n - A/\alpha$. Recall that $\|A\|_p^p = \operatorname{tr} A^p$ and $y = \alpha^p \left[n - H_t(\tilde{h}_m(I_n - \alpha^{-1}A)) \right]$. So, from the above:

$$\left| \|A\|_p^p - y \right| = \left| \operatorname{tr} A^p - \alpha^p \left[n - H_t(\tilde{h}_m(I_n - \alpha^{-1}A)) \right] \right| \leq \alpha^p \epsilon \operatorname{tr} B^p + \alpha^p \epsilon$$

$$\leq \epsilon \operatorname{tr} A^p + \epsilon(6\lambda_1(A))^p \leq \epsilon(1 + 6^p)\operatorname{tr} A^p = \epsilon(1 + 6^p)\|A\|_p^p,$$

with probability at least $1 - 2\delta$. We obtain the result by substituting ϵ with $\frac{\epsilon}{(1+6^p)}$, δ with $\frac{\delta}{2}$, and noting that $(1 + 6^p)^{1/p} \leq 7$. The bound on the running time follows from Theorem 5 and Lemma 1.

Acknowledgments. The author would like to thank anonymous reviewers and David Woodruff for helpful comments.

A   Extension of Hutchinson Estimator for $\tilde{h}_m(X^T X)$

To apply our methods to non-SPSD matrix X, it is necessary to apply Algorithm 1 to SPSD matrix $X^T X$. Algorithm 3 is a trivial extension of Algorithm 1 that computes $H_t(\tilde{h}_m(X^T X))$. We include it here for completeness.

Algorithm 3

1: **Input:** Matrix $X \in R^{l \times n}, t \geq 1$,   ▷ $\tilde{h}_m$ is defined in (7)
2: **Output:** $H_t(\tilde{h}_m(X^T X))$.
3: **Initialization:**
4: Let $g_1, g_2, \ldots, g_t \in \{-1, +1\}^n$ be i.i.d. random vectors
5:         whose entries are independent Rademacher variables.
6: **for** $i = 1, 2, \ldots, t$ **do**
7:     $w_1 = Xg_i, \quad v_1 = X^T w_1, \quad u_1 = g_i^T v_1$
8:     $a_1 = p$
9:     $S_i^1 = a_1 u_1$
10:     **for** $k = 2, \ldots, m$ **do**
11:         $w_k = X v_{k-1}$
12:         $v_k = X^T w_k$
13:         $u_k = g_i^T v_k$
14:         $a_k = a_{k-1} \frac{p-(k-1)}{k}$
15:         $S_i^k = S_i^{k-1} + (-1)^{k-1} a_k u_k$
16:     **end for**
17: **end for**
18: Return $y = \frac{1}{t} \sum_{i=1}^{t} S_i$.

Theorem 6. *Let $X \in R^{l \times n}$ be a matrix and let $t \geq 1$ be an integer. Then the output y of Algorithm 3 is $H_t(\tilde{h}_m(X^T X))$ and the running time of the algorithm is $O(t \cdot nnz(X))$.*

Proof. The proof follows directly from the proof of Theorem 5 by noting that for a fixed i and for all k, $v_k = (X^T X)^k g_i$ and $u_k = g_i^T (X^T X)^k g_i$. These claims can be verified by induction on k. Indeed, the case $k = 1$ follows since

$$u_1 = g_i^T v_1 = g_i^T X^T w_1 = g_i^T (X^T X) g_i,$$

and the case $k > 1$ follows since

$$u_k = g_i^T v_k = g_i^T X^T w_k = g_i^T X^T X v_{k-1} = g_i^T (X^T X)^k g_i,$$

where the last equality follows by induction.

References

1. Boutsidis, C., Drineas, P., Kambadur, P., Kontopoulou, E.-M., Zouzias, A.: A randomized algorithm for approximating the log determinant of a symmetric positive definite matrix. Linear Algebr. Appl. **533**, 95–117 (2017)
2. Han, I., Malioutov, D., Avron, H., Shin, J.: Approximating spectral sums of large-scale matrices using stochastic chebyshev approximations. SIAM J. Sci. Comput. **39**(4), A1558–A1585 (2017)
3. Hutchinson, M.: A stochastic estimator of the trace of the influence matrix for Laplacian smoothing splines. Commun. Stat. Simul. Comput. **19**(2), 433–450 (1990)
4. Musco, C., Netrapalli, P., Sidford, A., Ubaru, S., Woodruff, D.P.: Spectrum approximation beyond fast matrix multiplication: algorithms and hardness. CoRR, abs/1704.04163 (2017)
5. Musco, C., Netrapalli, P., Sidford, A., Ubaru, S., Woodruff, D.P.: Spectrum approximation beyond fast matrix multiplication: algorithms and hardness. In: Karlin, A.R. (ed.) 9th Innovations in Theoretical Computer Science Conference (ITCS 2018), volume 94 of Leibniz International Proceedings in Informatics (LIPIcs), Dagstuhl, Germany, pp. 8:1–8:21. Schloss Dagstuhl-Leibniz-Zentrum fuer Informatik (2018)
6. Roosta-Khorasani, F., Ascher, U.M.: Improved bounds on sample size for implicit matrix trace estimators. CoRR, abs/1308.2475 (2013)
7. Ubaru, S., Chen, J., Saad, Y.: Fast estimation of $tr(f(a))$ via stochastic lanczos quadrature. SIAM Journal on Matrix Analysis and Applications **38**(4), 1075–1099 (2017)

Nearly Linear Time Isomorphism Algorithms for Some Nonabelian Group Classes

Bireswar Das[(✉)] and Shivdutt Sharma[(✉)]

IIT Gandhinagar, Gandhinagar, India
{bireswar,shiv.sharma}@iitgn.ac.in

Abstract. The isomorphism problem for groups, when the groups are given by their Cayley tables is a well-studied problem. This problem has been studied for various restricted classes of groups. Kavitha gave a linear time isomorphism algorithm for abelian groups (JCSS 2007). Although there are isomorphism algorithms for certain nonabelian group classes, the complexities of those algorithms are usually super-linear. In this paper, we design linear and nearly linear time isomorphism algorithms for some nonabelian groups. More precisely,

- We design a linear time algorithm to factor Hamiltonian groups. This allows us to obtain an $\mathcal{O}(n)$ algorithm for the isomorphism problem of Hamiltonian groups, where n is the order of the groups.
- We design a nearly linear time algorithm to find a maximal abelian factor of an input group. As a byproduct we obtain an $\tilde{\mathcal{O}}(n)$ isomorphism for groups that can be decomposed as a direct product of a nonabelian group of bounded order and an abelian group, where n is the order of the groups.

1 Introduction

Two groups $(G, \cdot)$ and $(H, \times)$ are said to be isomorphic if there exists a bijective function $f : G \longrightarrow H$, which is a homomorphism i.e. $\forall a, b \in G, f(a \cdot b) = f(a) \times f(b)$. The decision version of this problem is to check whether two input groups $(G, \cdot)$ and $(H, \times)$ are isomorphic or not. There are multiple ways in which a group can be given as the input. Two of commonly used methods are by a generating set and by the Cayley table. The complexity of the group isomorphism problem varies with the input representation. In this paper we assume that input groups are given by their Cayley tables unless stated otherwise explicitly.

It is not known whether the group isomorphism problem (GrISO) is in P. If it is NP-complete then polynomial hierarchy collapses at the second level [4]. Tarjan (see e.g., [14]) gave an $n^{\log n + \mathcal{O}(1)}$ algorithm for GrISO. While this still remains the best upper bound for the general group isomorphism problem, progress has been made for restricted classes of groups. For solvable groups, Arvind and Torán showed that the problem is in NP ∩ co-NP under a reasonable complexity

© Springer Nature Switzerland AG 2019
R. van Bevern and G. Kucherov (Eds.): CSR 2019, LNCS 11532, pp. 80–92, 2019.
https://doi.org/10.1007/978-3-030-19955-5_8

theoretic assumption [1]. Rosenbaum and Wagner gave a $n^{1/2(\log n)+\mathcal{O}(1)}$ algorithm for the isomorphism problem of p-groups [17] and Rosenbaum gave an $n^{1/2(\log n)+\mathcal{O}(\log n/\log\log n)}$ time algorithm for solvable groups [16].

The isomorphism problem for various restricted classes of groups has been studied in the past [1,3,8,9,15,19,20]. Efficient polynomial time algorithms for the isomorphism problem of abelian groups were designed by Savag [18], and Vikas [20]. Kavitha gave a remarkable linear time algorithm for the isomorphism problem of abelian groups [19]. Kavitha's paper also provides us with a useful tool for computing the orders of all the elements in *any* group in linear time.

Polynomial time algorithms have been designed for some classes of nonabelian groups. For example, Le Gall designed an efficient algorithm for groups consisting of a semidirect product of an abelian group with a cyclic group of coprime order [7]. Later, Qiao, Sarma, and Tang gave a polynomial time algorithm for the isomorphism problem of groups with normal Hall subgroups [15]. Babai, Codenotti and Qiao gave a polynomial time algorithm for the class of groups with no normal abelian subgroups [2]. We believe that motivation behind these results on the isomorphism problem of nonabelian groups was to enlarge the family of group classes for which polynomial time algorithm is known. To the best of our knowledge the runtime of these algorithms are superquadratic.

Our goal in this paper is to design linear or nearly linear time algorithm for some nontrivial classes of nonabelian groups.

The isomorphism problem of nilpotent class 2 groups[1] is *not* known to be in polynomial time. A nonabelian group is Hamiltonian if all of its subgroups are normal. These groups are nilpotent class 2 groups. We design an $\mathcal{O}(n)$ algorithm for the recognition and the isomorphism problem of the Hamiltonian groups where n is the size of the input groups.

A Hamiltonian group is a direct product of the quaternion group Q_8 and an abelian group with certain structure [5]. This motivates us to study the class of groups that can be decomposed as a direct product of any arbitrary nonabelian group of bounded order and an abelian group without any specific structure. We design an $\tilde{\mathcal{O}}(n)$ algorithm for the recognition and the isomorphism problem of such groups.

Kayal and Nezhmetdinov gave an algorithm to factorize an input group into a direct product of indecomposable factors [12]. We note that this group factorization algorithm combined with any of the polynomial time isomorphism algorithms for abelian group gives us a polynomial time isomorphism test for the group classes considered in this paper. However, direct application of the result by Kayal and Nezhmetdinov only gives us superquadratic isomorphism algorithms. One of the contributions of this paper is to use the structure of the input groups to tweak and bypass some of the computation heavy steps of the algorithm by Kayal et al. [12].

The main results of this paper are stated below.

Theorem 1. *There exists an $\mathcal{O}(n)$ algorithm for the recognition and the isomorphism problem of Hamiltonian groups where n is the size of the input groups.*

---

[1] A group G is *nilpotent class 2* if $G/Z(G)$ is abelian.

Theorem 2. *There exists an $\tilde{O}(n)$ algorithm for the recognition and the isomorphism problem of groups that can be decomposed as a direct product of a nonabelian group of bounded order and an abelian group where n is the size of the input groups.*

2   Preliminaries

In this section, we describe some of the group-theoretic definitions and background used in the paper. For more details see [6,10,13,19]. For a group G, the number of elements in G or the *order* of G is denoted by $|G|$. Let $x \in G$ be an element of group G, then $\mathrm{ord}_G(x)$ denotes the order of the element x in G, which is the smallest power i of x such that $x^i = e$, where e is the identity element of the group G.

For a subset $S \subseteq G$, $\langle S \rangle$ denotes the subgroup generated by the set S. For a subgroup $A \leq G$, *the centralizer* of A, denoted $C_G(A)$, is the set $\{g \in G \mid ag = ga, \forall a \in A\}$. The *center* $Z(G)$ of group G is the subgroup with elements $\{g \in G \mid ga = ag, \forall a \in G\}$.

Given $H \leq G$, the *normal closure* of H in G is the smallest normal subgroup of G containing H and is denoted by $\langle H \rangle^G$. The *commutator* subgroup $[G, G]$ of a group G is the subgroup $\langle \{xyx^{-1}y^{-1} \mid \forall x, y \in G\} \rangle$.

Let G be a finite group and A, B be subgroups of G. Then G is a *direct product* of A and B, denoted $G = A \times B$, if (1) $A \trianglelefteq G$ and $B \trianglelefteq G$, (2) $|G| = |A||B|$, (3) $A \cap B = \{e\}$. We say that a group G is *decomposable* if there exist nontrivial subgroups A and B such that $G = A \times B$ and *indecomposable* otherwise. We say that a subgroup A of G is a *direct factor* (or *factor*) of G if there exists another subgroup B of G such that $G = A \times B$ and we will call B a *direct complement* (or *complement*) of A.

The fundamental theorem for finitely generated abelian groups implies that a finite group G can be decomposed as a direct product $G = G_1 \times G_2 \times \ldots \times G_t$, where each G_i is a cyclic group of order p^j for some prime p and integer $j \geq 1$. If a_i generates the cyclic group G_i for $i = 1, 2, 3, \ldots, t$ then the elements $a_1, a_2, \ldots, a_t$ are called a *basis* of G. An elementary abelian p-group is an abelian group in which every nontrivial element has order p. Chen and Fu [6], and Karagiorgos and Poulakis [11] gave linear time algorithms for finding a basis of abelian groups.

Theorem 3 (Remak-Krull-Schmidt, see e.g., [10]). *Let G be a finite group. If $G = G_1 \times G_2 \times \ldots \times G_s$ and $G = H_1 \times H_2 \times \ldots \times H_t$ with each G_i, H_j indecomposable, then $s = t$ and after reindexing $G_i \cong H_i$ for every i, and for any $r < t$, $G = G_1 \times \ldots \times G_r \times H_{r+1} \times \ldots \times H_t$.*

A *Remak-Krull-Schmidt decomposition* of a group G is a decomposition such that each direct factor of group G is indecomposable. The following lemma establishes a relationship between two different decompositions of a group G in which one of the factors is same.

Lemma 1 ([12]). *For a group G, suppose that $G = H \times K$. Then for a $K' \trianglelefteq G$, $G = H \times K'$ if and only if $K' = \{\alpha\phi(\alpha) \mid \alpha \in K\}$, where $\phi : K \longrightarrow Z(H)$ is a homomorphism.*

Model of Computation: Our model of computation is same as that of many of the algorithms for groups given by Cayley table (e.g., [6,19]). It is a RAM model where random access can be done in constant time. Each register and memory unit can store $\mathcal{O}(\log |G|)$ bits. The arithmetic, logic and comparison operations on $\mathcal{O}(\log |G|)$ bits take constant time. Unless stated otherwise we assume that the elements of the group are encoded as $1, 2, \ldots, |G|$.

Nearly Linear Time Algorithms: A group theoretic algorithm is *nearly linear time* if it has runtime $\mathcal{O}(|G| \log^{\mathcal{O}(1)} |G|)$. We hide the logarithmic factor by using the notation $\tilde{\mathcal{O}}(|G|)$. We list some useful nearly linear time algorithms for group theoretic problems for groups given by their Cayley tables in the next lemma. The ideas behind these results are either known as folklores or directly follows from easy observations.

Lemma 2. 1. *Given $S \subseteq G$, one can compute the elements of the subgroup $\langle S \rangle$ in $\mathcal{O}(|G| \log |G|)$ time.*
2. *Finding an $\mathcal{O}(\log |G|)$ sized generating set can be done in $\mathcal{O}(|G| \log |G|)$ time.*
3. *For a group G the center $Z(G)$ can be computed in $\mathcal{O}(|G| \log |G|)$ time.*
4. *Given $A \leq G$, one can check whether $A \trianglelefteq G$ in $\mathcal{O}(|G| \log |G|)$ time.*
5. *Given two subgroups A and B of G, one can check whether $G = A \times B$ in $\mathcal{O}(|G| \log |G|)$ time.*

Proof. We present a sketch of the proof for the statements *1* and *2*. The proof for the other statements are easy.

Consider a directed graph $X = (V, E)$, where $V = G$ and $E = \{(g, gs) | g \in G, s \in S\}$. Let $H = \langle S \rangle$. Finding H amounts to computing the set R of vertices reachable from the identity element $e \in G$. Notice that R is also a strongly connected component with $|H||S|$ edges. Thus, the runtime for computing the strongly connected component is $\mathcal{O}(|H||S|)$. This proves (*1*) if $|S| \leq \log |G|$. We use similar ideas as in the proof of *2* to handle the case when $|S| > \log |G|$. Hence, we first prove (*2*).

For *2*, we pick an element $a \in G \setminus \{e\}$ and set $S_1 = \{a\}$. The algorithm keeps on computing sets $S_1, S_2, \ldots$ in stages as follows. At the ith stage we have the set S_i. If we discover $G = \langle S_i \rangle$ we stop the algorithm and output S_i. Otherwise we pick $g \in G \setminus \langle S_i \rangle$ and let $S_{i+1} = S_i \cup \{g\}$. Note that $2|\langle S_i \rangle| \leq |\langle S_{i+1} \rangle|$ as $\langle S_{i+1} \rangle$ contains the disjoint cosets $\langle S_i \rangle$ and $\langle S_i \rangle g$. Thus, if the last set is S_r, then $r \leq \log |G|$. Let $\langle S_i \rangle = H_i$ and $n_i = |H_i|$. Computing H_i via finding a suitable strongly connected component in a graph as mentioned above takes time $\mathcal{O}(|H_i||S_i|) = \mathcal{O}(i|H_i|)$. Furthermore, we note that $n_r = |G|$ and $n_i \leq n_r/2^{r-i}$. Thus, computing all the H_is together takes time $\mathcal{O}(\sum_i i n_i)$ which is $\mathcal{O}(|G| \log |G|)$. Finding an element $g \in G \setminus \langle S_i \rangle$ takes time $\mathcal{O}(|G|)$ and this too happens at most $\log |G|$ times. Thus, the runtime of the algorithm is $\mathcal{O}(|G| \log |G|)$.

To complete the proof of *1*, we modify the algorithm for finding a logarithmic sized generating set by setting $S_1 = \{a\}$ for any $a \in S$ and then picking the new elements g from $S \setminus \langle S_i \rangle$ instead of $G \setminus \langle S_i \rangle$.

Organization of the Paper: In Sect. 3 we give algorithms to compute quotient groups in linear time. In Sect. 4 we discuss some nearly linear time algorithms for finding complements of certain groups. This results are then used in Sects. 5 and 6.

3   Algorithms in Quotient Groups

Suppose we have the list of elements of a group G and a black-box for the group multiplication. Let N be a normal subgroup of G (also given as a list or array). In this section we show how to construct the quotient structure G/N in linear time. More precisely, we describe a linear time algorithm to build a data structure that can serve as a black-box to compute multiplications in G/N. Once we have the data structure, a multiplication query in G/N can be processed in constant time with just one query to the black-box for G.

Many of the algorithms for groups given by their Cayley table work in the same running time if the algorithm has access to the list of group elements and a group multiplication black-box. As a consequence of this quotient black-box construction we can see that the algorithms by Kavitha [19], Chen and Fu [6], and Karagiorgos and Poulakis [11] still run in linear time in quotient groups.

Suppose the list of elements of a group G and a normal subgroup N of G are given along with a black-box for G. We construct lists L_i for $i = 1, \ldots, |G/N|$, each containing the elements of different cosets of N in G in Algorithm 1. We also compute the minimum elements m_i (in the input order) of the list L_i for each i.

Algorithm 1. Computing lists corresponding to the cosets of N in G.

1 **Input** : A group G and a normal subgroup N of G;
2 **Find** : Lists L_i's and the elements m_i's as described above;

3 Create an array *flag* indexed by the elements of G and set
$\quad flag[g] = 0, \forall g \in G$;
4 $i \leftarrow 0$;
5 **for** $g \in G$ **do**
6 $\quad$ **if** *flag*$[g] = 0$ **then**
7 $\quad\quad$ $i \leftarrow i + 1$;
8 $\quad\quad$ Prepare a list L_i of size $|G/N|$ with the elements of Ng;
9 $\quad\quad$ $m_i \leftarrow min\ L_i$;
10 $\quad\quad$ **for** $g_1 \in Ng$ **do**
11 $\quad\quad\quad$ $flag[g_1] = 1$;
12 $\quad\quad$ **end**
13 $\quad$ **end**
14 **end**

Run-Time Analysis of Algorithm 1: In the algorithm $flag[g] = 1$ line 6 indicates that we have already processed the coset containing g and no further action is required. If $flag[g] = 0$ then the algorithm spends $\mathcal{O}(|G/N|)$ time within the "if" condition. But in the process it also discovers all the elements of the coset Ng for which the "if" condition will *not* be executed in future. This shows that the run-time of Algorithm 1 is $\mathcal{O}(|G|)$. □

It is easy to see that in linear time we can compute an array S of size G and indexed by the elements of G such that $S[g] = i$, where L_i is the list produced by by Algorithm 1 containing the elements of Ng.

The data structure for the quotient group G/N consists of the lists L_i's, the sequence $m_1, m_2, \ldots, m_{|G/N|}$ and the array S along with an access to the black-box for G. The elements of G/N will be, as usual, $1, 2, \ldots, |G/N|$. The element i corresponds to the list L_i, which in turn corresponds to one of the cosets of N in G. If we need to compute the product of i and j, we first compute $m = m_i * m_j$ using the multiplication black-box for G and then return $S[m]$. By construction $S[m]$ is the index of the list containing the coset elements of the coset Nm.

We notice that any bounded number of repeated quotient construction can be done in linear time using the above method.

4   Algorithms for Finding Complements

Given a group G and a normal subgroup D of G, a complement of D in G is a subgroup B such that $G = D \times B$. It is important to note that a complement of a subgroup may or may not exist, and even if a complement exists it may not be unique. Kayal and Nezhmetdinov [12] gave an algorithm for finding a complement of a given normal subgroup D of G. Their algorithm is divided into two cases: G/D is abelian and G/D is nonabelian. The result for the first case can be stated as follows.

Theorem 4 ([12]). *There is an algorithm to check if a complement of a normal subgroup D of a group G exists in $\tilde{\mathcal{O}}(|G|)$ time, when G/D is abelian. The algorithm also returns a complement if it exists.*

Proof Sketch: A careful analysis of the complement finding algorithm given in [12] using the results from [6,19] shows that it takes $\tilde{\mathcal{O}}(|G|)$ time to find a complement of the subgroup D in G (if it exists). We use the linear time quotient construction techniques from Sect. 3 multiple times. Additionally, we use the facts that computing $Z(G)$ and testing normality can be done in $\tilde{\mathcal{O}}(|G|)$ time by Lemma 2.

In the second case when G/D is nonabelian, it is not clear how to make the algorithm by Kayal and Nezhmetdinov [12] run in nearly linear time in general (see [12]). Fortunately, for the purpose of this paper, as we would see in Sect. 6, we only need to deal with the subcase when D is a subgroup of the center $Z(G)$ of G. During its execution the algorithm in [12] computes a quotient group which can be done in linear time using results in Sect. 3.

The algorithm by Kayal and Nezhmetdinov computes a group $T = \langle \{ aga^{-1}g^{-1} \mid a \in C_G(D), g \in G \} \rangle$. One can verify that except for this, all other steps in the algorithm can be made to run in $\tilde{\mathcal{O}}(|G|)$ time without the assumption $D \leq Z(G)$. It is the computation of T where we use a different approach using the fact that $D \leq Z(G)$ to obtain the desired nearly linear runtime. We first mention an easy observation.

Observarion 1. *If $D \leq Z(G)$ then $C_G(D) = G$.*

From Observation 1, it is immediate that $T = \langle \{ aga^{-1}g^{-1} \mid a \in G, g \in G \} \rangle$ which is nothing else but the commutator subgroup $[G, G]$ of G. Lemma 3 gives us a way to compute T efficiently.

Lemma 3 (see e.g., [13]). *If $G = \langle S \rangle$ then $[G, G] = \langle [S, S] \rangle^G$, where S is a generating set of G and $[S, S] = \{ aga^{-1}g^{-1} \mid a, g \in S \}$.*

We can compute a generating set S of size $\mathcal{O}(\log |G|)$ of G in time $\mathcal{O}(|G| \log |G|)$ by Lemma 2. Again by Lemma 2 we can compute an $\mathcal{O}(\log |G|)$ sized generating set for the group $\langle [S, S] \rangle$ in $\mathcal{O}(|G| \log |G|)$ time. Let us denote this set by T_{gen}. Algorithm 2 given below computes a generating set for $[G, G]$ (see [13]).

Algorithm 2. Algorithm to find an $\mathcal{O}(\log |G|)$ sized generating set of $[G, G]$

---

1 **Input** : A group $G = \langle S \rangle$ and T_{gen} (defined above);
2 **Find** : An $\mathcal{O}(\log |G|)$ sized generating set of $[G, G]$;

3 Let $K \leftarrow \langle T_{gen} \rangle$;
4 **while** $\exists b \in T_{gen}, a \in S$ such that $a^{-1}ba \notin K$ **do**
5          $T_{gen} \leftarrow T_{gen} \cup \{a^{-1}ba\}$ and $K \leftarrow \langle T_{gen} \rangle$;
6 **return** T_{gen}

Runtime Analysis of Algorithm 2: Each time a new generator is added to T_{gen} the size of the group $K = \langle T_{gen} \rangle$ is at least doubled, which implies that the number of iterations of the while loop is $\mathcal{O}(\log |G|)$. We maintain the group K as an array A_K indexed by the group elements $g \in G$ such that $A_K[g] = 1$ if and only if $g \in K$. Thus, for any $a \in S$ and $b \in T_{gen}$, we can check if $a^{-1}ba \notin K$ in $\mathcal{O}(1)$ time. It takes $\mathcal{O}(|G| \log |G|)$ time to compute the group $\langle T_{gen} \rangle$. Now it is easy to verify that the overall runtime of Algorithm 2 is $\mathcal{O}(|G|(\log |G|)^3)$. $\square$

It is important to note that the inverse of an element $a \in G$ can be found in $\mathcal{O}(|G|)$ time (step 4). However since the number of iterations is only $\mathcal{O}(\log |G|)$, we would need to compute the inverse of $\mathcal{O}(\log |G|)$ many elements, which implies that the overall runtime to find inverses is $\mathcal{O}(|G| \log |G|)$.

Summarising the above discussion we obtain:

Theorem 5. *There exists an algorithm to find a complement of a subgroup D of the center $Z(G)$ of a groups G in time $\tilde{\mathcal{O}}(|G|)$ whenever a complement exists.*

5   Hamiltonian Group Recognition and Isomorphism

A Hamiltonian group is a nonabelian group all of whose subgroups are normal. Since every subgroup of such a group is normal, it follows that there is a unique Sylow subgroup of any fixed order. In this section we consider the following problem.

HAMILTONIAN GROUP RECOGNITION
Input : Given a group $(G, \cdot)$ by its Cayley table.
Find : Is G a Hamiltonian group?

The following structure theorem is one of the main ingredients for our result.

Theorem 6 ([5, page 114]). *Let G be a Hamiltonian group. Then*

– *G is the quaternion group Q_8; or,*
– *G is the direct product of Q_8 and B, or of Q_8 and A, or of Q_8, B and A, where A is an abelian group of odd order and B is an elementary 2-group. Moreover, every such direct product is a Hamiltonian group.*

We recall that the quaternion group Q_8 is a nonabelian group with eight elements and it is generated by two elements a and b with the conditions $a^4 = 1, a^2 = (ab)^2 = b^2$ (see e.g., [5]). An elementary 2-group is isomorphic to $\mathbb{Z}_2^k$ for some k. Thus, from Theorem 6 we can see that the Sylow 2-subgroup of a Hamiltonian group is $Q_8 \times \mathbb{Z}_2^k$ for some nonnegative integer k and the other Sylow subgroups are all abelian. The theorem also implies that Hamiltonian groups are nilpotent. The Sylow decomposition can be computed in $\mathcal{O}(|G|)$ time using methods[2] described in [6]. Next we decompose the Sylow 2-subgroup using Algorithm 3. If we find that the Sylow 2-subgroup is not of the form $Q_8 \times \mathbb{Z}_2^k$ for some k, then we can immediately conclude that the input group is not Hamiltonian. Otherwise, we can use the techniques developed by Kavitha [19] to test if the odd order Sylow subgroups are abelian. If that is the case then we know that the input group is Hamiltonian. Moreover, since the odd order Sylow subgroups are abelian, the algorithm given by Chen and Fu in [6] or Karagiorgos and Poulakis in [11] also give us a Remak-Krull-Schmidt decomposition of the odd order Sylow subgroups. The decomposition of the Sylow 2-subgroup obtained from Algorithm 3 along with the decomposition of the odd order abelian Sylow subgroups gives us a Remak-Krull-Schmidt decomposition of the input Hamiltonian group.

Given a Sylow 2-subgroup as input, Algorithm 3 checks if it is Hamiltonian and also returns a Remak-Krull-Schmidt decomposition isomorphic to $Q_8 \times \mathbb{Z}_2^k$ if the input is indeed Hamiltonian. We use the next lemma in the algorithm.

---

[2] We can compute the Sylow decomposition in $\mathcal{O}(|G|)$ *without* using the result given [6], if G is Hamiltonian 2-group. Note that in a Hamiltonian 2-group order of each non-trivial element will be either 2 or 4.

Lemma 4. *Any two non-commutating elements in a Hamiltonian 2-group generate a quaternion group which is also a direct factor.*

Proof. Let G be a Hamiltonian 2-group and let $g, g' \in G$ be two non-commutating elements. To show that $\langle g, g' \rangle \cong Q_8$ it is enough to show that $g^4 = 1, g^2 = (gg')^2 = g'^2$. As G is a Hamiltonian 2-group, $G = Q_8 \times \mathbb{Z}_2^k$ for some k. Thus, we can write $g = (a_1, b_1)$ and $g' = (a_2, b_2)$, where $a_1, a_2 \in Q_8$ and $b_1, b_2 \in \mathbb{Z}_2^k$. It is easy to verify that $g^4 = 1, g^2 = (gg')^2 = g'^2$. Now we prove that $\langle g, g' \rangle$ is also a factor of G. Let $C = \{a\phi(\alpha) \mid \alpha \in A\}$, where $\phi : Q_8 \longrightarrow \mathbb{Z}_2^k$ is a homomorphism that maps the generators a_1 and a_2 of Q_8 to b_1 and b_2 respectively. Now using Lemma 1, we can see that $G = C \times \mathbb{Z}_2^k$. Moreover, it is an easy verification to see that $C = \langle g, g' \rangle$. □

Algorithm 3. Algorithm for the recognition and decomposition of Hamiltonian 2-groups

---

1 **Input** : A group $(G, \cdot)$;
2 **Decide** : Is G a Hamiltonian 2-group?

3 **if** $G \cong Q_8$ **then** stop and return Hamiltonian 2-group, and G ;
4 $P = \{g \in G \mid \ ord_G(g) = 4\}$;
5 **if** $P = \emptyset$ **then** report "Not Hamiltonian 2-group";
6 Pick any $g \in P$;
7 Find an element $g' \in P$ such that $gg' \neq g'g$. If no such pair exists then
   report "Not Hamiltonian 2-group";
8 **if** $\langle g, g' \rangle \cong Q_8$ **then**
9 | Compute a complement C of $\langle g, g' \rangle$ in G;
10 | **if** C *exists and it is an elementary abelian 2-group* **then**
11 | | return C;
12 | **end**
13 **end**
14 **else**
15 | report "Not Hamiltonian 2-group";
16 **end**

We now prove the correctness of the algorithm and give the run-time analysis. Checking whether $G \cong Q_8$ or not can be done in $\mathcal{O}(1)$ time. From now on, we assume that $G \not\cong Q_8$. In a Hamiltonian 2-group, all non-central elements are of order 4 and constitutes the set P. Since we are interested in elements of order 4, P can be computed in linear time even without using results in [19].

Since the picked element $g \in P$ (Line 5) is non-central, there must exist an element $g' \in P$ such that $gg' \neq g'g$. If no such pair is found in P then G is not a Hamiltonian 2-group. Otherwise by Lemma 4, $\langle g, g' \rangle \cong Q_8$ and will also be direct factor of G. Thus, if the check $\langle g, g' \rangle \cong Q_8$ fails we conclude that G is not a Hamiltonian 2-group.

Using Kavitha's result given in [19], we can test whether C is abelian in time $\mathcal{O}(|G|)$ (Line 9). If C is abelian and all the elements of C have order 2, then we conclude that C is an elementary abelian 2-group and the algorithm returns the complement C.

Finally we argue that we can also compute a complement of $\langle g, g' \rangle$ in time $\mathcal{O}(|G|)$ in Line 8. We can use the result of Theorem 4 to find a complement. However, a direct application of Theorem 4 would only give us an $\tilde{\mathcal{O}}(|G|)$ upper bound. Below we show that the structure of Hamiltonian group could be used to get an $\mathcal{O}(|G|)$ upper bound.

The major time consuming computation tasks inside the complement finding algorithm of Theorem 4 are computing the quotient group $G/\langle g, g' \rangle$, computing the center and testing normality (see [12]).

Since $\langle g, g' \rangle$ is the quaternian group of order 8, testing its normality in time $\mathcal{O}(|G|)$ is trivial. We can compute $G/\langle g, g' \rangle$ in $\mathcal{O}(|G|)$ time using techniques discussed in Sect. 3. If G is a Hamiltonian 2-group, then $G/\langle g, g' \rangle$ will be an abelian group. The task of checking whether $G/\langle g, g' \rangle$ is abelian can be performed in $\mathcal{O}(|G|)$ time using the algorithm described in [19]. If G is a Hamiltonian 2-group, then the center of the group G consists of all order 2 elements along with the identity. One can find all these elements in $\mathcal{O}(|G|)$ time. If the original group is not a Hamiltonian 2-group then the final test, which is to confirm if we have actually computed a valid decomposition (see Section 4.1 of [12], last line) will identify any error that might have occurred in the computation of the center. The final test to verify the validity of the computed decomposition can be performed in linear time exploiting the structure of Hamiltonian 2-groups. These observations imply that Theorem 4 can be modified to find a complement of $\langle g, g' \rangle$ in G in $\mathcal{O}(|G|)$ time.

Once we have the Remak-Krull-Schmidt decompositions of two Hamiltonian groups the isomorphism test is trivial.

Theorem 7. *There exists an algorithm that given two Hamiltonian groups G and H tests if they are isomorphic in time $\mathcal{O}(|G|)$.*

6 Groups with a Bounded Nonabelian Direct Factor

Taking motivation from Hamiltonian groups, which are direct product of the nonabelian quaternion group Q_8 and an abelian group, we study the recognition and the isomorphism problem of a more general class of groups which can be decomposed as a direct product of a nonabelian group of bounded order and an abelian group. For a fixed d, let $\mathcal{G}_d = \{G \mid G = A \times B, \text{ where } |A| \leq d \text{ and } B \text{ is abelian}\}$. It is easy to see that the isomorphism problem for groups in $\mathcal{G}_d$ can be solved in linear time once we have a decomposition of each of the input groups as a direct product of a small nonabelian group with no cyclic factor and an abelian group.

In this section we show that given a nonabelian group G, it can be decomposed as a direct factor of a nonabelian group with *no cyclic factor* and an abelian group in nearly linear time. We note that for this algorithm we *do not need any upper bound* on the size of the nonabelian factor. The idea is to keep on peeling off direct cyclic factors from the given group as long as possible. Each time we factor out a cyclic group, the size of the other factor decreases by at least half. Thus, the process of factoring out cyclic groups can happen for at most $\log |G|$ iterations. Next we define the CYCLIC FACTOR problem below.

CYCLIC FACTOR
Input : A group $(G, \cdot)$ given by its Cayley table.
Find : A cyclic factor $\langle b \rangle$ and $H \trianglelefteq G$ (if they exist) such that $G = \langle b \rangle \times H$.

We show that the CYCLIC FACTOR problem can be solved in $\tilde{O}(|G|)$ time. From this result and the above discussion we can immediately obtain the following theorem.

Theorem 8. *There is an algorithm that takes the Cayley table of a nonabelian group G as input and in time $\tilde{O}(|G|)$ returns two groups A and B, such that $G = A \times B$ where A is a nonabelian group with no cyclic factor and B is abelian.*

In the rest of the section we focus on the CYCLIC FACTOR problem. The following lemma helps us to solve the problem.

Lemma 5. *If G has a cyclic factor then for any basis $\{b_1, b_2, \ldots, b_\ell\}$ of $Z(G)$, there is $i \in [\ell]$ such that $\langle b_i \rangle$ is a factor of G.*

Proof. Let $G = A \times B$, where A is nonabelian with no cyclic factor and B is abelian. Notice that $Z(G) = Z(A) \times B$. Let $Z(A) = \langle c_1 \rangle \times \ldots \times \langle c_r \rangle$ and $B = \langle d_1 \rangle \times \ldots \times \langle d_k \rangle$ be a basis decomposition of $Z(A)$ and B. This gives a basis decomposition $Z(G) = \langle c_1 \rangle \times \ldots \times \langle c_r \rangle \times \langle d_1 \rangle \times \ldots \times \langle d_k \rangle$. Let $c'_1, \ldots, c'_r, b'_1, \ldots, b'_k$ be an another basis of $Z(G)$, where c'_is and b'_js are ordered to satisfy the following conditions from Remak-Krull-Schmidt theorem:

(i) $\langle c'_i \rangle \cong \langle c_i \rangle, \forall i \in [r]$ and $\langle b'_j \rangle \cong \langle d_j \rangle, \forall j \in [k]$, and
(ii) $Z(G) = \langle c_1 \rangle \times \ldots \times \langle c_r \rangle \times B_p = Z(A) \times B_p$,

where $B_p = \langle b'_1 \rangle \times \ldots \times \langle b'_k \rangle$. By Lemma 1, we have $B_p = \{\alpha \phi(\alpha) \mid \alpha \in B\}$ for some homomorphism $\phi : B \longrightarrow Z(Z(A)) = Z(A)$. The same lemma can be used once more to show that $G = A \times B_p$. The result follows if we take $\{c'_1, \ldots, c'_r, b'_1, \ldots, b'_k\} = \{b_1, b_2, \ldots, b_\ell\}$. □

The above lemma immediately suggests an algorithm to solve the CYCLIC FACTOR problem for a given group G: (i) Find the center $Z(G)$ of the group G, (ii) Compute a basis $\{b_1, b_2, \ldots, b_\ell\}$ of $Z(G)$, and (iii) Try to find a complement of $\langle b_i \rangle$ in G for all $i = 1, 2, \ldots, \ell$.

In step (iii) of the algorithm if we find a complement then we have solved the problem. On the other hand, if G has a cyclic factor then Lemma 5 ensures that the algorithm would find a cyclic factor. This shows the correctness of the algorithm.

We now discuss the runtime of the algorithm. We can use Kavitha's result to check if G is abelian in linear time [19]. If G is abelian, then cyclic factors of G can be found in $\mathcal{O}(|G|)$ time using results from [6]. Let us assume that input group G is nonabelian.

We can compute the center of a group in nearly linear time using Lemma 2. Thus, step (i) of the algorithm takes $O(|G| \log |G|)$ time.

Since $Z(G)$ is abelian we can use the linear time algorithm of Chen and Fu [6] or Karagiorgos and Poulakis [11] for step (ii) of the algorithm. Notice that the number of basis elements of $Z(G)$ is at most $\log |G|$. Thus, the maximum number of iterations in step (iii) is at most $\log |G|$. In general, we do not know how to compute a complement of a subgroup in nearly linear time. However, the fact that each $\langle b_i \rangle$ is a subgroup of the center of the group allows us to use Theorem 5. Thus, the runtime of step (iii) as well as the whole algorithm is $\tilde{O}(|G|)$.

References

1. Arvind, V., Torán, J.: Solvable group isomorphism is (almost) in NP ∩ coNP. ACM Trans. Comput. Theory **2**(2), 4:1–4:22 (2011)
2. Babai, L., Codenotti, P., Qiao, Y.: Polynomial-time isomorphism test for groups with no abelian normal subgroups. In: Czumaj, A., Mehlhorn, K., Pitts, A., Wattenhofer, R. (eds.) ICALP 2012. LNCS, vol. 7391, pp. 51–62. Springer, Heidelberg (2012). https://doi.org/10.1007/978-3-642-31594-7_5
3. Babai, L., Qiao, Y.: Polynomial-time isomorphism test for groups with abelian sylow towers. In: STACS 2012 (29th Symposium on Theoretical Aspects of Computer Science), vol. 14, pp. 453–464. LIPIcs (2012)
4. Boppana, R.B., Hastad, J., Zachos, S.: Does co-NP have short interactive proofs? Inf. Process. Lett. **25**(2), 127–132 (1987)
5. Carmichael, R.D.: Introduction to the Theory of Groups of Finite Order. GINN and Company, Boston (1937)
6. Chen, L., Fu, B.: Linear and sublinear time algorithms for the basis of abelian groups. Theor. Comput. Sci. **412**(32), 4110–4122 (2011)
7. Gall, F.L.: Efficient isomorphism testing for a class of group extensions. In: 26th International Symposium on Theoretical Aspects of Computer Science, STACS 2009, February 26–28, 2009, Freiburg, Germany, Proceedings, pp. 625–636 (2009)
8. Garzon, M., Zalcstein, Y.: On isomorphism testing of a class of 2-nilpotent groups. J. Comput. Syst. Sci. **42**(2), 237–248 (1991)
9. Grochow, J.A., Qiao, Y.: Algorithms for group isomorphism via group extensions and cohomology. SIAM J. Comput. **46**(4), 1153–1216 (2017)
10. Hungerford, T.W.: Algebra. Graduate Texts in Mathematics, vol. 73. Springer, New York (1974)
11. Karagiorgos, G., Poulakis, D.: Efficient algorithms for the basis of finite abelian groups. Discrete Math. Algorithms Appl. **3**(04), 537–552 (2011)
12. Kayal, N., Nezhmetdinov, T.: Factoring groups efficiently. In: Albers, S., Marchetti-Spaccamela, A., Matias, Y., Nikoletseas, S., Thomas, W. (eds.) ICALP 2009. LNCS, vol. 5555, pp. 585–596. Springer, Heidelberg (2009). https://doi.org/10.1007/978-3-642-02927-1_49
13. Luks, E.M.: Lectures on polynomial-time computation in groups. University of Oregon, Department of Computer and Information Science (1990)
14. Miller, G.L.: On the $n^{\log n}$ isomorphism technique (a preliminary report). In: Proceedings of the Tenth Annual ACM symposium on Theory of computing, pp. 51–58. ACM (1978)
15. Qiao, Y.M., Sarma, J., Tang, B.S.: On isomorphism testing of groups with normal hall subgroups. J. Comput. Sci. Technol. **27**(4), 687–701 (2012)

16. Rosenbaum, D.J.: Breaking the $\mathcal{O}(n \log n)$ barrier for solvable-group isomorphism. In: Proceedings of The Twenty-fourth Annual ACM-SIAM Symposium on Discrete Algorithms, pp. 1054–1073. Society for Industrial and Applied Mathematics (2013)
17. Rosenbaum, D.J., Wagner, F.: Beating the generator-enumeration bound for p-group isomorphism. Theor. Comput. Sci. **593**, 16–25 (2015)
18. Savage, C.D.: An $\mathcal{O}(n^2)$ algorithm for abelian group isomorphism. Computer Studies [Program], North Carolina State University (1980)
19. Kavitha, T.: Linear time algorithms for abelian group isomorphism and related problems. J. Comput. Syst. Sci. **73**(6), 986–996 (2007)
20. Vikas, N.: An $\mathcal{O}(n)$ algorithm for abelian p-group isomorphism and an $\mathcal{O}(n \log n)$ algorithm for abelian group isomorphism. J. Comput. Syst. Sci. **53**(1), 1–9 (1996)

Belga B-Trees

Erik D. Demaine[1], John Iacono[2,3], Grigorios Koumoutsos[2(✉)],
and Stefan Langerman[2]

[1] Massachusetts Institute of Technology, Cambridge, USA
[2] Université Libre de Bruxelles, Brussels, Belgium
greg.koumoutsos@gmail.com
[3] New York University, New York, USA

Abstract. We revisit self-adjusting external memory tree data structures, which combine the optimal (and practical) worst-case I/O performances of B-trees, while adapting to the online distribution of queries.
Our approach is analogous to undergoing efforts in the BST model,
where *Tango Trees* (Demaine *et al.* 2007) were shown to be $O(\log \log N)$-
competitive with the runtime of the best offline binary search tree on
every sequence of searches. Here we formalize the B-Tree model as a
natural generalization of the BST model. We prove lower bounds for the
B-Tree model, and introduce a B-Tree model data structure, the Belga
B-tree, that executes any sequence of searches within a $O(\log \log N)$ factor of the best offline B-tree model algorithm, provided $B = \log^{O(1)} N$.
We also show how to transform any static BST into a static B-tree which
is faster by a $\Theta(\log B)$ factor; the transformation is randomized and we
show that randomization is necessary to obtain any significant speedup.

1 Introduction

Worst-case analysis does not capture the fact that some sequences of operations
on data structures, often typical ones, can be executed significantly faster than
worst case ones. Methods of analyzing algorithms whose performance depends
on more fine-grained characteristics of the input sequence other than the size
N have been coined *distribution sensitive data structures* [7,22]. Two general
methods to bound the performance of such a data structure exist. The first is
to explicitly bound the performance by some bound. For binary search trees
(BSTs) there is a rich set of such bounds (see e.g. [9,18]) like the sequential
access bound [29], the working set bound [21,28], the (weighted) dynamic finger
bound [11,12,24], the unified bound [3,21] and many others [5,10,20]. The other
method is to compare the performance of the data structure on a sequence of
operations to the performance of the best offline data structure in some model

The full version of this paper can be found at: https://arxiv.org/pdf/1903.03560.pdf.
J. Iacono—Research supported by the Fonds de la Recherche Scientifique-FNRS under
Grant no MISU F 6001 1.
J. Iacono—Research supported by NSF Grant CCF-1533564.
S. Langerman—Directeur de Recherches du F.R.S-FNRS.

© Springer Nature Switzerland AG 2019
R. van Bevern and G. Kucherov (Eds.): CSR 2019, LNCS 11532, pp. 93–105, 2019.
https://doi.org/10.1007/978-3-030-19955-5_9

on the same sequence. Such an analysis uses the language of competitive analysis introduced in [27], where the competitive ratio of an algorithm is the supremum ratio of the performance of the given algorithm to the offline optimal over all sequences of operations over a given length. A data structure which is $O(1)$-competitive in a particular model is said to be *dynamically optimal* [28]. In the BST model, the best known competitive ratio is $O(\log \log N)$, first achieved by Tango trees [15]. The existence of a dynamically optimal BST is one of the most intriguing and long-standing open problems in online algorithms and data structures (see [23] for a survey). The two prominent candidates to achieve dynamic optimality for BSTs are the *splay tree* of Sleator and Tarjan [28] and the *greedy* algorithm [14,25], but they are only known to be $O(\log N)$-competitive.

Disk-Access Model (DAM). The *external memory model*, or *disk-access model (DAM)* [2] is the leading way to theoretically model the performance of algorithms that can not fit all of their data in RAM, and thus must store it on a slower storage system historically known as *disk*. This model is parameterized by values M and B; the disk is partitioned into blocks of size B, of which M/B can be stored in memory at any given moment. The cost in the DAM is the number of transfers between memory and disk, called Input-Output operations (I/Os). The classic data structure for a comparison based dictionary in the DAM model, as well as in practice, is the B-Tree [4]. The B-Tree is a generalization of the BST, where each node stores up to $B - 1$ data items, for $B \geq 2$, and the number of children is one more than the number of data items. The B-Tree supports searches in time $O(\log_B N)$ in the DAM, a $\log B$ factor faster than traditional BSTs such as red-black trees [19] or AVL trees [1].

Dynamic Dictionaries in the DAM. Here, our goal is to explore dynamic dictionaries in the DAM and to obtain results similar to those known for BSTs.

Surprisingly, prior work in this direction is quite limited. One previous attempt was in the work of Sherk [26] where a generalization of splay trees to what we call the B-tree model was proposed, but without any strong results. Over ten years later, Bose et. al. [6] studied a self-adjusting version of skip-lists and B-Trees, where nodes can be split and merged to adapt to the query distribution by moving elements closer or farther from the root of the tree (here we call this model *classic self-adjusting* B-trees, see Sect. 2). They showed that dynamic optimality in this model is closely related to the working set bound. This bound captures temporal locality: for an access sequence $X = x_1, \ldots, x_m$, it is defined as $\mathrm{WS}(X) = \sum_{i=1}^{m} \log w_X(i)$, where $w_X(i)$ is the number of distinct elements accessed since the last access to the element x_i. In [6] the authors presented a data structure whose cost is upper bounded by $O(\mathrm{WS}(X)/\log B)$ and obtained a matching lower bound of $\Omega(\mathrm{WS}(X)/\log B)$ for this model, which implies that their structure is dynamically optimal.

Note that the lower bound of [6] shows a major limitation of B-trees with only split and merge operations: It implies there are sequences on which they are slower than BSTs. For example, repeatedly sequentially accessing all data items

$1, 2, \ldots, N$ requires $O(1)$ amortized time per search for BSTs like splay trees (this is the *sequential access bound* [29]) while the lower bound $\Omega(\mathrm{WS}(X)/\log B)$ implies an amortized cost $\Omega(\log_B N)$ in the classic self-adjusting model. In this work, we show that by adding just one more operation, an analogue of the rotation for B-Trees, we can overcome this limitation and obtain significant speedups with respect to standard B-trees.

Our Contribution. In this work we initiate a systematic study of dynamic B-trees. First, we formally define the (dynamic) B-Tree model of computation (§2). Second, we show how to produce lower bounds in the B-Tree model (§3). Then, we introduce a data structure, which we call the *Belga B-Tree*[1], which is $O(\log \log N)$ competitive with any dictionary in the B-Tree model of computation, when $B = O(\log^{O(1)} N)$ (§4).

More generally, we conjecture the following in §6: any BST-model algorithm can be transformed into a (randomized) B-Tree model algorithm with a $\Theta(\log B)$ factor cost savings. This would imply that BST model algorithms such as the splay tree [28] or greedy [14,25] would have B-Tree model counterparts, and that a dynamically optimal BST-model algorithm would imply a dynamically optimal algorithm in the B-Tree model. We leave this conjecture open, but in §5 we do resolve the case of a static (no rotations allowed) BSTs by showing a randomized transformation from a static BST to a static B-Tree such that any algorithm in the static BST model would have factor $\Theta(\log B)$ speedup in the B-Tree model. We also show that no $\omega(1)$-factor speedup is possible for a deterministic transformation in general.

2   The B-Tree Model of Computation

In this section, we define the tree models discussed in this paper. In all cases, we consider data structures supporting searches over a universe of N elements $\mathcal{U} = \{1, 2, \ldots, N\}$ which we refer to as *keys*. The input is a valid tree T_0 and request sequence of searches $X = x_1, x_2 \ldots, x_m$, where $x_i \in \mathcal{U}$ is the ith item to be searched.

2.1   The BST Model

In a Binary Search Tree (BST) data structure, each node stores a single key and three pointers, indicating its parent and its (left and right) children. The key value of a node is larger than all keys in its left subtree and smaller than all keys in its right subtree. To execute each request to search for element x_i, a BST algorithm initializes a single pointer at the root (at unit cost) then may perform any sequence of the following unit-cost operations:

---

[1] The Tango tree was invented on an overnight flight from JFK airport en route to Buenos Aires, Argentina. The work on the Belga B-Tree has been substantially completed at Cafe Belga, Ixelles, Belgium.

– Move the pointer to the parent or to the left or right child of the current node the pointer points to (if such a destination node exists).
– Perform a *rotation* of the edge between current node and its parent (if not the root).

Whenever the pointer moves to or it is initialized to a node v, we say that node v is *touched*. A BST-model search algorithm is correct if during each search, the element x_i that is being searched for is touched. The cost of a BST algorithm on the search sequence X equals the total number of unit-cost operations performed to execute the searches in the sequence. This model was formally defined in [15] and it is known to be equivalent up to constant factors to several alternative models which have been considered (e.g. [14,31]).

A BST data structure can be augmented such that each node stores $O(\log N)$ additional bits of information. The running time of such BST data structures in the RAM model is dominated by the number of unit-cost operations. A *static* BST is a restricted version of the BST model where rotations are not allowed and thus the shape of the tree never changes.

2.2   The B-Tree Model

We define the B-tree model to be a generalization of the BST model which allows more than one key to be stored in each node. The B-tree model is parameterized by a positive integer $B \geq 2$ which represents the maximum number of children of each node; in the case where $B = 2$ the B-tree model will be equivalent to the BST model. We denote by $n(v)$ the number of keys stored in a node v. Every node v has $n(v) \leq B - 1$ and $n(v) + 1$ child pointers (some of which could be null). A node v which stores exactly $n(v) = B - 1$ keys is called *full*.

Suppose $x_1, \ldots, x_{n(v)}$ are the keys stored at node v and $c_1, \cdots, c_{n(v)+1}$ are the children of v. Keys satisfy the in-order condition, i.e. $x_1 < \ldots < x_{n(v)}$ and for any key k_i stored in the subtree T_{c_i} rooted at c_i, we have that $k_1 < x_1 < \cdots < k_i < x_i < k_{i+1} < \cdots < k_{n(v)} < x_{n(v)} < k_{n(v)+1}$.

Similar to the BST model, to execute each search there is a single pointer initialized to the root of the tree at unit cost. To execute a search for x_i, a B-tree algorithm performs a sequence of the following unit-cost operations which are described formally later:

– Move the pointer to a child or to the parent of the current node.
– Split a node containing at least three keys.
– Join two sibling nodes storing no more than $B - 2$ keys in total.
– Rotate the edge between the current node and its parent.

B-tree model algorithms that only use the first type of operations are referred to as *static* as the shape of the B-tree does not change. We now fully describe the unit-cost operations of rotating, splitting and joining:

Rotations: Consider a (non-root) node u and let $p(u)$ be its parent. Let $P = \{p_1, \ldots, p_m\}$ be the union of all keys stored in u and $p(u)$. The keys stored at

u define an interval $[p_\ell, p_r]$ in P. A rotation of the edge $(p(u), u)$ essentially updates this interval to $[p_{\ell'}, p_{r'}]$, moving the keys as needed. Depending on the values of ℓ, ℓ' and r, r' we characterize a rotation as a promote/demote left—promote/demote right rotation. For example, a rotation of the type promote left k—demote right k' sets $\ell' = \ell + k$ (i.e. the k leftmost keys of u are promoted to $p(u)$) and $r' = r + k'$ (i.e. keys $p_{r+1}, \ldots, p_{r+k'}$ are demoted to u). Values k and k' should be non-negative and satisfy that after the rotation both u and $p(u)$ have at most $B - 1$ keys. Rotations of the type demote left—promote right, promote left—promote right and demote left—demote right can be defined analogously. As an example, Fig. 1 shows a rotation of type demote left - promote right.

Splitting a node: Let u be a node (except the root) containing at least three keys and let $p(u)$ be its non-full parent. Splitting node u at key u_m (which is not the smallest or the largest key stored at u) consists of promoting u_m to $p(u)$ and replacing u by 2 nodes u_L, u_R such that keys smaller than u_m are contained in u_L and keys larger than u_m are in u_R. To split the root (given that it stores at least three keys), we create an empty B-tree node, make it the parent of the root (i.e. the new root) and then perform a split operation as defined above.

Join: This operation is the inverse of a split. Let u and v be two sibling nodes and let p be their parent, such that there exists a unique key p_j in p such that p_j is larger than all keys stored at u and smaller than all keys stored at v. Joining nodes u and v (given that they store no more than $B - 2$ keys in total) consists of demoting p_j to u (and deleting it from p), adding all elements of v (including the pointers to children) to u and deleting v. Note that after a join operation p might become empty (in case p_j was the unique key of p). In that case, we set the parent of u to be the parent of p (if it exists) and we delete p. If p is empty and it is the root, then we just delete p and u becomes the new root of the tree.

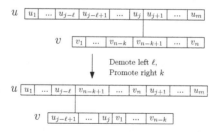

Fig. 1. A rotation of a B-tree edge (u, v) of the type demote left ℓ—promote right k: From the left of v, the ℓ neighboring keys of u, $u_{j-\ell+1}, \ldots, u_j$ are getting demoted to v. From the right, the k last elements of v, $v_{n-k+1}, \ldots, v_n$ are getting promoted to u.

A B-tree can be augmented with additional $O(B \log N)$ bits of information for each node. The performance of B-trees in the *external memory model* with blocks of size B, is within a constant factor of the sum of the unit-cost operations as we have defined them.

Relation with Other B-Tree Models. The classic structure of B-trees first appeared in [4]. In this framework, all leaves have the same depth and no join, split and rotate operations are performed during searches (to be precise, restricted versions of split and join were defined in order to support insertions and deletions and were not allowed for performing search operations, see [13] for an extensive treatment). We call this framework the *classic B-tree* model.

A more flexible model of B-trees was considered in [6]: We start with a classic B-tree and an algorithm is allowed to perform joins and splits, but not rotations. Note that by performing join and split operations, the property that all leaves of the tree have the same depth is maintained throughout the whole execution. This model was called "self-adjusting B-trees". To avoid confusion with our dynamic B-tree model, we call this model *classic self-adjusting* B-trees, in order to emphasize that all leaves have the same depth, as in classic B-trees. The self-adjustment relies on the fact that using joins and splits the algorithm might choose to bring an item closer to the root or demote it farther from the root. Also, note that the number of nodes in a B-tree on N keys is not fixed (as opposed to BSTs where we always have exactly N nodes) and the split/join operations might increase/decrease the number of nodes of the tree, changing thus its shape.

For the rest of this paper, whenever we use the term B-tree we refer to our B-tree model, unless stated otherwise.

3   Lower Bounds: Simulating Dynamic B-Trees Using BSTs

In this section we show how to simulate a dynamic B-tree algorithm using a BST-model algorithm with an $O(\log B)$ overhead in the cost. This will allow us to transform lower bounds from the BST model into lower bounds for the B-tree model.

Notation. For a search sequence X, we denote $\mathrm{OPT}_{\mathrm{BST}}(X)$ and $\mathrm{OPT}_{\text{B-Tree}}(X)$ the optimal (offline) cost to serve X using a BST-model and a B-tree-model data structure respectively.

Theorem 1. *For any search sequence X, $\mathrm{OPT}_{\mathrm{BST}}(X) = O(\mathrm{OPT}_{\text{B-Tree}}(X) \cdot \log B)$.*

Proof. (Sketch, formal proof is in the full version) We simulate a B-tree execution of X using a BST in the following way: Each node of the B-tree is simulated by a red-black tree of depth $O(\log B)$. Thus our BST is a tree of red-black trees. We also augment the red-black tree data structure such that each node stores a counter on the number of keys in its subtree. In the full version we show the details of how each unit-cost B-tree operation can be simulated in time $O(\log B)$ using our tree-of-trees BST data structure.   □

Theorem 1 implies that we can transform any lower bound for binary search trees to a lower bound for dynamic B-trees, as shown in the following corollary.

Corollary 2. *Let X be a search sequence and let $\mathrm{LB}(X)$ be any lower bound on the cost of executing X in the BST model. Then we have that $\mathrm{OPT}_{\text{B-Tree}}(X) = \Omega\left(\frac{\mathrm{LB}(X)}{\log B}\right).$*

Proof. Since $\mathrm{LB}(X)$ is a lower bound on $\mathrm{OPT}_{\text{BST}}(X)$, we have that $\mathrm{LB}(X) \le \mathrm{OPT}_{\text{BST}}(X) = O(\log B) \cdot \mathrm{OPT}_{\text{B-Tree}}(X)$, which implies $\mathrm{OPT}_{\text{B-Tree}}(X) = \Omega\left(\frac{\mathrm{LB}(X)}{\log B}\right).$ □

4   Belga B-Trees

In this section, we develop a dynamic B-tree data structure yclept *Belga B-tree* that achieves a competitive ratio of $O(\log \log N)$, for search sequences of length $\Omega(N)$, provided that $1 + \log_B \log N = O(\log_B \log N)$, i.e. $B = (\log N)^{O(1)}$. Our construction is built upon the ideas used in [15] to get a similar competitive ratio for binary search trees. Particularly, we crucially connect the cost of our algorithm to the *interleave lower bound*. For completeness, we present here the setup and the necessary background regarding this lower bound.

Interleave Lower Bound and Preferred Paths (See Fig. 2). Let $\{1, \ldots, N\}$ be the keys stored in our B-tree. Let P be a (fixed) complete binary search tree on those keys. For each internal node v in P, we define its left region to be v together with the subtree rooted at its left child and its right region to be the subtree rooted at its right child. Node v has a *preferred child*, which is left or right, depending on whether the last search for a node in its subtree was in its left or right region (if no node of the subtree rooted at v has been searched, then v has no preferred child).

We define a *preferred path* in P as follows: Start from a node that is not the preferred child of its parent (including the root) and perform a walk by following the preferred child of the current node, until reaching a node with no preferred child. Clearly, a preferred path contains $O(\log N)$ keys.

Note that during a search for a key, the preferred child of some nodes that are ancestors of the node with the key being searched might change. Each change of preferred child, changes also the preferred paths of P. For a search sequence X, the interleave lower bound $\mathrm{IB}(X)$ equals the total number of changes of preferred child from left to right or from right to left, over all nodes of P. We use the following lemma of [15], which is a slight variant of the first lower bound of [31]:

Lemma 1 (Lemma 3.2 in [15]**).** *The cost to execute X in the BST model is $\Omega(\mathrm{IB}(X))$ if $|X| = \Omega(N)$.*

High-Level Overview of Our Structure. We store each preferred path in a balanced classic B-tree. We call such classic B-trees *auxiliary trees*. Our dynamic B-tree will be a tree of classic B-trees. Recall that Lemma 1 essentially tells us that the number of preferred paths touched during a request sequence is a lower bound on the value of OPT_{BST}. The idea here is to show that for each preferred path touched, and thus unit of lower bound incurred, we can perform search and all update operations (cutting and merging preferred paths) with an overhead factor $O(\log_B \log N) = O(\frac{\log \log N}{\log B})$. This will imply that we have a dynamic B-tree with cost $O(\frac{\log \log N}{\log B} \cdot \text{IB}(X))$. This combined with Lemma 1 and Corollary 2 implies that the cost of our dynamic B-tree data structure is $O(\log \log N) \cdot \text{OPT}_{\text{B-Tree}}$.

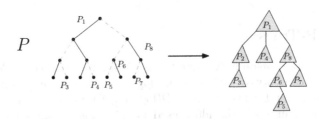

Fig. 2. Example of a reference tree P and the tree-of-trees representation of its preferred paths $P_1, \ldots, P_8$. Edges connecting different preferred paths are dashed gray.

Auxiliary Trees. Our auxiliary trees are augmented classic B-trees. Each auxiliary tree stores a preferred path. With each key x we also store its depth in the reference tree P. We call this value depth of key x. Also, each node stores the minimum and maximum depth of a key in its subtree. Last, a node may be marked or unmarked, depending on whether it is the root of an auxiliary tree or not. Note that P is just a reference tree used for the analysis. We do not need to store P explicitly in order to implement our algorithm. All necessary information about P is stored in our dynamic B-tree data structure.

During an execution of a search sequence we need to perform the following operations on a preferred path:

(i) Search for a key.
(ii) Cut the preferred path into two paths, one consisting of keys of depth at most d and the other of keys of depth greater than d.
(iii) Merge two preferred paths P_1 and P_2, where the bottom node of P_1 is the parent of the top node of P_2.

We will show that we can perform those operations using our auxiliary trees in time $O(1 + \log_B k)$, where k is the number of keys in the involved preferred paths. We defer this proof to the full version of this paper and we now proceed to the description and analysis of Belga B-trees, assuming that those operations can be done in time $O(1 + \log_B k)$. For the rest of this section, whenever we refer to cutting/merging operations on auxiliary trees, we mean the implementation of cutting/merging the corresponding preferred paths in our B-tree data structure.

Our Algorithm. A Belga B-tree is a tree of auxiliary classic B-trees, where each auxiliary tree stores a preferred path. Initially we transform the input tree T_0 to a valid Belga B-tree. Upon a request for a key x_i, we start from the root and search for x_i. Whenever we reach a marked node v (i.e. a root of an auxiliary tree), we have to update the preferred paths. Let Q be the preferred path stored in the auxiliary tree of the parent of v and R the preferred path in the auxiliary tree rooted at v. We update the preferred paths using the cut and merge operations of auxiliary trees. Particularly, if d is the minimum depth of a key of R (this value is stored at node v of our B-tree), we cut the auxiliary tree storing Q at depth $d-1$. This gives us two preferred paths Q_{d-} and Q_{d+}, where the first stores keys of Q of depth at most $d-1$ and the second keys of Q of depth greater than d. We mark the roots of the auxiliary trees corresponding Q_{d-} and Q_{d+}. We then merge the auxiliary tree storing Q_{d-} with the auxiliary tree rooted at v (which stores R). We mark the root of the new tree and continue the search for x_i.

Note that the only part where our algorithm needs to perform rotations is the initial step of transforming the input tree into a Belga B-tree.

Bounding the Cost. We now compare the cost of our Belga B-tree data structure to that of the optimal offline B-tree. The following lemma makes the essential connection between the number of preferred paths touched during a search and the cost of our algorithm.

Lemma 2. *Let ℓ be the number of preferred child changes during a search for key x_i. Then the cost of Belga B-tree for searching x_i is $O((\ell+1)(1+\log_B \log N))$.*

Proof. To search for x_i, we touch exactly $\ell + 1$ preferred paths. We account separately for the search cost and the update cost.

For each preferred path touched, the search cost is $O(\lceil \log_B \log N \rceil)$, since we are searching a balanced B-tree on $O(\log N)$ keys. Thus the total search cost is clearly $O((\ell+1)(1 + \log_B \log N))$.

We now account for the update cost. Recall that we can cut and merge preferred paths on k keys in time $O(1 + \log_B k)$. Since each preferred path has at most $O(\log N)$ keys, we can perform those updates in time $O(1+\log_B \log N)$. There are ℓ preferred path changes, and for each change we perform one cut and one merge operation, we get that the total time for merging and cutting is $O(\ell \cdot (1 + \log_B \log N))$. The lemma follows. □

Theorem 3. *For any search sequence of length $m = \Omega(N)$, Belga B-trees are $O(\log \log N)$-competitive.*

Proof. The total number of preferred path changes is at most $\mathrm{IB}(X) + N$. The additive N accounts for the fact that initially each node has no preferred child, so its first change from null to either left or right is not counted in $\mathrm{IB}(X)$. Using Lemma 2 and summing up over all search requests, we get that the cost of Belga B-trees is $O((\mathrm{IB}(X) + N + m)(1 + \log_B \log N))$. By our assumption on the value of B, we have that $1 + \log_B \log N = O(\log_B \log N)$, thus the cost is in $O((\mathrm{IB}(X) + N + m) \cdot \frac{\log \log N}{\log B})$. By Lemma 1 this is bounded by $(\mathrm{OPT}_{\mathrm{BST}}(X) + N + m) \cdot \frac{\log \log N}{\log B}$. Using Corollary 2 we get that cost of Belga B-tree is

$$O\left((\log B \cdot \mathrm{OPT}_{\text{B-Tree}} + N + m) \cdot \frac{\log \log N}{\log B}\right).$$

Note that for any request sequence $\mathrm{OPT}_{\text{B-Tree}} \geq m$. Since $m = \Omega(N)$, we have that $\log B \cdot \mathrm{OPT}_{\text{B-Tree}} + N + m = O(\log B \cdot \mathrm{OPT}_{\text{B-Tree}})$. We get that the total cost is upper bounded by

$$O\left(\log B \cdot \mathrm{OPT}_{\text{B-Tree}} \cdot \frac{\log \log N}{\log B}\right) = O(\mathrm{OPT}_{\text{B-Tree}} \cdot \log \log N).$$

5   Transforming Any Static BST into the B-Tree Model

In this section we focus on static trees, with the goal to simulate a static BST using a static B-tree and achieving a speedup by a factor of $\Theta(\log B)$. In the static BST and B-Tree models, all that is allowed in each operation is to move a single pointer around the tree, starting at the root, each time moving to a neighboring node, at unit cost per move. We refer to a sequence of moves of a single pointer as a *walk*. In particular, given a BST we wish to convert it to a B-Tree so that if a walk in the BST costs k, a walk in the B-Tree T_B that touches the same keys costs as little as possible in terms of k; k is clearly possible since a BST is a b-tree, but when can we achieve $o(k)$?

We note that the results of this section allow the pointer to move arbitrarily in a static BST/B-tree, i.e., it can visit nodes that are outside the path from the root to the searched node. In the case where only a search path of length D is considered, the worst-case cost has been completely characterized in [16] as $\Theta\left(\frac{D}{\lg(1+B)}\right)$ when $D = O(\lg N)$, $\Theta\left(\frac{\lg N}{\lg(1 + \frac{B \lg N}{D})}\right)$, when $D = \Omega(\lg N)$ and $D = O(B \lg N)$, and $\Theta\left(\frac{D}{B}\right)$ when $D = \Omega(B \lg N)$.

Block-Connected Mappings. The most natural approach to achieve our goal is to try to map a static BST T into a static B-tree T_B such that each node of T_B corresponds to a connected subtree of T. We call such a mapping $f : T \to T_B$, *block-connected*. Observe that in order to achieve a $\Omega(\log B)$ speedup for the B-tree model T_B, it is necessary that a block-connected mapping f should satisfy that every node at depth d in T is at depth $O(\frac{k}{\log B})$ in T_B. However, as we will see, this is not sufficient.

The next theorem shows that, perhaps surprisingly, this approach fails to give any super-constant factor improvement, given that the mapping is deterministic. Afterwards, we show how to achieve an $\Omega(\log B)$ factor speedup using randomization. The proofs of these theorems are in the full version.

Theorem 4. *There does not exist a block-connected mapping $f : T \to T_B$ such that any walk P on T of length k corresponds to a walk of length $o(k)$ in T_B.*

Randomized Construction. Theorem 4 above is based on an adversarial argument and relies crucially on the knowledge of the layout of the B-tree. To overcome this issue, we use randomization.

Theorem 5. *For any BST T, there is a randomized block-connected mapping which produces a static B-tree T_R such that for any walk of length k in T, there exists a corresponding walk in T_R with expected cost $O(k/\log B)$.*

6   Open Problems

We conclude with some open problems. The first is that our Belga B-trees are $O(\log \log N)$-competitive only when $B = \log^{O(1)} N$, and thus the case of large B where $B = \log^{\omega(1)} N$ remains open. The main impediment is to figure out how to fit multiple preferred paths into one block.

A more general open problem is to resolve the following conjecture: Is it possible to convert any BST-model algorithm into a B-Tree model algorithm such that if an algorithm costs $O(k)$ in the BST model, it costs $O(\frac{k}{\log B}+1)$ in the B-Tree model? Special cases of this theorem, when applied to, for example, splay trees and greedy future, would also be interesting should the general conjecture prove too difficult to resolve.

A third open problem is whether, given two B-tree model algorithms, can you achieve the runtime that is the minimum of them; this would be the B-Tree model analogue of the BST result of [17]. It would also allow one to then combine Belga B-trees with other B-tree model algorithms to get stronger results, like, for example [6] to add the working-set bound; in the BST model [30] gave a $O(\log \log N)$-competitive BST with the working set bound.

References

1. Adelson-Velskiĭ, G.M., Landis, E.M.: An algorithm for organization of information. Dokl. Akad. Nauk SSSR **146**, 263–266 (1962)
2. Aggarwal, A., Vitter, J.S.: The input/output complexity of sorting and related problems. Commun. ACM **31**(9), 1116–1127 (1988)
3. Badoiu, M., Cole, R., Demaine, E.D., Iacono, J.: A unified access bound on comparison-based dynamic dictionaries. Theor. Comput. Sci. **382**(2), 86–96 (2007)
4. Bayer, R., McCreight, E.M.: Organization and maintenance of large ordered indices. Acta Inf. **1**, 173–189 (1972)

5. Bose, P., Douïeb, K., Iacono, J., Langerman, S.: The power and limitations of static binary search trees with lazy finger. Algorithmica **76**(4), 1264–1275 (2016)

6. Bose, P., Douïeb, K., Langerman, S.: Dynamic optimality for skip lists and b-trees. In Symposium on Discrete Algorithms, SODA, pp. 1106–1114 (2008)

7. Bose, P., Howat, J., Morin, P.: A history of distribution-sensitive data structures. In: Brodnik, A., López-Ortiz, A., Raman, V., Viola, A. (eds.) Space-Efficient Data Structures, Streams, and Algorithms. LNCS, vol. 8066, pp. 133–149. Springer, Heidelberg (2013). https://doi.org/10.1007/978-3-642-40273-9_10

8. Brodnik, A., López-Ortiz, A., Raman, V., Viola, A. (eds.): Space-Efficient Data Structures, Streams, and Algorithms. LNCS, vol. 8066. Springer, Heidelberg (2013). https://doi.org/10.1007/978-3-642-40273-9

9. Chalermsook, P., Goswami, M., Kozma, L., Mehlhorn, K., Saranurak, T.: The landscape of bounds for binary search trees. CoRR, abs/1603.04892 (2016)

10. Chalermsook, P., Goswami, M., Kozma, L., Mehlhorn, K., Saranurak, T.: Multi-finger binary search trees. In 29th International Symposium on Algorithms and Computation, ISAAC, pp. 55:1–55:26 (2018)

11. Cole, R.: On the dynamic finger conjecture for splay trees. Part II: the proof. SIAM J. Comput. **30**(1), 44–85 (2000)

12. Cole, R., Mishra, B., Schmidt, J.P., Siegel, A.: On the dynamic finger conjecture for splay trees. Part I: splay sorting log n-block sequences. SIAM J. Comput. **30**(1), 1–43 (2000)

13. Cormen, T.H., Leiserson, C.E., Rivest, R.L., Stein, C.: Introduction to Algorithms, 3rd edn. MIT Press, Cambridge (2009)

14. Demaine, E.D., Harmon, D., Iacono, J., Kane, D.M., Patrascu, M.: The geometry of binary search trees. In: Symposium on Discrete Algorithms, SODA, pp. 496–505 (2009)

15. Demaine, E.D., Harmon, D., Iacono, J., Patrascu, M.: Dynamic optimality - almost. SIAM J. Comput. **37**(1), 240–251 (2007)

16. Demaine, E.D., Iacono, J., Langerman, S.: Worst-case optimal tree layout in external memory. Algorithmica **72**(2), 369–378 (2015)

17. Demaine, E.D., Iacono, J., Langerman, S., Özkan, Ö.: Combining binary search trees. In: Fomin, F.V., Freivalds, R., Kwiatkowska, M., Peleg, D. (eds.) ICALP 2013, Part I. LNCS, vol. 7965, pp. 388–399. Springer, Heidelberg (2013). https://doi.org/10.1007/978-3-642-39206-1_33

18. Elmasry, A., Farzan, A., Iacono, J.: On the hierarchy of distribution-sensitive properties for data structures. Acta Inf. **50**(4), 289–295 (2013)

19. Guibas, L.J., Sedgewick, R.: A dichromatic framework for balanced trees. In: Foundations of Computer Science (FOCS), pp. 8–21 (1978)

20. Howat, J., Iacono, J., Morin, P.: The fresh-finger property. CoRR, abs/1302.6914 (2013)

21. Iacono, J.: Alternatives to splay trees with o(log n) worst-case access times. In: Symposium on Discrete Algorithms (SODA), pp. 516–522 (2001)

22. Iacono, J.: Distribution Sensitive Data Structures. PhD thesis, Ph.D. Thesis. Rutgers, The State University of New Jersey (2001)

23. Iacono, J.: In pursuit of the dynamic optimality conjecture. In: Brodnik, A., López-Ortiz, A., Raman, V., Viola, A. (eds.) Space-Efficient Data Structures, Streams, and Algorithms. LNCS, vol. 8066, pp. 236–250. Springer, Heidelberg (2013). https://doi.org/10.1007/978-3-642-40273-9_16

24. Iacono, J., Langerman, S.: Weighted dynamic finger in binary search trees. In: Symposium on Discrete Algorithms, SODA, pp. 672–691 (2016)

25. Lucas, J.M.: Canonical forms for competitive binary search tree algorithms. Technical Report DCS-TR-250, Rutgers University (1988)
26. Sherk, M.: Self-adjusting k-ary search trees. J. Algorithms **19**(1), 25–44 (1995)
27. Sleator, D.D., Tarjan, R.E.: Amortized efficiency of list update and paging rules. Commun. ACM **28**(2), 202–208 (1985)
28. Sleator, D.D., Tarjan, R.E.: Self-adjusting binary search trees. J. ACM **32**(3), 652–686 (1985)
29. Endre, R.: Sequential access in play trees takes linear time. Combinatorica **5**(4), 367–378 (1985)
30. Wang, C.C., Derryberry, J., Sleator, D.D.: O(log log n)-competitive dynamic binary search trees. In: Symposium on Discrete Algorithms, SODA, pp. 374–383 (2006)
31. Wilber, R.E.: Lower bounds for accessing binary search trees with rotations. SIAM J. Comput. **18**(1), 56–67 (1989)

Eventually Dendric Shifts

Francesco Dolce[1]([⊠]) and Dominique Perrin[2]

[1] IRIF, Université Paris Diderot, Paris, France
dolce@irif.fr
[2] LIGM, Université Paris-Est Marne-la-Vallée, Champs-sur-Marne, France
dominique.perrin@esiee.fr

Abstract. We define a new class of shift spaces which contains a number of classes of interest, like Sturmian shifts used in discrete geometry. We show that this class is closed under conjugacy, a natural transformation obtained by sliding block coding.

Keywords: Formal languages · Symbolic dynamics · Sliding block codes

1 Introduction

Shift spaces are the sets of two-sided infinite words avoiding the words of a given language F denoted X_F. In this way the traditional hierarchy of classes of languages translates into a hierarchy of shift spaces. The shift space X_F is called of finite type when one starts with a finite language F and sofic when one starts with a regular language F. There is a natural equivalence between shift spaces called conjugacy. Two shift spaces are conjugate if there is a sliding block coding sending bijectively one upon the other (in this case the inverse map has the same form). Many basic questions are still open concerning conjugacy. For example, it is surprisingly not known whether the conjugacy of shifts of finite type is decidable. The complexity of a shift space X is the function $n \mapsto p(n)$ where $p(n)$ is the number of admissible blocks of length n in X. The complexities of conjugate shifts of linear complexity have the same growth rate (see [9, Corollary 5.1.15]).

In this paper, we are interested in shift spaces of at most linear complexity. This class is important for many reasons and includes the class of Stumian shifts which are by definition those of complexity $n + 1$, which play a role as binary codings of discrete lines. Several books are devoted to the study of such shifts (see [9] or [11] for example). We define a new class of shifts of at most linear complexity, called eventually dendric, which includes Sturmian shifts. It extends the class of dendric shifts introduced in [2] (under the name of *tree sets* given

This work was supported by the ANR project CODYS ANR–18–CE40–0007. We thank Valérie Berthé, Paulina Cecchi, Fabien Durand and Samuel Petite for useful conversations on this subject.

R. van Bevern and G. Kucherov (Eds.): CSR 2019, LNCS 11532, pp. 106–118, 2019.
https://doi.org/10.1007/978-3-030-19955-5_10

to their language) which themselves extend naturally episturmian shifts (also called Arnoux-Rauzy shifts) and interval exchange shifts. A *dendric shift* X is defined by introducing the extension graph of a word in the language $\mathcal{L}(X)$ of X and by requiring that this graph is a tree for every word in $\mathcal{L}(X)$. This kind of shifts has many interesting properties which involve free groups. In particular, in a dendric shift X on the alphabet A, the group generated by the set of return words to some word in $\mathcal{L}(X)$ is the free group on the alphabet and, in particular, has Card (A) free generators. This generalizes a property known for Sturmian (and episturmian) shifts whose link with automorphisms of the free group was noted by Arnoux and Rauzy. The class of eventually dendric shifts, introduced in this paper, is defined by the property that the extension graph of every word w in the language of the shift is a tree for every long enough word w.

The paper is organized as follows. In the first section, we introduce the definition of the extension graph and of an eventually dendric shift. In Sect. 3, we recall some mostly known properties on the complexity of a shift space and of left- or right-special words. We prove a result which characterizes eventually dendric shifts by the extension properties of left-special words (Proposition 4). This result shows us that asymptotically eventually dendric shifts behaves locally in a way similar to Sturmian shifts. In Sect. 4, we use the classical notion of asymptotic equivalence to give a second characterization of eventually dendric shifts (Proposition 1). In Sect. 5, we introduce the notion of a simple tree and we prove that for eventually dendric shift, the extension graph of every long enough word is a simple tree (Proposition 7), a property which holds trivially for every word in a Sturmian shift but that is quite surprising for this new larger class of shifts.. Finally, in Sect. 6 we use the previous results to prove the main result (Theorem 2), namely that the class of eventually dendric shifts is closed under conjugacy. This result shows the robustness of the class of eventually dendric shifts, giving a strong motivation for its introduction.

2 Eventually Dendric Shifts

Let A be a finite alphabet. We consider the set $A^{\mathbb{Z}}$ of bi-infinite words on A as a topological space for the product topology. The *shift map* $\sigma_A : A^{\mathbb{Z}} \to A^{\mathbb{Z}}$ is defined by $y = \sigma_A(x)$ if $y_i = x_{i+1}$ for every $i \in \mathbb{Z}$. A *shift space* on the alphabet A is a subset X of the set $A^{\mathbb{Z}}$ which is closed and invariant under the shift, that is such that $\sigma_A(X) = X$ (for more on shift spaces see, for instance, [9]).

We denote by $\mathcal{L}(X)$ the language of a shift space X, which is the set of finite factors of the elements of X. A language $\mathcal{L}$ on the alphabet A is the language of a shift if and only if it is *factorial* (that is contains the factors of its elements) and *extendable* (that is for any $w \in \mathcal{L}$ there are letters $a, b \in A$ such that $awb \in \mathcal{L}$). For $n \geq 0$ we denote $\mathcal{L}_n(X) = \mathcal{L}(X) \cap A^n$ and $\mathcal{L}_{\geq n}(X) = \cup_{m \geq n} \mathcal{L}_m(X)$. For $w \in \mathcal{L}(X)$ and $n \geq 1$, we denote $L_n(w, X) = \{u \in \mathcal{L}_n(X) \mid uw \in \mathcal{L}(X)\}$, $R_n(w, X) = \{v \in \mathcal{L}_n(X) \mid wv \in \mathcal{L}(X)\}$ and $E_n(w, X) = \{(u, v) \in L_n(w, X) \times R_n(w, X) \mid uwv \in \mathcal{L}(X)\}$. The *extension graph* of order n of w, denoted $\mathcal{E}_n(w, X)$, is the undirected graph with set of vertices the disjoint union of $L_n(w,X)$ and $R_n(w,X)$ and with edges the elements of $E_n(w,X)$. When the context is clear, we denote $L_n(w), R_n(w), E_n(w)$ and $\mathcal{E}_n(w)$ instead of $L_n(w,X),$ $R_n(w,X), E_n(w,X)$ and $\mathcal{E}_n(w,X)$ A path in an undirected graph is *reduced* if it does not contain successive equal edges (such a path is also known as *simple*). For any $w \in \mathcal{L}(X)$, since any vertex of $L_n(w)$ is connected to at least one vertex of $R_n(w)$, the bipartite graph $\mathcal{E}_n(w)$ is a tree if and only if there is a unique reduced path between every pair of vertices of $L_n(w)$ (resp. $R_n(w)$).

The shift X is said to be *eventually dendric* with threshold $m \geq 0$ if $\mathcal{E}_1(w)$ is a tree for every word $w \in \mathcal{L}_{\geq m}(X)$. It is said to be *dendric* if we can choose $m = 0$. The languages of dendric shifts were introduced in [2] under the name of tree sets. An important example of dendric shifts is formed by *episturmian shifts* (also called Arnoux-Rauzy shifts), which are by definition such that $\mathcal{L}(X)$ is closed by reversal and such that for every n there exists a unique $w_n \in \mathcal{L}_n(X)$ such that $\mathrm{Card}\,(R_1(w_n)) = \mathrm{Card}\,(A)$ and such that for every $w \in \mathcal{L}_n(X) \setminus \{w_n\}$ one has $\mathrm{Card}\,(R_1(w)) = 1$.

Example 1. Let X be the *Fibonacci shift*, which is generated by the morphism $a \mapsto ab, b \mapsto a$. It is well-known that the Fibonacci shift is a Sturmian shift (see, for example, [9]). The graphs $\mathcal{E}_1(a)$ and $E_3(a)$ are shown in Fig. 1.

Fig. 1. The graphs $\mathcal{E}_1(a)$ (on the left) and $\mathcal{E}_3(a)$ (on the right).

The class of *tree sets of characteristic* $c \geq 1$ introduced in [1,5] give an example of eventually dendric shifts of threshold 1.

Example 2. Let X be the shift generated by the morphism $a \mapsto ab, b \mapsto cda, c \mapsto cd, d \mapsto abc$ [4]. Its language is a tree set of characteristic 2 (see [1]).

3   Complexity of Shift Spaces

Let X be a shift space. For a word $w \in \mathcal{L}(X)$, we denote $\ell_k(w) = \mathrm{Card}\,(L_k(w))$, $r_k(w) = \mathrm{Card}\,(R_k(w))$, and $e_k(w) = \mathrm{Card}\,(E_k(w))$. For any $w \in \mathcal{L}(X)$, we have $1 \leq \ell_k(w), r_k(w) \leq e_k(w)$. The word w is *left-k-special* if $\ell_k(w) > 1$, *right-k-special* if $r_k(w) > 1$ and *k-bispecial* if it is both left-k-special and right-k-special. For $k = 1$, we use ℓ, r, e and we simply say special instead of k-special. Given a word w we define the quantity $m(w) = e(w) - \ell(w) - r(w) + 1$. We say that w is *strong* if $m(w) \geq 0$, *weak* if $m(w) \leq 0$ and *neutral* if $m(w) = 0$. It is clear that if $\mathcal{E}_1(w)$ is acyclic (resp. connected, resp. a tree), then w is weak (resp. strong, resp. neutral).

Proposition 1. *Let X be a shift space and let $w \in \mathcal{L}(X)$. If w is neutral, then*

$$\ell(w) - 1 = \sum_{b \in R_1(w)} (\ell(wb) - 1) \tag{1}$$

Set further $p_n(X) = \mathrm{Card}\,(\mathcal{L}_n(X))$, $s_n(X) = p_{n+1}(X) - p_n(X)$ and $b_n(X) = s_{n+1}(X) - s_n(X)$. The sequence $p_n(X)$ is called the *complexity* of X.

The following result is from [3] (see also [2, Lemma 2.12]).

Proposition 2. *We have for all $n \geq 0$,*

$$s_n(X) = \sum_{w \in \mathcal{L}_n(X)} (\ell(w)-1) = \sum_{w \in \mathcal{L}_n(X)} (r(w)-1) \quad and \quad b_n(X) = \sum_{w \in \mathcal{L}_n(X)} m(w).$$

In particular, the number of left-special (resp. right-special) words of length n is bounded by $s_n(X)$.

Proposition 3. *Let X be a shift space. If X is eventually dendric, then the sequence $s_n(X)$ is eventually constant.*

Proof. Let N be the threshold of X. By Proposition 2, we have that $b_n(X) = 0$ for all $n \geq N$. Thus $s_{n+1}(X) = s_n(X)$ for all $n \geq N$.

The converse of Proposition 3 is not true, as shown by the following example.

Example 3. Let X be the *Chacon ternary shift*, which is the substitutive shift space generated by the morphism $\varphi : a \mapsto aabc, b \mapsto bc, c \mapsto abc$. It is well known that the complexity of X is $p_n(X) = 2n + 1$ and thus that $s_n = 2$ for all $n \geq 0$ (see [9, Section 5.5.2]). The extension graphs of abc and bca are shown in Fig. 2.

Thus $m(abc) = 1$ and $m(bca) = -1$. Let now α be the map on words defined by $\alpha(x) = abc\varphi(x)$. Let us verify that if the extension graph of x is the graph of Fig. 2 on the left, the same holds for the extension graph of $y = \alpha(x)$. Indeed, since $axa \in \mathcal{L}(X)$, the word $\varphi(axa) = aabc\varphi(x)aabc = ayaabc$ is also in $\mathcal{L}(X)$ and thus $(a, a) \in \mathcal{E}_1(y)$. Since $cxa \in \mathcal{L}(X)$ and since a letter c is always preceded by a letter b, we have $bcxa \in \mathcal{L}(X)$. Thus $\varphi(bcxa) = bcyaabc \in \mathcal{L}(X)$ and thus $(c, a) \in \mathcal{E}_1(y)$. The proof of the other cases is similar. The same property holds for a word x with the extension graph on the right of Fig. 2. This shows that there is an infinity of words whose extension graph is not a tree and thus the Chacon set is not eventually dendric.

Fig. 2. The extension graphs $\mathcal{E}_1(abc)$ (on the left) and $\mathcal{E}_1(bca)$ (on the right).

Let X be a shift space. We define $LS_n(X)$ (resp. $LS_{\geq n}(X)$) as the set of left-special words of $\mathcal{L}(X)$ of length n (resp. at least n) and $LS(X) = \bigcup_{n \geq 0} LS_n(X)$.

The following result expresses the fact that eventually dendric shift spaces are characterized by an asymptotic property of left-special words which is a local version of the property defining Sturmian shift spaces.

Proposition 4. *A shift space X is eventually dendric if and only if there is an integer $n \geq 0$ such that any word w of $LS_{\geq n}(X)$ has exactly one right extension $wb \in LS_{\geq n+1}(X)$ with $b \in A$. Moreover, in that case one has $\ell(wb) = \ell(w)$.*

Proof. Assume first that X is eventually dendric with threshold m. Then any word w in $LS_{\geq m}(X)$ has at least one right extension in $LS(X)$. Indeed, since $L_1(w)$ has at least two elements and since the graph $\mathcal{E}_1(w)$ is connected, there is at least one element of $R_1(w)$ which is connected by an edge to more than one element of $L_1(w)$. Next, Eq. (1) shows that for any $w \in LS_{\geq m}(X)$ which has more than one right extension in $LS(X)$, one has $\ell(wb) < \ell(w)$ for each such extension. Thus the number of words in $LS_{\geq m}(X)$ which are prefix of one another and which have more than one right extension, is bounded by Card (A). This proves that there exists an $n \geq m$ such that for any $w \in LS_{\geq n}(X)$ there is exactly one $b \in A$ such that $wb \in LS(X)$. Moreover, one has then $\ell(wb) = \ell(w)$ by Eq. (1).

Conversely, assume that the condition is satisfied for some integer n. For any word w in $\mathcal{L}_{\geq n}(X)$, the graph $\mathcal{E}_1(w)$ is acyclic since all vertices in $R_1(w)$ except at most one have degree 1. Thus w is weak. Let N be the length of w. Then for every word u of length N and every $b \in R_1(u)$, one has $\ell(ub) = 1$ except for one letter b such that $\ell(ub) = \ell(u)$. Thus, by Proposition 2,

$$s_N(X) = \sum_{u \in \mathcal{L}_N(X)} (\ell(u) - 1) = \sum_{v \in \mathcal{L}_{N+1}(X)} (\ell(v) - 1) = s_{N+1}(X).$$

This shows that $b_N = 0$ for every $N \geq n$ and thus, by Proposition 2 again, all words in $\mathcal{L}_{\geq n}(X)$ are neutral. Since all graphs $\mathcal{E}_1(w)$ are moreover acyclic, this forces that these graphs are trees and thus that X is eventually dendric with threshold n.

Example 4. Let X be the *Tribonacci shift*, which is the episturmian shift generated by the substitution $\varphi : a \mapsto ab, b \mapsto ac, c \mapsto a$ and let α be the morphism $\alpha : a \mapsto a, b \mapsto a, c \mapsto c$. It can be verified that $\alpha(X)$ satisfies the condition of Proposition 4 with $n = 4$ and thus it is dendric with threshold at most 4. The threshold is actually 4 since $m(a^3) = 1$ in $\alpha(X)$.

4   Asymptotic Equivalence

The *orbit* of $x \in A^{\mathbb{Z}}$ is the equivalence class of x under the action of the shift transformation. Thus y is in the orbit of x if there is an $n \in \mathbb{Z}$ such that $x = \sigma^n(y)$. We say that x is a shift of y if they belong to the same orbit.

For $x \in A^{\mathbb{Z}}$, denote $x^- = \cdots x_{-2}x_{-1}$ and $x^+ = x_0x_1\cdots$ and $x = x^- \cdot x^+$. When X is a shift space, we denote X^+ the set of right infinite words u such that $u = x^+$ for some $x \in X$. A right infinite word $u \in A^{\mathbb{N}}$ is a *tail* of the two-sided infinite word $x \in A^{\mathbb{Z}}$ if $u = y^+$ for some shift y of x, that is $u = x_n x_{n+1} \cdots$ for some $n \in \mathbb{Z}$. The *right asymptotic equivalence* on a shift space X is the equivalence defined as follows. Two elements x, y of X are right asymptotically equivalent if there exists two shifts x', y' of x, y such that $x'^+ = y'^+$. In other words, x, y are right asymptotic equivalent if they have a common tail (see Fig. 3, where, for simplicity, we suppose $x = x'$ and $y = y'$). The classes of the right asymptotic equivalence not coinciding with only one orbit are called *right asymptotic classes* (they are called *asymptotic components* in [7]).

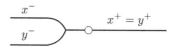

Fig. 3. Two right asymptotic sequences x, y.

Example 5. The Fibonacci shift X has only one right asymptotic class. It is formed of the orbits of the two elements $x, y \in X$ such that $x^+ = y^+ = \varphi^{\omega}(a)$ where $\varphi^{\omega}(a)$ is the Fibonacci word, that is the right infinite word having all $\varphi^n(a)$ for $n \geq 1$ as prefixes. Indeed, let $x, y \in X$ be such that $x^+ = y^+$ with $x \neq y$. Then all finite prefixes of $x^+ = y^+$ are left-special and thus are prefixes of $\varphi^{\omega}(a)$ (see, for instance, [9]). Thus $x^+ = y^+ = \varphi^{\omega}(a)$.

If C is a right asymptotic class, it is a union of orbits. The following result is proved in [7, Lemma 3.2] under a weaker hypothesis that we shall not need here. We give a proof for the sake of completeness.

Proposition 5. *Let X be a shift space such that the sequence $s_n(X)$ is bounded by k. The number of right asymptotic classes is finite and at most equal to k.*

Proof. Let $(x_1, y_1), \ldots, (x_h, y_h)$ be h pairs of distinct elements of X belonging to asymptotic classes $C_1, \ldots, C_h$ such that for all $1 \leq i \leq h$ one has $x_i^+ = y_i^+$ and $(x_i)_{-1} \neq (y_i)_{-1}$. For n large enough the prefixes of length n of the x_i^+ are h distinct left-special words and thus $h \leq s_n(X)$ since by Proposition 2 the number of left-special words is bounded by $s_n(X)$. This shows that the number of right asymptotic classes is finite and bounded by k.

Let X be a shift space. For a right asymptotic class C of X, we denote $\omega(C) = \mathrm{Card}\,(o(C)) - 1$ where $o(C)$ is the set of orbits contained in C. For a right infinite word $u \in X^+$, let $\ell_C(u) = \mathrm{Card}\,(a \in A \mid x^+ = au$ for some $x \in C)$. We denote by $LS(C)$ the set of right infinite words u such that $\ell_C(u) \geq 2$.

Proposition 6. *Let X be a shift space and let C be a right asymptotic class. Then*

$$\omega(C) = \sum_{u \in LS(C)} (\ell_C(u) - 1) \tag{2}$$

where both sides are simultaneously finite.

In order to prove Proposition 6, we use the following notion. A *cluster of trees* is a directed graph which is the union of a (non-trivial) cycle Γ and a family of disjoint trees (oriented from child to father) T_v with root v indexed by the vertices v on Γ (see Fig. 4). It is easy to verify that a finite connected graph is a cluster of trees if and only if every vertex has outdegree 1 and there is a unique strongly connected component. In a cluster of trees, the number of leaves (that is, the leaves of the trees T_v not reduced to their root) is equal to $\sum_u (d^-(u) - 1)$, where d^- stands for the indegree function and the sum runs over the set of internal nodes. Indeed, this is true for one cycle alone since there are no leaves and every internal node has indegree 1. The formula remains valid when suppressing a leaf in one of the trees not reduced to its root.

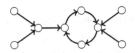

Fig. 4. A cluster of trees.

Proof (of Proposition 6). We first suppose that C does not contain periodic points which implies that $LS(C)$ does not contain periodic points either. It is easy to verify that if $u, v \in LS(C)$, there exist $n, m \geq 0$ such that $\sigma^n(u) = \sigma^m(v)$. We build a graph $T(C)$ as follows. The set of vertices of $T(C)$ is $o(C) \cup LS(C)$. There will be for each vertex u of $T(C)$ at most one edge going out of u, called its father.

Let first $x \in C$ and let u be the orbit of x. There is, up to a shift of x, at least one $y \in C$ with $x \neq y$ such that $y^+ = x^+$. Let $n \geq 0$ be the minimal integer such that $x_{-n} \neq y_{-n}$. Then $v = \sigma^{-n+1}(x)^+$ is in $LS(C)$ and depends only on the orbit u of x. We choose the vertex v as the father of u. Next, for every $u \in LS(C)$, we consider the minimal integer, if it exists, such that $v = \sigma^n(u)$ is in $LS(C)$. Then we choose v as the father of u.

Assume now that $\omega(C)$ is finite. Then $LS(C)$ is also finite and $T(C)$ is a finite tree. Indeed, if $u \in LS(C)$, there is at least one $x \in C$ such that $x^+ = u$ and thus such that u is an ancestor of the orbit of x. By the claim made above, any two elements of $LS(C)$ have a common ancestor. Since C does not contain periodic points, two vertices cannot be ancestors of one another. Thus there is a unique element of $LS(C)$ which has no father, namely the unique $u \in LS(C)$ with a maximal number of elements of $o(C)$ as descendants. Since it is an ancestor of all vertices of $T(C)$, this shows that $T(C)$ is a finite tree. Formula (2) now follows from the fact that in any finite tree with n leaves and and a set V of internal vertices, one has $n - 1 = \sum_{v \in V}(d^-(v) - 1)$.

Assume next that the right hand side of Eq. (2) is finite. Then the set $LS(C)$ is finite and thus $T(C)$ is again a tree with a finite number of internal nodes. Since the degree of each node is finite, it implies that it has also a finite number of leaves. Thus $w(C)$ is finite and Eq. (2) also holds.

Finally, assume that C contains a periodic point. It follows from the definition of a right asymptotic class that there is exactly one such periodic orbit, since two periodic points having a common tail are in the same orbit. The proof follows the same lines as in the first case, but this time $T(C)$ will be a cluster of trees instead of a tree. The set of leaves of $T(C)$ is, as above, the set $o(C)$ of non periodic orbits and the other vertices are the elements of $LS(C)$. The unique father of a vertex is defined in the same way as above. The fact that there is a unique strongly connected component is a consequence of the fact that there is a unique periodic orbit in C. Finally, Formula (2) holds with since the number of leaves is equal to $\sum(d^-(u) - 1) - 1$, where the sum runs over the set of internal nodes and the -1 corresponds to the unique periodic orbit.

Example 6. Consider again the image $\alpha(X)$ of the Tribonacci shift by the morphism $\alpha : a \mapsto a, b \mapsto a, c \mapsto c$ (Example 4). There is one right asymptotic class C made of three orbits represented in Fig. 5 on the left. The class is formed of the orbits of x, y, z where $x^+ = \alpha(\varphi^\omega(a))$ and $y^+ = z^+ = aax^+$. The tree $T(C)$ is shown on the right.

Fig. 5. The right asymptotic class C and the tree $T(C)$.

Using Proposition 6 we can give a characterization of eventually dendric shift spaces in terms of right asymptotic classes (for the proof, see [6]). We denote $w(X) = \sum w(C)$, where the sum is over the right asymptotic classes C of X.

Theorem 1. *A shift space X is eventually dendric if and only if:*

1. *The sequence $s_n(X)$ is eventually constant, and*
2. *We have $\lim s_n(X) = w(X)$.*

For example, the Tribonacci shift is such that $s_n(X) = 2$ for every $n \geq 0$ and $w(X) = 2$ since there is only one asymptotic class made of 3 orbits. Note that the Chacon shift X satisfies condition 1 of Theorem 1 but not condition 2. Indeed, one can verify that $s_n(X) = 2$ for all $n \geq 0$ but $w(X) = 1$.

5   Simple Trees

The *diameter* of a tree is the maximal length of simple paths. We call a tree *simple* if its diameter is at most 3. Note that if a simple tree is the extension graph $\mathcal{E}_n(w)$ in some shift space X of a bispecial word w, then the diameter of $\mathcal{E}_n(w)$ is at least 3, and it is exactly 3 if and only if any two vertices of $\mathcal{E}_n(w)$ on the same side (that is, both in $L_n(x)$ or both in $R_n(w)$) are connected to a common vertex on the opposite side. For example, if X is the Fibonacci shift, then $\mathcal{E}_1(a)$ is simple while $\mathcal{E}_3(a)$ is not (see Example 1).

We prove the following additional property of the graphs $\mathcal{E}_k(w)$.

Proposition 7. *Let X be an eventually dendric shift space. For any $k \geq 1$ there exists an $n \geq 1$ such that $\mathcal{E}_k(w)$ is a simple tree for every $w \in \mathcal{L}_{\geq n}(X)$.*

We first prove the following lemma.

Lemma 1. *Let X be an eventually dendric shift space. For every $k \geq 1$ there is an $n \geq 1$ such that if $p, w \in \mathcal{L}(X)$ with $|p| \leq k$ and $|w| \geq n$ are such that $pw, w \in LS(X)$, then pw, w have a unique right extension in $LS(X)$ for some letter $b \in A$ which is moreover such that $\ell(pwb) = \ell(pw)$ and $\ell(wb) = \ell(w)$.*

Proof. Consider two right asymptotic classes C, D and let $u \in LS(C)$, $v \in LS(D)$. If C, D are distinct, we cannot have $pu = v$ for some word p. Thus there is an integer n such that if w is the prefix of length n of u, then pw is not a prefix of v. Since there is a finite number of words p of length at most k, a finite number of right asymptotic classes (by Proposition 5) and since for each such class the set $LS(C)$ is finite (by Proposition 6), we infer that for every k there exists an n such that for every pair of right asymptotic classes C, D and any $u \in LS(C), v \in LS(D)$, if w is a prefix of u and pw a prefix of v, with $|p| \leq k$ and $|w| = n$, then $C = D$.

Next, assume that w is a prefix of u and pw a prefix of v with $u, v \in LS(C)$ for some right asymptotic class C. If $v \neq pu$, then there is a right extension w' of w such that pw' is not a prefix of v. By contraposition, if n is large enough, we have $v = pu$. We thus choose n large enough so that: all elements of $LS(C)$ for all right asymptotic components C have distinct prefixes of length n and such that for every pair of asymptotic classes C, D and any $u \in LS(C), v \in LS(D)$, if w is prefix of u and pw is prefix of v with $|p| \leq k$ and $|w| = n$ then $C = D$ and $pu = v$. We moreover assume that n is large enough so that the condition of Proposition 4 holds. Consider p, w with $|p| \leq k$ and $|w| = n$ such that pw, w are left-special. By condition 1, there are right asymptotic components C, D and elements $u \in LS(C)$ and $v \in LS(D)$ such that w is a prefix of u and pw a prefix of v. Because of condition 2, we must have $\sigma^k(v) = u$ (and in particular $C = D$). Thus there is a unique letter $b \in A$ such that $wb, pwb \in LS(X)$ which is moreover such that $\ell(wb) = \ell(w)$ and $\ell(pwb) = \ell(pw)$ by Proposition 4.

Proof (of Proposition 7). We choose n such that Proposition 4 and Lemma 1 hold. We prove by induction on h with $1 \leq h \leq k$ that for any $p, q \in L_h(w)$

there is an $r \in R_k(w)$ such that $pwr, qwr \in \mathcal{L}(X)$. This implies that every reduced path in the tree $\mathcal{E}_\ell(w)$ has length at most three, and thus that $\mathcal{E}_\ell(w)$ is a simple tree. The property is true for $h = 1$. Indeed, set $p = a$ and $q = b$. Apply iteratively Proposition 4 to obtain letters $c_1, \ldots, c_k$ such that $\ell(wc_1 \cdots c_i) = \ell(wc_1 \cdots c_i c_{i+1})$ and set $r = c_1 \cdots c_k$. Then $awr, bwr \in \mathcal{L}(X)$.

Assume next that the property is true for $h - 1$ and consider $ap, bq \in L_h(w)$ with $a, b \in A$. Replacing if necessary w by some longer word, we may assume that p, q end with different letters and thus that w is left-special. By the induction hypothesis, there is a word $r \in R_k(w)$ such that $pwr, qwr \in \mathcal{L}(X)$. By Lemma 1, the first letter of r is the unique letter c such that $\ell(pwc) = \ell(pw)$ and $\ell(qwc) = \ell(qw)$. Thus $apwc, bqwc \in \mathcal{L}(X)$. Applying Lemma 1 iteratively in this way, we obtain that $apwr, bqwr \in \mathcal{L}(X)$.

6   Conjugacy

Let A, B be two alphabets, and $X \subset A^{\mathbb{Z}}$ and $Y \subset B^{\mathbb{Z}}$ be two shift spaces. A map $\phi : X \to Y$ is called a *sliding block code* if there exists $m, n \in \mathbb{N}$ and a map $f : \mathcal{L}_{m+n+1}(X) \to B$ such that $\phi(x)_i = f(x_{i-m} \cdots x_{i+n})$ for all $i \in \mathbb{Z}$ and $x = (x_i) \in X$. It can be shown that a map $\phi : X \to Y$ is a sliding block code if and only if it is continuous and commutes with the shift, that is $\phi \circ \sigma_A = \sigma_B \circ \phi$ (see, for instance, [10]). Two shift spaces X, Y are said to be *conjugate* when there is a bijective sliding block code $\phi : X \to Y$. The following result shows that the property of being eventually dendric is a dynamical property, in the sense that it only depends on the class of a shift space under conjugacy.

Theorem 2. *The class of eventually dendric shift spaces is closed under conjugacy.*

We first treat the following particular case of conjugacy. Let X be a shift space on the alphabet A and let $k \geq 1$. Let $f : \mathcal{L}_k(X) \to A_k$ be a bijection from the set $\mathcal{L}_k(X)$ of blocks of length k of X onto an alphabet A_k. The map $\gamma_k : X \to A_k^{\mathbb{Z}}$ defined for $x \in X$ by $y = \gamma_k(x)$ if, for every $n \in \mathbb{Z}$, $y_n = f(x_n \cdots x_{n+k-1})$ is the k-th *higher block code* on X. The shift space $X^{(k)} = \gamma_k(X)$ is called the k-th *higher block shift space* of X. It is well known that the k-th higher block code is a conjugacy. We extend the bijection $f : \mathcal{L}_k(X) \to A_k$ to a map still denoted f from $\mathcal{L}_{\geq k}(X)$ to $\mathcal{L}_{\geq 1}(X^{(k)})$ by $f(a_1 a_2 \cdots a_n) = f(a_1 \cdots a_k) \cdots f(a_{n-k+1} \cdots a_n)$. Note that all nonempty elements of $\mathcal{L}(X^{(k)})$ are image by f of elements of $\mathcal{L}(X)$, that is, $\mathcal{L}(X^{(k)}) = \{f(w) \mid w \in \mathcal{L}_{\geq k}(X)\} \cup \{\varepsilon\}$.

Example 7. Let X be the Fibonacci shift. We show that the 2-block extension $X^{(2)}$ of X is eventually dendric with threshold 1. Set $A_2 = \{u, v, w\}$ with $f :$ $aa \mapsto u, ab \mapsto v, ba \mapsto w$. Since X is dendric, the graph $\mathcal{E}_1(w)$ is a tree for every word $w \in \mathcal{L}(X^{(2)})$ of length at least 1 (but not for $w = \varepsilon$). Thus $X^{(2)}$ is eventually dendric. It is actually a tree shift space of characteristic 2 since the graph $\mathcal{E}_1(\varepsilon)$ is the union of two trees (see Fig. 6).

Fig. 6. The extension graphs $\mathcal{E}_1(\varepsilon)$ (on the left) and $\mathcal{E}_1(vw)$ (on the right).

Lemma 2. *For every $k \geq 1$, the k-th higher block shift space $X^{(k)}$ is eventually dendric if and only if X is eventually dendric.*

Proof. We define for every $w \in \mathcal{L}_{\geq k}(X)$ a map from $\mathcal{E}_1(w)$ to $\mathcal{E}_1(f(w))$ as follows. To every $a \in L_1(w)$, we associate the first letter $\lambda(a)$ of $f(aw)$ and to every $b \in R_1(w)$, we associate the last letter $\rho(b)$ of $f(wb)$. Then, since $f(awb) = \lambda(a)f(w)\rho(b)$, the pair (a, b) is in $E_1(w)$ if and only if $(\lambda(a), \rho(b))$ is in $E_1(f(w))$. Thus, the maps λ, ρ define an isomorphism from $\mathcal{E}_1(w)$ onto $\mathcal{E}_1(f(w))$.

Thus we conclude that X is eventually dendric with threshold m if and only if $X^{(k)}$ is eventually dendric with threshold M with $0 \leq M \leq \max(1, m-k+1)$.

Example 8. Let X be the Fibonacci shift. For all $k \geq 2$, $X^{(k)}$ is an eventually dendric shift space with threshold 1.

A morphism $\alpha : A^* \to B^*$ is called *alphabetic* if $\alpha(A) \subseteq B$.

Lemma 3. *Let X be an eventually dendric shift space on the alphabet A and let $\alpha : A^* \to B^*$ be an alphabetic morphism which induces a conjugacy from X onto a shift space Y. Then Y is eventually dendric.*

Proof. Since α is invertible, there exists map $f : \mathcal{L}_{2r+1}(Y) \to A$, with $r \geq 0$, such that for $x = (x_k)_{k \in \mathbb{Z}}$ and $y = (y_k)_{k \in \mathbb{Z}}$, one has $y = \alpha(x)$ if and only if for every $k \in \mathbb{Z}$, one has $x_k = f(y_{k-r} \cdots y_{k-1}y_ky_{k+1} \cdots y_{k+r})$. We extend the definition of f to a map from $\mathcal{L}_{\geq 2r+1}$ to A: for $w = b_{1-r} \cdots b_{n+r} \in \mathcal{L}_{\geq 2r+1}(Y)$, set $f(w) = a_1 \cdots a_n$ where $a_i = f(b_{i-r} \cdots b_i \cdots b_{i+r})$. Note that if $u = f(w)$ and $w = svt$ with $s, t \in \mathcal{L}_r(Y)$, then $v = \alpha(u)$. Let n be the integer given by Proposition 7 for $k = r + 1$. We claim that every graph $\mathcal{E}_1(w)$ for $|w| \geq n + 2r$ is a tree. Let indeed $s, t \in \mathcal{L}_r(Y)$ and $v \in \mathcal{L}_{\geq n}(Y)$ be such that $w = svt$. Let $u = f(svt)$. Let $E'_k(u) = \{(p, q) \in L_k(u) \times R_k(u) \mid \alpha(puq) \in BwB\}$ and let $L'_k(u)$ (resp. $R'_k(u)$) be the set of $p \in L_k(u)$ (resp. $q \in R_k(u)$) which are connected to $L_k(u)$ (resp. $R_k(u)$) by an edge in $E'_k(u)$. Let $\mathcal{E}'_k(u)$ be the subgraph of $\mathcal{E}_k(u)$ obtained by restriction to the set of vertices which is the disjoint union of $L'_k(u)$ and $R'_k(u)$ (and that thus has $E'_k(u)$ as set of edges).

The graph $\mathcal{E}'_k(u)$ is a simple tree. Indeed, by Proposition 7, the graph $\mathcal{E}_k(u)$ is a simple tree. We may assume that u is k-bispecial (otherwise, the property is obviously true). Let (p, q) be an edge of $\mathcal{E}'_k(u)$. Then (p, q) is an edge of $\mathcal{E}_k(u)$ and since the latter is a simple tree either p is the unique vertex in $L_k(u)$ such that pu is right-special or q is the unique vertex in $R_k(u)$ such that uq is left-special (both cases can occur simultaneously). Assume the first case, the other being proved in a symmetric way. If (p', q') is another edge of $\mathcal{E}'(u)$, then (p, q') is an edge of $\mathcal{E}_k(u)$. Since $\alpha(p) \in Bs$ and $\alpha(q) \in tB$, we have actually $(p, q') \in E'_k(u)$. Thus $\mathcal{E}'_k(u)$ contains the two vertices of $\mathcal{E}_k(u)$ connected to more than one other vertex and this implies that $\mathcal{E}'_k(u)$ is a simple tree. For $p \in L'_k(u)$, let $\lambda(p)$ be the first letter of $\alpha(p)$ and for $q \in R'_k(u)$, let $\rho(q)$ be the last letter of $\alpha(q)$.

The graph $\mathcal{E}_1(w)$ is the image by the maps λ, ρ of the graph $\mathcal{E}'_k(u)$. Indeed, one has $(a, b) \in E_1(w)$ iff there exist $(p, q) \in E'_k(u)$ such that $\lambda(p) = a$ and $\rho(q) = b$. Let us consider a graph homomorphism ϕ preserving bipartiteness and such that left vertices are sent to left vertices and right vertices to right ones: Then, it is easy to verify that the image of a simple tree by ϕ is again a simple tree. Thus $\mathcal{E}_1(w)$ is a simple tree, which concludes the proof.

Proof (of Theorem 2). Every conjugacy is a composition of a higher block code and an alphabetic morphism (see [10, Proposition 1.5.12]). Thus Theorem 2 is a direct consequence of Lemmas 2 and 3.

Example 9. The fact that the image of the Tribonacci shift by the morphism α given in Example 4 is eventually dendric is actually a consequence of Theorem 2. Indeed α is an alphabetic morphism and thus a conjugacy. Images of episturmian shift spaces by non trivial alphabetic morphisms have been investigated in [12].

An interesting open question is whether the class of eventually dendric shifts is closed under taking *factors*, that is, images by a sliding block code not necessarily bijective. It would also be interesting to know whether the conjugacy of effectively given eventually dendric shifts is decidable (the conjugacy of substitutive shifts was recently shown to be decidable [8]).

References

1. Berthé, V., et al.: Specular sets. Theoret. Comput. Sci. **684**, 3–28 (2017)
2. Berthé, V., et al.: Acyclic, connected and tree sets. Monatsh. Math. **176**(4), 521–550 (2015)
3. Cassaigne, J.: Complexité et facteurs spéciaux. Bull. Belg. Math. Soc. Simon Stevin **4**(1), 67–88 (1997). Journées Montoises (Mons, 1994)
4. Cassaigne, J.: Personal communication (2013)
5. Dolce, F., Perrin, D.: Neutral and tree sets of arbitrary characteristic. Theor. Comput. Sci. **658**(part A), 159–174 (2017)
6. Dolce, F., Perrin, D.: Eventually dendric subshifts (2018). https://arxiv.org/abs/1807.05124
7. Donoso, S., Durand, F., Maass, A., Petite, S.: On automorphism groups of low complexity subshifts. Ergod. Th. Dynam. Sys. **36**, 64–95 (2016)
8. Durand, F., Leroy, J.: Decidability of the isomorphism and the factorization between minimal substitution subshifts (2018)
9. Fogg, P.N.: Substitutions in dynamics, arithmetics and combinatorics. In: Berthé, V., Ferenczi, S., Mauduit, C., Siegel, A. (eds.) Lecture Notes in Mathematics, vol. 1794. Springer, Berlin (2002). https://doi.org/10.1007/b13861
10. Lind, D., Marcus, B.: An Introduction to Symbolic Dynamics and Coding. Cambridge University Press, Cambridge (1995)

11. Queffélec, M.: Substitution dynamical systems—spectral analysis. In: Lecture Notes in Mathematics, 2nd edn., vol. 1294. Springer, Berlin (2010). https://doi.org/10.1007/978-3-642-11212-6

12. Vesely, V.: Properties of morphic images of S-adic sequences. Master's thesis, Czech Technical University (2018)

On Decidability of Regular Languages Theories

Sergey Dudakov and Boris Karlov[✉]

CS Department, Tver State University, Zhelyabova Str., 33, Tver 170100, Russia
sergeydudakov@yandex.ru, bnkarlov@gmail.com

Abstract. This paper is dedicated to studying decidability properties of some regular languages theories. We prove that the regular languages theory with the Kleene star only is decidable. If we use union and concatenation simultaneously then the theory becomes both Σ_1- and Π_1-hard over the one-symbol alphabet. Finally, we prove that the regular languages theory over one-symbol alphabet with union and the Kleene star is equivalent to arithmetic. The Kleene star is definable with union and concatenation, hence, the previous theory is equivalent to arithmetic also.

Keywords: Regular languages · Theory · Union · Concatenation · Kleene star · Quantifier elimination · Arithmetic · Undecidability

1 Introduction

One of the most important problems in mathematical logic is to study algorithmic decidability properties of different theories. Some classical results in this area are undecidability of Peano arithmetic, of group and semigroup theories etc. Examples of decidable theories are Presburger arithmetic, abelian groups theory, Boolean algebras theory etc. Some more recent results are connected to word theories. In [3,4] it was proved that the words theory over two-symbol alphabet with concatenation is essentially undecidable. In [10,11] it was proved that a variant of Robinson arithmetic is interpretable is this theory.

In this article we study some variants of regular languages theory. This interest is due to importance of such languages for formal linguistic and automata theory. Also regular languages have a lot of practical applications in compiler design, text processing algorithms, and many others. Each regular language has good decidability properties itself, many natural problems are decidable for them: the equivalence, the membership, the emptiness, the infinity problem etc.

But what about the set of all regular languages? If we consider the set of all regular languages (over fixed alphabet) with language-wide operations as a universe then analogous decidability questions naturally appear. With correspondence $w \leftrightarrow \{w\}$ it is easy to reduce concatenation theory of words to concatenation theory of languages. Hence, the last is undecidable.

© Springer Nature Switzerland AG 2019
R. van Bevern and G. Kucherov (Eds.): CSR 2019, LNCS 11532, pp. 119–130, 2019.
https://doi.org/10.1007/978-3-030-19955-5_11

In our paper we consider regular languages theory with different sets of classical operations. Section 2 of this article contains some basic definitions. In Sect. 3 we study the theory T_1 of regular languages with the Kleene star operation only. We prove that T_1 admits effective quantifier elimination, hence, it is decidable. In Sect. 4 we consider the theory T_2 of regular languages over one-symbol alphabet with union and concatenation. It is proved that T_2 is both Σ_1- and Π_1-hard. In Sect. 5 we study the theory T_3 of regular languages over one-symbol alphabet with union and the Kleene star. We establish that elementary arithmetic can be interpreted in T_3. Also we prove that theory T_4 of regular languages with all operations (union, concatenation, Kleene star) can be interpreted in arithmetic. Thus, the theories T_3, T_4 and arithmetic are algorithmically equivalent. The Kleene star can be expressed via union and concatenation, therefore, T_2 is equivalent to arithmetic also.

2   Preliminaries

An alphabet is a finite set of symbols. A word over an alphabet Σ is a finite sequence of symbols from Σ. The length of w is denoted $|w|$. The empty word is a word of zero length, it is denoted ε. Concatenation of two words u and v is a word which is obtained by appending v is the end of u. Concatenation of words u and v is denoted $u \cdot v$ or simply uv. The i-th power of the word w is the word $ww \ldots w$ where w is repeated i times. In particular, $w^0 = \varepsilon$ for every word w.

A language over an alphabet Σ is an arbitrary set of words over Σ. A union of two languages L_1 and L_2 is a usual union of the sets L_1 and L_2. Concatenation of languages L_1 and L_2 is the language $L_1 \cdot L_2 = \{ uv : u \in L_1, v \in L_2 \}$. A Kleene star of the language L is the language L^* consisting of concatenations of all possible sequences of words from L, i.e. $L^* = \{ w_1 w_2 \ldots w_n : n \geq 0, w_i \in L$ for every $i \}$. Since the case $n = 0$ is possible so L^* always contains the empty word ε. Obviously, $(L^*)^* = L^*$. L^* is finite if and only if $L = \emptyset$ or $L = \{ \varepsilon \}$, in both cases $L^* = \{ \varepsilon \}$.

The language L over the alphabet $\Sigma = \{ a_1, a_2, \ldots, a_n \}$ is regular if it can be obtained from the languages $\emptyset$, $\{ \varepsilon \}$, $\{ a_1 \}, \ldots, \{ a_n \}$ in finitely many steps with union, concatenation, and the Kleene star (for precise definitions see [1]). A deterministic finite automaton (DFA) is a quintuple $M = (Q, \Sigma, \delta, q_0, F)$, where Q is a finite set of states, Σ is an input alphabet, $q_0 \in Q$ is an initial state, $F \subseteq Q$ is a set of final states, and $\delta : Q \times \Sigma \to Q$ is a program. A word $w = a_1 a_2 \ldots a_n$ is accepted by a DFA M if there exists a sequence of states $q_0, q_1, \ldots, q_n$ such that $q_n \in F$, $\delta(q_i, a_{i+1}) = q_{i+1}$ for every $0 \leq i < n$. A DFA M recognizes a language of all words accepted by M. It is known (see [1,5,6]) that the language L is regular if and only if it is accepted by some DFA.

A theory T is a set of first-order formulas closed under logic inference. Formulas φ and ψ are equivalent in the theory T if $(\varphi \leftrightarrow \psi) \in T$. This is denoted $\varphi \equiv_T \psi$. A theory T admits quantifier elimination if for every formula φ there exists a quantifier-free formula $\psi \equiv_T \varphi$. For quantifier elimination it is enough to eliminate an existential quantifier from all formulas of kind $(\exists x)\varphi$ where

φ is an elementary conjunction (see [2]). A n-ary relation P is definable in a theory T if there exists a formula φ such that φ does not contain P and $P(x_1, \ldots, x_n) \equiv_T \varphi(x_1, \ldots, x_n)$. Similarly, the n-ary function f is definable in the theory T if there exists a formula φ such that φ does not contain f and $f(x_1, \ldots, x_n) = y \equiv_T \varphi(x_1, \ldots, x_n, y)$. The theory of a structure $\mathfrak{A}$ is the set of all formulas which are true in $\mathfrak{A}$.

3   Decidability of Regular Languages Theory with the Kleene Star

In this section we study the regular languages theory T_1 over an arbitrary alphabet $\Sigma = \{a_1, \ldots, a_n\}$ with constant $\emptyset^{(0)}$ (empty language) and operation $*^{(1)}$ (the Kleene star).

Let us note some properties of the Kleene star.

Lemma 1. *(1) There exists a language L_1 such that $L = L_1^*$ if and only if $L = L^*$.*
(2) If $L = L^*$ and $L \neq \{\varepsilon\}$ then there exist infinitely many languages L_i such that $L_i^* = L$.
(3) There exist infinitely many languages which are Kleene stars.
(4) There exists an infinite family of languages L_i such that $L_i^* \neq L_i$ and $L_i^* \neq L_j^*$ for all $i \neq j$.

Proof. (1) If $L_1^* = L$ then $L^* = (L_1^*)^* = L_1^* = L$.
(2) Let w be an arbitrary nonempty word from L. Let $L_i = L \setminus \{w^i\}$ for $i \geq 2$. Then $L_i^* = L$, and all languages L_i are different.
(3) All languages $\{a_1^i\}^*$ are different.
(4) Let $L_i = \{w \in \Sigma^* : |w| \geq i\}$, $i \geq 1$. All these languages are different, $L_i \neq L_i^*$ because $\varepsilon \in L_i^* \setminus L_i$. If $i < j$ then the shortest word from L_j is longer than the shortest word form L_i. Therefore, $L_i^* \neq L_j^*$. $\square$

Theorem 1. *The theory T_1 admits quantifier elimination.*

Proof. It is enough to eliminate a quantifier from $(\exists x)\varphi$, where the elementary conjunction φ can contain formulas of the forms $x = t$, $x^* = t$, $x = x^*$, $x \neq t$, $x^* \neq t$, and $x \neq x^*$. Until the proof end symbols r, s, t (possibly with indices) denote terms without the variable x. We suppose that φ has no terms of the form $(y^*)^*$ because $(y^*)^* = y^*$. We consider several possible cases.

Case 1. If φ is $x = t \wedge \varphi'$ then $(\exists x)(x = t \wedge \varphi') \equiv_{T_1} (\varphi')_t^x$, where $(\varphi')_t^x$ is obtained from φ by replacing x with t. Therefore, we assume in the following that φ doesn't contain $x = t$.

Case 2. Let φ contain both $x = x^*$ and $x^* = t$: $(\exists x)(x = x^* \wedge x^* = t \wedge \varphi')$. Then

$$(\exists x)\varphi \equiv_{T_1} (\exists x)(x = t \wedge x = x^* \wedge x^* = t \wedge \varphi') \equiv_{T_1} t = t^* \wedge (\varphi')_t^x.$$

Case 3. Let φ contain $x^* = t$ but no $x = x^*$. Then $(\exists x)\varphi$ is

$$(\exists x)(x^* = t \wedge \bigwedge_{i=1}^{k} x^* = t_i \wedge \bigwedge_{i=1}^{l} x \neq r_i \wedge \bigwedge_{i=1}^{m} x^* \neq s_i \wedge x \neq x^*).$$

If one of k, l, or m is zero then the corresponding conjunction is missing. The part $x \neq x^*$ may also be missing. Substituting t instead of x^* we have

$$(\exists x)\varphi \equiv_{T_1} (\exists x)(x^* = t \wedge \bigwedge_{i=1}^{p} x \neq r_i) \wedge \bigwedge_{i=1}^{k} t = t_i \wedge \bigwedge_{i=1}^{m} t \neq s_i.$$

Here $p = l$ if φ doesn't contain $x \neq x^*$, and $p = l + 1$, $r_{l+1} = t$ otherwise.

If $t^* = t$ and $t \neq \{\varepsilon\}$ then by Lemma 1 t is a Kleene star result for infinitely many languages. Hence, t is a Kleene star of some language rather than $r_1, \ldots, r_p$, and the conditions $x \neq r_i$ may be omitted. Hence, $(\exists x)x^* = t \equiv_{T_1} t = t^*$ by Lemma 1.

If $t = \{\varepsilon\}$ then x can be only either $\emptyset$ or $\{\varepsilon\}$. Then either all languages $r_1, \ldots, r_p$ must be different from $\emptyset$, or they must be different from $\{\varepsilon\}$.

Combining both variants we obtain the following equivalence:

$$(\exists x)\varphi \equiv_{T_1} \left(t = \emptyset^* \rightarrow \left(\bigwedge_{i=1}^{p} r_i \neq \emptyset \vee \bigwedge_{i=1}^{p} r_i \neq \emptyset^* \right) \right) \wedge t = t^* \wedge \bigwedge_{i=1}^{k} t = t_i \wedge \bigwedge_{i=1}^{m} t \neq s_i.$$

Case 4. Let φ contain subformula $x = x^*$ but no $x^* = t$: $(\exists x)\varphi$ is

$$(\exists x)(x = x^* \wedge \bigwedge_{i=1}^{k} x \neq s_i \wedge \bigwedge_{i=1}^{l} x^* \neq r_i).$$

Substituting x instead of x^* we obtain

$$(\exists x)\varphi \equiv_{T_1} (\exists x)(x = x^* \wedge \bigwedge_{i=1}^{k} x \neq s_i \wedge \bigwedge_{i=1}^{l} x \neq r_i).$$

By Lemma 1 there are infinitely many languages which are Kleene stars, hence, the last formula is true in T_1.

Case 5. Finally, let us suppose that φ contains only inequalities: $(\exists x)\varphi$ is

$$(\exists x)(\bigwedge_{i=1}^{k} x \neq r_i \wedge \bigwedge_{i=1}^{l} x^* \neq s_i \wedge x \neq x^*).$$

Again $x \neq x^*$ may be missing. In this case $(\exists x)\varphi$ is true also due to Lemma 1: there are infinitely many languages which are not Kleene stars. $\qquad\square$

Corollary 1. *The theory T_1 is decidable.*

4   Undecidability of Regular Languages Theory with Union and Concatenation

In this section we study the regular languages theory T_2 over the alphabet $\{0\}$ with a constant $0^{(0)}$ (the language $\{0\}$), operations $+^{(2)}$ (union) and $\cdot^{(2)}$ (concatenation). The following relations and functions are definable in T_2 with $+$:

- inclusion: $x \subseteq y \equiv_{T_2} x + y = y$;
- intersection: $x \cap y = z \equiv_{T_2} z \subseteq x \wedge z \subseteq y \wedge (\forall u)((u \subseteq x \wedge u \subseteq y) \rightarrow u \subseteq z)$;
- the empty language $\emptyset$: $x = \emptyset \equiv_{T_2} (\forall y)x + y = y$;
- the set of all words over alphabet $\{0\}$: $x = 0^* \equiv_{T_2} (\forall y)x + y = x$;
- complement $\bar{x}$: $\bar{x} = y \equiv_{T_2} x + y = 0^* \wedge x \cap y = \emptyset$;
- subtraction: $x \setminus y = z \equiv_{T_2} x \cap \bar{y} = z$;
- cardinality test $\mathrm{Card}_k(x)$ for every fixed natural k. The formula $\mathrm{Card}_k(x)$ says that the language x contains exactly k words:

$$\mathrm{Card}_1(x) \equiv_{T_2} x \neq \emptyset \wedge (\forall y)(y \subseteq x \rightarrow (y = \emptyset \vee y = x));$$
$$\mathrm{Card}_{k+1}(x) \equiv_{T_2} (\exists y)(y \subseteq x \wedge \mathrm{Card}_k(x \setminus y) \wedge \mathrm{Card}_1(y)).$$

The constant ε denoting the language $\{\varepsilon\}$ can be defined with concatenation:

$$x = \varepsilon \equiv_{T_2} (\forall y)x \cdot y = y.$$

Let the relation $\mathrm{Mult}_k(x, y)$ denote the following: the language x contains one word 0^n, $n \neq 0$, and y contains one word 0^{kn} for some natural n. This relation is definable in T_2 as follows:

$$\mathrm{Mult}_k(x, y) \equiv_{T_2} \mathrm{Card}_1(x) \wedge x \neq \varepsilon \wedge y = \underbrace{x \cdot x \cdot \ldots \cdot x}_{k \text{ times}}.$$

We prove undecidability of T_2 by reducing to it the halting problem for two-counter machines. A two-counter machine (see [7]) has a finite set of states $Q = \{q_0, q_1, \ldots, q_{n-1}\}$, an initial state q_0, a final state q_f, two counters a and b, and an instruction set which are of the form:

- the increment instruction $q_i\ c = c + 1$; q_j means that in the state q_i the machine increases its counter c by one and moves to the state q_j;
- the decrement instruction q_i if $c \neq 0$ $then$ $c = c - 1$; $q_j\, else\, q_k$ means that in the state q_i the machine decreases its counter c by one and moves to the state q_j if $c > 0$, otherwise it does not change c and moves to the state q_k.

Configuration of a counter machine is a triple (q_i, a, b) where q_i is a state, a and b are naturals. $(q_i, a, b) \vdash_M (q_j, a', b')$ means that the machine moves from the configuration (q_i, a, b) to the configuration (q_j, a', b') in one step. $\vdash_M^*$ denotes reflexive and transitive closure of the relation $\vdash_M$. The machine M halts on an input (a, b) if $(q_0, a, b) \vdash_M^* (q_f, a', b')$ for some a' and b'. Remember that Σ_1 is a class of recursive enumerable sets and Π_1 is a class of its complements (see [8]). It is known (see [7]) that halting problem and its complement are respectively

Σ_1- and Π_1-hard for two-counter machines for initial configurations of the form $(q_0, 0, 0)$.

In the proof of the next theorem we show the method of modeling a counter machine using regular languages. In the next section we will prove stronger version of this theorem but the proof will be technically more difficult.

Theorem 2. *The theory T_2 is Σ_1-hard and Π_1-hard.*

Proof. We reduce the halting problem and its complement to T_2. Let M be any two-counter machine with the set of states $Q = \{q_0, q_1, \ldots, q_{m-1}\}$ and the set of instructions $P = \{p_1, \ldots, p_s\}$. Let us suppose that the state q_0 is initial and the state q_1 is final. We encode the configuration (q_i, a, b) by the word of length $2^i 3^a 5^b$.

Now for every l, $1 \le l \le s$, we construct a formula $\mathrm{Step}_l(x, y)$ expressing the following property:

- both languages x and y contain exactly one word each;
- if x contains the code of configuration c_1 then y contains the code of configuration c_2 such that $c_1 \vdash_M c_2$ according to the instruction p_l.

Note that we do not check whether x and y indeed contain the correct code because we cannot express the property that the word length is of the form $2^i 3^a 5^b$.

We use two auxiliary formulas $\mathrm{State}_i(x)$ and $\mathrm{ChangeState}_{i,j}(x, y)$. The formula $\mathrm{State}_i(x)$ says that the configuration of code x has the state q_i:

$$(\exists z)\mathrm{Mult}_{2^i}(z, x) \wedge \neg(\exists u)\mathrm{Mult}_{2^{i+1}}(u, x).$$

The formula $\mathrm{ChangeState}_{i,j}(x, y)$ says that x and y encode two equal configurations except the state, x has the state q_i and y has the state q_j:

$$(\exists z)(\mathrm{Mult}_{2^i}(z, x) \wedge \mathrm{Mult}_{2^j}(z, y)) \wedge \neg(\exists u)\mathrm{Mult}_{2^{i+1}}(u, x).$$

If $p_l \in P$ is q_i $a = a + 1$; q_j then we construct a formula $\mathrm{Step}_l(x, y)$ as

$$(\exists z)(\mathrm{ChangeState}_{i,j}(x, z) \wedge \mathrm{Mult}_3(z, y)).$$

If $p_l \in P$ is q_i *if* $a \ne 0$ *then* $a = a - 1$; q_j *else* q_k then a formula $\mathrm{Step}_l(x, y)$ is

$$\Big(\neg(\exists z)\mathrm{Mult}_3(z, x) \rightarrow \mathrm{ChangeState}_{i,k}(x, y)\Big) \wedge$$

$$\wedge \Big((\exists z)\mathrm{Mult}_3(z, x) \rightarrow (\exists u)(\mathrm{Mult}_3(u, x) \wedge \mathrm{ChangeState}_{i,j}(u, y))\Big).$$

For the counter b formulas $\mathrm{Step}_l(x, y)$ are obtained by replacing 3 with 5.

The formula $\mathrm{Step}(x, y)$ says that the machine M moves in one step from the configuration of code x to the configuration of code y:

$$\bigvee_{l=1}^{s} \mathrm{Step}_l(x, y).$$

The formula $\text{Closed}(x)$ says: if the language x contains the code of a configuration c then it contains the codes of all configurations reachable from c. I.e. $\text{Closed}(x)$ is:

$$(\forall u)(\forall v)((u \subseteq x \wedge \text{Step}(u, v)) \rightarrow v \subseteq x).$$

The formula $\text{Reachable}(x)$ says that the language x contains the word 0 encoding the initial configuration $(q_0, 0, 0)$, the codes of all configurations reachable from $(q_0, 0, 0)$, and no other word, i.e. that x is the smallest closed language containing 0:

$$0 \subseteq x \wedge \text{Closed}(x) \wedge (\forall y)((0 \subseteq y \wedge \text{Closed}(y)) \rightarrow x \subseteq y).$$

Now we can describe the reductions. Let φ and ψ be the next formulas correspondingly:

$$(\exists x)(\text{Reachable}(x) \wedge (\exists y)(y \subseteq x \wedge \text{State}_1(y))),$$
$$(\exists x)(\text{Reachable}(x) \wedge \neg(\exists y)(y \subseteq x \wedge \text{State}_1(y))).$$

The formula φ says that the final configuration is reachable from $(q_0, 0, 0)$, and ψ says that it is not reachable. Then $\varphi \in T_2$ if and only if M halts on $(q_0, 0, 0)$, and $\psi \in T_2$ if and only if M does not halt on $(q_0, 0, 0)$. Therefore, the theory T_2 is both Σ_1- and Π_1-hard. □

5 Undecidability of Regular Languages Theory with Union and the Kleene Star

In this section we study the regular languages theory T_3 over the alphabet $\{0\}$ with the constant $0^{(0)}$ (the language $\{0\}$) and operations $+^{(2)}$ (union) and $*^{(1)}$ (the Kleene star).

Since the signature of T_3 contains union, all relations and functions from the beginning of the previous section are also definable in T_3, because concatenation wasn't used. The constant ε denoting the language $\{\varepsilon\}$ can be defined with Kleene star as $\emptyset^*$.

The following lemma is evident.

Lemma 2. *If the relation* $\text{Mult}_{k_i}(x, y)$ *is definable in* T_3 *for naturals* $k_1, \ldots, k_n$ *then* $\text{Mult}_m(x, y)$ *is also definable for the product* $m = k_1 \ldots k_n$.

Our main technical result is the following.

Lemma 3. *The relations* $\text{Mult}_k(x, y)$ *are definable in* T_3 *for* $k = 2, 3, 5, 7$.

Proof. Let the relation $\text{Mult}_{k_1, \ldots, k_m}(x, y)$ mean that the language x contains a single word 0^p, and the language y contains exactly the words $0^{k_1 p}, \ldots, 0^{k_m p}$ for some natural $p \neq 0$. Obviously, if $\text{Mult}_{k_1}(x, y), \ldots, \text{Mult}_{k_m}(x, y)$ are definable, then $\text{Mult}_{k_1, \ldots, k_m}(x, y)$ is also definable:

$$(\exists z_1) \ldots (\exists z_m)(\text{Mult}_{k_1}(x, z_1) \wedge \cdots \wedge \text{Mult}_{k_m}(x, z_m) \wedge y = z_1 + \cdots + z_m).$$

The formula $\mathrm{Mult}_{2,3}(x, y)$ can be defined as

$$\mathrm{Card}_1(x) \wedge \mathrm{Card}_2(y) \wedge x^* \setminus x = y^*.$$

Indeed, let $L = \{0^p\}$ where p is some natural, $L_1 = L^* \setminus L$. Then $L_1 = \{\varepsilon, 0^{2p}, 0^{3p}, 0^{4p}, \dots\} = \{0^{2p}, 0^{3p}\}^*$. Thus, L_1 is a Kleene star of some language containing exactly two words. On the other hand, every such language must contain both words 0^{2p} and 0^{3p} because they cannot be obtained from shorter words.

Now we define $\mathrm{Mult}_2(x, y)$ and $\mathrm{Mult}_3(x, y)$ correspondingly:

$$\mathrm{Card}_1(y) \wedge (\exists z)(\mathrm{Mult}_{2,3}(x, z) \wedge y \subseteq z) \wedge \neg(\exists u)(\mathrm{Card}_2(u) \wedge x^* \setminus (x + y) = u^*);$$

$$\mathrm{Card}_1(y) \wedge (\exists z)(\mathrm{Mult}_{2,3}(x, z) \wedge y \subseteq z) \wedge \neg\mathrm{Mult}_2(x, y).$$

If we remove 0^{3p} from L_1 then we get $L_2 = \{\varepsilon, 0^{2p}, 0^{4p}, 0^{5p}, 0^{6p}, \dots\}$. This language is a Kleene star of $\{0^{2p}, 0^{5p}\}$. Removing 0^{2p} from L_1 we get $L_3 = \{\varepsilon, 0^{3p}, 0^{4p}, 0^{5p}, 0^{6p}, \dots\}$. If L_3 is a Kleene star of some language L_3' then L_3' must contain at least three words $0^{3p}, 0^{4p}, 0^{5p}$ because neither of these words can be represented as concatenation of two shorter nonempty words from L_3.

The construction of $\mathrm{Mult}_k(x, y)$ for $k = 5, 7$ follows the same idea. The formula $\mathrm{Mult}_{5,7}(x, y)$ is

$$(\exists z)(\exists u)(\exists v)(\mathrm{Mult}_{2,3,4}(x, u) \wedge \mathrm{Mult}_{6,8,9}(x, v) \wedge$$
$$\wedge \mathrm{Card}_5(z) \wedge x^* \setminus (x + u) = z^* \wedge y = z \setminus v).$$

Here z must be $\{0^{5p}, 0^{6p}, 9^{7p}, 0^{8p}, 0^{9p}\}$ and $y = \{0^{5p}, 0^{7p}\}$. Then we use the equality $\{\varepsilon, 0^{5p}, 0^{6p}, 0^{8p}, 0^{9p}, \dots\} = \{0^{5p}, 0^{6p}, 0^{8p}, 0^{9p}\}^*$. On the other hand, any language L_4 such that $L_4^* = \{\varepsilon, 0^{6p}, 0^{7p}, 0^{8p}, 0^{9p}, \dots\}$ must include at least $\{0^{6p}, 0^{7p}, 0^{8p}, 0^{9p}, 0^{10p}, 0^{11p}\}$:

$$\mathrm{Mult}_5(x, y) \equiv_{T_3} \mathrm{Card}_1(y) \wedge (\exists z)(\exists u)(\mathrm{Mult}_{5,7}(x, z) \wedge \mathrm{Mult}_{2,3,4}(x, u) \wedge$$
$$\wedge y \subseteq z \wedge \neg(\exists v)(\mathrm{Card}_4(v) \wedge x^* \setminus (x + u + y) = v^*)),$$

$$\mathrm{Mult}_7(x, y) \equiv_{T_3} \mathrm{Card}_1(y) \wedge (\exists z)(\exists u)(\mathrm{Mult}_{5,7}(x, z) \wedge \mathrm{Mult}_{2,3,4}(x, u) \wedge$$
$$\wedge y \subseteq z \wedge (\exists v)(\mathrm{Card}_4(v) \wedge x^* \setminus (x + u + y) = v^*). \qquad \square$$

Corollary 2. *The constants 0^2, 0^3, 0^5, 0^7 are definable in T_3.*

Lemma 4. *The following relation $\mathrm{Join}(x, y, z)$ is definable in T_3: x, y, and z contain one word each, and if $x = \{0^p\}$ and $y = \{0^q\}$ for some co-primes p and q then $z = \{0^{pq}\}$.*

Proof. Let us define $\mathrm{Join}(x, y, z)$ as

$$\mathrm{Card}_1(x) \wedge \mathrm{Card}_1(y) \wedge \mathrm{Card}_1(z) \wedge z^* = x^* \cap y^*.$$

If $L_1 = \{0^p\}$ and $L_2 = \{0^q\}$ then $L_1^* \cap L_2^* = \{0^{pqi} : i \geq 0\}$ because p and q are co-prime. Therefore, $L_1^* \cap L_2^* = \{0^{pq}\}^*$. $\qquad \square$

Theorem 3. *For every arithmetical formula φ one can construct a formula ψ such that φ is true in arithmetic if and only if $\psi \in T_3$.*

Proof. It is known (see [9]) that two numbers can be multiplied by a counter machine with only three counters. We use next three-counters machines:

- M_1 moves a number from the first counter into the second one:

$$(q_0, a, 0, 0) \vdash^* (q_1, 0, a, 0);$$

- M_2 computes the sum of the first two counters:

$$(q_0, a, b, 0) \vdash^* (q_1, a + b, 0, 0);$$

- M_3 computes the product of the first two counters:

$$(q_0, a, b, 0) \vdash^* (q_1, ab, 0, 0).$$

Like in the proof of Theorem 2 we can construct the formulas $\mathrm{Closed}_i(x)$ saying the following: if $0^c \in x$ where $c = 2^j 3^a 5^b 7^d$ is code of M_i configuration (q_j, a, b, d) then x contains codes of all reachable configurations. Concatenation was used only to construct the formulas $\mathrm{Mult}_k(x, y)$. But since each M_i has only three counters we only need the formulas $\mathrm{Mult}_k(x, y)$ where k is either power of 2 or one of 3, 5, 7. All such relations are definable in T_3 by Lemmas 2 and 3.

The formula $\mathrm{Primes}_{p_1, \ldots, p_m}(x)$ says that $x = \{\, 0^n \,\}$ where $n = p_1^{\alpha_1} \ldots p_m^{\alpha_m}$ for some primes $p_1, \ldots, p_m \in \{\, 2, 3, 5, 7 \,\}$:

$$\mathrm{Primes}_{p_1, \ldots, p_m}(x) \equiv_{T_3} x = 0 \vee \Big(x \neq \varepsilon \wedge \mathrm{Card}_1(x) \wedge$$

$$\wedge (\forall y)((\mathrm{Card}_1(y) \wedge x \subseteq y^*) \rightarrow (y = 0 \vee y \subseteq (0^{p_1})^* \vee \cdots \vee y \subseteq (0^{p_m})^*)) \Big).$$

In particular, the formula $\mathrm{Primes}_p(x)$ says $x = \{\, 0^{p^\alpha} \,\}$ for some α.

The formula $\mathrm{Conf}(x)$ says that $x = \{\, 0^c \,\}$ where c encodes some configuration:

$$\mathrm{Conf}(x) \equiv_{T_3} \mathrm{Primes}_{2,3,5,7}(x).$$

The formula $\mathrm{InitConf}_1(x) \equiv_{T_3} \mathrm{Primes}_3(x)$ says that $x = \{0^c\}$ where c encodes an initial configuration of M_1, $\mathrm{InitConf}_2(x) \equiv_{T_3} \mathrm{InitConf}_3(x) \equiv_{T_3} \mathrm{Primes}_{3,5}(x)$ says that $x = \{\, 0^c \,\}$ where c encodes an initial configuration of M_2 or M_3.

Analogously formulas $\mathrm{FinalConf}_i(x)$ says that x encodes a final configuration: $\mathrm{State}_1(x) \wedge \mathrm{Primes}_{2,5}(x)$ and $\mathrm{State}_1(x) \wedge \mathrm{Primes}_{2,3}(x)$ correspondingly.

The formula $\mathrm{Result}_i(x, y)$ says that x encodes an initial configuration of M_i and y encodes a corresponding final configuration:

$$\mathrm{InitConf}_i(x) \wedge \mathrm{FinalConf}_i(y) \wedge (\forall u)((x \subseteq u \wedge \mathrm{Closed}_i(u)) \rightarrow y \subseteq u).$$

We represent a natural n by $\{0^{3^n}\}$. The formula $\mathrm{Add}(x, y, z)$ says that the language z represents the sum of numbers represented by x and y:

$$\mathrm{Primes}_3(x) \wedge \mathrm{Primes}_3(y) \wedge (\exists u)(\exists v)(\exists w)(\exists t)(\mathrm{Result}_1(y, u) \wedge$$
$$\wedge \mathrm{Mult}_2(v, u) \wedge \mathrm{Join}(x, v, w) \wedge \mathrm{Result}_2(w, t) \wedge \mathrm{Mult}_2(y, t)).$$

Let $x = \{0^{3^a}\}$ and $y = \{0^{3^b}\}$. Then $u = \{0^{2 \times 5^b}\}$, $v = \{0^{5^b}\}$, $w = \{0^{3^a 5^b}\}$, $t = \{0^{2 \times 3^{a+b}}\}$, and $y = \{0^{3^{a+b}}\}$.

Using the same method we can construct the formula $\mathrm{Mult}(x, y, z)$ for multiplication:

$$\mathrm{Primes}_3(x) \wedge \mathrm{Primes}_3(y) \wedge (\exists u)(\exists v)(\exists w)(\exists t)(\mathrm{Result}_1(y, u) \wedge$$
$$\wedge \mathrm{Mult}_2(v, u) \wedge \mathrm{Join}(x, v, w) \wedge \mathrm{Result}_3(w, t) \wedge \mathrm{Mult}_2(y, t)).$$

Now we can describe reduction from arithmetic to T_3. Let φ be an arbitrary arithmetical formula. We may suppose that all atomic subformulas of φ are of the forms $x + y = z$ or $x \times y = z$ where x, y, z are variables. For every arithmetical formula θ we construct its translation $\mathrm{T}(\theta)$ by induction:

- $\mathrm{T}(x + y = z)$ is $\mathrm{Add}(x, y, z)$;
- $\mathrm{T}(x \times y = z)$ is $\mathrm{Mult}(x, y, z)$;
- $\mathrm{T}(\chi \circ \theta)$ is $\mathrm{T}(\chi) \circ \mathrm{T}(\theta)$ for $\circ \in \{\wedge, \vee, \rightarrow\}$;
- $\mathrm{T}(\neg\theta)$ is $\neg\mathrm{T}(\theta)$;
- $\mathrm{T}((\exists x)\theta)$ is $(\exists x)(\mathrm{Primes}_3(x) \wedge \mathrm{T}(\theta))$;
- $\mathrm{T}((\forall x)\theta)$ is $(\forall x)(\mathrm{Primes}_3(x) \rightarrow \mathrm{T}(\theta))$.

Let ψ be $\mathrm{T}(\varphi)$. Then φ is true in arithmetic if and only if $\psi \in T_3$.   □

Finally, let us consider the regular languages theory T_4 over an arbitrary alphabet $\Sigma = \{a_1, \ldots, a_n\}$ with constants $a_1^{(0)}, \ldots, a_n^{(0)}$ and operations $+^{(2)}$, $\cdot^{(2)}$, and $*^{(1)}$. The constant a_i is interpreted as the language $\{a_i\}$, other symbols have previous meaning.

Theorem 4. *For every formula φ one can construct an arithmetical formula ψ such that $\varphi \in T_4$ if and only if ψ is true in arithmetic.*

Proof. We suppose that all atomic subformulas of φ are of the forms $x + y = z$, $x \cdot y = z$, $x^* = y$, or $x = a_i$.

Let us fix some "reasonable" numeration of all DFA and let M_n be a DFA of number n. We may suppose that for every number n there exists a DFA M with the number n. There are algorithms which given DFA M_1 and M_2 construct DFA recognizing the languages $L(M_1) \cup L(M_2)$, $L(M_1) \cdot L(M_2)$, and $L(M_1)^*$ (see [1,5]). Then the relations $L(M_x) \cup L(M_y) = L(M_z)$, $L(M_x) \cdot L(M_y) = L(M_z)$, and $L(M_x)^* = L(M_y)$ are recursive. But every recursive relation is representable in arithmetic (see [2]). Let corresponding formulas be $\mathrm{Union}(x, y, z)$,

Concat(x, y, z), and Star(x, y). Let n_i be the number of some DFA recognizing $\{a_i\}$. For every formula φ we construct its translation by induction:

- $T(x = a_i)$ is $x = \underbrace{1 + \cdots + 1}_{n_i \text{ times}}$;
- $T(x + y = z)$ is Union(x, y, z);
- $T(x \cdot y = z)$ is Concat(x, y, z);
- $T(x^* = y)$ is Star(x, y);
- $T(\chi \circ \theta)$ is $T(\chi) \circ T(\theta)$ for $\circ \in \{\wedge, \vee, \rightarrow\}$;
- $T(\neg\theta)$ is $\neg T(\theta)$;
- $T((\exists x)\theta)$ is $(\exists x)T(\theta)$;
- $T((\forall x)\theta)$ is $(\forall x)T(\theta)$.

Let ψ be $T(\varphi)$. Then $\varphi \in T_4$ if and only if ψ is true in arithmetic. $\qquad\square$

Corollary 3. *The theories T_3, T_4 and arithmetic are recursively isomorphic.*

Corollary 4. *The theory T_2 and arithmetic are recursively isomorphic.*

Proof. It is enough to define the Kleene star with union and concatenation:

$$x^* = y \equiv_{T_4} x \subseteq y \wedge y \cdot y = y \wedge (\forall z)((x \subseteq z \wedge z \cdot z = z) \rightarrow y \subseteq z). \qquad\square$$

Corollary 5. *The theories T_2, T_3, and T_4 are undecidable.*

6   Conclusion

We have considered classical operations on languages: union, concatenation and the Kleene star. We have proved that the regular language theory is decidable with the Kleene star only but undecidable with union and other operations.

Some interesting questions remain open.

- What is the exact place of the regular languages theory with concatenation only in arithmetical hierarchy?
- Does an answer to the previous question depend on the alphabet size?
- Is the regular languages theory with union only decidable?
- Are there other operations on regular languages with decidable theory?

References

1. Aho, A.V., Ullman, J.D.: The Theory of Parsing, Translation and Compiling. Volume 1: Parsing. Prentice-Hall Inc., Englewood Cliffs (1972)
2. Boolos, G.S., Burgess, J.P., Jeffrey, R.C.: Computability and Logic, 5th edn. Cambridge University Press, New York (2007)
3. Grzegorczyk, A.: Undecidability without arithmetization. Stud. Logica **79**(2), 163–230 (2005)

4. Grzegorczyk, A., Zdanowski, K.: Undecidability and concatenation. In: Ehrenfeucht, A., Marek, V.W., Srebrny, M. (eds.) Andrzej Mostowski and Foundational Studies, pp. 72–91. IOS Press, Amsterdam (2008)
5. Hopcroft, J.E., Motwani, R., Ullman, J.D.: Introduction to Automata Theory, Languages, and Computation, 3rd edn. Pearson, Harlow (2013)
6. Kleene, S.C.: Representation of events in nerve nets and finite automata. In: Shannon, C., McCarthy, J. (eds.) Automata Studies, pp. 3–42. Princeton University Press, Princeton (1951)
7. Minsky, M.L.: Computation: Finite and Infinite Machines. Prentice-Hall, Inc., Englewood Cliffs (1967)
8. Rogers, H.: Theory of Recursive Functions and Effective Computability. McGraw-Hill Education, New York (1967)
9. Schroeppel, R.: A Two-Counter Machine Cannot Calculate 2^N. Technical Report 257, Massachusetts Institute of Technology, A. I. Laboratory (1973)
10. Švejdar, V.: On interpretability in the theory of concatenation. Notre Dame J. Formal Logic $50(1)$, 87–95 (2009)
11. Visser, A.: Growing commas. A study of sequentiality and concatenation. Notre Dame J. Formal Logic $50(1)$, 61–85 (2009)

Minimizing Branching Vertices in Distance-Preserving Subgraphs

Kshitij Gajjar[✉] and Jaikumar Radhakrishnan

Tata Institute of Fundamental Research, Mumbai, India
{kshitij.gajjar,jaikumar}@tifr.res.in

Abstract. It is NP-hard to determine the minimum number of branching vertices needed in a single-source distance-preserving subgraph of an undirected graph. We show that this problem can be solved in polynomial time if the input graph is an interval graph.

In earlier work, it was shown that every interval graph with k terminal vertices admits an all-pairs distance-preserving subgraph with $O(k \log k)$ branching vertices [13]. We extend this result to bi-interval graphs; these are graphs that can be expressed as the strong product of two interval graphs. We present a polynomial time algorithm that takes a bi-interval graph with k terminal vertices as input, and outputs an all-pairs distance-preserving subgraph of it with $O(k^2)$ branching vertices. This bound is tight.

1 Introduction

Distance-preserving minors were introduced by Krauthgamer and Zondiner [19]. They showed that every undirected graph with k terminals admits a distance-preserving minor with $O(k^4)$ vertices and edges.

Distance-preserving subgraphs are closely related to distance-preserving minors. In fact, for many graph classes, distance-preserving minors are obtained by first constructing distance-preserving subgraphs and then contracting edges adjacent to vertices of degree 2.

Definition 1 (Distance-preserving subgraph). *Given an undirected graph* $G = (V, E)$ *and two subsets of vertices* $S \subseteq V, T \subseteq V$, *we say that a subgraph* $H(V, E')$ *of* G *is distance-preserving for* (G, S, T) *if for all pairs of vertices* $(u, v) \in S \times T$, *we have* $d_G(u, v) = d_H(u, v)$, *where* d_G *and* d_H *denote the distances in* G *and* H *respectively.*

If $S = T$, *then the distance-preserving subgraph is called an all-pairs distance-preserving subgraph. If* $|S| = 1$, *then the distance-preserving subgraph is called a* single-source *distance-preserving subgraph. In both cases, the vertices of* T *are called* terminals.

K. Gajjar—Supported by a DAE scholarship.

R. van Bevern and G. Kucherov (Eds.): CSR 2019, LNCS 11532, pp. 131–142, 2019.
https://doi.org/10.1007/978-3-030-19955-5_12

In such cases, the size of the distance-preserving minor depends solely on the number of vertices of degree 3 or more in the distance-preserving subgraph. We call such vertices *branching vertices*. It is therefore natural to minimize the number of branching vertices while constructing distance-preserving subgraphs. An *optimal* distance-preserving subgraph is one which has the minimum number of branching vertices.

The $O(k^4)$ upper bound of Krauthgamer and Zondiner is obtained through distance-preserving subgraphs. In subsequent work, Krauthgamer, Nguyên and Zondiner showed that this bound is tight by exhibiting a planar graph on k terminals for which every distance-preserving subgraph has $\Omega(k^4)$ branching vertices [18, Section 5].

Gajjar and Radhakrishnan [13] study the problem of constructing distance-preserving subgraphs of interval graphs and showed that every interval graph with k terminals has an optimal all-pairs distance-preserving subgraph with $O(k \log k)$ branching vertices. They also showed that this bound is tight. We later present a setting (borrowed from [13, Section 1.2]) in which this problem may be applied to interval graphs.

In this work we consider the algorithmic version of this problem; namely the efficiency of computing optimal distance-preserving subgraphs. This problem was already shown to be NP-complete for general graphs [13]. We observe that the same proof shows that even the single-source version of the problem is NP-complete for general graphs (see Theorem 3). We show the following result.

Theorem 1 (Single-source interval graphs). *There exists a polynomial time algorithm that, given an interval graph G with a source s and k terminals $t_1, t_2, \ldots, t_k$ as input, constructs an optimal shortest path tree which preserves the distances between each (s, t_i) pair.*

In other words, there is a polynomial time algorithm that takes an interval graph as input and outputs an optimal single-source distance-preserving subgraph of it.

As stated above, the upper bound for interval graphs is significantly better than the lower bound for general graphs. We ask if similar better upper bounds can be established for super-classes of interval graphs. Since interval graphs are precisely the graphs of boxicity one, a natural candidate is graphs of boxicity two (intersection graphs of axis-parallel rectangles). However, it can be shown that there are graphs of boxicity two with k terminals for which every all-pairs distance-preserving subgraph requires $\Omega(k^4)$ branching vertices (see Theorem 4).

A strict sub-class of graphs of boxicity two is bi-interval graphs. These are intersection graphs of axis-parallel rectangles, where the rectangles arise from the cross product of two families of intervals (see Definition 3).

Theorem 2 (All-pairs bi-interval graphs).

(a) *(Upper bound) There is a polynomial time algorithm that, given a bi-interval graph with k terminals as input, produces an all-pairs distance-preserving subgraph of it with $O(k^2)$ branching vertices.*

(b) *(Lower bound) For every $k \geq 4$, there is a bi-interval graph G_{diag} on k terminals such that every all-pairs distance-preserving subgraph of G_{diag} has $\Omega(k^2)$ branching vertices.*

1.1 Related Work

The problem of constructing small distance-preserving subgraphs bears close resemblance to several well-studied problems in the area of graph algorithms: graph compression [10], graph spanners [1–3,6,23,26], Steiner point removal [11, 16,17], vertex sparsification [4,9,21], graph homeomorphisms [12,20], graph sparsification [15,25], graph contractions [7], etc.

Distance-preserving subgraphs are also used in several other scenarios. All of them have their own set of objectives that they are trying to optimize. To avoid confusing them with our work, we briefly mention them over here. Djoković [8] (see also Chepoi [5]) characterizes distance-preserving subgraphs of hypercubes, Nussbaum *et al.* [22] partition a graph into disjoint distance-preserving subgraphs for clustering purposes, Yan *et al.* [27] present algorithms for processing distance-preserving subgraph queries with applications to road networks, and Sadri *et al.* [24] provide a heuristic for constructing distance-preserving subgraphs and run an experiment using the New York City road network as an example.

1.2 Our Techniques

In this section, we explain the main ideas behind our algorithms and proofs.

Our first result concerns the construction of single-source distance-preserving subgraphs in interval graphs with the minimum number of branching vertices. Constructing a single-source shortest path tree in an interval graph is straightforward: the simple greedy algorithm does the job. However, this method is not guaranteed to produce a tree with the *minimum* number of branching vertices. We do not know of a simple modification of the greedy method that helps us minimize the number of branching vertices. We instead proceed as follows. Using a breadth-first search, we partition the graph into layers based on their distance from the source vertex. Within a layer, we order vertices based on the position of their respective intervals. We observe that with respect to this ordering an optimal solution can be assumed to have a very special structure: some vertices are directly connected by paths to the source, and the others via a single special vertex. Once this structure is established, a combination of network flows and dynamic programming allows us to determine the optimal solution. Since finding a single-source shortest path tree in interval graphs is simple, it is not clear why we need a somewhat sophisticated solution if we need to minimize the number of branching vertices. We describe our solution in detail assuming that the source vertex is the leftmost interval. This case already contains most of the ideas; the algorithm for the general case requires some more analysis.

Our second result concerns a 2D generalization of interval graphs. In earlier work [13], the worst-case number of branching vertices for distance-preserving subgraphs of interval graphs with k terminals was determined to be $\Theta(k \log k)$. In particular, the lower bound of $\Omega(k \log k)$ was obtained by considering certain interval graphs with regularly placed intervals. A natural 2D generalization of such graphs would be regularly placed rectangles, a special case of bi-interval graphs (Definition 4). We present an example (see [14, Section 3.6]) where such graphs with k terminals require $\Omega(k^2)$ branching vertices. To show that this is tight, we observe that shortest paths in such graphs can be constructed with two components: (i) ones that proceed *diagonally*, (ii) ones that proceed *horizontally* or *vertically* (we call such paths *straight* paths). We first generate the diagonal paths from all terminals; with some care one can argue that these paths meet only at $O(k^2)$ vertices. However, we need to complete these partial paths by adding straight segments. We observe that this second phase of our plan can be mapped to a certain problem for constructing distance-preserving subgraphs in interval graphs. By borrowing some ideas from earlier work [13], we show how these problems can be solved efficiently in such a way that the total number of branching vertices remains $O(k^2)$.

1.3   An Example

The following stylized example (taken from [13]) illustrates a setting where distance-preserving subgraphs of interval graphs arise naturally when considering a shipping problem.

A freight container needs to be delivered from seaport X to seaport Y, but there is no ship that travels from X to Y. In such cases, the container is typically first transported from port X to a central hub H, and transferred through a series of ships arriving at H until it is finally picked up by a ship that is destined for port Y. Thus, the container reaches its final destination via some "intermediate" ships at port H.[1]

However, there is a cost associated with transferring the container from one ship to another. There is also an added cost if an intermediate ship receives containers from multiple ships, or sends containers to multiple ships. Thus, given the docking times of ships at H, and a small subset of these ships that require a transfer of containers between each other, our goal is to devise a transfer strategy that meets the following objectives: (i) Minimize the number of transfers for each container; (ii) Minimize the number of ships that have to deal with multiple transfers.

Representing each ship's visit to the port as an interval on the time line, this problem can be modelled using distance-preserving subgraphs of interval graphs. In this setting, a shortest path from an earlier interval to a later interval corresponds to a valid sequence of transfers across ships that moves forward in time. Objective (i) corresponds to minimizing pairwise distances between terminals and objective (ii) corresponds to minimizing the number of branching vertices in the subgraph.

---

[1] The container cannot be left at the warehouse/storage unit of port H itself beyond a certain limited period of time.

In prior work [13], it was shown that $O(k \log k)$ branching vertices are sufficient for preserving pairwise distances between k terminals, and that this is the
best possible bound. In the next section, we focus on the problem of determining
this number exactly. For general graphs this problem is NP-complete, even when
restricted to one source. We do not know if the all-pairs version of the problem
is NP-complete for interval graphs. We show that the problem can be solved
efficiently for interval graphs when there is only one source.

2   Single-Source Distance-Preserving Subgraphs

In this section, we investigate the computational complexity of finding optimal distance-preserving subgraphs, or distance-preserving subgraphs with the
minimum number of branching vertices. Let $\mathcal{A}$ be the following decision
problem.

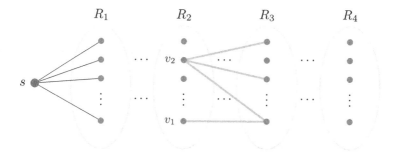

Fig. 1. An instance of G_{BFS}. The source vertex s is the leftmost interval in G. The
layer R_i is the set of vertices at a distance of i from s, and the vertices within each
layer are arranged in increasing order from bottom to top. For instance, v_1 is below
v_2 in R_2, which means that the right end point of the interval v_1 is to the left of the
right end point of the interval v_2. Thus, the neighbourhood of v_1 in R_3 is a subset of
the neighbourhood of v_2 in R_3.

Input: An unweighted graph G with k target vertices, one source vertex s, and
   a positive integer m.
Output: Yes, if there is a single-source distance-preserving subgraph (for source
   s) of G with at most m branching vertices; No, otherwise.

The main theorem for this section is the following.

Theorem 1 (Single-source interval graphs). *There exists a polynomial time
algorithm that, given an interval graph G with a source s and k terminals
$t_1, t_2, \ldots, t_k$ as input, constructs an optimal shortest path tree which preserves
the distances between each (s, t_i) pair.*

We will see a proof of Theorem 1 shortly. First, we investigate what happens
in the general setting.

2.1 General Graphs

In this section, we will prove that it is NP-complete to find a distance-preserving subgraph of a general graph with the minimum number of branching vertices.

Theorem 3. *The decision problem $\mathcal{A}$ is* NP-*complete.*

Proof. In [13, Section 2.1] (see the full version), it is shown that finding an optimal all-pairs distance-preserving subgraph is NP-hard for unweighted graphs (or graphs with unit edge weights) in general. In fact, the same proof carries through for the single-source case as well. Consider the graph G_{set} in [13, Theorem 6] (full version). Let $S = \{t_1\}$ be the source vertex and let $T = \mathcal{U} \cup \{t_0, t_1\}$ be the set of target vertices. Observe that the proof essentially shows that even preserving distances from t_1 to T is NP-hard. This completes the proof.

Since there is no hope of solving the problem for the general case (unless $P = NP$), we move on to interval graphs.

2.2 Interval Graphs

Every interval graph has two representations: one using vertices and edges (called the *graph* representation), and one using a set of intervals on the real line (called the *interval* representation).

Let G be an interval graph and let $\mathfrak{I}_G$ be the interval representation of G. For every vertex $v \in V(G)$, let $\mathsf{left}(v)$ and $\mathsf{right}(v)$ be the left and right end points, respectively, of its corresponding interval in $\mathfrak{I}_G$. For simplicity, we assume that all the end points of the intervals have distinct values in $\mathfrak{I}_G$. Define a relation "$\prec$" on the vertices of G such that $v_1 \prec v_2$ in G if and only if $\mathsf{right}(v_1) < \mathsf{right}(v_2)$ in $\mathfrak{I}_G$. In other words, the intervals are ordered according to their right end points. The relation "$\preceq$" is similarly defined. When talking about interval graphs, we use the terms interval and vertex interchangeably, and when talking about bi-interval graphs, we use the terms rectangle and vertex interchangeably.

It is well known that one method of constructing shortest paths in interval graphs is the following *greedy* algorithm. Suppose we need to construct a shortest path from interval u to interval v in G (assume $u \prec v$). The greedy algorithm starts at vertex u. The second vertex on the greedy shortest path from u to v is the interval with the maximum right value of all the intervals that overlap with u. In this way, each step of the greedy algorithm chooses the next vertex as the interval with the maximum right value of all the intervals that overlaps with the current interval. The algorithm terminates when the current interval overlaps with v. It is easy to prove (via induction) that the path thus obtained is a shortest path from u to v. Let $P_G^{\text{gr}}(u, v)$ be the shortest path produced by this greedy algorithm.

This is where our algorithm heavily uses the interval graph property. This greedy algorithm specifically applies to interval graphs and gives them a nice layered graph representation (see Fig. 1). We defer the detailed proof of Theorem 1 to the full version of the paper [14, Section 2.3]. We now move on to all-pairs distance-preserving subgraphs.

3   All-Pairs Distance-Preserving Subgraphs

Interval graphs are "one-dimensional" by nature. It is therefore interesting to consider two-dimensional analogues of interval graphs. There are several different ways to generalize interval graphs into two (or more) dimensions, and many of them have been studied in literature. The main theorem for this section is the following.

Theorem 2 (All-pairs bi-interval graphs).

(a) (Upper bound) There is a polynomial time algorithm that, given a bi-interval graph with k terminals as input, produces an all-pairs distance-preserving subgraph of it with $O(k^2)$ branching vertices.

(b) (Lower bound) For every $k \geq 4$, there is a bi-interval graph G_{diag} on k terminals such that every all-pairs distance-preserving subgraph of G_{diag} has $\Omega(k^2)$ branching vertices.

We will see a proof of this shortly. First, let us see analyze a few other 2D analogues of interval graphs.

3.1   Two-Dimensional Analogues of Interval Graphs

A bi-interval graph G is determined by two families of intervals $\mathcal{I}_X$ and $\mathcal{I}_Y$. There is a vertex $v_{a,b}$ in $V(G)$ for each pair $(a, b) \in \mathcal{I}_X \times \mathcal{I}_Y$. The criterion used

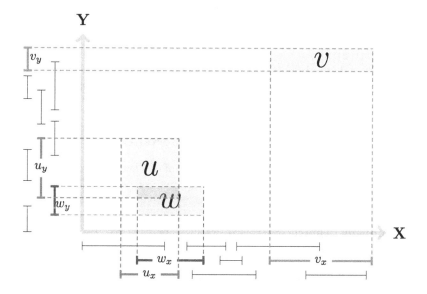

Fig. 2. Three vertices u, v, w, represented as rectangles in a bi-interval graph. Their corresponding intervals in the interval graphs X and Y are u_x, v_x, w_x and u_y, v_y, w_y, respectively. Note that (u, w) is an edge in the bi-interval graph since the rectangles u and w overlap, but (u, v) and (v, w) are not.

to connect two vertices by an edge has a natural geometric interpretation: two vertices are adjacent if the rectangles associated with them intersect (see Fig. 2). Formally, (a_1, b_1) and (a_2, b_2) are connected by an edge if and only if $(a_1 \times b_1) \cap (a_2 \times b_2) \neq \emptyset$. Thus, bi-interval graphs are rectangle intersection graphs where the rectangles that appear have a product structure induced by two families of intervals.

Bi-interval graphs are a natural generalization of interval graphs in the sense that they have boxicity two[2], and interval graphs by definition are precisely the graphs having boxicity one. Thus, it might be interesting to study the class of graphs having boxicity two (or higher). However, the following theorem shows that not much can be done for such graphs.

Theorem 4. *For every $k \geq 4$, there exists a graph having boxicity two on k terminals for which every all-pairs distance-preserving subgraph has $\Omega(k^4)$ branching vertices.*

Proof (Proof sketch). It is easy to see that the weighted planar graph G presented in [18, Section 5] can be made unweighted (by subdividing the edges) so that every distance-preserving subgraph of G has $\Omega(k^4)$ branching vertices. Now that G is unweighted and planar, it is possible to replace its vertices by axis-parallel rectangles, so that G now has boxicity two. This completes the proof.

In this paper, we study the class of unweighted bi-interval graphs. It is also reasonable to consider bi-interval graphs with non-negative real edge weights. Using earlier work by Krauthgamer *et al.* [18], it can be shown that for every $k \geq 4$, there exists a *weighted* bi-interval graph[3] on k terminals for which every distance-preserving subgraph has $\Omega(k^4)$ branching vertices (see the full version of [13, Corollary 8(b)]).

None of these lower bounds can be improved beyond constant factors, since an $O(k^4)$ upper bound exists for both weighted and unweighted graphs [18, Section 2.1].

3.2   Proof of the Upper Bound for Bi-Interval Graphs: Idea

In this section, we give a rough outline of our proof of Theorem 2(a). The detailed proof can be found in the full version of the paper [14, Section 3.3].

The proof is constructive by nature. However, we do not provide any approximation guarantees for our solution with respect to an optimal solution. Let us begin by formally defining bi-interval graphs. For this, we need the notion of strong graph products.

Definition 2 (Strong graph product). *Given two graphs G_1 and G_2, the strong graph product of G_1 and G_2, denoted by $G_1 \boxtimes G_2$, is defined as follows.*

---

[2] The boxicity of a graph is the minimum dimension in which a given graph can be represented as an intersection graph of axis-parallel boxes.

[3] Every interval graph is a bi-interval graph.

- $V(G_1 \boxtimes G_2) = V(G_1) \times V(G_2)$.
- Let (u_1, u_2) and (v_1, v_2) be two vertices of $G_1 \boxtimes G_2$ such that $(u_1, u_2) \neq (v_1, v_2)$. Then $((u_1, u_2), (v_1, v_2)) \in E(G_1 \boxtimes G_2)$ if and only if one of the following is true:
  1. $u_1 = v_1$ in G_1 and $(u_2, v_2) \in E(G_2)$.
  2. $u_2 = v_2$ in G_2 and $(u_1, v_1) \in E(G_1)$.
  3. $(u_1, v_1) \in E(G_1)$ and $(u_2, v_2) \in E(G_2)$.

In other words, two distinct vertices (u_1, u_2) and (v_1, v_2) are adjacent in $G_1 \boxtimes G_2$ if and only if they are adjacent or equal in each coordinate.

Definition 3 (Bi-interval graph). *A graph G is a bi-interval graph if $G = G_1 \boxtimes G_2$, where G_1 and G_2 are interval graphs.*

We now define a sub-class of bi-interval graphs known as king's graphs. King's graphs are used in our proof of the lower bound (Theorem 2(b)). Note that a path graph is an interval graph. When X and Y are both path graphs, then $X \boxtimes Y$ resembles a chessboard. The vertices are the squares of the chess board, and there is an edge between two vertices of the graph if and only if a king can go from one square to the other in a single move. Such a bi-interval graph is therefore called a king's graph.

Definition 4 (King's graph). *A graph G is a king's graph if $G = G_1 \boxtimes G_2$, where G_1 and G_2 are path graphs.*

See [14, Section 3.6] for a proof of Theorem 2(b) using Definition 4. Let us now move on to our proof sketch of Theorem 2(a).

In order to describe our construction of distance-preserving subgraphs of bi-interval graphs, it will be helpful to fix a method of constructing shortest paths in such graphs. It is well known and easy to prove that a greedy algorithm works for finding shortest paths in interval graphs. In a nutshell, we run this greedy algorithm on two interval graphs and put the resulting paths together.

Suppose $G = X \boxtimes Y$ is a bi-interval graph (see Definition 3), and we need to construct a shortest path between the vertices $u = (u_x, u_y)$ and $v = (v_x, v_y)$ in G (assume that $u_y \preceq v_y$). $P_X^{gr}(u_x, v_x) = (u_x = i_0, i_1, i_2, \ldots, i_p = v_x)$ and $P_Y^{gr}(u_y, v_y) = (u_y = j_0, j_1, j_2, \ldots, j_q = v_y)$ be the greedy shortest paths from u_x to v_x in X and from u_y to v_y in Y, respectively. Thus, $d_X(u_x, v_x) = p$ and $d_Y(u_y, v_y) = q$. Suppose $p \leq q$ (otherwise, exchange X and Y). Then, $P_G^{gr}(u, v) = ((i_0, j_0), (i_1, j_1), (i_2, j_2), \ldots, (i_p, j_p), (i_p, j_{p+1}), (i_p, j_{p+2}), \ldots, (i_p, j_{q-1}), (i_p, j_q))$ is a shortest path between u and v in G. This method can be visualized geometrically: starting from the rectangle u, embark on the greedy shortest paths in both X and Y simultaneously until the current rectangle lies in the same row or column as the destination rectangle (that is, move *diagonally*); then, move optimally within the row or column in the corresponding interval graph[4] to reach the destination (*horizontally* or *vertically up*; we call such paths *straight* paths).

---

[4] Note that when restricted to a fixed row (or a fixed column), a bi-interval graph is simply an interval graph.

Since we assumed that $u_y \leq v_y$, the diagonal segments of these paths either go northeast or northwest (but never southeast or southwest[5]). Our subgraph will consist of such paths for all pairs of terminals. First, we make provisions for the diagonal parts of the shortest paths, by independently moving northeast and northwest along greedy paths. Two paths, both proceeding northeast (or both proceeding northwest) from different terminals can meet, but once they have met they move in unison, never to diverge again. Furthermore, a path proceeding northeast and another proceeding northwest meet at most once and diverge immediately, never to meet again. Thus, by introducing at most $O(k^2)$ branching vertices, we succeed in making provision for all diagonal segments of shortest paths out of terminals.

Two tasks still remain: (i) we must identify vertices at which these diagonal segments branch off into a vertical or a horizontal segment (potentially introducing a branching vertex) and arrive at a vertex (which we call *pseudo-terminal*) in the row or column of another terminal; (ii) ensure that every pseudo-terminal is connected to every terminal in the row or column (with appropriate straight paths). The first task is straightforward for there are no choices to be made. The second task requires some care; we might have multiple terminals in the same column, which need shortest paths to all the pseudo-terminals in that column. To keep the number of branching vertices small, we borrow some ideas from the earlier analyses of distance-preserving subgraphs in interval graphs [13]. The total count of branching vertices breaks down as follows: (i) $O(k^2)$ when diagonal paths are added; (ii) $O(k^2)$ when diagonal paths branch off to join a straight path. Let us explain (ii) briefly. There are at most $O(k)$ pseudo-terminals in any row or column. We argue that within a row (or column) with p terminals and q pseudo-terminals, shortest paths can be provided by introducing $O(p(p + q))$ additional branching vertices. Since $p + q = O(k)$ and the total number of terminals is k, we need at most $O(k^2)$ branching vertices for (ii). In the full version of our paper, we show that this bound is tight [14, Section 3.6].

4   Conclusion

Finding an optimal all-pairs distance-preserving subgraph is NP-complete for general graphs, and a single-source distance-preserving subgraph is in P for interval graphs. However, the question remains: what about all-pairs distance-preserving subgraphs for interval graphs? Is it also in P? Is it NP-complete? If so, is it fixed-parameter tractable (with parameter k, the number of terminals)? This question is wide open and any progress toward solving it would be interesting.

---

[5] Anti-parallel paths are two greedy shortest paths in opposite directions. We do not consider southeast or northwest paths because southeast is anti-parallel to northwest, and southwest is anti-parallel to northeast, and there does not seem to be any straightforward method to get a handle on the number of branching vertices in terms of k by using anti-parallel paths (see [14, Section 3.7] for a more thorough explanation of this).

Acknowledgment. We thank Suhail Sherif for the "unique diagonal paths" idea [14, Section 3.6] that eventually led to our proof of Theorem 2(b). We are also grateful to the anonymous reviewers of this paper for their helpful suggestions and comments.

References

1. Baswana, S., Kavitha, T., Mehlhorn, K., Pettie, S.: New constructions of (α, β)-spanners and purely additive spanners. In: Proceedings of the sixteenth annual ACM-SIAM symposium on Discrete algorithms. pp. 672–681. Society for Industrial and Applied Mathematics (2005)
2. Bodwin, G.: Linear size distance preservers. In: Proceedings of the Twenty-Eighth Annual ACM-SIAM Symposium on Discrete Algorithms, pp. 600–615. Society for Industrial and Applied Mathematics (2017)
3. Bollobás, B., Coppersmith, D., Elkin, M.: Sparse distance preservers and additive spanners. SIAM J. Discrete Math. **19**(4), 1029–1055 (2005)
4. Charikar, M., Leighton, F.T., Li, S., Moitra, A.: Vertex sparsifiers and abstract rounding algorithms. In: 2010 IEEE 51st Annual Symposium on Foundations of Computer Science, pp. 265–274, October 2010. https://doi.org/10.1109/FOCS.2010.32
5. Chepoi, V.: Distance-preserving subgraphs of Johnson graphs. Combinatorica **37**(6), 1–17 (2015)
6. Coppersmith, D., Elkin, M.: Sparse sourcewise and pairwise distance preservers. SIAM J. Discrete Math. **20**(2), 463–501 (2006)
7. Däubel, K., Disser, Y., Klimm, M., Mütze, T., Smolny, F.: Distance-preserving graph contractions. CoRR abs/1705.04544 (2017). http://arxiv.org/abs/1705.04544
8. Djoković, D.Ž.: Distance-preserving subgraphs of hypercubes. J. Comb. Theory Ser. B **14**(3), 263–267 (1973)
9. Englert, M., Gupta, A., Krauthgamer, R., Rcke, H., Talgam-Cohen, I., Talwar, K.: Vertex sparsifiers: new results from old techniques. SIAM J. Comput. **43**(4), 1239–1262 (2014). https://doi.org/10.1137/130908440
10. Feder, T., Motwani, R.: Clique partitions, graph compression and speeding-up algorithms. J. Comput. System Sci. **51**(2), 261–272 (1995). https://doi.org/10.1006/jcss.1995.1065
11. Filtser, A.: Steiner point removal with distortion $O(\log k)$. In: Proceedings of the Twenty-Ninth Annual ACM-SIAM Symposium on Discrete Algorithms, pp. 1361–1373. Society for Industrial and Applied Mathematics (2018)
12. Fortune, S., Hopcroft, J., Wyllie, J.: The directed subgraph homeomorphism problem. Theor. Comput. Sci. **10**(2), 111–121 (1980)
13. Gajjar, K., Radhakrishnan, J.: Distance-preserving Subgraphs of Interval Graphs. In: 25th Annual European Symposium on Algorithms (ESA 2017), Leibniz International Proceedings in Informatics (LIPIcs), vol. 87, pp. 39:1–39:13. Schloss Dagstuhl-Leibniz-Zentrum fuer Informatik, Dagstuhl (2017). https://doi.org/10.4230/LIPIcs.ESA.2017.39. http://drops.dagstuhl.de/opus/volltexte/2017/7879. http://arxiv.org/abs/1708.03081
14. Gajjar, K., Radhakrishnan, J.: Minimizing branching vertices in distance-preserving subgraphs. CoRR abs/1810.11656 (2018). http://arxiv.org/abs/1810.11656
15. Goranci, G., Henzinger, M., Peng, P.: Improved guarantees for vertex sparsification in planar graphs. arXiv preprint arXiv:1702.01136 (2017)

16. Gupta, A.: Steiner points in tree metrics don't (really) help. In: Proceedings of the twelfth annual ACM-SIAM symposium on Discrete algorithms, pp. 220–227. Society for Industrial and Applied Mathematics (2001)

17. Kamma, L., Krauthgamer, R., Nguyên, H.L.: Cutting corners cheaply, or how to remove steiner points. SIAM J. Comput. **44**(4), 975–995 (2015)

18. Krauthgamer, R., Nguyên, H., Zondiner, T.: Preserving terminal distances using minors. SIAM J. Discrete Math. **28**(1), 127–141 (2014). https://doi.org/10.1137/120888843

19. Krauthgamer, R., Zondiner, T.: Preserving terminal distances using minors. In: Czumaj, A., Mehlhorn, K., Pitts, A., Wattenhofer, R. (eds.) ICALP 2012. LNCS, vol. 7391, pp. 594–605. Springer, Heidelberg (2012). https://doi.org/10.1007/978-3-642-31594-7_50

20. LaPaugh, A.S., Rivest, R.L.: The subgraph homeomorphism problem. J. Comput. Syst. Sci. **20**(2), 133–149 (1980)

21. Leighton, F.T., Moitra, A.: Extensions and limits to vertex sparsification. In: Proceedings of the Forty-second ACM Symposium on Theory of Computing, STOC 2010, pp. 47–56. ACM, New York (2010). https://doi.org/10.1145/1806689.1806698. https://doi.acm.org/10.1145/1806689.1806698

22. Nussbaum, R., Esfahanian, A.H., Tan, P.N.: Clustering social networks using distance-preserving subgraphs. In: Özyer, T., Rokne, J., Wagner, G., Reuser, A. (eds.) The Influence of Technology on Social Network Analysis and Mining. LNSN, vol. 6, pp. 331–349. Springer, Vienna (2013). https://doi.org/10.1007/978-3-7091-1346-2_14

23. Peleg, D., Schäffer, A.A.: Graph spanners. J. Graph Theory **13**(1), 99–116 (1989). https://doi.org/10.1002/jgt.3190130114

24. Sadri, A., Salim, F.D., Ren, Y., Zameni, M., Chan, J., Sellis, T.: Shrink: distance preserving graph compression. Inf. Syst. **69**, 180–193 (2017)

25. Spielman, D.A., Teng, S.H.: Spectral sparsification of graphs. SIAM J. Comput. **40**(4), 981–1025 (2011)

26. Thorup, M., Zwick, U.: Spanners and emulators with sublinear distance errors. In: Proceedings of the seventeenth annual ACM-SIAM symposium on Discrete algorithm, pp. 802–809. Society for Industrial and Applied Mathematics (2006)

27. Yan, D., Cheng, J., Ng, W., Liu, S.: Finding distance-preserving subgraphs in large road networks. In: 2013 IEEE 29th International Conference on Data Engineering (ICDE), pp. 625–636. IEEE (2013)

On Tseitin Formulas, Read-Once Branching Programs and Treewidth

Ludmila Glinskih and Dmitry Itsykson[(✉)]

St. Petersburg Department of V.A. Steklov Institute of Mathematics of the Russian Academy of Sciences, Fontanka 27, St. Petersburg, Russia
lglinskih@gmail.com, dmitrits@pdmi.ras.ru

Abstract. We show that any nondeterministic read-once branching program that decides a satisfiable Tseitin formula based on an $n \times n$ grid graph has size at least $2^{\Omega(n)}$. Then using the Excluded Grid Theorem by Robertson and Seymour we show that for arbitrary graph $G(V, E)$ any nondeterministic read-once branching program that computes a satisfiable Tseitin formula based on G has size at least $2^{\Omega(\mathrm{tw}(G)^{\delta})}$ for all $\delta < 1/36$, where $\mathrm{tw}(G)$ is the treewidth of G (for planar graphs and some other classes of graphs the statement holds for $\delta = 1$). We also show an upper bound of $O(|E|2^{\mathrm{pw}(G)})$, where $\mathrm{pw}(G)$ is the pathwidth of G.

We apply the mentioned results to the analysis of the complexity of derivations in the proof system OBDD($\wedge$, reordering) and show that any OBDD($\wedge$, reordering)-refutation of an unsatisfiable Tseitin formula based on a graph G has size at least $2^{\Omega(\mathrm{tw}(G)^{\delta})}$.

Keywords: Tseitin formula · Read-once branching program · Treewidth · Grid

1 Introduction

This paper continues the study of representation of satisfiable Tseitin formulas by read-once branching programs.

A Tseitin formula $\mathrm{TS}_{G,c}$ [20] is defined for every undirected graph $G(V, E)$ and labelling function $c : V \to \{0, 1\}$. We introduce a propositional variable for every edge of G. The Tseitin formula $\mathrm{TS}_{G,c}$ represents a linear system over the field $\mathrm{GF}(2)$ that for every vertex $v \in V$ states that the sum of all edges adjacent to v equals $c(v)$. It is well known that a Tseitin formula is satisfiable if and only if the sum of values of the labeling function for all vertices in every connected component is even [21].

In 2017 Itsykson et al. [13] showed that any OBDD representing satisfiable Tseitin formulas based on d-regular expanders on n vertices has size at least

The research was supported by Russian Science Foundation (project 16-11-10123).

R. van Bevern and G. Kucherov (Eds.): CSR 2019, LNCS 11532, pp. 143–155, 2019.
https://doi.org/10.1007/978-3-030-19955-5_13

$2^{\Omega(n)}$. Then Glinskih and Itsykson [9] extended this lower bound to nondeterministic read-once branching programs (1-NBP).

In this paper we consider an $n \times n$ grid and study the complexity of representation of Tseitin formulas based on it by read-once branching programs. In Theorem 2 we prove that any 1-NBP computing a satisfiable Tseitin formula based on an $n \times n$ grid has size $2^{\Omega(n)}$. Although an $n \times n$ grid graph has some edge-expansion properties, we could not prove the lower bound based only on these properties; our proof requires careful analysis and we use the geometric properties of the grid.

As an important corollary we establish a connection between the complexity of 1-NBP representation of a satisfiable Tseitin formula and the treewidth of the underlying graph. The treewidth is one of the most important structural measures of a graph and it is one of the main parametrizations for computational graph problems. Theorem 3 states that any 1-NBP computing a satisfiable Tseitin formula $\text{TS}_{G,c}$ has size at least $2^{\Omega(\text{tw}(G)^\delta)}$ for all $\delta < 1/36$, where $\text{tw}(G)$ denotes the treewidth of the graph G. The proof is based on the Excluded Grid Theorem by Robertson and Seymour [19]: there is a function g such that if a graph G has treewidth at least $g(t)$, then G contains a grid of size $t \times t$ as a minor. Recent results of Chekuri and Chuzhoy [4] and [5] give polynomial upper bound on the function g. Hence we know that every graph G has a $t \times t$ grid as a minor, where $t = \Omega(\text{tw}(G)^\delta)$ for all $\delta < 1/36$. For several classes of graphs it is possible to improve the value of δ, for example, for planar graphs $\delta = 1$ [11,18]. Thus Theorem 3 is followed by Lemma 8 stating that if H is a minor of G, then for every S and for every 1-NBP of size S that computes a satisfiable Tseitin formula $\text{TS}_{G,c}$ there is an 1-NBP that computes a satisfiable Tseitin formula $\text{TS}_{H,c'}$ of size at most S. This lemma is proved separately for every operation: an edge deletion, a vertex deletion and an edge contraction. We use the non-determinism in the case of an edge contraction: we replace nodes labelled with contracted edges by guessing nodes.

In Theorem 4 we show that for every satisfiable Tseitin formula based on a graph $G(V, E)$ has an OBDD of size $O(|E|2^{\text{pw}(G)})$, where $\text{pw}(G)$ is the pathwidth of G (note that the pathwidth differs from the treewidth by at most a logarithmic factor: $\text{tw}(G) \leq \text{pw}(G) \leq O(\text{tw}(G) \log |V|)$). Since the pathwidth of an $n \times n$ grid is $O(n)$, our upper and lower bounds for grids match up to a constant in the exponent.

There are several known approaches to defining the treewidth of CNF formulas. Ferrara, Pan and Vardi [7] considered a graph on variables of a CNF formula where two variables are connected iff they share a common clause. They proved that if a graph associated with CNF formula has the treewidth t, then the formula has an OBDD of size $n^{O(t)}$ (it is very similar to Theorem 4 but it uses another notion of treewidth). Razgon [17] showed that this bound is tight and there is a family of CNF formulas with the treewidth at most k that requires 1-NBP of size $n^{\Omega(k)}$. In the case of a Tseitin formula $\text{TS}_{G,c}$, the associated graph is the edge-graph of G, where the vertices are the edges of G and two edges are connected iff they are incident to the same vertex of G. For example, if G is a star on $n+1$ vertices (a star is a tree and, hence, it has the treewidth 1), then the edge-graph is the complete graph K_{n-1} and it has the treewidth $n-2$.

Applications to Proof Complexity. The interest of the study of Tseitin formulas comes from the propositional proof complexity; unsatisfiable Tseitin formulas are one of the basic examples of hard formulas for many proof systems.

The study of representations of satisfiable Tseitin formulas by read-once branching program was motivated by the study of proof systems based on OBDDs introduced by Atserias, Kolaitis and Vardi [2]. Itsykson et al. [13] studied the proof system OBDD($\wedge$, reordering); a proof of unsatisfiability of a CNF formula φ in this proof system is a sequence of OBDDs: $D_1, D_2, \ldots, D_s$ such that D_s is a constant false OBDD and for all $i \in [s]$, D_i either represents a clause of φ, or represents the conjunction of $D_k \wedge D_\ell$, where $k, \ell < i$ and D_i, D_k and D_ℓ use the same order, or represents the same function as D_ℓ but using another order, where $\ell < i$. The paper [13] gives an exponential lower bound on size of OBDD($\wedge$, reordering)-refutations of unsatisfiable Tseitin formulas based on constant degree expanders. The lower bound proof is organized as follows: for any refutation of Tseitin formula $TS_{G,c}$ of size S it is possible to construct an OBDD of size at most S^2 representing a satisfiable Tseitin formula $TS_{G',c'}$, where G' is a graph obtained from G by the deletion of several edges. Thus it is sufficient to prove lower bound on the size of OBDD representation of $TS_{G',c'}$. We adapt this approach and show in Theorem 6 that our results imply that any OBDD($\wedge$, reordering)-refutation of an unsatisfiable Tseitin formula $TS_{G,c}$ has size at least $2^{\Omega(\text{tw}(G)^\delta)}$, where δ is a constant as above. In particular we get a lower bound $2^{\Omega(n)}$ on the complexity of OBDD($\wedge$, reordering)-refutations of Tseitin formulas based on the $n \times n$ grid.

The recent paper by Buss et al. [3] shows that this proof system cannot be polynomially simulated by Resolution and even by Cutting Planes. The paper shows that any Resolution proof of Tseitin formula based on the complete graph on $\log n$ vertices $K_{\log n}$ has size at least $2^{\Omega(\log^2 n)}$, while it has an OBDD($\wedge$, reordering)-refutation of polynomial size. It is well known that the size of the shortest regular Resolution proof of any unsatisfiable CNF formula ϕ equals the size of the minimal read-once branching program for the following search problem Search_ϕ: given an assignment of variables of ϕ, find a clause that is refuted by this assignment [15,16]. Our upper bound implies that satisfiable $TS_{K_{\log n},c}$ can be computed by an OBDD of size $poly(n)$. Thus we have that computing of $\text{Search}_{TS_{K_{\log n},c}}$ for an unsatisfiable $TS_{K_{\log n},c}$ is superpolynomially harder than computing of a satisfiable $TS_{K_{\log n},c'}$ for read-once branching programs.

Tseitin formulas based on the grid graphs were studied in proof complexity. The first superpolynomial lower bound for regular resolution was proved for grid graphs in 1968 by Tseitin [20]. In 1987 Urquhart proved a lower bound for Tseitin formulas based on expanders in unrestricted Resolution [21] but tight lower bounds for grids were proved by Dantchev and Riis only in 2001 [6]. In the recent paper [12] Hastad proved lower bound on Bounded depth Frege refutations for Tseitin formulas based on $n \times n$ grid graphs that implies that polynomial size Frege proofs of such formulas should use formulas with almost logarithmic depth.

The treewidth was also studied in the context of resolution refutations of Tseitin formulas. Alekhnovich and Razborov [1] considered a hypergraph that corresponds to every CNF formula, where variables are vertices and clauses as sets of variables form hyperedges. For Tseitin formulas the branch-width of this hypergraph is up to a constant factor equal to the Resolution width [1]. For constant degree graphs the treewidth is equal to the branch-width of the hypergraph up to a multiplicative constant. Galesi, Talebanfard and Torán in the recent paper [8] consider cop-robber games on graphs that is very similar to games characterising the treewidth, they used such games in an analysis of the complexity parameters of resolution refutations of Tseitin formulas.

Proofs omitted due to space constraints can be found in [10].

2　Preliminaries

Branching Programs. A deterministic branching program (BP) is a form of representation of Boolean functions. A Boolean function $f(x_1, x_2, \ldots, x_n)$ is represented by a directed acyclic graph with exactly one source and two sinks. All nodes except sinks are labeled with a variable; every internal node has exactly two outgoing edges: one is labeled with 1 and the other is labeled with 0. One of the sinks is labeled with 1 and the other is labeled with 0. The value of the function for given values of variables is evaluated as follows: we start a path from the source such that for every node on its path we go along the edge that is labeled with the value of the corresponding variable. This path will end in a sink. The label of this sink is the value of the function.

A nondeterministic branching program (NBP) differs from a deterministic in the way that we also allow guessing nodes that are unlabeled and have two outgoing unlabeled edges. So nondeterministic branching program may have three types of nodes: guessing nodes, nodes labeled with a variable (we call them just labeled nodes) and two sinks; the source is either a guessing node or a labeled node. The result of a function represented by a nondeterministic branching program for given values of variables equals 1, if there exists at least one path from the source to the sink labeled with 1 such that for every node labeled with a variable on its path we go along an edge that is labeled with the value of the corresponding variable (for guessing nodes we are allowed to choose any of two outgoing edges). Note that deterministic branching programs constitute a special case of nondeterministic branching programs.

A deterministic or nondeterministic branching program is (syntactic) read-k (k-BP or k-NBP) if every path from the source to a sink contains at most k occurrences of each variable.

Let π be a permutation of the set $\{1, \ldots, n\}$ (an order). A π-ordered binary decision diagram is a 1-BP such that on every path from the source to a sink variable $x_{\pi(i)}$ can not appear before $x_{\pi(j)}$ if $i > j$. An ordered binary decision diagram (OBDD) is a π-ordered binary decision diagram for some π.

Lemma 1 ([22]). *Assume that Boolean functions f_1 and f_2 have π-ordered OBDDs of sizes k_1 and k_2 respectively, then (1) $f_1 \wedge f_2$ has a π-ordered OBDD of size at most $k_1 k_2$; (2) for any partial substitution ρ, $f_1|_\rho$ has a π-ordered OBDD of size at most k_1.*

Tseitin Formulas. Let $G(V, E)$ be an undirected graph without loops but possibly with multiple edges, $c : V \to \{0, 1\}$ be a labeling function that matches every vertex with a Boolean value. We associate every edge $e \in E$ with a propositional variable x_e. A Tseitin formula $\mathrm{TS}_{G,c}$ based on a graph G and a labeling function c is the conjunction of the following conditions: for every vertex v the sum of variables x_e for all edges e that are incident to v equals $c(v)$ modulo 2. More formally: $\bigwedge_{v \in V} \left(\sum_{e \text{ is incident to } v} x_e = c(v) \bmod 2 \right)$.

Usually, Tseitin formulas are written in the CNF. If the maximal degree of a graph is upper bounded by a constant d, then a sum modulo 2 can be written as a d-CNF of size at most 2^d, hence the size of CNF representation of $\mathrm{TS}_{G,c}$ is at most $O(2^d n)$.

We will use the following criterion of the satisfiability of Tseitin formulas:

Lemma 2 ([21]). *A Tseitin formula $\mathrm{TS}_{G,c}$ is satisfiable if and only if for every connected component of the graph G the sum of values of the function c for all of the vertices is even. I.e., for every connected component U the following holds: $\sum_{v \in U} c(v) = 0 \bmod 2$.*

Remark 1. Note that a substitution of a value to a variable $x_e := \alpha$ transforms Tseitin formula $\mathrm{TS}_{G,c}$ to a Tseitin formula $\mathrm{TS}_{G',c'}$, where graph G' is obtained from the graph G by deleting the edge e, c' equals c in every vertex except two vertices that are incident to edge e. On these two vertices the values of c and c' differ by α.

For a graph $G(V, E)$ let $k_G(l)$ be the maximal number of connected components that can be obtained from G by the deletion of l edges. The following lower bound on the size of 1-NBP for satisfiable Tseitin formula is known:

Lemma 3 ([9], **Corollary 20**). *For every connected graph $G(V, E)$ and arbitrary $1 \le l \le |E|$ any 1-NBP evaluating a satisfiable Tseitin formula $\mathrm{TS}_{G,c}$ has size at least $2^{|V| - k_G(l) - k_G(|E| - l) + 1}$.*

Proof (sketch). If a graph $H(U, F)$ consists of k connected components, then a satisfiable Tseitin formula $\mathrm{TS}_{H,f}$ has exactly $2^{|F| - |U| + k}$ satisfying assignments ([9], Lemma 2). For every l we estimate the number of nodes of an 1-NBP for $\mathrm{TS}_{G,c}$ on the level l. The graph G is connected, hence $\mathrm{TS}_{G,c}$ has exactly $2^{|E| - |V| + 1}$ satisfying assignments. For every satisfying assignment of $\mathrm{TS}_{G,c}$ we consider an accepting path of the 1-NBP corresponding to it. We consider the beginnings of these paths of length l. The number of accepting paths with the same beginning of length l is at most $2^{|E| - l - |V| + k_G(|E| - l)}$ (it is an upper bound on the number of satisfying assignments for Tseitin formulas on subgraph of G

with $|E| - l$ edges). Thus, the number of different beginnings of length l of the accepting paths is at least $2^{l-k_G(|E|-l)+1}$. The number of different beginnings of length l of accepting paths that go through a fixed vertex on the level l is at most $2^{l-|V|+k_G(l)}$ (it is an upper bound on the number of satisfying assignments for Tseitin formulas on subgraph of G with l edges). Finally, the number of vertices on the l-th level is at least $2^{|V|-k_G(l)-k_G(|E|-l)+1}$. □

Lemma 4. *Let G be an undirected graph, c and c' be such that Tseitin formulas $\mathrm{TS}_{G,c}$ and $\mathrm{TS}_{G,c'}$ are satisfiable. Then sizes of minimum-size 1-NBPs (1-BPs and OBDDs) for $\mathrm{TS}_{G,c}$ and $\mathrm{TS}_{G,c'}$ are equal.*

Treewidth, Pathwidth and Minors. A *tree decomposition* of an undirected graph $G(V, E)$ is a tree $T = (V_T, E_T)$ such that every vertex $u \in V_T$ corresponds to a set $X_u \subseteq V$ and it satisfies the following properties: 1. The union of X_u for $u \in V_T$ equals V. 2. For every edge $(a, b) \in E$ there exists $u \in V_T$ such that $a, b \in X_u$. 3. If a vertex $a \in V$ is in the sets X_u and X_v for some $u, v \in V_T$, then it is also in X_w for all w on the path between u and v in T.

If a tree T is a path, then this representation is a *path decomposition*. The *width* of a tree decomposition is the maximum $|X_u|$ for $u \in V_T$ minus one. A *treewidth* of a graph G is the minimal value of the width among all tree decompositions of the graph G. We denote it as $\mathrm{tw}(G)$. The *pathwidth* of a graph G is the minimal value of the width among all path decompositions of a graph G. We denote it as $\mathrm{pw}(G)$.

Lemma 5 ([14]). *For every graph G on n vertices $\mathrm{pw}(G) = O(\log(n) \cdot \mathrm{tw}(G))$.*

A *minor* of an undirected graph G is a graph that can be obtained from a graph G by a sequence of edge contractions, edge deletions and vertex deletions.

Theorem 1 ([5]). *For every constant $\delta < 1/36$ every graph G contains a $t \times t$ grid as a minor, where $t = \Omega(\mathrm{tw}(G)^\delta)$.*

3   Lower Bound for Grids

In this section we prove the following Theorem.

Theorem 2. *Let T_n be an $n \times n$ grid graph. Then if a Tseitin formula $\mathrm{TS}_{T_n,c}$ is satisfiable, then every 1-NBP that computes $\mathrm{TS}_{T_n,c}$ has size at least $2^{\Omega(n)}$.*

Proof. T_n contains $(n + 1)^2$ vertices and $2n(n + 1)$ edges. In order to prove this theorem we use Lemma 3 for $l = n(n + 1)$ (so l is the half of the number of edges). So we have to prove that if we delete half of the edges of T_n, then the resulting graph will have at most $\frac{(n+1)^2}{2} - \varepsilon \cdot n$ connected components for some constant $\varepsilon > 0$. Hence, by Lemma 3, every 1-NBP for $\mathrm{TS}_{T_n,c}$ has size at least $2^{2\varepsilon n+1}$.

We call a subgraph of T_n *optimal* if it contains l edges and has the maximal number of connected components. The plan of the proof is the following. At first we show that there exists an optimal subgraph H that has one connected component that contains all edges and all other connected components are isolated vertices. Then we estimate the number of connected components of H.

Lemma 6. *There is an optimal subgraph of T_n that has exactly one connected component with at least two vertices.*

Proof. Consider all optimal subgraphs of T_n. Choose among them a subgraph H that contains a connected component M with the maximal number of edges. If M contains all edges of H, then the lemma is proved. Further we assume that not all edges are in M.

Consider the properties of the chosen graph H.

1. All the edges of the grid T_n between vertices of M are in M. Indeed, otherwise we can delete an edge from another connected component and add it to M. After this operation the number of connected components does not decrease, but the number of edges in M is strictly increased. This is a contradiction since M has the maximal number of edges among all the optimal subgraphs.
2. Every connected component, except M, is edge-biconnected (i.e. it is impossible to increase the number of connected components by the deletion of an edge from it). Indeed, assume that for some connected component except M it is possible to delete an edge from it such that the number of connected component increases. Then we delete this edge and add an edge of the grid that connects M with a vertex out of M. In this case the number of connected components is not changed but the number of edges in the maximal component would be increased. This is a contradiction.
3. There is no vertex v of T_n such that it is not in M but there are at least two edges between v and vertices of M in T_n. Proof by contradiction, assume that such a vertex exists. Consider a connected component K that differs from M and has edges. Consider a set of the lowest vertices in K and let u be the leftmost vertex among them. There are no edges to the left or down from the vertex u in the graph H. Since the connected component K has edges, there is at least one edge that is incident to u. By the previous property, K is edge-biconnected, hence, u has precisely two incident edges. Let us delete the two edges incident to u from K and add two edges that connect the vertex v and M. The number of connected components doesn't decrease, but the number of edges in the maximal component increases. This is a contradiction.
4. Every 1×1 square of the grid T_n contains 0, 1 or 4 edges from M. A 1×1 square cannot contain exactly 3 edges because it contradicts the property 1. Let an 1×1 square contains exactly 2 edges from M. If these are two incident edges then we get a contradiction with the property 3 or the property 1. If these are two opposite edges, then we get a contradiction with the property 1.
5. For every $u, v \in M$, the minimal rectangle of the grid that contains both u and v (with all interior edges) is a subgraph of M (one of the sides of the rectangle could be of zero length, in that case it's just a line of the grid). It can be easily shown by the induction on the length of the shortest path between u and v using the property 4.

6. M is a rectangle of the grid with all edges of this rectangle. Consider the maximal rectangle of the grid that is fully contained (with all edges) in M. If there are vertices in M that are not in this rectangle, then we could increase this rectangle using the property 5.

So M is a rectangle of the grid. We say that M can be moved one step to the left (right, down or up) if all left (right, down or up) neighbours of the left (right, down or up) border of M are isolated vertices in H. Such a move doesn't change the number of connected components and the number of edges in them. Consider some connected component K that differs from M and contains edges. We move M one step closer to K in the way of decreasing the distance between them while it is possible. By the distance we understand the minimal L_1-distance between two vertices from M and K. After some step it is not possible anymore; it means that one of the borders of M (w.l.o.g. it is the upper border) has upper neighbours from connected components that consist of more than one vertex.

Let M be a rectangle $x \times y$, where x, y are non-negative integers. That means that every horizontal line of M contains $x + 1$ vertices. Assume that among upper neighbours of the upper border of M there are m vertices that are in some connected component that consists of more than one vertex. Let these m vertices be in k connected components. Assume that there are r edges of the graph H between $x + 1$ upper neighbours of M (see an example on the left part of Fig. 1). Since every edge between upper neighbours of M decreases the number of connected components, the following inequality holds $k \leq m - r$. Obviously, $r \leq x$.

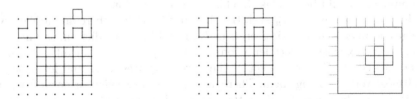

Fig. 1. Example: $x = 6, y = 4, r = 3, k = 3, m = 7$ **Fig. 2.** Spines

Consider the following modification of the graph H: we move the rectangle M one step up and add edges down from r vertices on the bottom border of M (see example on the right part of Fig. 1). The number of edges after this transformation is not changed since r edges overlapped and we added r edges. Now we estimate the number of connected components. On the bottom border we add $(x+1)-r$ new connected components. On the upper border $(x+1)-m+k$ connected components disappeared (were merged to one). Finally, the number of connected components increased by $m - k - r$, that is at least zero since $k \leq m - r$. But the number of edges in the maximal connected component increases, this contradicts the choice of the graph H. So we get a contradiction with the assumption that there are more than one connected components with at least one edge. $\qquad\square$

So we may assume that there is an optimal graph H that has one connected component M with at least two vertices and t isolated vertices. Notice that M is not necessary a rectangle now (in Lemma 6 we only show that it has to be a rectangle under the assumption that the statement of the lemma is wrong). We are going to estimate $k_{T_n}(l) = t + 1$ from the above.

For every connected component K we define a set of *spines* that go out of it. We assume that T_n is a part of the infinite grid. An edge e of the infinite grid is a spine of K if it connects a vertex u from K with a vertex out of K and u is the right or bottom endpoint of e (see Fig. 2). Note that spines can go outside of the square T_n if u is on the upper or left border of T_n. If a component is an isolated vertex, then it has exactly two spines.

The same edge cannot be a spine for two different connected components since we choose only the edges that go up or to the left from the component. Let h be the total number of all spines for all of the connected components of H. Every spine is either an edge of the square T_n that is not in H or is among $2(n+1)$ edges that go outside the square T_n. Hence, $2(n+1) + n(n+1) \geq h$.

Let X be the number of spines of the connected component M, then we get that $h = 2t + X$. Using the previous inequality we estimate t as follows: $t \leq (n+1)(n+2)/2 - X/2 = (n+1)^2/2 + (n+1-X)/2$. Hence, we need to show that $X \geq (1+\epsilon)n$ for some constant ϵ.

Consider the minimal grid rectangle that contains M. Assume it has size $(a-1) \times (b-1)$, where a, b are natural numbers. Then there are spines of M in every of a vertical lines and in every of b horizontal lines. Then we get that $X \geq a + b \geq 2\sqrt{ab}$. On the other hand, the component M contains exactly $n(n+1)$ edges, they need to be embedded into a rectangle $(a-1) \times (b-1)$ that contains $a(b-1) + b(a-1)$ edges. Then we can estimate $2ab > a(b-1) + b(a-1) \geq n(n+1)$ and we get $X \geq \sqrt{2n(n+1)} > \sqrt{2}n$.

Using the upper bound on t, the number of connected components can be estimated as: $k_{T_n}(l) = t + 1 \leq (n+1)^2/2 - (\sqrt{2}-1)n + \frac{3}{2}$. Then by Lemma 3, every 1-NBP for a satisfiable Tseitin formula $\mathrm{TS}_{T_n,c}$ has size at least $2^{2(\sqrt{2}-1)n-2}$. □

4 Treewidth

The main goal of this section is to prove the following theorem.

Theorem 3. *Let* $\mathrm{TS}_{G,c}$ *be a satisfiable Tseitin formula. Then every 1-NBP for* $\mathrm{TS}_{G,c}$ *has size at least* $2^{\Omega(\mathrm{tw}(G)^\delta)}$ *for all* $\delta < 1/36$.

Lemma 7. *Let* D *be a 1-NBP computing a Boolean function* $f : \{0,1\}^n \to \{0,1\}$. *If we change every node in* D *labeled with the variable* x_1 *by a guessing node and remove all labels of all its outgoing edges, then we obtain a valid 1-NBP that computes* $\exists x_1 f(x_1, x_2, \ldots, x_n)$.

Lemma 8. *Let* H *be a minor of an undirected graph* G *and Tseitin formulas* $\mathrm{TS}_{G,c}$ *and* $Ts_{H,c'}$ *be satisfiable. Then for every* S *and every 1-NBP of size* S *that computes* $\mathrm{TS}_{G,c}$, *there is a 1-NBP that computes* $\mathrm{TS}_{H,c'}$ *of size at most* S.

Proof. It suffices to prove the statement of the lemma for the case when H is obtained from G by the application of one operation. Let us consider all types of operations separately.

1. H is obtained from G by the deletion of an edge e. Let σ be a satisfying assignment of $\mathrm{TS}_{G,c}$. We apply the partial assignment $x_e := \sigma(x_e)$ to $\mathrm{TS}_{G,c}$ and get the satisfiable formula $\mathrm{TS}_{H,c''}$. It is well known that the application of a substitution does not increase size of the 1-NBP. By Lemma 4, sizes of the minimal 1-NBPs for $\mathrm{TS}_{H,c''}$ and $\mathrm{TS}_{H,c'}$ are equal.
2. H is obtained from G by the deletion of a vertex v. Since all the variables of Tseitin formulas are associated with edges, this case can be considered as a sequence of edge deletions.
3. A graph $H(V_H, E_H)$ is obtained from a graph $G(V, E)$ by the contraction of an edge $e = (u, v)$. Let us define a labeling function $c'' : V_H \to \{0, 1\}$ as follows: for all the vertices that differ from the joined vertex $\{u, v\}$ it has the same value as in the labeling function c and $c''(\{u, v\}) = c(u) + c(v) \bmod 2$. By the construction, every connected component U of H corresponds to a connected component U' of G with the same sum of the labeling functions: $\sum_{w \in U} c''(w) = \sum_{w \in U'} c(w) \bmod 2$. Hence, by Lemma 2, $\mathrm{TS}_{H,c''}$ is satisfiable.

$\square$

Lemma 9. *Formulas $\mathrm{TS}_{H,c''}$ and $\exists x_e \mathrm{TS}_{G,c}$ define the same function.*

By Lemma 7, the minimal size of a 1-NBP for $\exists x_e \mathrm{TS}_{G,c}$ (that is by Lemma 9 equivalent to $\mathrm{TS}_{H,c''}$) is at most the minimal size of a 1-NBP for $\mathrm{TS}_{G,c}$. By Lemma 4, minimal sizes of 1-NBPs for $\mathrm{TS}_{H,c''}$ and for $\mathrm{TS}_{H,c'}$ are equal. $\square$

Proof (Proof of Theorem 3).
By Theorem 1, the graph G contains a $t \times t$ grid graph as a minor, where $t = \Omega(\mathrm{tw}(G)^\delta)$. The theorem follows from Theorem 2 and Lemma 8. $\square$

We also obtain an upper bound:

Theorem 4. *Every satisfiable Tseitin formula $\mathrm{TS}_{G,c}$ can be represented as OBDD of size $|E|2^{\mathrm{pw}(G)+1} + 2$.*

Corollary 1. *Any satisfiable Tseitin formula based on a graph $G(V, E)$ can be represented as OBDD of size $O(|E||V|^{O(\mathrm{tw}(G))})$.*

Proof. Follows from Theorem 4 and Lemma 5. $\square$

5   Lower Bound in the Proof System OBDD($\wedge$, reordering)

In this section we show that any refutation of an unsatisfiable Tseitin formula $\mathrm{TS}_{G,c}$ in the proof system OBDD($\wedge$, reordering) has size at least $2^{\Omega(\mathrm{tw}(G)^\delta)}$ for all $\delta < 1/36$.

If F is a formula in CNF, we say that a sequence $D_1, D_2, \ldots, D_t$ of OBDDs is an OBDD$(\wedge, \text{reordering})$-refutation of F if D_t is an OBDD that represents the constant false function, and for all $1 \le i \le t$, D_i is an OBDD that represents a clause of F or can be obtained from the previous D_j's by one of the following inference rules: (conjunction or join) D_i is an OBDD with order π, that represents $D_k \wedge D_l$ for $1 \le l, k < i$, where D_k, D_l have the same order π; (reordering) D_i is an OBDD that is equivalent to an OBDD D_j with $j < i$ (note that D_i and D_j may have different orders).

We say that a graph H is t-good if H is connected and every OBDD-representation of any satisfiable Tseitin formula $\mathrm{TS}_{H,c}$ has size at least t. The following theorem can be proved using ideas from [13]:

Theorem 5 (cf. [13]). *Let $G(V, E)$ be a connected graph and degrees of all vertices of G be bounded by a constant. Assume that the graph G has the following properties: (1) if we delete any vertex from G, we get a t-good graph; (2) for every two vertices u and v of G there is a path p between them such that if we delete all vertices from p, we get a t-good graph. And if we delete from G vertices u and v and the edges of the path p, we also get a connected graph.*
Then any OBDD$(\wedge, \text{reordering})$-refutation of an unsatisfiable Tseitin formula $\mathrm{TS}_{G,c}$ has size at least $\Omega(\sqrt{t})$.

Proof (sketch). We consider the last step of the OBDD$(\wedge, \text{reordering})$-refutation: the conjunction of OBDDs F_1 and F_2 is the identically false function but both F_1 and F_2 are satisfiable. Both F_1 and F_2 are conjunctions of several clauses of $\mathrm{TS}_{G,c}$.

Since G remains connected after removing of any single vertex, F_1 and F_2 together contain all clauses of $\mathrm{TS}_{G,c}$. Assume that there are two nonadjacent vertices u and v such that F_1 does not contain a clause C_u that corresponds to the vertex u and F_2 does not contain a clause C_v that corresponds to v (if this assumption is false, the proof is rather straightforward). We consider two partial substitutions ρ_1 and ρ_2 that are both defined on the edges adjacent to u and v and on the edges of the path p between u and v. The substitutions ρ_1 and ρ_2 assign opposite values to edges of the path p and are consistent on all other edges. The substitution ρ_1 satisfies C_v and refutes C_u and ρ_2 satisfies C_u and refutes C_v.

By the construction $F_1|_{\rho_1} \wedge F_2|_{\rho_2}$ is almost a satisfiable Tseitin formula based on the graph that is obtained from G by deletion of the vertices u and v and all edges from the path p. However, it is also possible that this formula does not contain some clauses for the vertices from p. Thus we make additional partial substitution τ that substitute values from a satisfying assignment for all remaining edges for vertices from p. $(F_1|_{\rho_1} \wedge F_2|_{\rho_2})|_{\tau}$ is satisfiable Tseitin formula based on the graph that is obtained from G by deletion of all vertices from the path p. The size of an OBDD representation of such a formula is at least t by the condition of the theorem. Hence by Lemma 1 we get that either F_1 or F_2 has size at least $\Omega(\sqrt{t})$ in the given order. $\qquad\square$

Theorem 5 is used in the proof of the main result of this section:

Theorem 6. *Let G be a connected graph and $\mathrm{TS}_{G,c}$ be an unsatisfiable formula. Then every $\mathrm{OBDD}(\wedge, \text{reordering})$-refutation of $\mathrm{TS}_{G,c}$ has size at least $2^{\Omega\left(\mathrm{tw}(G)^{\delta}\right)}$ for all $\delta < 1/36$.*

Acknowledgements. The authors are grateful to Fedor Fomin, Alexander Kulikov, Alexander Knop, Dmitry Sokolov and anonymous reviewers for discussions and useful comments. The second author is a Young Russian Mathematics award winner and would like to thank sponsors and jury of the contest.

References

1. Alekhnovich, M., Razborov, A.A.: Satisfiability, branch-width and tseitin tautologies. Comput. Complex. **20**(4), 649–678 (2011)
2. Atserias, A., Kolaitis, P.G., Vardi, M.Y.: Constraint propagation as a proof system. In: Wallace, M. (ed.) CP 2004. LNCS, vol. 3258, pp. 77–91. Springer, Heidelberg (2004). https://doi.org/10.1007/978-3-540-30201-8_9
3. Buss, S., Itsykson, D., Knop, A., Sokolov, D.: Reordering rule makes OBDD proof systems stronger. In: CCC-2018, pp. 16:1–16:24 (2018)
4. Chekuri, C., Chuzhoy, J.: Polynomial bounds for the grid-minor theorem. J. ACM **63**(5), 40:1–40:65 (2016)
5. Chuzhoy, J.: Excluded grid theorem: improved and simplified. In: Proceedings of STOC-2015, pp. 645–654 (2015)
6. Dantchev, S.S., Riis, S.: "planar" tautologies hard for resolution. In: FOCS-2001, pp. 220–229 (2001)
7. Ferrara, A., Pan, G., Vardi, M.Y.: Treewidth in verification: local vs. global. In: Sutcliffe, G., Voronkov, A. (eds.) LPAR 2005. LNCS (LNAI), vol. 3835, pp. 489–503. Springer, Heidelberg (2005). https://doi.org/10.1007/11591191_34
8. Galesi, N., Talebanfard, N., Torán, J.: Cops-robber games and the resolution of Tseitin formulas. In: Beyersdorff, O., Wintersteiger, C.M. (eds.) SAT 2018. LNCS, vol. 10929, pp. 311–326. Springer, Cham (2018). https://doi.org/10.1007/978-3-319-94144-8_19
9. Glinskih, L., Itsykson, D.: Satisfiable Tseitin formulas are hard for nondeterministic read-once branching programs. In: MFCS-2017, pp. 26:1–26:12 (2017)
10. Glinskih, L., Itsykson, D.: On Tseitin formulas, read-once branching programs and treewidth. Technical Reports. TR-19-020, ECCC (2019)
11. Gu, Q.P., Tamaki, H.: Improved bounds on the planar branchwidth with respect to the largest grid minor size. Algorithmica **64**(3), 416–453 (2012)
12. Håstad, J.: On small-depth Frege proofs for Tseitin for grids. In: FOCS-2017, pp. 97–108 (2017)
13. Itsykson, D., Knop, A., Romashchenko, A., Sokolov, D.: On OBDD-based algorithms and proof systems that dynamically change order of variables. In: STACS-2017, pp. 43:1–43:14 (2017)
14. Korach, E., Solel, N.: Tree-width, path-width, and cutwidth. Discrete Appl. Math. **43**(1), 97–101 (1993)
15. Krajíček, J.: Bounded Arithmetic, Propositional Logic and Complexity Theory. Cambridge University Press, Cambridge (1995)

16. Lovász, L., Naor, M., Newman, I., Wigderson, A.: Search problems in the decision tree model. SIAM J. Discrete Math. **8**(1), 119–132 (1995)
17. Razgon, I.: On the read-once property of branching programs and CNFs of bounded treewidth. Algorithmica **75**(2), 277–294 (2016)
18. Robertson, N., Seymour, P., Thomas, R.: Quickly excluding a planar graph. J. Comb. Theory Ser. B **62**(2), 323–348 (1994)
19. Robertson, N., Seymour, P.D.: Graph minors. V. Excluding a planar graph. J. Comb. Theory Ser. B **41**(1), 92–114 (1986)
20. Tseitin, G.S.: On the complexity of derivation in the propositional calculus. Zapiski nauchnykh seminarov LOMI **8**, 234–259 (1968)
21. Urquhart, A.: Hard examples for resolution. JACM **34**(1), 209–219 (1987)
22. Wegener, I.: Branching Programs and Binary Decision Diagrams. SIAM, Philadelphia (2000)

Matched Instances of Quantum Satisfiability (QSat) – Product State Solutions of Restrictions

Andreas Goerdt[(✉)]

Fakultät für Informatik, Technische Universität Chemnitz,
Straße der Nationen 62, 09111 Chemnitz, Germany
goerdt@informatik.tu-chemnitz.de
http://www.tu-chemnitz.de/informatik/TI

Abstract. Matched instances of the quantum satisfiability problem have the following property: They have a product state solution. This is a mere existential statement and the problem is to find such a solution efficiently. Recent work by Gharibian and coauthors has made first progress on this question: They give an efficient algorithm which works for instances whose interaction hypergraph is restricted in a certain way.

We continue this line of research and give two results: First, an efficient algorithm is presented which works when the constraints themselves are restricted (the interaction hypergraph is not restricted). The restriction is that each constraint has at most 2 additive terms. Second, over the field of real numbers the problem of solving matched instances of QSat by product state solutions becomes NP-hard.

Keywords: Quantum satisfiability · Product states ·
Efficient algorithms · QMA$_1$

1 Introduction

Generalities. The quantum complexity class QMA (Quantum Merlin Arthur) is defined analogously to NP: Given a quantum state, a quantum computer can efficiently check (with sufficiently high probability) that this state solves a given instance of the problem considered, given the instance has a solution. The theory of Hamiltonian Complexity, see for example [7], is currently being developed to deal with the relevant questions. Many complete problems for this class have been found, and one of them is QSat, the quantum satisfiability problem (in fact it is complete for the restriction QMA$_1$ where the check succeeds with probability 1.) QSat has been introduced by Bravyi in [5]. It is a natural generalization of the classical satisfiability problem to the quantum setting. Note that problems in QMA seem much more difficult than those in NP, as the witnessing quantum state in general is a sum of exponentially many terms. Therefore it cannot even be described classically in any efficient way.

© Springer Nature Switzerland AG 2019
R. van Bevern and G. Kucherov (Eds.): CSR 2019, LNCS 11532, pp. 156–167, 2019.
https://doi.org/10.1007/978-3-030-19955-5_14

We focus on QSat. In particular we are interested in efficiently solvable restricted versions. Recently remarkable results have been obtained in this area: Already Bravyi shows that the quantum analogue of 2-Sat, called 2-QSat, is efficiently solvable. It took 10 years until it was noted that 2-QSat can even be solved in linear time [2, 4], extending a well known classical result for 2-Sat [3, 6]. More recently Gharibian and coauthors [1] have looked at QSat instances where each qubit (corresponding to a variable of the classical case) occurs at most twice and obtained an efficient algorithm. Again this extends a previously known classical result [11], problem 9.5.4 and [8]. The present paper contributes to the program to find more restrictions of QSat with an efficient algorithm.

Notation. States $|0\rangle$ and $|1\rangle$ of a single qubit are the states of our fixed computational basis. They directly correspond to classical boolean values. The computational basis of k qubits consists of the states $|b_1 \ldots b_k\rangle$ with b_i a classical bit $0, 1$.

An instance of the quantum satisfiability problem QSat on n qubits $1, \ldots, n$ is a conjunction of constraints $C_1, \ldots, C_m$. A constraint C is a quantum state of k qubits $i_1, \ldots, i_k$ from these n qubits:

$$C = \sum_{b_1, \ldots, b_k} \alpha_{b_1 \ldots b_k} |b_1 \ldots b_k\rangle_{i_1 \ldots i_k} \tag{1}$$

with b_i ranging over the two classical bits $0, 1$. States $|qb_1\rangle, \ldots, |qb_n\rangle$ with $|qb_i\rangle = \overline{a_{i,0}} |0\rangle + \overline{a_{i,1}} |1\rangle$ (with $\overline{a}$ the complex conjugate of a) are a solution of C iff

$$\sum_{b_1, \ldots, b_k} \alpha_{b_1, \ldots, b_k} \cdot a_{i_1, b_1} \cdot \ldots \cdot a_{i_k, b_k} = 0.$$

That is the state $|qb_{i_1}\rangle \otimes \cdots \otimes |qb_{i_k}\rangle$ is orthogonal to the state C. In spite of the fact that C is simply a state of the qubits $i_1, \ldots, i_k$, we use the terminology that $i_1, \ldots, i_k$ are the qubits on which C *acts*. If each constraint acts on (at most) k out of n qubits we have an instance of the problem k–QSat.

We only consider product state solutions $|qb_1\rangle \otimes \cdots \otimes |qb_n\rangle$. By the way this avoids the problem of general solutions which cannot be efficiently described. Note that constraints can still be entangled and thus really quantum. An instance $C_1, \ldots, C_m$ is satisfiable iff we have states $|qb_i\rangle$ of our qubits which solve all C_j simultaneously.

For $C = |b_1 \ldots b_k\rangle$ a constraint from the computational basis, the states $|qb_1\rangle, \ldots, |qb_k\rangle$ are a solution iff at least one $|qb_i\rangle$ is a basis state with $|qb_i\rangle = |\neg b_i\rangle$. Interpreting a basis state constraint, for example $|10\rangle$, as the clause $\neg x_1 \vee x_2$, we see that the classical satisfiability problem is included in QSat. Note that $x_1 = 1, x_2 = 0$ is the unique assignment falsifying the clause.

An instance of QSat is *matched* iff for each constraint we can pick one among the qubits on which it acts (we call it the matched qubit) such that no qubit is picked twice. Collecting the qubits of each constraint in one set (hyperedge) a QSat instance induces a hypergraph (with multiple edges) on vertices $1, \ldots, n$. We call this the interaction hypergraph. Thus an instance is matched iff its interaction hypergraph has an SDR (system of distinct representatives.) Note that an SDR can be found efficiently by bipartite matching techniques. The graph considered is: One part the vertices $1, \ldots, n$, the other part the hyperedges and each hyperedge is adjacent to the vertices it consists of. Then an SDR is a matching in which each hyperedge occurs.

For a matched instance of the classical satisfiability problem a solution can easily be found by assigning the matched variables the right truth value.

Motivation. In [10] (see also [12]) the following result is proven by non-constructive means (of basic Algebraic Geometry): Each matched instance of QSat has a *product state* solution. The motivating question for this work is obvious: Can we find such a solution efficiently? In seminal work on this problem Gharibian and coauthors [1] have given a positive answer to the question, provided the interaction hypergraph has certain restricting properties (and the constraints are generic.)

Results and Techniques. A constraint has l additive terms iff exactly l among the $\alpha_{b_1 \ldots b_k}$ (see (1)) are non-zero. An instance has l terms iff each of its constraints has at most l additive terms.

With $l = 1$ we only have constraints from the computational basis and QSat in this case is the classical satisfiability problem. Our first result concerns the natural next step, matched QSat instances with 2 terms.

Theorem 1. *The following problem has an efficient algorithm. Input: A matched QSat instance with 2 additive terms. Output: A product state solution to this instance.*

Following usage of at least part of the literature, for example [2], we disregard all questions of numerical precision here. Our algorithms are based on simple algebraic calculations, except that they use the natural logarithm of numbers from the input and one exponentiation to get the final output.

Example 1. We observe that the case $l = 2$ cannot be directly reduced to a classical constraint satisfaction problem as the case $l = 1$. A constraint with 2 terms, for example, is $(|000\rangle + |111\rangle)_{i,j,k}$. A product state solution consisting of *basis states* only must assign $|0\rangle$ to one qubit among i, j, k and $|1\rangle$ to another one. This gives us a 2–coloring of the hyperedge $\{i, j, k\}$.

Note that 2–coloring means: Not all vertices of a hyperedge receive the same color. Even in a matched instance of the hypergraph 2–coloring problem a solution cannot be found by simply assigning the matched vertex of each hyperedge the right color (as in the case of classical satisfiability.)

Consider the Fano Plane, see for example [1]. It is a 3–uniform hypergraph (that is each edge consists of exactly 3 vertices) with the following properties: First, it is matched (has an SDR) and second, it is not 2–colorable.

We consider the QSat instance which has for each edge $\{i, j, k\}$ of the Fano plane the constraint $(|000\rangle + |111\rangle)_{i,j,k}$. The instance is matched and therefore must have a product state solution. It cannot have one of basis states only because the Fano plane is not 2–colorable. As the instance has 2 additive terms our theorem applies. A solution is $|qb_i\rangle = |0\rangle - |1\rangle$ for all qubits i under consideration.

The proof of Theorem 1 is based on a simple observation: Constructing a product state solution of the constraint $|000\rangle + |111\rangle$, for example, means looking for complex numbers a_i, b_i, c_i for $i = 0, 1$ such that $a_0 \cdot b_0 \cdot c_0 = -a_1 \cdot b_1 \cdot c_1$. Taking logarithms we reduce this to a linear equation. Actually, as we work over the complex numbers, we choose to introduce 2 linear equations. Thus for an instance of QSat (with 2 terms) we get 2 systems of linear equations. If both systems have a solution, we get a solution to the instance.

We will see that the systems are non-homogeneous. So they may well be unsolvable. In this case we resort to an old idea of Seymour from the theory of 2–colorability of hypergraphs, see [13] (we learned it from [9].) We consider the homogeneous version of the system (just setting each right-hand-side to 0.) As the instance is matched the number of equations is at most the number of variables. Therefore the homogeneous system must have a non-trivial solution now. This allows us to assign basis states to some qubits in order to solve some constraints. A smaller matched instance remains.

Clearly, this approach is limited to $l = 2$ because otherwise taking logarithms does not yield a linear system.

In general it seems difficult to find a product state solution to a matched instance efficiently. Therefore we look for indicators of algorithmic hardness. We have a result in this direction, too.

Theorem 2. *(a) A matched QSat instance over the reals does not always have a solution with only real coefficients.*
(b) The following problem is NP-hard. Input: A matched QSat instance over the reals. Output: A product state solution only with real coefficients, provided the instance has such a solution.

Theorem 2 (a) reflects that the reals are not algebraically closed.

2   Proof of Theorem 1

2.1   An Algorithm for Symmetric Instances

A symmetric constraint C has the form

$$C = \alpha \cdot |b_1 \ldots b_k\rangle + \beta \cdot |\neg b_1 \ldots \neg b_k\rangle \quad \text{with} \quad \alpha \neq 0 \text{ and } \beta \neq 0.$$

We normalize one of the coefficients α or β to 1. A symmetric instance consists of symmetric constraints only. A basis state solution to a symmetric constraint like C corresponds to a solution in the sense of Not-all-equal satisfiability of the clause corresponding to $|b_1 \ldots b_k\rangle$ (or $|\neg b_1 \ldots \neg b_k\rangle$), that is the clause falsified by assigning $b_1, \ldots, b_k$ to its variables (or $\neg b_1, \ldots, \neg b_k$).

Definition 1. *Let $C = (|b_1 \ldots b_k\rangle + \alpha |\neg b_1 \ldots \neg b_k\rangle)_{i_1 \ldots i_k}$ be a symmetric constraint acting on qubits $i_1, \ldots, i_k$ from qubits $1, \ldots, n$. For each qubit we pick a corresponding variable, x_i for qubit i. Then the left-hand-side of C is*

$$LHS(C) = \sum_{j, b_j = 1} x_{i_j} - \sum_{j, b_j = 0} x_{i_j}$$

with j ranging over $1, \ldots, k$. For $\alpha = 1$ this is only unique up to multiplication with -1. In this case we pick one of the two possibilities as $LHS(C)$.

Example 2. We consider the symmetric constraint $|110\rangle + \alpha |001\rangle$ acting on qubits i_1, i_2, i_3. Then the left-hand-side is $x_{i_1} + x_{i_2} - x_{i_3}$. Note that the constraint $|110\rangle - \alpha |001\rangle = |110\rangle + \exp(\mathrm{i}\pi)\alpha |001\rangle$ has the same left-hand-side, but is generally not solved by the same states. The constraint $\beta |110\rangle + |001\rangle$ has the left-hand-side $-x_{i_1} - x_{i_2} + x_{i_3}$. This is the previous left-hand-side multiplied with -1.

Definition 2. *Let* $C = (|b_1 \ldots b_k\rangle + \alpha |\neg b_1 \ldots \neg b_k\rangle)_{i_1 \ldots i_k}$ *be a symmetric constraint with* $\alpha = r \cdot \exp(\mathrm{i} \cdot \phi), r > 0$.

(a) The radius equation of C is $LHS(C) = \ln r$.
(b) The phase equation of C is $LHS(C) = \pi + \phi$.

Example 3. Consider the constraint

$$|110\rangle + r \cdot \exp(\mathrm{i}\phi) |001\rangle, \quad r > 0$$

on qubits $1, 2, 3$ for simplicity. We look for solutions which consist only of non-basis states. We make the general ansatz:

$$|qb_j\rangle = |0\rangle + r_j \exp(\mathrm{i}\psi_j) |1\rangle, \quad r_j > 0, \ j = 1, 2, 3. \tag{2}$$

Qubits $1, 2, 3$ are a solution of the constraint iff

$$r_1 \exp(\mathrm{i}\psi_1) \cdot r_2 \exp(\mathrm{i}\psi_2) = -r \exp(\mathrm{i}\phi) \cdot r_3 \exp(\mathrm{i}\psi_3).$$

As we have $r, r_j > 0$ the preceding equation is true iff both of the following equations hold:

$$r_1 \cdot r_2 = r \cdot r_3 \quad \text{and} \quad \exp(\mathrm{i}(\psi_1 + \psi_2)) = - \exp(\mathrm{i}(\phi + \psi_3)) = \exp(\mathrm{i}(\pi + \phi + \psi_3)).$$

Taking logarithms, the first equation holds iff

$$\ln r_1 + \ln r_2 - \ln r_3 = \ln r. \tag{3}$$

The second is implied by (the π could also be $-\pi$ for example)

$$\psi_1 + \psi_2 - \psi_3 = \pi + \phi. \tag{4}$$

The radius and phase equations of the constraint are

$$x_1 + x_2 - x_3 = \ln r \quad \text{and} \quad x_1 + x_2 - x_3 = \pi + \phi.$$

Comparing with (3, 4) we see that real solutions to the radius and phase equations give us non-basis states solving the constraint via (2).

The next proposition collects the observations of the example. Its proof is by simple calculation following the example.

Proposition 1. *Let $C_1, \ldots, C_m$ be a symmetric QSat instance over qubits $1, \ldots, n$. Let $t_1, \ldots, t_n$ be a real solution for $x_1, \ldots, x_n$ of the linear system of radius equations. Let $\psi_1, \ldots, \psi_n$ be a real solution for $x_1, \ldots, x_n$ of the linear system of phase equations.*

Then the states

$$|qb_j\rangle = |0\rangle + r_j \cdot \exp(\mathbf{i}\psi_j)|1\rangle \quad \text{with } r_j = \exp(t_j), \ j = 1, \ldots, n$$

are a solution to $C_1, \ldots, C_m$

Note that the preceding proposition only says: If we have solutions to both linear systems we have states solving all constraints.

Example 4. Consider the symmetric instance

$$|00\rangle + |11\rangle, \ |00\rangle - |11\rangle = |00\rangle + \exp(\mathbf{i}\pi)|11\rangle$$

with both constraints on the same two qubits, say $1, 2$. The instance is clearly matched. The left-hand-side can be picked as $-x_1 - x_2$ for both constraints.

The system of radius equations is

$$-x_1 - x_2 = 0 \text{ and } -x_1 - x_2 = 0$$

both equations are equal. Note that the radius is 1 in both cases (and recall $\ln 1 = 0$.) The system of phase equations is

$$-x_1 - x_2 = \pi \text{ and } -x_1 - x_2 = \pi + \pi$$

We can solve the radius equations with $x_2 = -x_1$ giving $r_1 = \exp(x_1)$ and $r_2 = \exp(-x_1)$, their product must be 1. The phase equations are not solvable. However, we have a solution of basis states, one qubit must be $|0\rangle$ and the other one $|1\rangle$. We see that the associated homogeneous system $-x_1 - x_2 = 0$ has a non trivial solution, one $x_i > 0$ and the other one < 0. Proposition 2 is the general version of this observation.

Note, that adding the third constraint $|01\rangle + |10\rangle$ we have an instance which is not any more solvable by product states. But, the instance is not any more matched.

Proposition 2 is similar to Corollary 2 in section 2 of [13].

Proposition 2. *Let $C_1, \ldots, C_m$ be a symmetric instance of QSat over qubits $1, \ldots, n$. Let $a_1, \ldots, a_n$ be a non-trivial real solution for $x_1, \ldots, x_n$ of the homogeneous linear system*

$$LHS(C_1) = 0, \ \ldots, \ LHS(C_m) = 0.$$

Assign qubit j with

$$|qb_j\rangle = |0\rangle \ \text{if } a_j < 0 \quad \text{and} \quad |qb_j\rangle = |1\rangle \ \text{if } a_j > 0,$$

and let $|qb_j\rangle$ be an arbitrary state if $a_j = 0$.

Then we have: The state $|qb_1\rangle \otimes \cdots \otimes |qb_n\rangle$ is a solution to any constraint of the instance which acts on at least one qubit j with $a_j \neq 0$.

Proof. Let $C = |b_1 \ldots b_k\rangle + \alpha |\neg b_1 \ldots \neg b_k\rangle$ be a constraint of the instance. For simplicity we assume it acts on qubits $1, \ldots, k$. Let j with $1 \leq j \leq k$ be such that $a_j \neq 0$. As

$$LHS(C) = \sum_{j, b_j = 1} x_j - \sum_{j, b_j = 0} x_j$$

there must be another $j' \neq j$, $1 \leq j' \leq k$, with $a_{j'} \neq 0$ in order that $LHS(C) = 0$ is solved. Assume that $a_j > 0$ and $b_j = 1$. Then another $a_{j'} < 0$ and $b_{j'} = 1$ or an $a_{j'} > 0$ and $b_{j'} = 0$.

Assume the first alternative applies. Then $|qb_j\rangle = |1\rangle$ and $|qb_{j'}\rangle = |0\rangle$. The term $|\neg b_1 \ldots \neg b_k\rangle$ evaluates to 0 regardless of the states of the remaining qubits as $\neg b_j = 0$. The second term $|b_1 \ldots b_k\rangle$ evaluates to 0 as $b_{j'} = 1$. Thus C is solved regardless of the remaining qubits.

Assume that the second alternative applies. Then we have $|qb_j\rangle = |qb_{j'}\rangle = |1\rangle$. The term $|b_1 \ldots b_k\rangle$ evaluates to 0 because $b_{j'} = 0$. The term $|\neg b_1 \ldots \neg b_k\rangle$ evaluates to 0 as $\neg b_j = 0$.

The remaining cases are: First, an $a_j < 0$ with $b_j = 1$ and second, all a_j with $b_j = 1$ are assigned 0. These cases are easily treated in the same way.

The following algorithm subsumes the consideration behind Propositions 1 and 2.

Algorithm 1

```
Input: A matched symmetric instance of QSat on qubits 1,...,n.
Output: A solution to this instance.
Set I := the input instance.
1. Set up the linear systems of radius and phase equations of I.
If they both have a solution:
    a. Assign the qubits of I as described in Proposition 1.
    b. Assign qubits still unassigned arbitrarily.
    c. Output: The constructed assignment of states. End.
       Comment: Proposition 1 is only used once in the algorithm.
       I acts only on qubits which have not been assigned.
2. Construct the homogeneous system with equations LHS(C) = 0 for
each constraint C from I.
3. Obtain a non-trivial solution to this system.
    a. Assign basis states to the qubits as in Proposition 2.
    b. Leave the remaining qubits unassigned.
4. Set I := those constraints of I which do not act on any qubit
assigned by now.
       Comment: Only basis states have been assigned by now.
5. If I has no constraints:
    a. Assign the qubits unassigned by now arbitrarily.
    b. Output: The assignment of states. End.
6. Goto 1.
```

Concerning correctness: First, I always consists of those constraints of the input which act only on unassigned qubits. As such I is a matched instance throughout. If 1. applies we are done by Proposition 1.

If 1. does not apply 3. must be successful because I is always matched. Note, for a matched instance the number of qubits on which it acts must be at least as large as the number of its constraints. The linear systems considered in 1. or 2. are represented by a matrix with entries $1, -1, 0$. And this matrix has at least as many non-zero columns as it has rows. Each column corresponds to a qubit on which I acts and each row to a constraint.

If one of the non-homogeneous systems cannot be solved (over the reals) the column rank of the matrix is strictly less than the number of rows which in turn is at most the number of non-zero columns. Therefore the kernel of the linear mapping must be non-trivial and the algorithm finds a solution in 3. At least 1 constraint is solved by the assignment in 3. a., regardless of how the remaining qubits will be assigned (this is by Proposition 2). Then I obtained in 4. has at least one constraint less than before.

We emphasize that the instance I still to be solved always is a subset of the original constraints of the input. The constraints themselves are not changed.

Concerning running time: $O(n)-$times solving linear equations yields $O(n^4)$.

2.2   Reduction to Symmetric Instances

We need a basic transformation step:

$$\text{Assign } |qb_i\rangle = |b\rangle, \ b \ = \ 0, 1.$$

First, this transformation decomposes each constraint C which acts on qubit i and at least one additional qubit as

$$C \ = \ |b\rangle_i \otimes |\phi_1\rangle \ + \ |\neg b\rangle_i \otimes |\phi_2\rangle.$$

Second, it substitutes C with $|\phi_1\rangle$. If the first term is not present ($|\phi_1\rangle = 0$) we delete the constraint. For this transformation we have: If we have no constraint acting only on i then any solution of the transformed instance yields a solution to the original instance after assigning $|qb_i\rangle = |b\rangle$.

Now we have a matched instance of QSat with two terms. We transform it into a matched symmetric instance such that any solution to the transformed instance yields a solution to the original instance.

To begin with, we get rid of the non-symmetric constraints with two terms. Non-symmetric constraints with two terms can be written as

$$|b\rangle \otimes (|b_1 \ldots b_k\rangle \ + \ \alpha |c_1 \ldots c_k\rangle), \quad \alpha \neq 0, \ b \ = \ 0, \text{ or } b \ = \ 1.$$

We use the following transformation rules:

1. Elimination of the constraint

$$|b\rangle_i \otimes (|b_1 \ldots b_k\rangle \ + \ \alpha |c_1 \ldots c_k\rangle)$$

where i is not the matched qubit of the constraint. In this case we substitute the constraint with

$$|b_1 \ldots b_k\rangle + \alpha\,|c_1 \ldots c_k\rangle\,.$$

Note that qubit i may well occur in the new instance. The instance remains matched and any solution to the new instance is a solution to the original instance regardless of the state qubit i gets.

2. Elimination of the constraint

$$|b\rangle_i \otimes (|b_1 \ldots b_k\rangle + \alpha\,|c_1 \ldots c_k\rangle)$$

where i is the matched qubit of the constraint. In this case we first delete the constraint. Then we use the rule Assign $|qb_i\rangle = |\neg b\rangle$. We observe that the only constraint acting only on qubit i can be the constraint considered as each constraint acting on i must act on its matched qubit, too. The instance remains matched.

We apply 1. as long as it applies and then 2. Now we have a matched instance in which all constraints with 2 terms are symmetric. We still need to eliminate constraints with only 1 term. We apply the following transformation rule to the instance.

3. Elimination of the constraint $|b_1 \ldots b_k\rangle_{i_1,\ldots,i_k}$. Let i_j be the matched qubit. First, we delete the constraint, then we do Assign $|qb_{i_j}\rangle = |\neg b_{i_j}\rangle$. The new instance is matched. Note that an application of this step may well generate new constraints with 1 term.

We iterate 3. as long as we have constraints with 1 term only. This yields a symmetric matched instance with two terms or an instance without constraints. In this case we assign any qubits still left over arbitrarily.

3   Proof of Theorem 2

We give a translation of a classical 3–Sat formula into a matched instance of QSat. For this translation we need several subinstances as building blocks.

First the so called bit instance consisting 2 constraints:

$$|00\rangle_{i,j}\,,\quad |11\rangle_{i,j}\,.$$

This is a matched instance (matching the first constraint to i and the other one to j) with the following property: A product state is a solution to this instance iff $|qb_i\rangle = |b\rangle$ and $|qb_j\rangle = |\neg b\rangle$, $b = 0, 1$. The instance allows to restrict attention to states which are basis states and thus classical boolean values. (Note that the instance is solved by the state $|01\rangle + |10\rangle$, but this is not a product state and therefore is not of relevance here.)

Second, the conjunction constraint:

$$|011\rangle + |1\rangle \otimes (|01\rangle + |10\rangle + |00\rangle) \text{ on qubits } i, j, k.$$

This constraint computes the logical $\wedge$ in the following sense: If $|qb_j\rangle = |b\rangle$ and $|qb_k\rangle = |c\rangle$ where $b, c = 0, 1$ then $|qb_i\rangle \otimes |qb_j\rangle \otimes |qb_k\rangle$ is a solution to this constraint iff $|qb_i\rangle = |b \wedge c\rangle$.

For disjunction the constraint:

$$|0\rangle \otimes (|01\rangle + |10\rangle + |11\rangle) + |100\rangle \text{ on qubits } i, j, k.$$

For negation we get $|00\rangle + |11\rangle$. If a product state is a solution we have: If one of the qubits is a basis state, the other one must its negation. (We have included this only for didactic purposes, we do not need this constraint.) Note that the two constraints above $|00\rangle$, $|11\rangle$ enforce that both qubits are basis states.

These building blocks can be assembled to get a matched instance which determines the truth value of a 3–Sat formula with boolean variables $x_1, \ldots, x_n$ given the truth values of the variables.

Initialization subinstance:

$$|00\rangle_{1,n+1}, |11\rangle_{1,n+1}, |00\rangle_{2,n+2}, |11\rangle_{2,n+2}, \ldots, |00\rangle_{n,2n}, |11\rangle_{n,2n}.$$

This is clearly matched. For any solution we have that qubits $1, \ldots, n$ are basis states and qubits $n+1, \ldots, 2n$ the corresponding negations. Moreover, any combination of basis states of qubits $1, \ldots, n$ can occur in a solution.

Given truth values for qubits $1, \ldots, n$ we can calculate the truth value of a clause with 3 literals with two disjunction constraints. The first qubit (the result) of these constraints always is a new qubit. This new qubit is (and must be) the matched qubit of the constraint.

After this we use $m - 1$ conjunction constraints to calculate the conjunction of the values computed for the m clauses of the 3–Sat formula before. The first qubit of each conjunction constraint is a new qubit. Again it is the matched qubit.

Clearly the instance constructed by now is matched. Let r be the qubit which contains the final result. With the additional constraint $|0\rangle_r$ we can ensure that $|qb_r\rangle = |1\rangle$. Thus any solution of the instance constructed gives a satisfying assignment. But with this constraint the instance is not any more matched. (We see here that constructing product state solutions, provided they exist, of QSat instances with one more constraint than the number of variables is NP-hard.)

But, restricting to the reals the test $|qb_r\rangle = |1\rangle$ is possible without violating the matching condition. The building block for this is:

$$(|00\rangle + |11\rangle)_{t,s} , \quad (|01\rangle - |10\rangle)_{t,s}$$

on two qubits t, s. Clearly by itself this is matched. A product state solution is

$$(a_0 |0\rangle + \overline{a_1} |1\rangle) \otimes (b_0 |0\rangle + \overline{b_1} |1\rangle) \text{ with } a_0 = b_0 = 1, \ a_1 = b_1 = i$$

as $a_0 b_0 = 1$ and $a_1 b_1 = -1$ and $a_0 b_1 = a_1 b_0 = i$.

It is easy to see that we have no real product state solution: First, each product state solution is such that a_0, a_1, b_0, b_1 must all be $\neq 0$. For if $a_0 = 0$, for example, we must have that $b_1 = 0$, in order to have a solution of the first constraint. This means $b_0 = 1$ and then the state does not solve the second constraint.

Now

$$a_0 b_0 + a_1 b_1 = 0 \text{ and } a_0 b_1 - a_1 b_0 = 0$$

implies $a_0 = -a_1 b_1 / b_0$ and then with the second equation $- a_1 b_1^2 - a_1 b_0^2 = 0$. This implies $-b_1^2/b_0^2 = 1$ and $b_1/b_0 = i$. Thus we have no solution of real product states for the building block. (Note that $|00\rangle - |11\rangle$ is a real solution, but non-product.)

Now we add to the instance the following two test constraints:

$$((|00\rangle + |11\rangle) \otimes |0\rangle)_{t,s,r} \; , \; ((|01\rangle - |10\rangle) \otimes |0\rangle)_{t,s,r}$$

with r the result qubit above and t, s new qubits.

Clearly the whole instance is still matched. A product state solution over the reals must have $|qb_r\rangle = |1\rangle$ to solve the test constraints. If $|qb_r\rangle = |0\rangle$ we must have a solution to $|00\rangle + |11\rangle$, $|01\rangle - |10\rangle$ on qubits s, t. Thus qubits $1, \ldots, n$ of a solution over the reals must be a satisfying assignment of the original 3–Sat instance.

4 Conclusion

The whole paper centers around the question of constructing product state solutions. For Theorem 1 this is the obvious task as a solution is promised to exist. But for Theorem 2 this is not the only way to address the question: One might ask for the complexity of the mere decision version, and in this case the promise in the definition of QMA_1 for instances without solution must be accounted for. Such a result seems to be stronger than the one obtained here. We ask: Can our techniques be adapted to get hardness for the decision question over the reals?

Much broader is the following: Do their exist more (classical) constraint satisfaction problems such that matched instances are guaranteed to have a solution, but is not clear how to find one?

References

1. Aldi, M., de Beaudrap, N., Gharibian, S., Saeedi, S.: On efficiently solvable cases of quantum k-SAT. In: 43rd International Symposium on Mathematical Foundations of Computer Science, MFCS 2018, 27–31 August, Liverpool, UK. pp. 38:1–38:16 (2018), https://doi.org/10.4230/LIPIcs.MFCS.2018.38
2. Arad, I., Santha, M., Sundaram, A., Zhang, S.: Linear time algorithm for quantum 2SAT. In: 43rd International Colloquium on Automata, Languages, and Programming, ICALP 2016, 11–15 July, Rome, Italy, pp. 15:1–15:14 (2016). https://doi.org/10.4230/LIPIcs.ICALP.2016.15

3. Aspvall, B., Plass, M.F., Tarjan, R.E.: A linear-time algorithm for testing the truth of certain quantified boolean formulas. Inf. Process. Lett. **8**(3), 121–123 (1979). https://doi.org/10.1016/0020-0190(79)90002-4

4. de Beaudrap, J.N., Gharibian, S.: A linear time algorithm for quantum 2-SAT. In: 31st Conference on Computational Complexity, CCC 2016, 29 May–1 June, Tokyo, Japan, pp. 27:1–27:21 (2016). https://doi.org/10.4230/LIPIcs.CCC.2016.27

5. Bravyi, S.: Efficient algorithm for a quantum analogue of 2-SAT. In: Cross Disciplinary Advances in Quantum Computing, University of Tyer, Texas, 1–4 October, pp. 33–48 (2009). https://arxiv.org/abs/quant-ph/0602108

6. Even, S., Itai, A., Shamir, A.: On the complexity of timetable and multicommodity flow problems. SIAM J. Comput. **5**(4), 691–703 (1976). https://doi.org/10.1137/0205048

7. Gharibian, S., Huang, Y., Landau, Z., Shin, S.W.: Quantum hamiltonian complexity. Found. Trends Theor. Comput. Sci. **10**(3), 159–282 (2015). https://doi.org/10.1561/0400000066

8. Johannsen, J.: Satisfiability problems complete for deterministic logarithmic space. In: Diekert, V., Habib, M. (eds.) STACS 2004. LNCS, vol. 2996, pp. 317–325. Springer, Heidelberg (2004). https://doi.org/10.1007/978-3-540-24749-4_28

9. Kullmann, O., Zhao, X.: Bounds for variables with few occurrences in conjunctive normal forms. CoRR abs/1408.0629 (2014). http://arxiv.org/abs/1408.0629

10. Laumann, C.R., Läuchli, A.M., Scardicchio, A., Sondhi, S.L.: On product, generic and random generic quantum satisfiability. CoRR abs/0910.2058 (2009). http://arxiv.org/abs/0910.2058

11. Papadimitriou, C.H.: Computational Complexity. Addison-Wesley, Reading (1994)

12. Parthasarathy, K.R.: On the maximal dimension of completely entangled subspace for finite level quantum systems. Proc. Indian Acad. Sci. (Math. Sci.) 114, 364–375 (2004). https://arxiv.org/abs/quant-ph/0405077

13. Seymour, P.D.: On the two-colouring of hypergraphs. Q. J. Math. **25**, 303–312 (1974)

Notes on Resolution over Linear Equations

Svyatoslav Gryaznov$^{(\boxtimes)}$ (iD)

St. Petersburg Department of Steklov Mathematical Institute of Russian
Academy of Sciences, 27 Fontanka, St. Petersburg 191023, Russia
svyatoslav.i.gryaznov@gmail.com

Abstract. We consider the proof system Res($\oplus$) introduced by Itsykson
and Sokolov [8] which is an extension of Resolution proof system and
operates with disjunctions of linear equations over $\mathbb{F}_2$. In this paper we
prove exponential lower bounds on tree-like Res($\oplus$) refutations for Ordering and Dense Linear Ordering principles by Prover-Delayer games.

We also consider the following problem: given two disjunctions of linear equations over ring R decide whether all Boolean satisfying assignments of one of them satisfy another. Part and Tzameret conjectured
that for rings $R \neq \mathbb{F}_2$ this problem is coNP-hard, but proved it only for
rings with char$(R) = 0$ and char$(R) \geq 5$ [10]. We completely prove the
conjecture.

1 Introduction

The resolution proof system is among most studied and well-known proof systems. Some formulas that encode unsatisfiable systems of linear equations over
$\mathbb{F}_2$ are hard for resolution and even for bounded-depth Frege [3]. Itsykson and
Sokolov [8] suggested an extension Res($\oplus$) which operates with disjunctions of
linear equations over $\mathbb{F}_2$. This extension has short proofs for unsatisfiable systems
of linear equations over $\mathbb{F}_2$. Res($\oplus$) is an interesting partial case of bounded-
depth Frege with parity gates. However, there are no known super-polynomial
lower bounds on the size of derivations in Res($\oplus$). We are primarily concerned
with tree-like Res($\oplus$). For an unsatisfiable CNF formula ϕ consider a problem
$Search_\phi$: given an assignment of variables of ϕ find a clause of ϕ that is falsified
by this assignment. It is known that tree-like resolution refutations for ϕ are
equivalent to decision trees for $Search_\phi$. Itsykson and Sokolov [8] proved a similar result: tree-like Res($\oplus$) refutations of ϕ are equivalent to the parity decision
trees for $Search_\phi$.

We emphasize three main methods for proving lower bounds on tree-like
Res($\oplus$) refutations:

S. Gryaznov: The research is supported by Russian Science Foundation (project 18-71-
10042).

R. van Bevern and G. Kucherov (Eds.): CSR 2019, LNCS 11532, pp. 168–179, 2019.
https://doi.org/10.1007/978-3-030-19955-5_15

1. Lower bounds from communication complexity. A tree-like $\text{Res}(\oplus)$ refutation of ϕ of size s can be transformed into a randomized communication protocol for the problem $Search_\phi$ with the randomized communication complexity $\log(s)$ [8]. Hence, the size of any $\text{Res}(\oplus)$ refutation of ϕ can be obtained from the lower bound on the randomized communication complexity [2,6]. However, lower bounds on communication complexity of $Search_\phi$ are only known for specifically constructed formulas which often involve lifting. This is a disadvantage of this method since it most probably cannot be used for particular formulas.

2. Lower bounds from Polynomial Calculus over $\mathbb{F}_2$. For an unsatisfiable k-CNF formula ϕ tree-like $\text{Res}(\oplus)$ refutation of ϕ of size s can be transformed into a Polynomial Calculus refutation of ϕ with degree at most $O(\log(s)) + k$. Hence, the size of tree-like $\text{Res}(\oplus)$ refutations of ϕ follows from the lower bounds on the degree of Polynomial Calculus refutations of ϕ. It was first noted by Garlík and Kołodziejczyk [5]. For a detailed overview of this method one may refer to [10]. This method gives lower bounds on random 3-CNF formulas [10]. One of the restrictions of this method is that it requires ϕ to be in a k-CNF for a small k.

3. Lower bounds from Prover-Delayer games. Prover-Delayer games were originally proposed for tree-like resolutions by Pudlák and Impagliazzo [11] and extended to $\text{Res}(\oplus)$ by Itsykson and Sokolov [7,8]. Itsykson and Sokolov [8] used this technique to prove lower bound $2^{\frac{n-1}{2}}$ on the size of proofs for Pigeonhole Principle (PHP_n^m). The lower bounds on Pigeonhole principle can also be obtained from the lower bound on the degree of Graph-Pigeonhole Principle refutation in Polynomial Calculus, but it requires the number of pigeons m to be polynomial in the number of holes n [9]. The proof which uses Prover-Delayer games does not apply any restrictions on m. Until this moment, Pigeonhole was the only non-trivial example of this technique for tree-like $\text{Res}(\oplus)$. In this paper we give another two natural examples of hard formulas for tree-like $\text{Res}(\oplus)$ which we prove using Prover-Delayer games. This method is relatively easier than the previous two and the proofs are more explicit.

1.1   Our Results

In Sect. 3 we prove lower bounds on the size of $\text{Res}(\oplus)$ refutations of two families of unsatisfiable CNF formulas using Prover-Delayer games.

In Sect. 3.2 we consider the encoding of the negation of Ordering principle which states that there are no linear orderings without minimum and give $2^{n-O(1)}$ lower bound on the size of any tree-like $\text{Res}(\oplus)$ refutation. It is also possible to prove a similar (but slightly worse) lower bound $2^{n/4-O(1)}$ using the lower bound on the degree of derivations in Polynomial Calculus of Graph-Ordering principle [4,9].

In Sect. 3.3 we consider the encoding of the negation of Dense Linear Ordering principle which states that there are no dense linear orderings and give $2^{n/3-O(1)}$ lower bound on the size of any tree-like $\text{Res}(\oplus)$ refutation. This principle was considered by Atserias and Dalmau [1]. There are no known lower bounds in

Polynomial Calculus on graph variant of this principle (unlike the case of Graph-Ordering). This proof is more involved than the previous one: it uses ideas from both the Ordering principle and Pigeonhole principle lower bound proofs.

Both Ordering and Dense Linear Ordering principles have resolution refutations of polynomial size [1]. Thus, we give natural examples that are hard for tree-like $Res(\oplus)$ and easy for resolution.

Itsykson and Knop described in [7] how to construct hard *satisfiable* formulas for drunken $DPLL(\oplus)$ algorithms based on hard unsatisfiable formulas such that their hardness is proved by Prover-Delayer games. Thus, hard satisfiable formulas for drunken $DPLL(\oplus)$ algorithms can be also constructed from Ordering and Dense Linear Ordering principles.

Part and Tzameret considered an extension of $Res(\oplus)$ over an arbitrary ring R [10]. They used syntactic variant of weakening rule and conjectured that verifying the correctness of application of semantic weakening rule on Boolean cube is hard for rings $R \neq \mathbb{F}_2$. They proved it for rings with $char(R) = 0$ and $char(R) \geq 5$ and left the remaining cases as an open question. In Sect. 4 we prove the conjecture completely by showing it for rings with $char(R) = 3$ and $char(R) = 2, R \neq \mathbb{F}_2$.

1.2   Further Research

It is an open question whether Graph-Dense Linear Ordering principle requires large degree for Polynomial Calculus proofs.

Part and Tzameret [10] generalized the lower bound for Pigeonhole principle from [8] to the case of an arbitrary ring. For Ordering principle and Dense Linear Ordering principle their approach fails. It is interesting whether it is possible to generalize the proofs for this principles for the case of an arbitrary ring.

2   Preliminaries

2.1   Resolution over Linear Equations

In this section we describe a proof system that we use throughout the paper. We focus on the case $R = \mathbb{F}_2$ and give a definition for the system $Res(\oplus)$ as given in [8]. For other choices of the ring R one may refer to [10].

A *linear clause* is a disjunction $\bigvee_{i=1}^{m} (f_i = \alpha_i)$ of linear forms $\{f_i\}_{i=1}^{m}$ over $\mathbb{F}_2$ where $\alpha_i \in \mathbb{F}_2, i = 1, 2, \ldots, m$. We say that the linear clause D is semantically implied by linear clauses $C_1, \ldots, C_s$ if every satisfying assignment of the variables of $\bigwedge_{i=1}^{s} C_i$ also satisfies D. We denote it as $C_1, \ldots, C_s \vDash D$.

This proof system has two rules:

- Resolution rule that allows to derive the linear clause $A \vee B$ from linear clauses $A \vee (f = 0)$ and $B \vee (f = 1)$ where f is a linear form over $\mathbb{F}_2$.

– (Semantic) Weakening rule that allows to derive the linear clause D from the linear clause C if D is semantically implied by C.

A proof of the unsatisfiability of a CNF formula is a derivation of the empty clause from the clauses of the initial formula using one of the rules above. If a proof can be arranged as a binary tree with the empty clause in the root, the initial clauses in the leaves, and every clause in an internal node being a result of application of the rules to the children of that node then the proof is *tree-like*.

Itsykson and Sokolov [8] proved that the described system is indeed a proof system. We emphasize that it is not the case for different choices of underlying ring R, as noted in [10]. For instance, for $R = \mathbb{F}_2$ the following lemma holds while for other rings this problem is coNP-hard. This is covered in Sect. 4 of the paper.

Lemma 2.1 ([8]). *Application of semantic weakening rule can be verified in polynomial time: there is an algorithm that for linear clauses C and D verifies whether D is semantically implied by C in time $poly(|C| + |D|)$.*

2.2   Prover-Delayer Games

Pudlák and Impagliazzo introduced in [11] Prover-Delayer games for proving lower bounds on tree-like resolution refutations. We incorporate a straightforward extension of such games (see [7] for a detailed description of the method).

The game is played by two players: Prover and Delayer. They both have an unsatisfiable CNF formula ϕ. Delayer pretends to know a satisfying assignment of ϕ and Prover wants to find a falsified clause by asking Delayer about the values of linear forms over $\mathbb{F}_2$. The game consists of two steps:

1. Prover chooses the linear form g over $\mathbb{F}_2$ and asks Delayer for the value of g on the assignment.
2. Delayer either answers with 0 or 1 or with $*$ allowing Prover to choose any value. In the latter case Delayer earns a coin.

The game ends when there is a clause C of ϕ which is falsified by every solution of a Prover's system.

Lemma 2.2 ([7,8]). *If Delayer has a strategy that guarantees winning of at least k coins against any Prover in any game the unsatisfiable formula ϕ then the size of any tree-like $\mathrm{Res}(\oplus)$ proof of the unsatisfiability of ϕ is at least 2^k*

3   Framework for Proving Lower Bounds Based on Prover-Delayer Games

For our case it is crucial that for field $\mathbb{F}_2$ the disequality $f(x) \neq \alpha$ is equivalent to the equality $f(x) = 1 - \alpha$. Due to this fact it is more convenient to work with the *negation* of the linear clause $C = \bigvee_{i=1}^{m} (f_i = \alpha_i)$ being the linear system

$\neg C = \bigwedge_{i=1}^{m} (f_i = 1 - \alpha_i)$. The correspondence between them is the following: all solutions of the system $\neg C$ are exactly the variable assignments that falsify the clause C.

For Pigeonhole principle PHP_n^m we can find a certain property about the structure of its solutions. This originally was proved in [8] for tree-like Res($\oplus$) and then generalized by Part and Tzameret in [10] for arbitrary rings. We apply similar ideas to Ordering and Dense Linear Ordering principles.

Consider an unsatisfiable CNF formula ϕ. We split its clauses into two groups $\phi = F \wedge E$ where F and E are CNF subformulas of ϕ. We say that the solution σ of the linear system $Av = b$ in the variables of E (and possibly other variables) is E-respectful if for every clause C in E the solution σ satisfies C.

Theorem 3.1 (cf. [8], Lemma 5.1 and Theorem 5.1). *Suppose that for the function f for every linear system $Av = b$ over $\mathbb{F}_2$ in the variables of $\phi_n = F \wedge E$ with less than $f(n)$ equations in the system we have the following property: if the system has E-respectful solution that falsifies some clause C in F then it also has another E-respectful solution that does not falsify C. Then the size of any tree-like Res($\oplus$) proof is at least $2^{f(n)}$.*

Proof. We construct a strategy for Delayer that guarantees winning of at least $f(n)$ coins. Then the statement follows from Lemma 2.2. We fix n and drop index n from ϕ_n until the end of the proof. Let Φ be the system of all linear forms from Prover's questions together with its assigned values (Delayer's answers or Prover's choices for $*$). Initially, the system is empty. We keep the invariant while Delayer earned less than $f(n)$ coins: the system has at least one E-respectful solution and there is no clause C in F that is falsified by all E-respectful solutions.

The round of the game starts with the linear form g from Prover. If $g(x) = \alpha$ is E-respectfully implied by Φ for some $\alpha \in \mathbb{F}_2$ (i.e. all E-respectful solutions of Φ satisfy $g(x) = \alpha$) then Delayer simply answers α. Otherwise, there are some E-respectful solutions with $g(x) = 0$ and some with $g(x) = 1$. In that case Delayer says $*$ and lets Prover choose the value arbitrarily.

Consider three possible conclusions of the game:

1. Φ does not have any solutions. This case is impossible because by the invariant it has an E-respectful solution.
2. Φ only has solutions that falsify clauses in E. It is not the case because Φ has at least one E-respectful solution.
3. Φ has at least one solution that falsifies a clause C in F. Consider the linear system $\Phi' \subset \Phi$ which corresponds only to the answers $*$ of Delayer. By construction all lines in $\Phi \setminus \Phi'$ are E-respectfully implied by Φ', hence Φ' does not have any new E-respectful solutions. Therefore, C in F also falsifies every E-respectful solution of Φ', and by assumption Φ' must have more than $f(n)$ lines (and that is exactly how many coins Delayer earned).

The described strategy allows Delayer to earn at least $f(n)$ coins, therefore the size of any tree-like Res($\oplus$) proof is at least $2^{f(n)}$ by Lemma 2.2. $\qquad\square$

3.1 Pigeonhole Principle

In this case E consists of all hole-clauses (that state that no more than one pigeon can be in any hole) and F consists of pigeon-clauses (that state that each pigeon sits in at least one hole).

Lemma 3.2 (cf. [8], Lemma 5.1). *Let $Ax = b$ be the linear system in the variables of PHP_n^m with at most $\frac{n-1}{2}$ equations that has an E-respectful solution. Then for every clause $C \in F$ there is the E-respectful solution τ of the system $Ax = b$ that satisfies C.*

Theorem 3.3 ([8]). *The size of any tree-like $\mathrm{Res}(\oplus)$ refutation of PHP_n^m is at least $2^{\frac{n-1}{2}}$.*

Proof. Follows from application of Lemma 3.2 to Theorem 3.1.   □

3.2 Ordering Principle

Ordering principle states that for any positive n there are no linear orderings of n elements without minimal element. For any n we encode the negation of this statement by the unsatisfiable CNF formula $Ordering_n$ with variables $\{x_{ij}\}_{i \neq j \in [n]}$ where $x_{ij} = 1$ if element i is less than element j (we denote it as $i \prec j$) and $x_{ij} = 0$ otherwise. The clauses of $Ordering_n$ are of four types:

1. Anti-symmetry: $\neg x_{ij} \vee \neg x_{ji}$ for all $i \neq j \in [n]$.
2. Totality: $x_{ij} \vee x_{ji}$ for all $i \neq j \in [n]$.
3. Transitivity: $\neg x_{ij} \vee \neg x_{jk} \vee x_{ik}$ for all distinct $i, j, k \in [n]$.
4. Non-minimality (i.e. non-existence of the minimal element):
   $$C_i = \bigvee_{j \in [n] \setminus \{i\}} x_{ji} \quad \text{for all } i \in [n].$$

We separate all clauses into two groups E and F (with its meaning as in Theorem 3.1) so that E consists of all clauses that encode a property of solution being a linear ordering (types 1–3) and F consists of the non-minimality clauses $\{C_i\}$.

Lemma 3.4. *Let $Ax = b$ be the linear system in the variables of $Ordering_n$ with at most $n - 2$ equations that has an E-respectful solution. Then for every clause $C \in F$ there is the E-respectful solution τ of the system $Ax = b$ that satisfies C.*

Proof. Every E-respectful solution of $Ax = b$ encodes a linear ordering, hence exactly one of the clauses of F must be refuted by each solution (the clause corresponding to the non-minimality of the minimum element in this particular ordering).

Let σ be the E-respectful solution of $Ax = b$. Without loss of generality we may assume that σ encodes the natural ordering $(1, 2, \ldots, n)$. We also replace all occurrences of x_{ji} where $j > i$ with $1 - x_{ij}$. As we are only interested in E-respectful solutions, the operation is safe.

We prove the statement by induction on n.

The base case $n = 2$ is trivial. The system $Ax = b$ is empty so every linear ordering on n elements can be encoded as an E-respectful solution of the system. Clearly, for $n > 1$ there are at least two of them with different minimal elements.

Induction step: among all variables x_{ij} that appear in $Ax = b$ with non-zero coefficient choose i, j so that no x_{kl} appear in the system for all k, l such that $i \leq k < l \leq j$ and $(k, l) \neq (i, j)$. If there are more than one such a choice, choose one with the greatest value of i. We can always choose such i as any x_{ij} in the system with smallest value of $j - i$ satisfies such conditions.

Suppose that $i = 1$. By the construction (we chose i to be maximal possible) only variables $x_{12}, \ldots, x_{1n}$ can appear in the system. Consider $Ax = b$ as a system of these variables. It has a solution by the assumption, and the number of lines in the system is $n-2$ that is fewer then the number of variables. Hence, the system must have another solution τ'. This partial solution of the original system can be extended to any E-respectful solution with fixed position of element 1 (as we fix only $\{x_{1k}\}_{k=2}^{n}$). Arbitrarily extend τ' to τ for all other variables such that τ is still an E-respectful solution, and 1 is not a minimal element in corresponding ordering: our choice of the values of $\{x_{1k}\}_{k=2}^{n}$ is different from the σ (which set all of them to 1), hence there is some k, s.t. $x_{1k} = 0$ in τ, i.e $k \prec 1$.

Otherwise $i > 1$. In this case we "glue" two specifically chosen elements (i and j) of the order encoded by σ and then reduce the problem to the induction hypothesis.

Firstly, consider any line of $Ax = b$ with x_{ij} and add it to other lines with x_{ij}. This way x_{ij} appears only in one equation of the resulting system (which is clearly equivalent to $Ax = b$ since we only added some equations together). We denote this equation as $f(x) = \alpha$. Let $Bx = c$ be the rest of the system.

We "glue" elements i and j together in $Bx = c$ by replacing all of the occurrences of j by i. This operation is correct as there are no x_{kl} for all $i \leq k < l \leq j$ in $Bx = c$ by the choice of i, j. Let $B'x' = c$ be the "glued" system. It has an E-respectful solution $\sigma' = (1, 2, \ldots, i, \ldots, j - 1, j + 1, \ldots, n)$.

The system $B'x' = c$ has at most $n - 3$ lines, depends on the variables corresponding to the ordering of $n-1$ elements, and has the E-respectful solution σ'. Therefore we can apply the induction hypothesis and get the different E-respectful solution τ' corresponding to the ordering $(\tau_1, \ldots, \tau_{n-1})$ with $\tau_1 \neq 1$.

Let $i = \tau_k$ for some k. We consider two orderings

$$(\tau_1, \ldots, \tau_{k-1}, i, j, \tau_{k+1}, \ldots, \tau_{n-1}) \tag{1}$$

and

$$(\tau_1, \ldots, \tau_{k-1}, j, i, \tau_{k+1}, \ldots, \tau_{n-1}). \tag{2}$$

Each of them corresponds to the E-respectful solution $\tau^{(t), t=1,2}$ of $Bx = c$ and only one of them satisfies $f(x) = \alpha$. Therefore, for some $t \in \{1, 2\}$ we have $\tau = \tau^{(t)}$ that is an E-respectful solution of $Ax = b$, and σ and τ have different minimal elements in corresponding orderings. $\qquad \square$

Theorem 3.5. *The size of any tree-like* Res($\oplus$) *refutation of Ordering$_n$ is at least* 2^{n-2}.

Proof. Follows from application of Lemma 3.4 to Theorem 3.1. $\square$

3.3   Dense Linear Ordering Principle

Dense Linear Ordering (DLO) principle states that there are no dense linear orderings of n elements. We encode the negation of this statement as the CNF formula DLO_n with variables $\{x_{ij}\}_{i \neq j \in [n]}$ where $x_{ij} = 1$ iff i goes before j ($i \prec j$) and $\{z_{ikj}\}_{i \neq j \neq k \in [n]}$ where if $i \prec j$ and $z_{ikj} = 1$ then $i \prec k \prec j$ (we call such k a "witness" of $i \prec j$). DLO_n has five types of clauses:

1. Anti-symmetry: $\neg x_{ij} \vee \neg x_{ji}$ for all $i \neq j \in [n]$.
2. Totality: $x_{ij} \vee x_{ji}$ for all $i \neq j \in [n]$.
3. Transitivity: $\neg x_{ij} \vee \neg x_{jk} \vee x_{ik}$ for all distinct $i, j, k \in [n]$.
4. Semantics for z's: $\neg z_{ikj} \vee x_{ik}$ and $\neg z_{ikj} \vee x_{kj}$ for all distinct $i, j, k \in [n]$.
5. Density (if $i \prec j$ then there must be some k between them):
   $C_{ij} = \neg x_{ij} \vee (\bigvee_{k \in [n] \backslash \{i,j\}} z_{ikj})$ for all $i \neq j \in [n]$.

We separate clauses into two groups: E is a set of clauses of type 1–4, and F is the rest, i.e. $\{C_{ij}\}$.

Lemma 3.6. *Let $Av = b$ be the linear system in the variables of DLO_n with at most $M = \frac{n-3}{3}$ equations that has an E-respectful solution. Then for every clause C_{st} in F there is an E-respectful solution of the system $Av = b$ that satisfies this clause.*

Firstly, we prove a restricted auxiliary version of this lemma:

Claim. Let $Av = b$ be the linear system in the variables of DLO_n with at most $n - 3$ equations that has the E-respectful solution σ which sets the values of all z's to 0. Then for every clause C_{st} in F there is the E-respectful solution τ of the system $Av = b$ that satisfies this clause and τ can set a z variable to 1 only if it has the form z_{skt} for some k.

Proof. Without loss of generality assume that the system has the E-respectful solution σ that corresponds to the natural ordering $(1, 2, \ldots, n)$. We replace all occurrences of $x_{ji}, j > i$ in the system with $1 - x_{ij}$.

Let C_{st} be the clause of F that is falsified by σ.

We prove the statement by induction on n.

Base case $n = 3$ is trivial since the system is empty and for every clause in F can be satisfied by some ordering. We can either choose an ordering with $s \succ t$ or choose an ordering with $s \prec w \prec t$ for some w (which exists since $n \geq 3$), and set $z_{swt} = 1$.

Induction step. Choose the pair (i, j) such that $t \leq i < j$, x_{ij} appears in $Av = b$ with non-zero coefficient and no other x_{kl} appears in the system for all

k, l satisfying $i \leq k < l \leq j, (k, l) \neq (i, j)$. If there are more than one such a choice, choose the pair with the maximal value of i. Suppose that such a pair exists. We "glue" i and j into i in a similar way as we did in the proof of Lemma 3.4.

Modify the system by leaving x_{ij} only in one equation $f(v) = \alpha$ by adding this equation to any other line of the system containing x_{ij}. Let $Bv = c$ be the rest of the system. Replace each occurrence of j in $Bv = c$ by i and denote the resulting system as $B'v = c'$. System $B'v = c'$ has an E-respectful solution τ by the induction hypothesis. Now extend τ by adding j and placing to the one of the sides of i: choose an order for i and j to satisfy $f(v) = \alpha$.

If we cannot choose such pair, then t does not compare with any other $w > t$ and we can modify σ so that t is a maximal element in the ordering encoded by σ: we can set $x_{wt} = 1$ for all $w > t$ and it will remain a solution as x_{wt} does not appear in the system.

Similarly try to choose the pair (i, j) to the left of s: $i < j \leq s$ such that x_{ij} appears in the system with non-zero coefficient and no x_{kl} appears in the system for all k, l satisfying $i \leq k < l \leq j, (k, l) \neq (i, j)$. "Glue" them into j and apply the induction hypothesis if successful, otherwise make s the minimum element.

Otherwise, we have the E-respectful solution with s as the minimum and t as the maximum, hence any other element can be taken as a witness of $s \prec t$. Assign the values from σ for all variables except z_{skt} for all k. The resulting system has $n - 2$ variables, at most $n - 3$ equations, and has a zero solution, hence it has at least one other solution with some z_{skt} set to 1. We can change σ according to that solution and get the resulting E-respectful solution τ which has $\tau(z_{skt}) = 1$ for some k. ∎

Proof (Lemma 3.6). Every E-respectful solution encodes a linear ordering.

Let σ be the E-respectful solution of the system with minimal number of 1's assigned to the z variables. We claim that σ sets at most M ones to the z variables. Indeed, set all x's from σ to $Av = b$ and view the resulting system as a system in z variables. There are at most M lines, hence there are at most M basic variables, and all other variables are free. We can set all free variables to 0, and only the basic ones can be set to 1. Here we stress the fact that if σ is an E-respectful solution with $\sigma(z_{ijk}) = 1$ for some i, j, k, then we can set $\sigma(z_{ijk}) = 0$ and it will remain E-respectful, i.e. changing 1 to 0 in z's does not falsify any clause in E.

Let us define the system $Cv = d$: it contains all equations from $Av = b$ with z_{ijk} replaced by 1 if $\sigma(z_{ijk}) = 1$; also, for every z_{ijk} with $\sigma(z_{ijk}) = 1$ it contains two additional equations $x_{ij} = 1$ and $x_{jk} = 1$.

By the construction, the resulting system $Cv = d$ has at most $3M \leq n - 3$ equations and for every z_{ijk} which appears in $Cv = d$ we have $\sigma(z_{ijk}) = 0$. Therefore, the system has E-respectful solution σ' which coincides with σ on x's and sets all z's to 0.

We can apply Lemma 3.3 to the system $Cv = d$ with the solution σ' for the clause C_{st} and get the E-respectful solution τ of $Cv = d$ that satisfies C_{st}. Since for every z_{ijk} in $Av = b$ with $\sigma(z_{ijk}) = 1$ we have $\tau(x_{ij}) = 1$ and $\tau(x_{jk}) = 1$, we can set $\tau(z_{ijk}) = 1$ and it will remain E-respectful. The modified τ is an E-respectful solution of $Av = b$ because of that and it satisfies C_{st}. ☐

Theorem 3.7. *The size of any tree-like* $\mathrm{Res}(\oplus)$ *refutation of* DLO_n *is at least* $2^{(n-3)/3}$.

Proof. Follows from application of Lemma 3.6 to Theorem 3.1. ☐

We emphasize here that this result is new and to the moment cannot be obtained using different techniques.

4   Weakening in Rings $R \neq \mathbb{F}_2$ is Hard to Verify

Let R be a commutative ring with identity. A *linear clause* over R is a disjunction $\bigvee_{i=1}^{s} f_i(x) = \alpha_i$ of linear equations $f_i(x) = \alpha_i$ over R. We say that the linear clause D is semantically implied on Boolean cube by the linear clause C if every 0/1 assignment of the variables that satisfies C also satisfies D. Semantic weakening rule on Boolean cube allows to derive the linear clause D from the linear clause C if D is semantically implied by C on Boolean cube.

Part and Tzameret conjectured [10] that verifying application of semantic weakening rule on Boolean cube is a coNP-hard problem if $R \neq \mathbb{F}_2$ (recall that for $R = \mathbb{F}_2$ it can be done in polynomial time by Lemma 2.1). Part and Tzameret only proved this statement for R with $\mathrm{char}(R) = 0$ or $\mathrm{char}(R) \geq 5$:

Proposition 4.1 ([10], Proposition 5). *Let R be a ring with* $\mathrm{char}(R) = 0$ *or* $\mathrm{char}(R) \geq 5$. *Then the following problem is* coNP-*complete: given a linear clause over the ring R decide whether it is a tautology on Boolean cube.*

We consider the remaining cases: $R \neq \mathbb{F}_2$ with $\mathrm{char}(R) = 2$ or $\mathrm{char}(R) = 3$, hence completing the proof of the conjecture.

Theorem 4.2. *Let R be a ring, $R \neq \mathbb{F}_2$. Then the following problem is* coNP-*complete: given a linear clause over the ring R decide whether it is a tautology on Boolean cube.*

Proof. For the case of rings of characteristic 0 one may refer to the proof of Proposition 5 by Part and Tzameret in [10]. We prove the statement for the ring R with $\mathrm{char}(R) > 0, R \neq \mathbb{F}_2$.

We show how to encode any given 3-DNF formula $\phi = \phi(x_1, x_2, \ldots, x_n)$ as a disjunction of linear equations $\Phi = \Phi(y_1, y_2, \ldots, y_m)$ so that ϕ is a tautology iff Φ is a tautology on Boolean cube $\{0,1\}^m$. The plan of the proof is the following. Firstly, we transform ϕ into an equivalent disjunction of "all equal" predicates (denoted as ae). Then we show that such predicates can be encoded as linear equations in R on Boolean cube.

Definition 4.3. *The predicate "all equals" is defined on* R^3 *and* $\mathrm{ae}(x,y,z) = 1$ *iff* $x = y = z$ *in* R.

To perform the transformation we use the idea from the proof that NAE-3-SAT is NP-complete [12]: the clause $x \lor y \lor z$ holds iff the formula $\exists a. \neg \mathrm{ae}(0, x, \neg a) \land \neg \mathrm{ae}(y, z, a)$ holds (we prove the similar statement below).

Without loss of generality we assume that every term (a conjunction of width at most 3) of the 3-DNF formula has width exactly 3: we can simply repeat an arbitrary literal of this term several times (e.g. replace $x \land y$ with $x \land y \land x$). For DNF formulas we use a similar statement: $x \land y \land z$ is equivalent to $\forall a. \mathrm{ae}(0, \neg x, \neg a) \lor \mathrm{ae}(\neg y, \neg z, a)$. The proof is straightforward: for $a = 0$ the second condition forces $y = z = 1$ and for $a = 1$ the first condition forces $x = 1$. We apply the transformation to every term of ϕ, introducing a fresh variable a_i for (a in the example above) for every term t_i of the system. Since we are checking the system for being a tautology on Boolean cube, we can simply remove all $\forall$ quantifiers for freshly-introduced variables. We denote the resulting system as Φ.

Remark 4.4. We assume that every variable can only be set to either 0 or 1, hence for every variable a in R we can safely use $\neg a$ to denote $1 - a$ in R.

Claim. ϕ is a tautology iff Φ is a tautology.

Proof. Let σ be the assignment to the variables of ϕ.

Assume that ϕ is a tautology. Hence some term $x \land y \land z$ is satisfied by σ. By the construction after application of σ we get $\mathrm{ae}(0, 0, \neg a) \lor \mathrm{ae}(0, 0, a)$ which concludes that either $\neg a = 0$ or $a = 0$, which is clearly true since $a \in \{0, 1\}$. Therefore, any extension of σ to the variables of Φ satisfies Φ.

The opposite direction is similar. Assume that Φ is a tautology. Consider system $\Psi = \Phi|_\sigma$. Since Φ is a tautology on Boolean cube, Ψ is also a Boolean tautology consisting of only fresh variables a_i introduced during the transformation (recall that for every term t_i we introduced variable a_i which appears exactly in two clauses). By the construction and after application of σ, every variable a_i appears in at most two clauses of the form $a_i = \beta$ for some $\beta \in R$, and does not appear in any other clauses. Hence, Ψ is a Boolean tautology iff for some i there are clauses $a_i = 0$ and $a_i = 1$ in Ψ. We can get these equations only if the assignment σ satisfies the original term (it can be checked by considering all possible assignments to the variables of the term). ∎

Claim. A Boolean restriction of the predicate "all equals" ae can be expressed in any ring R of positive characteristic and $R \neq \mathbb{F}_2$ by a single linear equation in R.

Proof. For R with $\mathrm{char}(R) = 2, R \neq \mathbb{F}_2$: let α be any element of R that is not 0 or 1. Then $\mathrm{ae}(x, y, z) = 1$ iff $x + \alpha y + (1 + \alpha)z = 0$ for Boolean x, y, z.

For simplicity, we start with the definition of this operator for $\mathrm{char}(R) = 3$. In this case $\mathrm{ae}(x, y, z) = 1$ iff $x + y + z = 0$ on Boolean cube. It can be easily checked by considering all possibilities for Boolean x, y, and z.

In case of arbitrary char$(R) = k \geq 3$ repeat any variable in the sum so that the total number of summed variables is k. For the ring R with characteristic $k \geq 3$ and Boolean variables $x_1, x_2, \ldots, x_k$ the linear equation $x_1 + x_2 + \cdots + x_k = 0$ holds iff all x_i are equal. Hence, we can encode the "all equals" predicate $ae(x, y, z)$ as $x + y + z + \underbrace{x + \cdots + x}_{k-3} = 0$ in R.   ■ □

Acknowledgements. The author is grateful to Dmitry Itsykson, Dmitry Sokolov, and Anastasia Sofronova for fruitful discussions. The author also thanks Dmitry Itsykson for the statement of the problem and Dmitry Sokolov for telling about Dense Linear Ordering principle and related problems.

References

1. Atserias, A., Dalmau, V.: A combinatorial characterization of resolution width. J. Comput. Syst. Sci. **74**(3), 323–334 (2008)
2. Beame, P., Pitassi, T., Segerlind, N.: Lower bounds for LovÁsz-Schrijver systems and beyond follow from multiparty communication complexity. SIAM J. Comput. **37**(3), 845–869 (2007)
3. Ben-Sasson, E.: Hard examples for the bounded depth Frege proof system. Comput. Complex. **11**(3), 109–136 (2002)
4. Galesi, N., Lauria, M.: Optimality of size-degree tradeoffs for polynomial calculus. ACM Trans. Comput. Logic **12**(1), 4:1–4:22 (2010)
5. Garlík, M., Kołodziejczyk, L.A.: Some subsystems of constant-depth frege with parity. ACM Trans. Comput. Logic **19**(4), 1–34 (2018)
6. Göös, M., Pitassi, T.: Communication lower bounds via critical block sensitivity. In: Proceedings of the 46th Annual ACM Symposium on Theory of Computing - STOC 2014. ACM Press (2014)
7. Itsykson, D., Knop, A.: Hard satisfiable formulas for splittings by linear combinations. In: Gaspers, S., Walsh, T. (eds.) SAT 2017. LNCS, vol. 10491, pp. 53–61. Springer, Cham (2017). https://doi.org/10.1007/978-3-319-66263-3_4
8. Itsykson, D., Sokolov, D.: Lower bounds for splittings by linear combinations. In: Csuhaj-Varjú, E., Dietzfelbinger, M., Ésik, Z. (eds.) MFCS 2014. LNCS, vol. 8635, pp. 372–383. Springer, Heidelberg (2014). https://doi.org/10.1007/978-3-662-44465-8_32
9. Mikša, M., Nordström, J.: A Generalized Method for Proving Polynomial Calculus Degree Lower Bounds. In: Proceedings of the 30th Conference on Computational Complexity, CCC 2015, pp. 467–487. Schloss Dagstuhl-Leibniz-Zentrum fuer Informatik, Portland, Oregon (2015)
10. Part, F., Tzameret, I.: Resolution with counting: lower bounds over different moduli. In: Electronic Colloquium on Computational Complexity (ECCC), vol. 25, p. 117 (2018)
11. Pudlák, P., Impagliazzo, R.: A lower bound for DLL algorithms for k-SAT (preliminary version) (2000)
12. Schaefer, T.J.: The complexity of satisfiability problems. In: Proceedings of the Tenth Annual ACM Symposium on Theory of Computing, STOC 1978, pp. 216–226. ACM, San Diego (1978)

Undecidable Word Problem in Subshift Automorphism Groups

Pierre Guillon[1(✉)], Emmanuel Jeandel[2], Jarkko Kari[3], and Pascal Vanier[4]

[1] Université d'Aix-Marseille, CNRS, Centrale Marseille I2M, UMR 7373,
13453 Marseille, France
pierre.guillon@math.cnrs.fr

[2] Université de Lorraine, CNRS, Inria LORIA, 54000 Nancy, France
emmanuel.jeandel@loria.fr

[3] Department of Mathematics and Statistics, University of Turku,
20014 Turku, Finland
jkari@utu.fi

[4] Laboratoire d'Algorithmique, Complexité et Logique, Université de Paris-Est,
LACL, UPEC, Paris, France
pascal.vanier@lacl.fr

Abstract. This article studies the complexity of the word problem in groups of automorphisms (or reversible cellular automata) of subshifts. We show in particular that for any computably enumerable Turing degree, there exists a (two-dimensional) subshift of finite type whose automorphism group contains a subgroup whose word problem has exactly this degree. In particular, there are such subshifts of finite type where this problem is uncomputable. This remains true in a large setting of subshifts over groups.

Subshifts are sets of colorings of a group G avoiding some family of forbidden finite patterns. They have first been introduced, for $G = \mathbb{Z}$, as a way of discretizing dynamical systems on compact spaces. SFTs correspond to the particular case when only finitely many patterns are forbidden; they are used in information theory to model data streams with coding constraints. When $G = \mathbb{Z}^2$, SFTs turn out to be, up to recoding, the sets of colorings defined by some Wang tiles, and a tool to study decidability questions. When G is the free group, subshifts can be seen as sets of colorings of a tree; the case of the free monoid is known to correspond to the so-called tree languages, and SFTs to tree automata [1–3].

Subshifts are hence both a means to model complex systems, and to provide complete problems for a wide range of complexity and computability classes.

An automorphism of a subshift X is a shift-invariant continuous bijection from X onto X, or equivalently a reversible cellular automaton on X. Understanding the automorphism group of a subshift can be seen as a way to understand how constraints over the "physical space" (the possible configurations) restrict the interactions between the cellular automata that act on them.

Little is known about automorphism groups of subshifts in general, besides that they are countable. As an example of our ignorance, it is a long-standing

© Springer Nature Switzerland AG 2019
R. van Bevern and G. Kucherov (Eds.): CSR 2019, LNCS 11532, pp. 180–190, 2019.
https://doi.org/10.1007/978-3-030-19955-5_16

open problem whether the automorphism groups of the 2-symbol full shift and of the 3-symbol full shift are isomorphic.

Many results have nevertheless been recently reached, for $G = \mathbb{Z}$, in two kinds:

- The automorphism group of some *large* subshifts (positive-entropy SFTs, ...) is rich [4]: it contains all finite groups, finitely generated abelian groups, countable free and free abelian groups, ... This means that when you pick some reversible cellular automata over these subshifts, they can have very complex interactions. In [5], it is proved that periodicity of cellular automata is undecidable, which can be interpreted as the torsion problem for the automorphism group of these subshifts.
- The automorphism group of some *small* subshifts (small complexity function, substitutive, ...) is poor [6–8]: in the most extreme case, it is proven to be virtually $\mathbb{Z}$, which means that every reversible cellular automaton is essentially the shift (up to finitely many local permutations).

With $G = \mathbb{Z}^d$ when $d \geq 2$, computability has played a central role in the study of SFTs. From a computability point of view, it is noted in [9] that their automorphism groups have a computably enumerable word problem (which is formalized in a general setting in Theorem 2). The word problem essentially corresponds to picking up a reversible cellular automaton rule over this subshift, and asking whether it is equal to the identity. We show that it can be arbitrarily complex: for any given computably enumerable degree, one can construct an SFT the automorphism group of which has a word problem with this degree (Corollary 2).

1   Preliminaries

By countable set, we mean injectable in $\mathbb{N}$. Let λ denote the empty word. For $\mathcal{A}$ a countable alphabet, we note $\mathcal{A}^* := \bigsqcup_{n \in \mathbb{N}} \mathcal{A}^n$ the set of finite words over $\mathcal{A}$. We also note $\mathcal{A}^{\leq r} := \bigsqcup_{n \leq r} \mathcal{A}^n$ for $r \in \mathbb{N}$.

Let us note X^C the complement of set X. $W \Subset X$ means that $W \subset X$ and W is finite. $V \sqcup W$ means $V \cup W$ assuming that $V \cap W = \emptyset$.

1.1   Computability

Computability problems are naturally defined over $\mathbb{N}$, but can easily be extended through subsets of it, cartesian products or disjoint union (by canonically injecting $\mathbb{N}$ in sets of tuples). For example, if $\mathcal{G} \subset \mathbb{N}$, then the set $\mathcal{G}^*$ of tuples admits a simple injection into $\mathbb{N}$. Let us fix a (computable) countable set I, that we can identify to integers.

Definition 1. *Let us define the following reducibility notions, for $X, Y \subset I$:*

1. X is Turing-reducible *to Y, $X \leq_T Y$, if: one can compute X with oracle Y.*

2. X *is* enumeration-reducible *to* Y, $X \leq_e Y$, *if: from any* x *and any integer* $i \in \mathbb{N}$, *one can compute a finite set* $Y_i(x)$ *such that* $x \in X$ *if and only if* $\exists i \in \mathbb{N}, Y_i(x) \subset Y$.

3. X *is* positive-reducible *to* Y, $X \leq_p Y$, *if: from any* x, *one can compute finitely many finite sets* $Y_0(x), \ldots, Y_{n-1}(x)$ *such that* $x \in X$ *if and only if* $\exists i < n, Y_i(x) \subset Y$.

4. X *is* many-one-reducible *to* Y, $X \leq_m Y$, *if: from any* x, *one can compute some* $\phi(x)$ *such that* $x \in X$ *if and only if* $\phi(x) \in Y$.

5. X *is* one-one-reducible *to* Y, $X \leq_1 Y$, *if,* $X \leq_m Y$ *and the corresponding* ϕ *is one-to-one.*

One-one reducibility implies many-one reducibility, which in turns implies positive-reducibility, which implies both Turing-reducibility and enumeration-reducibility.

Each reducibility $\leq_r$ induces a notion of equivalence $\equiv_r$: $A \equiv_r B$ iff $A \leq_r B$ and $B \leq_r A$. And each notion of equivalence $\equiv_r$ induces a notion of degree $\deg_r$: the *degree of a set* A is its equivalence class for $\equiv_r$.

The *join* $A \oplus B$ of A and B is the set C such that $2n + 1 \in C$ iff $n \in A$ and $2n \in C$ iff $n \in B$. It has the property that $A \leq_r A \oplus B$ and $B \leq_r A \oplus B$ for any reducibility $\leq_r$ previously defined.

See [10] for a reference on computability-theoretical reductions.

1.2 Monoids and Groups

We will deal with countable monoids $\mathbb{M} = \mathcal{G}^*/R$, where $\mathcal{G} \subset \mathbb{N}$, $\mathcal{G}^*$ is the free monoid generated by symbols from $\mathcal{G}$ and R is a monoid congruence[1]. The monoid is always implicitly endowed with its generating set $\mathcal{G}$ (later, some problems may depend on the presentation). Each element of the monoid is represented by a word $u \in \mathcal{G}^*$, but the representation is not one-to-one (except for the free monoid itself). We note $i =_\mathrm{M} j$ if $\pi(i) = \pi(j)$ and $\pi : \mathcal{G}^* \to \mathbb{M}$ is the natural quotient map.

It is also clear that the concatenation map, which from any two words $i, j \in \mathcal{G}^*$ outputs $i \cdot j$, which is one representative of the corresponding product, is computable. We say that $\mathbb{M}$ is an *effective group* if, additionally, there is a computable map $\psi : \mathcal{G}^* \to \mathcal{G}^*$ such that $i \cdot \psi(i) =_\mathrm{M} \psi(i) \cdot i =_\mathrm{M} \lambda$.

The *equality problem* of $\mathbb{M}$, endowed with generating family $\mathcal{G}$, is the set of pairs $\{ (i,j) \in (\mathcal{G}^*)^2 \,|\, i =_\mathrm{M} j \}$, endowed with a natural enumeration so that we can consider it as a computability problem.

Remark 1

1. It is clear that the *word problem* $\{ i \in \mathcal{G}^* \,|\, i =_\mathrm{M} \lambda \}$ is one-one-reducible to the equality problem.

2. If $\mathbb{M}$ is an effective group, then the word problem is actually many-one-equivalent to the equality problem.

---

[1] We could deal in the same way with semigroups, by prohibiting the empty word.

3. The equality problems for $\mathbb{M}$ endowed with two distinct finite generating sets are one-one-equivalent.
4. If $\mathbb{M}'$ is a submonoid of $\mathbb{M}$ endowed with a generating set which is included in that of $\mathbb{M}$, then the equality problem in $\mathbb{M}'$ is one-one-reducible to that of $\mathbb{M}$.
5. In particular, the equality problem in any finitely generated submonoid is one-one-reducible to that of $\mathbb{M}$.

Nevertheless, there are countable groups whose word problem is computable when endowed with one generating family, and uncomputable when endowed with another one.

The word problem is known to be decidable if and only if the group is *computable* (see [11] for a proof in the finitely generated case), that is, it can be seen as a computable subset of $\mathbb{N}$ over which the composition rule is a computable function (this implies that inversion is also a computable map).

1.3 Subshifts

Let $\mathcal{A}$ be a finite alphabet with at least two letters, and $\mathbb{M}$ a group (most of the following should be true if $\mathbb{M}$ is a cancellative monoid though). In a first reading, the reader is encouraged to think of $\mathbb{M}$ as being $\mathbb{Z}$: the results are not significantly simpler in that specific setting (except those that mention 1D SFT). A finite *pattern* w over $\mathcal{A}$ with *support* $W = \mathcal{S}(w) \Subset \mathcal{G}^*$ is a map $w = (w_i)_{i \in W} \in \mathcal{A}^W$. An element of $\mathcal{A}^{\mathbb{M}}$ is called a *configuration*. Configurations can be seen as colorings of the Cayley graph by the letters of $\mathcal{A}$ and patterns can be seen as finite configurations. Depending on the context, note that, for $g \in \mathcal{S}(w)$, w_g may either be an element of $\mathcal{A}$ or a subpattern with support $\{g\}$. If $g \in \mathcal{G}^*$ and w is a pattern, we will denote by $\sigma^g(w)$ the pattern with support $W \cdot g$ such that $\sigma^g(w)_{i \cdot g^{-1}} = w_i$ for all $i \in \mathcal{S}(w)$.

We are interested in $\mathcal{A}^{\mathbb{M}}$, which is a Cantor set, when endowed with the prodiscrete topology, on which $\mathbb{M}$ acts continuously by (left) shift: we note $\sigma^i(x)_j = x_{i \cdot j}$ for $i, j \in \mathbb{M}$ and $x \in \mathcal{A}^{\mathbb{M}}$. $\mathcal{A}^{\mathbb{M}}$ is thus called the *full shift* on alphabet $\mathcal{A}$. A *subshift* is a closed σ-invariant subset $X \subset \mathcal{A}^{\mathbb{M}}$. Equivalently, X can be defined as the set $X_{\mathcal{F}} := \{ x \in \mathcal{A}^{\mathbb{M}} \mid \forall i \in \mathbb{M}, \forall w \in \mathcal{F}, \exists j \in \mathcal{S}(w), x_{i \cdot j} \neq w_j \}$ avoiding a language $\mathcal{F} \subset \bigsqcup_{W \Subset \mathcal{G}^*} \mathcal{A}^W$, which is then called a (defining) *forbidden language*. If $\mathcal{F}$ can be chosen finite, the subshift is called *of finite type* (SFT); if it can be chosen computably enumerable, it is called *effective*. Figure 1 shows an example of forbidden language and configuration of the associated subshift.

The *language* with *support* $W \Subset \mathcal{G}^*$ of subshift X is the set $\mathcal{L}_W(X) := \{ (x_{\pi(i)})_{i \in W} \mid x \in X \}$; the *language* of X is $\mathcal{L}(X) = \bigsqcup_{W \Subset \mathcal{G}^*} \mathcal{L}_W(X)$, and its *colanguage* is the complement of it. The latter is a possible defining forbidden language. If $u \in \mathcal{L}_W(X)$, we define the corresponding *cylinder*

$$[u] = \{ x \in X \mid \forall i \in W, x_{\pi(i)} = u_i \}.$$

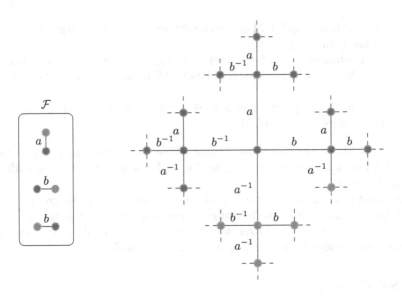

Fig. 1. If $\mathbb{M}$ is the free group on two elements $\{a, b\}$ and the set of forbidden patterns is on the left, then the configuration on the right is in $X_{\mathcal{F}}$.

Remark 2. π induces a natural covering $\Pi : \mathcal{A}^{\mathbb{M}} \to \mathcal{A}^{\mathcal{G}^*}$ by $\Pi(x)_i = x_{\pi(i)}$. Its image set $\Pi(\mathcal{A}^{\mathbb{M}})$ is a subshift over the free monoid. One can note the following.

1. $X = X_{\mathcal{L}(X)^C}$.
2. The colanguage of the full shift $\mathcal{A}^{\mathbb{M}}$ is the same as that of the subshift $\Pi(\mathcal{A}^{\mathbb{M}})$: the set

$$\mathcal{L}(\mathcal{A}^{\mathbb{M}})^C = \bigsqcup_{W \in \mathcal{G}^*} \{ w \in \mathcal{A}^W \mid \exists i, j \in W, i =_{\mathbb{M}} j, w_i \neq w_j \}$$

of patterns that do not respect the monoid congruence.
3. Nevertheless, $\emptyset$ is a forbidden language defining $\mathcal{A}^{\mathbb{M}}$.
4. The colanguage of every subshift $X_{\mathcal{F}} \subset \mathcal{A}^{\mathbb{M}}$ is the set of patterns $w \in \mathcal{A}^W$, $W \Subset \mathcal{G}^*$, whose all extensions to configurations $x \in [w]$ involve as a subpattern a pattern of either $\mathcal{F}$, or $\mathcal{L}(\mathcal{A}^{\mathbb{M}})^C$. In that case, by compactness, there exists $V \supset W$ (which depends only on W) such that $w \in \mathcal{A}^W$ is in the colanguage iff every $v \in \mathcal{A}^V$ such that $v_{|W} = u$ involves a subpattern from $\mathcal{F}$ or $\mathcal{L}(\mathcal{A}^{\mathbb{M}})^C$.

Remark 3. Let $\mathbb{M}$ be a monoid.

1. The equality problem in $\mathbb{M}$ is positive-equivalent (and one-one-reducible) to the colanguage of the full shift.
2. The colanguage of any subshift X is enumeration-reducible to the join of any defining forbidden language for X and the equality problem of $\mathbb{M}$.

Proof

1. one-one-reducibility: one can computably map each word $(i, j) \in (\mathcal{G}^*)^2$ to a unique pattern over $\{i, j\}$ involving two different symbols. By Point 2 of Remark 2, this pattern is in the colanguage of the full shift if and only if $i =_M j$.

   positive-reducibility (with all Y_is being singletons): from each pattern $w \in \mathcal{A}^{\mathcal{G}^*}$, one can compute the set of pairs $(i, j) \in \mathcal{S}(w)^2$ such that $w_i \neq w_j$. By Point 2 of Remark 2, w is in the colanguage if and only if one of these pairs is an equality pair in M.

2. Consider the set Z of *locally inadmissible* patterns, that involve a subpattern either from the forbidden language or from $\mathcal{L}(\mathcal{A}^M)^C$. From any pattern w, one can enumerate all of its subpatterns and all of their shifts, *i.e.* all patterns v such that there exists $i \in \mathcal{G}^*$ with $\mathcal{S}(v) \cdot i \subset \mathcal{S}(w)$ and $w_{j \cdot i} = v_j$ for every $j \in \mathcal{S}(v)$. This shows that Z is enumeration-reducible to the join of the forbidden language and $\mathcal{L}(\mathcal{A}^M)^C$, the latter being equivalent to the equality problem, by the previous point. It remains to show that the colanguage of X is enumeration-reducible to Z.

   From any pattern $w \in \mathcal{A}^{\mathcal{G}^*}$ and any $i \in \mathbb{N}$, one can compute some $V_i \Subset \mathcal{G}^*$ including $\mathcal{S}(w)$, in a way that $V_{i+1} \supset V_i$ and $\bigcup_{i \in \mathbb{N}} V_i = \mathcal{G}^*$ (for example take the union of $\mathcal{S}(w)$ with balls in the Cayley graph). Then, one can compute the set Y_i of extensions of w to V_i, *i.e.* patterns with support V_i whose restriction over $\mathcal{S}(w)$ is w. By Point 4 of Remark 2, $w \in \mathcal{L}(X)^C$ if and only if there exists $V \Subset \mathcal{G}^*$ with $V \supset \mathcal{S}(w)$ such that all extensions of w to V are in Z; and in particular this should happen for some V_i, which precisely means that $Y_i \subset Z$. $\square$

It results that, in some sense, one expects most subshifts to have a colanguage at least as complex as the equality problem in the underlying monoid.

1.4   Homomorphisms

Let $X \subset \mathcal{A}^M$ and $Y \subset \mathcal{B}^M$ be subshifts. Denote $\mathcal{E}nd(X, Y)$ the set of *homomorphisms* (continuous shift-commuting maps) from X to Y, and $\mathcal{A}ut(X, Y)$ the set of bijective ones (*conjugacies*). We also note $\mathcal{E}nd(X) = \mathcal{E}nd(X, X)$ the monoid of *endomorphisms* of X, and $\mathcal{A}ut(X) = \mathcal{A}ut(X, X)$ the group of its *automorphisms*.

If M is finitely generated, then homomorphisms correspond to block maps (and endomorphisms to cellular automata), thanks to a variant of the Curtis-Hedlund-Lyndon theorem [12].

Theorem 1. *Let M be finitely generated. A map Φ from subshift $X \subset \mathcal{A}^M$ into subshift $Y \subset \mathcal{B}^M$ is a homomorphism if and only if there exist a radius $r \in \mathbb{N}$ and a block map $\phi : \mathcal{A}^{\mathcal{G}^{\leq r}} \to \mathcal{B}$ such that for every $x \in \mathcal{A}^M$ and $i \in \mathcal{G}^*$, $\Phi(x)_{\pi(i)} = \phi(x_{|\pi(i \cdot \mathcal{G}^{\leq r})})$ (where the latter has to be understood with the obvious reindexing of the argument).*

See Fig. 2 for an example of block map and associated homomorphism. Let us order the block maps $\phi : \mathcal{A}^{\mathcal{G}^{\leq r}} \to \mathcal{B}$ by increasing radius $r \in \mathbb{N}$, and then by lexicographic order, so that we have a natural bijective enumeration $\mathbb{N} \to \bigsqcup_{r \in \mathbb{N}} \mathcal{B}^{\mathcal{A}^{\mathcal{G}^{\leq r}}}$ (because $\mathcal{A}$, $\mathcal{B}$ and $\mathcal{G}$ are finite). This gives in particular a surjective enumeration $\mathbb{N} \to \mathcal{E}nd(\mathcal{A}^{\mathcal{G}^*}, \mathcal{B}^{\mathcal{G}^*})$ and in general, a partial surjective enumeration $\mathbb{N}' \subset \mathbb{N} \to \mathcal{E}nd(X, Y)$.

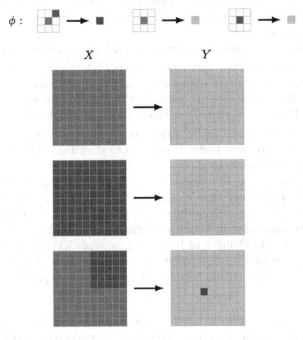

Fig. 2. The block map ϕ takes the $\mathbb{Z}^2$ subshift X to the $\mathbb{Z}^2$ subshift Y by applying it locally at each position.

Let us discuss briefly the situation of this enumeration, which is not the main topic of the present paper, but sheds light on the difficulties for an effective representation of homomorphisms. In general, $\mathbb{N}' \neq \mathbb{N}$. It is a nontrivial problem to ask whether $\mathbb{N}'$ is computable (this would mean that we can decide whether a block map sends X into Y), like for the full shift. Similarly, the domain $\mathbb{N}'$ of an enumeration of $\mathcal{A}ut(X, Y)$ need not be computable: it is already uncomputable for $\mathcal{A}ut(\mathcal{A}^{\mathbb{Z}^2})$, because it corresponds to the reversibility problem for two-dimensional cellular automata (over the full shift) [13].

On the other hand, obtaining a bijective enumeration for $\mathcal{E}nd(\mathcal{A}^{\mathcal{G}^*}, \mathcal{B}^{\mathcal{G}^*})$ would be easily achieved by representing each block map only for its smallest possible radius. Nevertheless, trying to achieve a bijective enumeration in general for $\mathcal{E}nd(X, Y)$, or even for $\mathcal{E}nd(\mathcal{A}^M, \mathcal{B}^M)$, is a process that would depend on the colanguage of the subshift (we want to avoid two block maps that differ only over the colanguage), which may be uncomputable.

For the rest of the paper, let us assume that $\mathbb{M}$ is an effective group. More precisely, all results could be interpreted as reductions to a join with a problem representing the composition map of the group, and sometimes to an additional join with a problem representing the inversion.

2   Equality Problem Is Not Too Hard

Remark 4.   Two distinct block maps $\phi, \psi : \mathcal{A}^{\mathcal{G}^{\leq r}} \to \mathcal{A}$ representing endomorphisms of X actually represent the same endomorphism if and only if for every pattern $u \in \mathcal{A}^{\mathcal{G}^{\leq r}}$, $\phi(u) \neq \psi(u) \Rightarrow u \in \mathcal{L}(X)^C$.

This remark allows to establish that the equality problem is at most as complex as knowing whether a pattern is in the colanguage.

Theorem 2. *The equality problem in $\mathcal{E}nd(X)$ is positive-reducible to $\mathcal{L}(X)^C$.*

Proof. One can directly apply Remark 4, by noting that it is easy to transform each block map into an equivalent one, so that the resulting two block maps have the same radius (the original maximal one, by ignoring extra symbols). □

Of course, this remains true for the equality problem in $\mathcal{A}ut(X)$. Since positive-reducibility implies both Turing-reducibility and enumeration-reducibility, we get the following for the lowest classes of the arithmetic hierarchy (which was already known; see [9]).

Corollary 1

1. *The equality problem is decidable, in the endomorphism monoid of any subshift with computable language (for instance 1D sofic subshift, 1D substitutive subshift, minimal effective subshift, two-way space-time diagrams of a surjective cellular automaton...).*
2. *The equality problem is computably enumerable, in the endomorphism monoid of any effective subshift (for instance multidimensional sofic subshift, substitutive subshift, limit set of cellular automaton...).*

3   Automorphism Groups with Hard Equality Problem

The purpose of this section is to prove a partial converse to Theorem 2: we can build a subshift X for which the two problems involved are equivalent, however complex they are.

Let $X \subset \mathcal{A}^{\mathbb{M}}$ and $Y \subset \mathcal{B}^{\mathbb{M}}$ be subshifts. For $\alpha : \mathcal{B} \to \mathcal{B}$ and $u \in \mathcal{A}^{\mathbb{M}}$, let us define the *controlled map* $C_{u,\alpha}$ as the homomorphism over $X \times Y$ such that $C_{u,\alpha}(x,y)_0 = (x_0, \alpha(y_0))$ if $x \in [u]$; (x_0, y_0) otherwise. Informally, $C_{u,\alpha}(x,y)$ applies α somewhere in y iff it sees u at the corresponding position in x. Denote also π_1 the projection to the first component, and σ_1^g the shift of the first component with respect to element $g \in \mathbb{M}$: $\sigma_1^g(x,y)_0 = (x_g, y_0)$ for every $(x, y) \in X \times Y$.

Remark 5

1. $\pi_1 C_{u,\alpha} = \pi_1$.
2. If $\mathbb{M}$ is a group and $g \in \mathbb{M}$, then $C_{u,\alpha} = \sigma_1^g C_{\sigma^{g-1}(u),\alpha} \sigma_1^{-g}$.
3. $C_{u,\alpha} \in \mathcal{E}nd(X \times Y, X \times \mathcal{B}^{\mathbb{M}})$.
4. $C_{u,\alpha}$ is injective if and only if α is a permutation.
5. $C_{u,\alpha} \in \mathcal{E}nd(X \times Y)$ if Y is (locally) α-*permutable, i.e.* for all $y \in Y$, if we define z by $z_0 = \alpha(y_0)$, $z_i = y_i$ for $i \neq 0$, then $z \in Y$.
6. From Remark 4, $C_{u,\alpha}$ is the identity over $X \times Y$ if and only if $u \notin \mathcal{L}(X)$, or α is the trivial permutation over letters appearing in Y.

Example 1. Examples of α-permutable subshifts are the full shift on $\mathcal{B}$ or, if $\mathcal{B} = \mathcal{B}' \sqcup \{\bot\}$ and $\alpha(\bot) = \bot$, the $\mathcal{B}'$-*sunny-side-up* defined by forbidding every pattern which involves two occurences of $\mathcal{B}'$. We have seen that the colanguage of the former is positive-equivalent to the word problem in $\mathbb{M}$. The language of the latter can be easily proven to be many-one-equivalent to the word problem in $\mathbb{M}$ (as essentially noted in [14, Prop 2.11]), hence yielding a kind of jump for the colanguage.

If $a, b, c \in \mathcal{B}$, let us denote $\alpha_{abc} : \mathcal{B} \to \mathcal{B}$ the 3-*cycle* that maps a to b, b to c, c to b, and any other symbol to itself. The following lemma corresponds essentially to [15, Lemma 18] and shows that controlled permutations, no matter the size of the control pattern u, can be expressed with a finite number of generators.

Lemma 1. *Suppose $\mathcal{B}$ has at least 5 distinct elements a, b, c, d, e. Let $u \in \mathcal{A}^{\mathcal{S}(u)}$ be a pattern, $g \in \mathcal{S}(u)$, and $v = u_{|\mathcal{S}(u) \setminus \{g\}}$. Then $C_{u,\alpha_{abc}} = (\Psi\Phi)^2$, where $\Phi = \sigma_1^g C_{u_g,\alpha_{ade}} C_{u_g,\alpha_{bad}} \sigma_1^{g-1}$ and $\Psi = C_{v,\alpha_{bde}} C_{v,\alpha_{cbd}}$.*

Proof. If $x_g = u_g$, then $\Phi(x,y)_0 = (x_0, \phi(y_0))$, where ϕ is the involution that swaps a and b on the one hand, d and e on the other hand; otherwise $\Phi(x,y)_0 = (x_0, y_0)$. If $x \in [v]$, then $\Psi(x,y)_0 = (x_0, \psi(y_0))$, where ψ is the involution that swaps b and c on the one hand, d and e on the other hand; otherwise $\Psi(x,y)_0 = (x_0, y_0)$. Since $\phi^2 = \psi^2 = \mathrm{id}$, one can see that if $x \notin [u]$, then $(\Psi\Phi)^2(x,y)_0 = (x_0, y_0)$. Now if $x \in [u]$, then we see that $\Psi\Phi(x,y)_0 = (x_0, \psi\phi(y_0))$, and $\psi\phi = \alpha_{acb}$, so that we get the stated result. $\square$

Theorem 3. *Let $X \subset \mathcal{A}^{\mathbb{M}}$ be a subshift and $Y \subset \mathcal{B}^{\mathbb{M}}$ an α_{abc}-permutable subshift for every $a, b, c \in \mathcal{B}' \subset \mathcal{B}$, where $|\mathcal{B}'| \geq 5$. Then $\mathcal{L}(X)^C$ is one-one-reducible to the word problem in the subgroup of automorphisms of $X \times Y$ generated by σ_1^g and $C_{u_0,\alpha_{abc}}$ for $g \in \mathcal{G}$, $a, b, c \in \mathcal{B}'$ and $u_0 \in \mathcal{A}$.*

Proof. From an induction and Lemma 1, we know that this subgroup includes every $C_{u,\alpha_{abc}}$ for every $a, b, c \in \mathcal{B}'$ and $u \in \mathcal{A}^*$. From Point 6 of Remark 5, an automorphism $C_{u,\alpha_{abc}}$ is equal to the identity if and only if $u \notin \mathcal{L}(X)$. $\square$

Consequently, subshifts can have finitely generated automorphism subgroups with equality problem as complex as their colanguage, as formalized by the following corollary. In that case, the equality problem of the whole automorphism group is as complex also.

Corollary 2

1. If X and Y are as in Theorem 3, then $\mathcal{L}(X)^C$ is one-one-equivalent to the word problem in (a finitely generated subgroup of) $\mathcal{A}ut(X \times Y)$.
2. For every subshift X over a finitely generated group $\mathbb{M}$, there exists a countable-to-one extension $X \times Y$ such that $\mathcal{L}(X)^C$ is one-one-equivalent to the word problem in (a finitely generated subgroup of) $\mathcal{A}ut(X \times Y)$.
3. For every subshift X over a finitely generated group $\mathbb{M}$, there exists a full extension $X \times \mathcal{B}^{\mathbb{M}}$ such that $\mathcal{L}(X)^C$ is one-one-equivalent to the word problem in (a finitely generated subgroup of) $\mathcal{A}ut(X \times \mathcal{B}^{\mathbb{M}})$.
4. Every Σ_1^0 Turing degree contains the word problem in (a finitely generated subgroup of) $\mathcal{A}ut(X)$, for some 2D SFT X.
5. There exists a 2D SFT X for which the word problem in (a finitely generated subgroup of) $\mathcal{A}ut(X)$ is undecidable.

Point 5 answers [9, Problem 5].

Proof

1. Just use Point 5 of Remark 1. For the converse reduction in the one-one-equivalence, simply apply Theorem 2 and Point 1 of Remark 1.
2. We use Theorem 3 with Y being the $\{0,1,2,3,4\}$-sunny-side-up.
3. We use Theorem 3 with $Y = \{0,1,2,3,4\}^{\mathbb{M}}$. Remark that $\mathcal{L}(X)^C$ and $\mathcal{L}(X \times \{0,1,2,3,4\}^{\mathbb{M}})^C$ are one-one-equivalent.
4. Every Σ_1^0 degree contains the colanguage of a 2D SFT, thanks to constructions from [16,17]. Then its product with the full shift $\{0,1,2,3,4\}^{\mathbb{Z}^2}$ is still an SFT, and we conclude by the previous point.
5. Apply the previous point with any uncomputable Σ_1^0 degree. □

Note that the number of generators can be decreased if we want to reduce only the language whose support is spanned by a subgroup. For instance 2D SFTs are already known to have (arbitrarily Σ_1^0) uncomputable 1D language. Indeed, our automorphisms do not alter the X layer, so that their parallel applications to all traces with respect to a subgroup is still an automorphism.

Among the open questions, we could wonder whether there is a natural class of SFT (irreducible, with uncomputable language, at least over $\mathbb{Z}^2$) whose colanguage could be proven reducible to the word problem in the automorphism group. This could require to encode the whole cartesian product of Theorem 3 inside such subshifts. Another question would be to adapt our construction while controling the automorphism group completely so that it is finitely generated.

Acknowledgements. This research was supported by the Academy of Finland grant 296018.

We thank Ville Salo for some discussions on commutators, on the open questions, and for a careful reading of this preprint.

References

1. Aubrun, N., Béal, M.-P.: Decidability of conjugacy of tree-shifts of finite type. In: Albers, S., Marchetti-Spaccamela, A., Matias, Y., Nikoletseas, S., Thomas, W. (eds.) ICALP 2009. LNCS, vol. 5555, pp. 132–143. Springer, Heidelberg (2009). https://doi.org/10.1007/978-3-642-02927-1_13

2. Comon, H., et al.: Tree automata techniques and applications (2007). http://www.grappa.univ-lille3.fr/tata. Accessed 12 Oct 2007

3. Thomas, W.: Automata on infinite objects. In: van Leeuwen, J. (ed.) Formal Models and Semantics, Handbook of Theoretical Computer Science, pp. 133–191. Elsevier, Amsterdam (1990)

4. Boyle, M., Lind, D.A., Rudolph, D.J.: The automorphism group of a shift of finite type. Trans. Am. Math. Soc. **306**(1), 71–114 (1988)

5. Kari, J., Ollinger, N.: Periodicity and immortality in reversible computing. In: Ochmański, E., Tyszkiewicz, J. (eds.) MFCS 2008. LNCS, vol. 5162, pp. 419–430. Springer, Heidelberg (2008). https://doi.org/10.1007/978-3-540-85238-4_34

6. Coven, E., Yassawi, R.: Endomorphisms and automorphisms of minimal symbolic systems with sublinear complexity (2014)

7. Cyr, V., Kra, B.: The automorphism group of a shift of linear growth: beyond transitivity (2014)

8. Donoso, S., Durand, F., Maass, A., Petite, S.: On automorphism groups of low complexity subshifts (2015)

9. Hochman, M.: Groups of automorphisms of SFTs. Open problems (2017). http://math.huji.ac.il/~mhochman/problems/automorphisms.pdf

10. Rogers Jr., H.: Theory of Recursive Functions and Effective Computability. MIT Press, Cambridge (1987)

11. Rabin, M.O.: Computable algebra, general theory and theory of computable fields. Trans. Am. Math. Soc. **95**, 341–360 (1960)

12. Hedlund, G.A.: Endomorphisms and automorphisms of the shift dynamical system. Math. Syst. Theory **3**, 320–375 (1969)

13. Kari, J.: Reversibility and surjectivity problems of cellular automata. J. Comput. Syst. Sci. **48**(1), 149–182 (1994)

14. Aubrun, N., Barbieri, S., Sablik, M.: A notion of effectiveness for subshifts on finitely generated groups. Theor. Comput. Sci. **661**, 35–55 (2017)

15. Boykett, T., Kari, J., Salo, V.: Finite generating sets for reversible gate sets under general conservation laws. Theor. Comput. Sci. **701**, 27–39 (2017)

16. Simpson, S.G.: Medvedev degrees of 2-dimensional subshifts of finite type. Ergodic Theory Dyn. Syst. **34**(November 2012), 665–674 (2014)

17. Hanf, W., Myers, D.: Non recursive tilings of the plane II. J. Symbolic Logic **39**(2), 286–294 (1974)

Parameterized Complexity
of Conflict-Free Set Cover

Ashwin Jacob[1]([✉]), Diptapriyo Majumdar[2], and Venkatesh Raman[1]

[1] The Institute of Mathematical Sciences, HBNI, Chennai, India
{ajacob,vraman}@imsc.res.in
[2] Royal Holloway, University of London, Egham, UK
diptapriyo.majumdar@rhul.ac.uk

Abstract. SET COVER is one of the well-known classical NP-hard problems. Following some recent trends, we study the *conflict-free* version of the SET COVER problem. Here we have a universe $\mathcal{U}$, a family $\mathcal{F}$ of subsets of $\mathcal{U}$ and a graph $G_{\mathcal{F}}$ on the vertex set $\mathcal{F}$ and we look for a subfamily $\mathcal{F}' \subseteq \mathcal{F}$ of minimum size that covers $\mathcal{U}$ and also forms an independent set in $G_{\mathcal{F}}$. Here we initiate a systematic study of the problem in parameterized complexity by restricting the focus to the variants where SET COVER is fixed-parameter tractable (FPT). We give upper bounds and lower bounds for conflict-free version of the SET COVER with and without duplicate sets along with restrictions to the graph classes of $G_{\mathcal{F}}$.

1 Introduction and Previous Work

Covering problems are problems in combinatorics that ask whether a certain structure "covers" another. Covering problems are very well-studied in theoretical computer science. Examples include VERTEX COVER, FEEDBACK VERTEX SET, CLUSTER VERTEX DELETION among others.

Several of these covering problems can be encapsulated by a a problem called SET COVER which is one of the well-studied classical NP-hard problems. In the SET COVER problem, we have a universe $\mathcal{U}$, a family $\mathcal{F}$ of subsets of $\mathcal{U}$ and an integer k and the goal is to find a subfamily $\mathcal{F}'$ of size at most k such that $\bigcup_{S \in \mathcal{F}'} S = \mathcal{U}$.

SET COVER is very well studied in a variety of algorithmic settings, especially in the realm of approximation algorithms and parameterized complexity. Unfortunately, SET COVER when parameterized by solution size k is $W[2]$-hard [5] and hence is unlikely to be fixed-parameter-tractable (FPT).

It has been seen in computational problems where some pairs of elements in the problem are in conflict with each other and hence cannot go in the solution together. This can be modeled by defining a graph on the elements and an edge (u, v) is added if elements u and v do not go into the solution together or in other words form a conflict. Hence a solution without conflicts will form an independent set in this graph. Conflict-free versions of classical problems in P like MAXIMUM FLOW [20], MAXIMUM MATCHING [6], SHORTEST PATH [13]

R. van Bevern and G. Kucherov (Eds.): CSR 2019, LNCS 11532, pp. 191–202, 2019.
https://doi.org/10.1007/978-3-030-19955-5_17

have been studied. Conflict-free version of problems like VERTEX COVER [12], FEEDBACK VERTEX SET [1], INTERVAL COVERING [2,3] are studied from the parameterized point of view very recently.

We look at the conflict-free version of SET COVER defined as follows:

CONFLICT-FREE SET COVER
Input: An universe $\mathcal{U}$, a family $\mathcal{F}$ of subsets of $\mathcal{U}$, a graph $G_{\mathcal{F}}$ with vertex set $\mathcal{F}$ and an integer k.
Goal: Is there a subfamily $\mathcal{F}' \subseteq \mathcal{F}$ of size at most k such that $\cup_{F \in \mathcal{F}'} F = \mathcal{U}$ and $\mathcal{F}'$ forms an independent set in G?

We assume that there are no duplicate sets in the family $\mathcal{F}$. Hence $|\mathcal{F}| \leq 2^{|\mathcal{U}|}$.

Note that if $G_{\mathcal{F}}$ is edgeless, the problem is equivalent to SET COVER as every subset of vertices of $G_{\mathcal{F}}$ forms an independent set. Hence CONFLICT-FREE SET COVER is $W[2]$-hard whenever SET COVER is $W[2]$-hard and the only interesting cases of CONFLICT-FREE SET COVER are those special instances or parameterizations where SET COVER is FPT. Banik et al. [3] introduced CONFLICT-FREE SET COVER and considering restrictions on $G_{\mathcal{F}}$ showed that

- CONFLICT-FREE SET COVER is W[1]-hard when $G_{\mathcal{F}}$ is from those classes of graphs where INDEPENDENT SET is W[1]-hard, and
- when $G_{\mathcal{F}}$ has bounded arboricity, CONFLICT-FREE SET COVER is FPT parameterized by k whenever SET COVER is FPT parameterized by k.

Our Results: We note that the reduction instance of CONFLICT-FREE SET COVER in the W[1]-hardness result above [3] contains duplicate sets.

- Our first result is an $f(k)|\mathcal{F}|^{o(k)}$ lower bound for CONFLICT-FREE SET COVER without any duplicate sets assuming the Exponential Time Hypothesis (ETH) even when the sets of $\mathcal{F}$ pairwise intersect in at most one element. The lower bound holds even when $G_{\mathcal{F}}$ is restricted to bipartite graphs where INDEPENDENT SET is polynomial-time solvable. Hence the result can be seen as generalizing the previous W[1]-hardness result.
- On the positive side, for CONFLICT-FREE SET COVER we give FPT algorithms parameterized by k whenever SET COVER is FPT and $G_{\mathcal{F}}$ belongs to graph classes which are sparse like graphs with bounded degeneracy or nowhere dense graphs using the recently introduced independence covering family [17]. On the other hand if $G_{\mathcal{F}}$ is a dense graph like split or co-chordal, we give an FPT algorithm. This algorithm works for a large class of graphs where the number of maximal independent sets is polynomial in the number of vertices (that are sets in the family in our case).

  Next we consider the problem parameterized by the universe size $|\mathcal{U}|$. Here, since the number of sets $|\mathcal{F}| \leq 2^{|\mathcal{U}|}$, CONFLICT-FREE SET COVER is FPT as the trivial brute-force algorithm of choosing at most k sets from $\mathcal{F}$ is of complexity bounded by $\binom{|\mathcal{F}|}{k} \leq \binom{2^{|\mathcal{U}|}}{|\mathcal{U}|} \leq 2^{|\mathcal{U}|^2}$.
- We give a matching lower bound of $2^{o(|\mathcal{U}| \log |F|)}$ for any value of $|\mathcal{F}|$ as well assuming the ETH.

We note that the problem does not have a polynomial kernel as when $G_{\mathcal{F}}$ is an empty graph, the problem becomes SET COVER parameterized by universe size which does not have a polynomial kernel unless $\mathsf{NP} \subseteq \mathsf{coNP/poly}$ [4].

Unlike the SET COVER problem, in CONFLICT-FREE SET COVER duplicate sets do play an important role as the neighbourhood sets in the graph can vary which matters in the independence requirement of the solution.

- For CONFLICT-FREE SET COVER with duplicate sets we give an $f(|\mathcal{U}|)|\mathcal{F}|^{o(|\mathcal{U}|)}$ lower bound assuming the ETH even when all pairs of sets intersect in at most one element and even when $G_{\mathcal{F}}$ is restricted to bipartite graphs where the INDEPENDENT SET problem can be solved in polynomial time.
- In addition, we give FPT algorithms when we restrict $G_{\mathcal{F}}$ to interval graphs via a dynamic programming algorithm using the perfect elimination ordering of graphs. We extend this idea and give an FPT algorithm for chordal graphs which is a superclass of interval graphs via dynamic programming on the clique-tree decomposition of the graph.

We also study the CONFLICT-FREE SET COVER problem where there is an underlying (linearly representable) matroid on the family of subsets, and we want the solution to be an independent set in the matroid. Banik et al. [3] studied this version for a specialization of SET COVER where the sets are intervals on a real line.

- We show that even the more general problem (where the sets in the family are arbitrary) is FPT when parameterized by the universe size, using the idea of dynamic programming over representative families [9].

We note that this result can be obtained as a corollary of a result by Bevern et al. [22] where they studied generalization called uncapacitated facility location problem with multiple matroid constraints. But our algorithm is simpler and has a better running time.

2   Preliminaries

We use $[n]$ to denote the set $\{1, \ldots, n\}$. We use the standard terminologies of the graph theory book by Diestel [7]. For a graph $G = (V, E)$ we denote n as the number of vertices and m as the number of edges. For $S \subseteq V(G)$, we denote $G[S]$ to be the subgraph induced on S. A *complement* of graph G is a graph H on the same vertices such that two distinct vertices of H are adjacent if and only if they are not adjacent in G. A set $S \subseteq V(G)$ is called an *independent set* if for all $u, v \in S, (u, v) \notin E(G)$.

An *interval* graph is an undirected graph formed from a family of intervals in the real line $\mathcal{I}$ with the vertex set as $\mathcal{I}$ and for intervals $u, v \in \mathcal{I}$, add edge (u, v) if the intervals u and v intersect. A chord in a cycle is an edge between two non-adjacent vertices of the cycle. A *chordal* graph is a graph in which any cycle of four or more vertices has a chord. A graph G is said to be *d-degenerate* if every subgraph of G has a vertex of degree at most d. The *arboricity* of an undirected graph is the minimum number of forests into which its edges can be partitioned.

We use the following conjecture to prove lower bounds.

Conjecture 1 (Exponential Time Hypothesis (ETH)([11])). 3-CNF-SAT cannot be solved in $\mathcal{O}^*(2^{o(n)})$[1] time where the input formula has n variables and m clauses.

3   CONFLICT-FREE SET COVER **Parameterized by** k

3.1   Hardness Results

Theorem 1. CONFLICT-FREE SET COVER *where every pair of sets in $\mathcal{F}$ intersect in at most one element is* W[1]*-hard with respect to solution size k when $G_{\mathcal{F}}$ is bipartite.*

Proof. We give a reduction from the W[1]-hard problem MULTICOLORED BICLIQUE [5] defined as follows:

MULTICOLORED BICLIQUE   **Parameter:** k
Input: A bipartite graph $G = (A \cup B, E)$, an integer k, a partition of A into k sets $A_1, A_2, \ldots, A_k$ and a partition of B into k sets $B_1, B_2, \ldots, B_k$.   **Question:** Does there exist a subgraph of G isomorphic to the biclique $K_{k,k}$ with one vertex from each of the sets A_i and B_i?

We note that while the $W[1]$-hardness of MULTICOLORED BICLIQUE can be easily shown from a reduction from MULTICOLORED CLIQUE, the complexity of the normal k-biclique problem was open for a long time and was only shown recently to be $W[1]$-hard [14].

Given an instance of $(G, A_1, \ldots, A_k, B_1, \ldots, B_k)$ of MULTICOLORED BICLIQUE with $V(G) = \{v_1, v_2, \ldots, v_n\}$, we construct an instance of CONFLICT-FREE SET COVER $(\mathcal{U}, \mathcal{F}, G_{\mathcal{F}}, k)$ without duplicates as follows.

We define the universe $\mathcal{U} = [2k] \cup V(G) \cup \{x\}$. Let S_{v_j} denote the set corresponding to vertex v_j. For $i \in [k]$, if $v_j \in A_i$, define $S_{v_j} = \{v_j, i\}$. For $i \in [2k] \setminus [k]$, if $v_j \in B_{i-k}$, define $S_{v_j} = \{v_j, i\}$. Define a set $D = V(G) \cup \{x\}$. We have $\mathcal{F} = \bigcup_{v \in V(G)} S_v \cup \{D\}$. The graph $G_{\mathcal{F}}$ is obtained by taking the complement of the graph G, making the sets A and B independent and making D as an isolated vertex. Note that the graph $G_{\mathcal{F}}$ remains bipartite.

Note that $\mathcal{F}$ is defined in such a way that all pairs of sets intersect in at most one element. Also there are no duplicate sets.

We claim that there is a multicolored biclique of size k in G if and only if there is a CONFLICT-FREE SET COVER of size $2k+1$ in the instance $(\mathcal{U}, \mathcal{F}, G_{\mathcal{F}})$.

Let $S = \{a_1, \ldots, a_k, b_1, \ldots, b_k\}$ be the vertices in G that form a multicolored biclique. Then $\mathcal{F}' = \{D, S_{a_1}, \ldots, S_{a_k}, S_{b_1}, \ldots, S_{b_k}\}$ covers $\mathcal{U}$ as D covers $V(G) \cup \{x\}$ and $i \in S_{a_i}$ for $i \in [k]$ and $i \in S_{b_{i-k}}$ for $i \in [2k] \setminus [k]$. Since the edges across A and B in G are non-edges in $G_{\mathcal{F}}$ and D is an isolated vertex, $\mathcal{F}'$ forms an independent set in $G_{\mathcal{F}}$. In the reverse direction, let $\mathcal{F}' = \{S_1, \ldots, S_{2k+1}\}$ be a

---

[1] $\mathcal{O}^*$ notation ignores polynomial factors of input.

solution of size $2k+1$ covering $\mathcal{U}$. The set D has to be part of the solution $\mathcal{F}'$ as only D contains element x. Now note that an element $i \in [k]$ can be covered only by sets S_v where $v \in A_i$. Similarly an element $i \in [2k] \setminus [k]$ can be covered only by sets S_v where $v \in B_{i-k}$. Hence the vertices of the sets in $\mathcal{F}'$ are such that there is at least one vertex from each of the sets A_i and B_i. Since the budget is limited to $2k$ after picking D, exactly one vertex from each of the sets A_i and B_i is contained in $\mathcal{F}'$. Since the vertices $\mathcal{F}' \setminus D$ form an independent set in $G_{\mathcal{F}}$, the corresponding vertices form a biclique in G. □

Since MULTICOLORED BICLIQUE cannot be solved in time $f(k)|\mathcal{F}|^{o(k)}$ for solution size k assuming ETH [19], we have the following corollary.

Corollary 1. CONFLICT-FREE SET COVER *where every pair of sets in $\mathcal{F}$ intersect in at most one element cannot be solved in time $f(k)|\mathcal{F}|^{o(k)}$ for solution size k in bipartite graphs for any computable function f assuming the ETH.*

3.2 Upper Bounds

In the following results, we restrict the graph $G_{\mathcal{F}}$.

Graphs with Bounded Number of Maximal Independent Sets

Theorem 2. *When $G_{\mathcal{F}}$ is restricted to a graph where the number of maximal independent sets is polynomial in $|\mathcal{F}|$ and can be enumerated in time polynomial in $|\mathcal{F}|$, if SET COVER can be solved in $\mathcal{O}^*(f(k))$ time, CONFLICT-FREE SET COVER can be solved in $\mathcal{O}^*(f(k))$ time.*

Proof. For each maximal independent set I of G, we run the $\mathcal{O}^*(f(k))$ algorithm for SET COVER with the family $\mathcal{F}$ containing sets corresponding to the vertices in I. Since the solution X of CONFLICT-FREE SET COVER is an independent set, $X \subseteq I'$ for some maximal independent set I'. So if the SET COVER algorithm returns YES for any I, return YES, else return NO.

Note that although SET COVER problem is W-hard when parameterized by k, there are variants of SET COVER like when the size of the intersection of sets in $\mathcal{F}$ is bounded [21] where the problem can be solved in $\mathcal{O}^*(f(k))$ time.

As the number of maximal independent sets in split graphs (since at most one vertex of the clique can be in the independent set), co-chordal graphs [10] and $2K_2$-free graphs [8] are polynomial in the number of vertices, we have the following corollary.

Corollary 2. *If SET COVER can be solved in $\mathcal{O}^*(f(k))$ time, CONFLICT-FREE SET COVER can be solved in $\mathcal{O}^*(f(k))$ time when $G_{\mathcal{F}}$ is restricted to split graphs, co-chordal graphs or $2K_2$-free graphs.*

Graphs with Bounded Degeneracy

We use the notion of k-Independence Covering Family introduced by [17] defined as follows:

Definition 1 (k-Independence Covering Family). *For a graph G and integer k, a family of independent sets of G is called an independence covering family for (G, k), denoted by $\mathscr{F}(G, k)$, if for any independent set X in G of size at most k, there exists an independent set $Y \in \mathscr{F}(G, k)$ such that $X \subseteq Y$.*

Lemma 1. *(Deterministic Independence Covering Lemma [17]) Given a d-degenerate graph G and an integer k, there is an algorithm that runs in time $O^*(\binom{k(d+1)}{k} \cdot 2^{o(k(d+1)} \cdot (n + m) \log n)$ and outputs a k-independence covering family for (G, k) of size at most $O^*(\binom{k(d+1)}{k} \cdot 2^{o(k(d+1))} \cdot \log n)$.*

Theorem 3. CONFLICT-FREE SET COVER *has an algorithm with running time $O^*(f(k)\binom{k(d+1)}{k} \cdot 2^{o(k(d+1))})$ with solution size k when $G_{\mathcal{F}}$ is a d-degenerate graph if* SET COVER *can be solved in $O^*(f(k))$ time.*

Proof. We use Lemma 1 on $G_{\mathcal{F}}$ to get a k-independence covering family $\mathscr{F}(G_{\mathcal{F}}, k)$. For each independent set $Y \in \mathscr{F}(G_{\mathcal{F}}, k)$, we run the algorithm for SET COVER for the instance $(\mathcal{U}, Y, k)$ in $O^*(f(k))$ time. If for any of the sets Y, $(\mathcal{U}, Y, k)$ is a yes instance, we return yes. Otherwise we return no.

Let X be the solution of size k. There is a set Y in $\mathscr{F}(G_{\mathcal{F}}, k)$ such that $X \subseteq Y$. Hence when we run the algorithm for SET COVER in instance $(\mathcal{U}, Y, k)$, since $G[Y]$ is an independent set, the algorithm will return X. $\square$

Note that graphs with bounded degeneracy contain many other graph classes such as planar graphs, graphs with bounded arboricity and graphs with bounded treewidth. We note that a similar result as Theorem 3 has been proven in graphs with bounded arboricity [3] (through a different argument) from which a result for graphs with bounded degeneracy follows as the degeneracy of a graph is also bounded when the arboricity is bounded.

Nowhere Dense Graphs

Nowhere dense graphs contains a number of classes of graphs that are not contained in the class of graphs with bounded degeneracy, including graphs with bounded local treewidth and graphs that locally exclude a fixed minor. In [17], the authors construct a k-independence covering family for nowhere dense graphs.

Lemma 2. *([17]). Let G be a nowhere dense graph and k be an integer. There is a deterministic algorithm that runs in time*

$$O\left(f(k, \frac{1}{k}) \cdot n^{1+o(1)} + g(k) \cdot \binom{k^2}{k} \cdot 2^{o(k^2)} \cdot n(n+m) \log n\right)$$

and outputs a k-independence covering family for (G, k) of size $O(g(k)\binom{k^2}{k} \cdot 2^{o(k^2)} \cdot n \log n)$ where f is a computable function and $g(k) = (f(k, \frac{1}{k}))^k$.

We get the following theorem whose proof is similar to Theorem 3.

Theorem 4. CONFLICT-FREE SET COVER *has an algorithm with running time* $\mathcal{O}^*(h(k)g(k)\binom{k^2}{k} \cdot 2^{o(k^2)})$ *for nowhere dense graphs with solution size k and a computable function g if* SET COVER *can be solved in* $\mathcal{O}^*(h(k))$ *time.*

4   CONFLICT-FREE SET COVER **Parameterized by** $|\mathcal{U}|$

4.1   Lower Bounds When $\mathcal{F}$ Has No Duplicates

We define the following variant of MULTICOLORED BICLIQUE.

SMALL MULTICOLORED BICLIQUE   **Parameter:** k
Input: A bipartite graph $G = (A \cup B, E)$, an integer k, a partition of A into k sets $A_1, A_2, \ldots, A_k$ and a partition of B into k sets $B_1, B_2, \ldots, B_k$ such that $|A_i| = |B_i| = s$ where $k \leq s \leq 2^k/2k$.
Question: Does there exist a subgraph of G isomorphic to the biclique $K_{k,k}$ with one vertex from each of the sets A_i and B_i?

We first note that the reduction from 3-COLORING used in [15] can be modified so that we get the following lower bound for SMALL MULTICOLORED BICLIQUE.

Theorem 5. [2]SMALL MULTICOLORED BICLIQUE *cannot be solved in time* $2^{o(k \log s)}$ *under the ETH*

Theorem 6. CONFLICT-FREE SET COVER *without duplicates when $G_{\mathcal{F}}$ is bipartite cannot be solved in time* $2^{o(|\mathcal{U}| \log |\mathcal{F}|)}$ *under ETH.*

Proof. Given an instance of $(G, A_1, \ldots, A_k, B_1, \ldots, B_k)$ of SMALL MULTICOLORED BICLIQUE with $V(G) = \{v_1, v_2, \ldots, v_n\}$, we construct an instance of CONFLICT-FREE SET COVER $(\mathcal{U}, \mathcal{F}, G_{\mathcal{F}}, 2k + 1)$ without duplicates as follows:

Let us define sets $Z = \{z_1, z_2, \ldots z_{\lceil \log n \rceil}\}$ and $O = \{o_1, o_2, \ldots o_{\lceil \log n \rceil}\}$.

We define the universe $\mathcal{U} = [2k] \cup Z \cup O \cup \{x\}$.

Let us look at vertex $v_j \in V$ and construct sets $S_{v_j} \in \mathcal{F}$. Let us map j to its binary representation $b_1, b_2, \ldots, b_{\lceil \log n \rceil}$. We create a set T_j as follows: for all $i \in [[\log n]]$, when $b_i = 0$, add z_i to T_j, else add o_i to T_j. For $i \in [k]$, if $v_j \in A_i$, define $S_{v_j} = \{i\} \cup T_j$. For $i \in [2k] \setminus [k]$, if $v_j \in B_{i-k}$, define $S_{v_j} = \{i\} \cup T_j$. Define the extra set $D = Z \cup O \cup \{x\}$. The graph $G_{\mathcal{F}}$ is obtained by taking the complement of the graph G, making the sets A and B independent and making D as an isolated vertex. Note that the graph $G_{\mathcal{F}}$ remains bipartite.

Note that the construction is almost exactly the same as in Theorem 1 but the vertices are encoded in binary form. The correctness of the reduction can then easily be seen after noting that the set D has to go in the solution as it is the only set containing element x.

Note that in the SMALL MULTICOLORED BICLIQUE instance, $n = 2k \cdot s \leq 2^k$. Since $\log n \leq k$, $|U| \leq 4k + 1$.

---
[2] Proof in full version.

Now suppose CONFLICT-FREE SET COVER has an algorithm with running time $2^{o(|\mathcal{U}| \log |\mathcal{F}|)}$. Since $s = \frac{|\mathcal{F}|}{2k}$ and $|\mathcal{U}| \le 4k + 1$, we have a running time of $2^{o(4k \log(2k \cdot s))} = 2^{o(k(\log s + \log k))} = 2^{o(k \log s)}$ for SMALL MULTICOLORED BICLIQUE violating the ETH. □

4.2 Lower Bounds When $\mathcal{F}$ Has Duplicates

We have the following hardness result [3].

Theorem 7 ([3]). *If for a subclass of graphs $\mathcal{G}$, finding an independent set of size k is W[1]-hard, then* CONFLICT-FREE SET COVER *parameterized by $|\mathcal{U}|$ is W[1]-hard when $G_{\mathcal{F}}$ is restricted to the class $\mathcal{G}$.*

Bipartite Graphs

Bipartite graphs is one class of graphs where the INDEPENDENT SET problem can be solved in polynomial time. In contrast to Theorem 7, we show that CONFLICT-FREE SET COVER on bipartite graphs is W[1]-hard. Note that in Theorem 1 proven previously, the size of the universe can be much larger than the solution size k and hence the hardness result does not follow from it.

Lemma 3. CONFLICT-FREE SET COVER *parameterized by $|\mathcal{U}|$ is W[1]-hard on bipartite graphs.*

Proof. We again give a reduction from the W[1]-hard problem MULTICOLORED BICLIQUE. The construction is very similar to that in Theorem 1, the difference being the vertex v is not added to sets S_v.

Given an instance of MULTICOLORED BICLIQUE, we construct an instance of CONFLICT-FREE SET COVER as follows: $\mathcal{U} = [2k]$. Let S_v denote the set corresponding to vertex v we add to $\mathcal{F}$. For $i \in [k], i \in S_v$ if $v \in A_i$. For $i \in [2k] \setminus [k], i \in S_v$ if $v \in B_{i-k}$. The graph G' is obtained by complementing the graph G and making the sets A and B independent. The graph G' remains bipartite.

The correctness proof easily follows. □

4.3 Upper Bounds When $\mathcal{F}$ Has Duplicates

Interval Graphs

Interval graphs have the property that its vertices can be ordered as $v_1, \ldots, v_n$ such that for each v_i, $N[v_i] \cap \{v_i, \ldots, v_n\}$ is present consecutively in the ordering where $N[v_i]$ is the closed neighbourhood set of v_i. Such an ordering is actually the *perfect elimination ordering* of the graph and can be obtained in time polynomial in $|V(G)|$ by arranging the corresponding intervals in order of their leftmost endpoint. We make use of this ordering to give a dynamic programming algorithm for CONFLICT-FREE SET COVER with duplicates on interval graphs.

Theorem 8. [3] CONFLICT-FREE SET COVER *with duplicate sets when $G_{\mathcal{F}}$ is restricted to interval graphs can be solved in $\mathcal{O}^*(2^{|\mathcal{U}|})$ time.*

---

[3] Proof in full version.

Now we give a $\mathcal{O}^*(3^{|\mathcal{U}|})$-time dynamic programming algorithm for chordal graphs which is a superclass of interval graphs.

Chordal Graphs

A *clique tree decomposition* is a nice tree decomposition T where for all nodes $i \in V(T)$, the vertices of in the bag X_i are such that $G[X_i]$ forms a clique. All chordal graphs have clique-tree decompositions and can be found in polynomial time [10]. Given a clique tree decomposition, it can be converted to a nice clique tree decomposition in polynomial time as well [5].

In the theorem below, we give an algorithm for CONFLICT-FREE SET COVER with duplicates on chordal graphs using dynamic programming on the nice clique tree decomposition of the graph.

Theorem 9. [4]CONFLICT-FREE SET COVER *with duplicates on chordal graphs can be solved in* $O^*(3^{|\mathcal{U}|})$ *running time.*

5 Matroidal Conflict-Free Set Cover

Let us define the MATROIDAL CONFLICT-FREE SET COVER problem.

MATROIDAL CONFLICT-FREE SET COVER
Input: A universe $\mathcal{U}$, a family $\mathcal{F}$ of subsets of $\mathcal{U}$, a linear matroid $M = (\mathcal{F}, \mathcal{I})$ and an integer k.
Goal: Is there a subfamily $\mathcal{F}' \subseteq \mathcal{F}$ of size at most k such that $\bigcup_{F \in \mathcal{F}'} F = \mathcal{U}$ and $\mathcal{F}'$ forms an independent set in M?

We give a dynamic programming algorithm for MATROIDAL CONFLICT-FREE SET COVER containing duplicate sets using computation of representative sets noting that the similar ideas used in [3] for INTERVAL COVERING can be extended to MATROIDAL CONFLICT-FREE SET COVER .

For $W \subseteq \mathcal{U}$, let $\mathcal{B}^W$ denote the collection of subfamilies X of $\mathcal{F}$ of size at most k such that X covers W and forms a independent set in the matroid M.

$$\mathcal{B}^W = \{X \subseteq \mathcal{F} \mid |X| \leq k, W \subseteq \bigcup_{S \in X} S \text{ and } X \in \mathcal{I}\}$$

Note that $\mathcal{B}^{\mathcal{U}}$ contains all the solutions of size at most k of MATROIDAL CONFLICT-FREE SET COVER . Hence we solve the MATROIDAL CONFLICT-FREE SET COVER problem by checking whether $\mathcal{B}^{\mathcal{U}}$ is empty or not.

Definition 2 (q-representative family [18]). *Let* $M = (E, \mathcal{I})$ *be a matroid and* $\mathcal{A}$ *be a family of sets of size p in M. For sets $A, B \subseteq E$, we say that A fits B if $A \cap B = \phi$ and $A \cup B \in \mathcal{I}$. A subfamily $\hat{\mathcal{A}} \subseteq \mathcal{A}$ is said to q-represent $\mathcal{A}$ if for every set B of size q such that there is an $A \in \mathcal{A}$ that fits B, there is an $\hat{A} \in \hat{\mathcal{A}}$ that also fits B. We use $\hat{\mathcal{A}} \subseteq^q_{rep} \mathcal{A}$ to denote that $\hat{\mathcal{A}}$ q-represents $\mathcal{A}$.*

---
[4] Proof in full version.

Lemma 4 ([9]). *For a matroid $M = (E, \mathcal{I})$ and $S \subseteq E$, if $S_1 \subseteq^q_{rep} S$ and $S_2 \subseteq^q_{rep} S_1$, then $S_2 \subseteq^q_{rep} S$.*

Note that $\mathcal{B}^{\mathcal{U}}$ is nonempty if and only if $\hat{\mathcal{B}}^{\mathcal{U}} \subseteq^0_{rep} \mathcal{B}^{\mathcal{U}}$ is nonempty. Let us define $\mathcal{B}^{Wj}$ as the subset of $\mathcal{B}^W$ containing sets of size exactly j. We denote $\hat{\mathcal{B}}^W \subseteq^{1,\ldots,k}_{rep} \mathcal{B}^W$ to denote that $\hat{\mathcal{B}}^W$ contains the union of all the i-representative families of $\mathcal{B}^W$ where $1 \leq i \leq k$. In other words,

$$\hat{\mathcal{B}}^W = \bigcup_{j=1}^{k} \left(\hat{\mathcal{B}}^{Wj} \subseteq^{k-j}_{rep} \mathcal{B}^{Wj} \right)$$

Lemma 5 ([16]). *Let $M = (E, \mathcal{I})$ be a linear matroid of rank n and S be a family of t independent sets of size p. Let A be a $n \times |E|$ matrix representation of M over a field $\mathbb{F}$ where $\mathbb{F} = \mathbb{F}_{p^\ell}$ or $\mathbb{F}$ is $\mathbb{Q}$. Then there is a deterministic algorithm to compute $\hat{S} \subseteq^q_{rep} S$ size $np\binom{p+q}{p}$ in $\mathcal{O}\left(\binom{p+q}{p} tp^3 n^2 + t\binom{p+q}{p}^{\omega-1}(pn)^{\omega-1}\right) + (n + |E|)^{O(1)})$ operations over $\mathbb{F}$ where ω is the matrix multiplication exponent.*

Theorem 10. MATROIDAL CONFLICT-FREE SET COVER *can be solved in* $\mathcal{O}^*(2^{(\omega+1)\cdot|\mathcal{U}|})$ *time where ω is the matrix multiplication exponent.*

Proof. Let $\mathcal{D}$ be an array of size $2^{|\mathcal{U}|}$ with $\mathcal{D}[W]$ storing the family $\hat{\mathcal{B}}^W \subseteq^{1,\ldots,k}_{rep} \mathcal{B}^W$. We compute the entries of $\mathcal{D}$ in the increasing order of subsets of $\mathcal{U}$. To do so we compute the following:

$$\mathcal{N}^W = \bigcup_{S_i \in \mathcal{F}} (\mathcal{D}[W \setminus S_i] \bullet S_i) \cap \mathcal{I} \tag{1}$$

where $\mathcal{A} \bullet \mathcal{B} = \{A \cup B \mid A \in \mathcal{A} \text{ and } B \in \mathcal{B} \text{ and } A \cap B = \phi\}$.

We show that $\mathcal{N}^W \subseteq^{1\ldots k}_{rep} \mathcal{B}^W$. Let $S \in \mathcal{B}^{Wj}$ and Y be a set of size $k - j$ such that $S \cap Y = \phi$ and $S \cup Y \in \mathcal{I}$. We give a set $\hat{S} \in \mathcal{N}^{Wj}$ such that $\hat{S} \cap Y = \phi$ and $\hat{S} \cup Y \in \mathcal{I}$.

Let $S = \{S_1, S_2, \ldots, S_j\}$. Let $S' = S \setminus S_j$. Let $Y' = Y \cup S_j$. $|S'| = j - 1$ and $|Y'| = k-j+1$. Since S' covers $W \setminus S_j$, $S' \in \mathcal{B}^{(W \setminus S_j)(j-1)}$. By definition, $\mathcal{D}[W \setminus S_j]$ contains $\hat{\mathcal{B}}^{(W \setminus S_j)(j-1)} \subseteq^{k-j+1}_{rep} \mathcal{B}^{(W \setminus S_j)(j-1)}$ and hence a set $S^* \in \mathcal{D}[W \setminus S_j]$ such that $S^* \cap Y' = \phi$ and $S^* \cup Y' \in \mathcal{I}$. From Eq. (1), $S^* \cup S_j \in \mathcal{N}^W$. The set $\hat{S} = S^* \cup S_j$ is such that $\hat{S} \cap Y = \phi$ and $\hat{S} \cup Y \in \mathcal{I}$. Hence $\mathcal{N}^W \subseteq^{1\ldots k}_{rep} \mathcal{B}^W$.

We store $\hat{\mathcal{N}}^W \subseteq^{1\ldots k}_{rep} \mathcal{N}^W$ in $\mathcal{D}[W]$. The sets $\hat{\mathcal{N}}^{Wj}$ are computed using Lemma 5. We have $\hat{\mathcal{N}}^{Wj} \subseteq^{k-j}_{rep} \mathcal{N}^{Wj} \subseteq^{k-j}_{rep} \mathcal{B}^{Wj}$ for all $1 \leq j \leq k$. Hence from Lemma 4, we have $\mathcal{D}[W] = \hat{\mathcal{N}}^W \subseteq^{1\ldots k}_{rep} \mathcal{B}^W$.

We now focus on the running time to compute $\mathcal{D}[W]$ and the size of $\mathcal{D}[W]$. Assume that $\mathcal{D}[Y]$ is precomputed for all subsets $Y \subseteq W$. We have $|\mathcal{D}[Y]| = |\hat{\mathcal{N}}^Y| = \sum_{j=1}^{k} |\hat{\mathcal{N}}^{Yj}|$. From Lemma 5, $|\hat{\mathcal{N}}^{Yj}| \leq |\mathcal{F}| \cdot k \cdot \binom{k}{j}$. Hence from Eq. (1), putting $Y = W \setminus S_i$, we have $|\mathcal{N}^{Wj}| \leq |\mathcal{F}|^2 \cdot k \cdot \binom{k}{j}$. Using Lemma 5, the time to compute $\hat{\mathcal{N}}^{Wj} \subseteq_{rep}^{k-j} \mathcal{N}^{Wj}$ is $\mathcal{O}^* \left(\binom{k}{j}^2 + \binom{k}{j}^\omega \right)$ where ω is the exponent for matrix multiplication. Hence the total time to compute $\mathcal{D}[W]$ is $\sum_{j=1}^{k} \mathcal{O}^* \left(\binom{k}{j}^\omega \right) = \mathcal{O}^*(2^{\omega k})$. The size of $\mathcal{D}[W]$ is $\mathcal{O}(|\mathcal{F}| \cdot k \cdot \sum_{j=1}^{k} \binom{k}{j}) = \mathcal{O}(2^k \cdot k \cdot |\mathcal{F}|)$.

The overall running time to check if $\mathcal{D}[U]$ is empty or not is bounded by $\mathcal{O}^*(2^{|\mathcal{U}|} \cdot 2^{\omega k}) = \mathcal{O}^*(2^{\omega |\mathcal{U}| + |\mathcal{U}|}) = \mathcal{O}^*(10.361^{|\mathcal{U}|})$. $\qquad\square$

6   Conclusion

We have initiated a systematic study of CONFLICT-FREE SET COVER with various parameterizations with restrictions to $G_\mathcal{F}$. One open question is to identify a general characterization for the graph classes of $G_\mathcal{F}$ when CONFLICT-FREE SET COVER becomes FPT parameterized by k or by $|\mathcal{U}|$.

Acknowledgements. The authors thank Fahad Panolan and Saket Saurabh for several valuable discussions on the theme of the paper.

References

1. Agrawal, A., Jain, P., Kanesh, L., Lokshtanov, D., Saurabh, S.: Conflict free feedback vertex set: a parameterized dichotomy. In: LIPIcs-Leibniz International Proceedings in Informatics, vol. 117. Schloss Dagstuhl-Leibniz-Zentrum fuer Informatik (2018)
2. Arkin, E.M., et al.: Conflict-free covering. In: Conference on Computational Geometry, p. 17 (2015)
3. Banik, A., Panolan, F., Raman, V., Sahlot, V., Saurabh, S.: Parameterized complexity of geometric covering problems having conflicts. Algorithms and Data Structures. LNCS, vol. 10389, pp. 61–72. Springer, Cham (2017). https://doi.org/10.1007/978-3-319-62127-2_6
4. Cygan, M., et al.: On problems as hard as CNF-SAT. ACM Trans. Algorithms (TALG) **12**(3), 41 (2016)
5. Cygan, M., et al.: Parameterized Algorithms. Springer, Cham (2015). https://doi.org/10.1007/978-3-319-21275-3
6. Darmann, A., Pferschy, U., Schauer, J., Woeginger, G.J.: Paths, trees and matchings under disjunctive constraints. Discrete Appl. Math. **159**(16), 1726–1735 (2011)
7. Diestel, R.: Graph Theory. Springer, Heidelberg (2006)
8. Farber, M.: On diameters and radii of bridged graphs. Discrete Math. **73**(3), 249–260 (1989)
9. Fomin, F.V., Lokshtanov, D., Panolan, F., Saurabh, S.: Efficient computation of representative families with applications in parameterized and exact algorithms. J. ACM (JACM) **63**(4), 29 (2016)
10. Golumbic, M.C.: Algorithmic Graph Theory and Perfect Graphs, vol. 57. Elsevier, Amsterdam (2004)
11. Impagliazzo, R., Paturi, R., Zane, F.: Which problems have strongly exponential complexity? J. Comput. Syst. Sci. **63**(4), 512–530 (2001)

12. Jain, P., Kanesh, L., Misra, P.: Conflict free version of covering problems on graphs: classical and parameterized. In: Fomin, F.V., Podolskii, V.V. (eds.) CSR 2018. LNCS, vol. 10846, pp. 194–206. Springer, Cham (2018). https://doi.org/10.1007/978-3-319-90530-3_17

13. Kann, V.: Polynomially bounded minimization problems which are hard to approximate. In: Lingas, A., Karlsson, R., Carlsson, S. (eds.) ICALP 1993. LNCS, vol. 700, pp. 52–63. Springer, Heidelberg (1993). https://doi.org/10.1007/3-540-56939-1_61

14. Lin, B.: The parameterized complexity of the k-biclique problem. J. ACM (JACM) **65**(5), 34 (2018)

15. Lokshtanov, D., Marx, D., Saurabh, S.: Slightly superexponential parameterized problems. In: Proceedings of the Twenty-Second Annual ACM-SIAM Symposium on Discrete Algorithms, pp. 760–776. Society for Industrial and Applied Mathematics (2011)

16. Lokshtanov, D., Misra, P., Panolan, F., Saurabh, S.: Deterministic truncation of linear matroids. ACM Trans. Algorithms (TALG) **14**(2), 14 (2018)

17. Lokshtanov, D., Panolan, F., Saurabh, S., Sharma, R., Zehavi, M.: Covering small independent sets and separators with applications to parameterized algorithms. In: Proceedings of the Twenty-Ninth Annual ACM-SIAM Symposium on Discrete Algorithms, pp. 2785–2800. Society for Industrial and Applied Mathematics (2018)

18. Marx, D.: A parameterized view on matroid optimization problems. Theor. Comput. Sci. **410**(44), 4471–4479 (2009)

19. Marx, D., Salmasi, A., Sidiropoulos, A.: Constant-factor approximations for asymmetric tsp on nearly-embeddable graphs. In: LIPIcs-Leibniz International Proceedings in Informatics, vol. 60. Schloss Dagstuhl-Leibniz-Zentrum fuer Informatik (2016)

20. Pferschy, U., Schauer, J.: The maximum flow problem with conflict and forcing conditions. In: Pahl, J., Reiners, T., Voß, S. (eds.) INOC 2011. LNCS, vol. 6701, pp. 289–294. Springer, Heidelberg (2011). https://doi.org/10.1007/978-3-642-21527-8_34

21. Raman, V., Saurabh, S.: Short cycles make w-hard problems hard: FPT algorithms for w-hard problems in graphs with no short cycles. Algorithmica **52**(2), 203–225 (2008)

22. van Bevern, R., Tsidulko, O.Y., Zschoche, P.: Fixed-parameter algorithms for maximum-profit facility location under matroid constraints. In: Proceedings of the 11th International Conference on Algorithms and Complexity (2019)

Forward Looking Huffman Coding

Shmuel Tomi Klein[1], Shoham Saadia[2], and Dana Shapira[2(✉)]

[1] Department of Computer Science, Bar Ilan University, 52900 Ramat Gan, Israel
tomi@cs.biu.ac.il
[2] Department of Computer Science, Ariel University, 40700 Ariel, Israel
shohamsaadia@gmail.com, shapird@ariel.ac.il

Abstract. Huffman coding is known to be optimal, yet its dynamic version may yield smaller compressed files. The best known bound is that the number of bits used by dynamic Huffman coding in order to encode a message of n characters is at most larger by n bits than the number of bits required by static Huffman coding. In particular, dynamic Huffman coding can also generate a larger encoded file than the static variant, though in practice the file might often, but not always, be smaller. We propose here a new dynamic Huffman encoding approach, that provably always performs at least as good as static Huffman coding, and may be better than the standard dynamic Huffman coding for certain files. This is achieved by reversing the direction for the references of the encoded elements to those forming the model of the encoding, from pointing backwards to looking into the future.

1 Introduction

Huffman coding is one of the cornerstones of data compression algorithms, and enjoys popularity in spite of almost seven decades since its invention, probably because of its well-known optimality. Given is a text $T = x_1 \cdots x_n$ over some alphabet $\Sigma = \{\sigma_1, \ldots, \sigma_m\}$ such that σ_i occurs w_i times in T. The problem is to assign binary codewords of lengths ℓ_i bits to the characters σ_i, such that the set of codewords forms a *prefix code* and such that the total size of the encoded file in bits, $\sum_{i=1}^{m} w_i \ell_i$, is minimized.

The alphabet will often consist of a set of characters, and it is convenient in many applications to use just the basic ASCII set of 128 or 256 letters, but Huffman's algorithm works just as well for larger sets, such a bigrams, k-grams or even words, and ultimately, any set $\mathcal{S}$ of substrings of the text, as long as there is a well defined way to parse the text into a sequence of elements of $\mathcal{S}$. We shall use the term *alphabet* in this broader sense.

Huffman's seminal paper [4] solves this problem optimally, but it should be remembered that Huffman codes are optimal under the following constraints:

1. the alphabet Σ is given and fixed in advance;
2. the codeword lengths ℓ_i are integers.

© Springer Nature Switzerland AG 2019
R. van Bevern and G. Kucherov (Eds.): CSR 2019, LNCS 11532, pp. 203–214, 2019.
https://doi.org/10.1007/978-3-030-19955-5_18

In particular, the second condition seems self-evident when one considers the binary codewords of known codes like ASCII, Huffman, Shanon-Fano or others, though it may be circumvented by *arithmetic coding* [12]. By encoding the entire input file as a single element rather than each character individually, an arithmetic encoder effectively assigns to each occurrence of the character σ_i of probability $p_i = w_i/n$ an encoding of exactly $-\log_2 p_i$ bits, without rounding. The average codeword length thus reaches entropy and is always at least as good as that of Huffman coding based on the same alphabet.

There is, however, another implicit constraint, which is rarely mentioned because it seems obvious, that

3. the encoding is static, i.e., the same codeword is used for a given character throughout the encoding process.

Data compression algorithms are often classified into static or adaptive techniques. The static ones base the coding procedures on a model of the distribution of the encoded elements that is either assumed in advance or gathered in a first pass over the data, whereas the adaptive methods learn the model details incrementally. Therefore, the static models would seem to be better and produce smaller files as they have the ability to exploit more knowledge: not only the distribution of the elements up to the current point, but global occurrence statistics in the entire file. However, in practice, adaptive compression is sometimes better, in particular when there is much variability in the occurrence patterns of the different elements on which the model is based. Furthermore, if the model is not learned adaptively, a prelude consisting of the details of the chosen model should be prepended to the compressed file, allowing the decoder to be synchronized with the encoder. For the adaptive methods, transmitting the details of the model to the decoder is superfluous, as the model gets updated identically on both encoder and decoder ends. The adaptive methods are often referred to as dynamic ones, and we shall use these terms interchangeably.

Static Huffman coding can use a known distribution of the alphabet corresponding to the nature of the file, for example, English text with its well known character distribution, or use accurate probabilities for the specific input file that are computed via a preprocessing stage. An advantage of using known statistics is saving the preprocessing stage, however, one then relies on the assumption that the given file fits the assumed model, which is not always the case. If there is too much discrepancy between the conjectured and the actual model, there might be a loss in the compression gain.

In the adaptive mode of Huffman coding, the encoder and decoder maintain identical copies of a varying Huffman tree, which, at each point, corresponds to the frequencies of the elements processed so far. The trivial solution of reconstructing the Huffman tree from scratch after each processed character, is obviously very wasteful, since in most cases, the Huffman tree is not altered: only one of the frequencies is incremented by 1, and the others remain unchanged. This motivated the development of efficient adaptive Huffman coding procedures by Faller [1] and Gallager [3], who propose essentially the same one-pass solution.

Knuth extends Gallager's work and also suggests that the frequencies may be decreased as well as increased [6]. These independent adaptive Huffman coding methods are known as the FGK algorithm.

A further enhancement by Vitter [11] also minimizes the external path length $\Sigma_{i=1}^m \ell_i$ and the height $\max\{\ell_i\}$ among all Huffman trees. Vitter proved that the number of bits used in his adaptive Huffman procedure in order to encode a message of n characters, is bounded by the size of the compressed file resulting from the optimal two-pass static Huffman algorithm, plus n. That is, the dynamically produced file may be *larger* than the static counterpart, and examples can be given for which this actually happens, though empirical results often show that on the contrary, there might be an improvement in the compression rate of the dynamic version as compared to the static one. One may thus conclude that in certain cases, though not in all, adaptive Huffman coding is *better than the optimal* static Huffman coding! The contribution of this paper is a new dynamic Huffman encoding which *always* performs at least as good as static Huffman coding.

In the following section, we describe the main idea of the new approach, followed, in Sect. 3, by bring the details and an extended example. In Sect. 4 we first present an extreme case in which the new approach can be twice as efficient as the dynamic algorithm used so far, and then formally prove that the forward looking variant does not hurt the compression performance of static coding, a property that does not hold for classical dynamic Huffman coding. Section 5 brings empirical results.

2   Forward Looking Dynamic Huffman Coding

The traditional dynamic Huffman coding, and in fact, practically all adaptive compression algorithms, update the model according to what has already been seen in the file processed so far. The underlying assumption is that the past is a good approximation of the future, according to the biblical postulate that

> The thing that hath been, it is that which shall be
> *Ecclesiastes* 1:9, King James Version

More precisely, the distribution of the elements in the prefix of the file up to the current point, serves as an estimate for the distribution of these elements from the current point onwards. Such an assumption seems to be justifiable, especially for homogeneous texts written in some natural language, but there is of course no guarantee that it holds for all possible input files.

The algorithm we suggest corresponds to a scenario in which the exact statistics of the element's occurrences are known, yet we prefer using a dynamic model that adapts itself while processing the file. However, contrarily to the classic dynamic methods, which base their current model on what has already been seen in the *past*, our algorithm uses the model's knowledge of what is still to come, i.e., it looks into the *future*. The idea of looking forwards rather than backwards in adaptive compression has already been proposed in a completely different context for files compressed by LZSS [9]: instead of pointing backwards to reoccurring strings, the locations of the pointers were moved and their direction was altered to point forwards, in order to enable streaming compressed pattern matching [5].

The motivation of the suggested amendment is a different approach to what should be done in order to produce a more economical encoding. Traditional dynamic Huffman coding concentrates only on the character that is currently processed: its frequency is incremented, which tends to shorten its codeword length for future usage, but it ignores the fact that these savings may come at the price of having certain other codewords lengthened. This "selfish" behavior is counterbalanced by the more altruistic approach of the forward looking variant, where the frequency of the currently processed element is *decreased*, even though, as a consequence, the corresponding codeword can only become longer, yet this action may shorten the codewords of other symbols that are still present in the tree, yielding an overall gain.

To allow a fair comparison, one has to take into account that the models on which the methods rely require different amounts of storage for their encodings. For instance, static Huffman coding does not need the exact frequencies of the m characters; if a canonical [8] Huffman tree is used, it suffices to transmit its *quantized source* $\langle n_1, n_2, \ldots, n_k \rangle$ as defined in [2], where n_i is the number of codewords of length i, for $1 \leq i \leq k$, and k is the longest codeword length. For example, the quantized source of the canonical tree equivalent to that in Fig. 1(a) would be $\langle 0, 1, 2, 6, 4 \rangle$. By using a canonical tree, one can thus save the transmission of the frequencies, but the sequence of characters must then be sorted by frequency order. If, on the other hand, a non-canonical tree is acceptable, the order of the characters may be implicit, e.g., lexicographic, but then we need to know the length of each of the m codewords.

The forward looking dynamic Huffman algorithm requires the exact frequencies of the elements, and not just the corresponding codeword lengths. The standard dynamic Huffman method does not need any frequencies, since they are incrementally learned by both encoder and decoder.

We have therefore included the size of a header describing the model in our experiments below. When the size n of the text to be compressed is large relative to the size m of the alphabet, the amount of storage required to encode the model is often negligible. For larger alphabets, for example when words, instead of just characters, are the elements to be encoded [7], the additional overhead of the forward looking variant may not be justifiable, unless the text to be encoded is very large. For certain applications, like large full-text Information Retrieval Systems, this overhead can be ignored, since the list of different words and their frequencies are usually stored anyway as part of the standard auxiliary data called *inverted files* [13].

3   Implementation Details and Extended Example

For the ease of description, we shall assume that the alphabet of the entire file is known in advance to both encoder and decoder. In practice, adaptive Huffman coding may reserve leaves in the Huffman tree only to characters that have already been processed, plus one leaf, often called NYT for Not Yet Transmitted, to enable the extension of the alphabet whenever a character c that has not been seen previously is encountered. In this case, the codeword of NYT is sent to the output, followed by some standard encoding of c, e.g., in ASCII (though this limits the size of the alphabet to 256, which is often not enough, for example when we wish to encode words rather than single letters). The leaf for NYT is then split, i.e., transformed into a node with two leaf children, one for c and one for NYT.

Our assumption is thus that we initialize a Huffman tree with m leaves. For the classical dynamic Huffman coding, all nodes are assigned a frequency of 0 (or all 1 if zero frequencies may cause trouble), and the frequencies are incremented at each step. The Huffman tree at the end is one that would have been produced by static Huffman coding. For the forward looking approach, the initial frequencies are those corresponding to the entire file, and they are decremented after the processing of each character. If one of the frequencies reaches 0, we know that no further occurrences of the corresponding character will appear in the file, so that its leaf may be removed from the tree.

The main tool for updating the Huffman tree, rather than reconstructing it from scratch after each character, is the *sibling property*, which guarantees that the tree is a Huffman tree, and is defined as follows:

Sibling Property: A weighted binary tree with m leaves is said to have the sibling property if and only if

1. the m leaves have nonnegative weights $w_1, \ldots, w_m$, and the weight of each internal node is the sum of the weights of its two children; and
2. the nodes can be numbered in nondecreasing order by weight, so that the nodes indexed $2j - 1$ and $2j$ in this numbering are siblings, for $1 \le j \le m - 1$.

It has been shown in [3] that a tree is a Huffman tree if and only if it has the sibling property. As example, consider the tree in Fig. 1(a), in which the internal nodes contain their weights, the leaf for character x contains the pair $(x, f(x))$, where $f(x)$ is its frequency, and the indexes in the above mentioned numbering are written above the nodes. We use a bottom-up, left to right numbering, but any other one is plausible, as long as it complies with the sibling property.

Given a Huffman tree, the following update procedure is used by the standard dynamic Huffman coding. If the currently processed character is c, the weights of all the nodes on the path from the leaf corresponding to c up to the root have to be incremented by 1. For example, referring again to Fig. 1(a), suppose the next character is $c = $ r whose leaf is the node indexed 16, then the weights of the nodes indexed 16, 20, 23 and 25 are incremented to 3, 5, 9 and 24, respectively. Note that for this example, the sibling property still holds after the updates (with the same numbering), so no further action is needed. However, in other cases, the increments may disrupt the monotonicity of the numbering.

Consider the case in which the next character is $c = $ A, rather than r. The path from its leaf to the root is emphasized in Fig. 1(a), and the weights of the nodes indexed 1, 11, 18, 21, 24 and 25 are incremented to 2, 3, 5, 8, 16 and 24, respectively. If we were to keep the same numbering, the weights of nodes 10–13 would be 1, 3, 2, 2, which is not a monotone sequence. In fact, with the present layout of the tree, no numbering can fulfill the second condition of the sibling property.

The difference between the two examples lies in the fact that in the first one, all the nodes that get updates have indexes which are maximal for their given weights before the increments, a property which does not hold for the second example. Indeed, there are many nodes with weight 2, all those with indexes 11 to 16, and the leaf (r,2) is indexed 16; nodes 18 to 20 have weight 4, nodes 22 and 23 have weight 8, and only node 25 has weight 23. On the other hand, for the second example, the leaf (A,1) is indexed 1, but there are also other nodes with weight 1 and that have higher indexes (up to 10); node 18 has weight 4, and node 20, which has a higher index, has also weight 4.

Clearly, this rule holds in general and not only for the examples, since a non-decreasing sequence of integers will remain such, even if the highest ranking elements within the sub-sequences of identical integers are incremented by 1. For example, $\cdots 6\,6\,7\,7\,7\,7\,9\,9\cdots$ may turn into $\cdots 6\,\mathbf{7}\,7\,7\,7\,\mathbf{8}\,9\,9\cdots$, which is still non-decreasing. To ensure that only such highest ranking nodes are updated, the dynamic Huffman algorithm exploits another property of Huffman trees, namely that nodes with identical weights may be interchanged. More precisely, since swapping the nodes is actually implemented by swapping the pointers to them, not just the nodes are interchanged, but the entire sub-trees rooted by these nodes. As a result, the shape of the tree might change, which yields a different set of codewords, but the weighted total path-length $\sum_{i=i}^{m} w_i \ell_i$ remains the same, so that the transformed tree is also a legitimate Huffman tree minimizing this sum, which represents the size of the compressed file.

Our algorithm adapts the dynamic Huffman procedure to the forward looking variant. After each encoded character, the number of its occurrences is *decremented* by 1. Before doing so, each node on the path from the updated leaf to the root is swapped, if necessary, with the *smallest* numbered node of identical weight. Then the weights of these nodes can be decremented without violating the sibling property. To continue the previous example, $\cdots 6\,6\,7\,7\,7\,7\,9\,9\cdots$ may now turn into $\cdots 6\,\mathbf{6}\,6\,7\,7\,7\,\mathbf{8}\,9\cdots$.

Given a text $T = x_1 \cdots x_n$ to be compressed, FORWARD-HUFFMAN compresses T using initially the Huffman tree of the static encoding. After every read character x_k, the corresponding codeword is output to the encoded file, and the Huffman tree gets modified to correspond to the frequencies within $\{x_{k+1} \cdots x_n\}$. The formal algorithm is given in Algorithm 1. The decoding algorithm is symmetrical.

Algorithm 1. *forward looking dynamic Huffman encoding*

FORWARD-HUFFMAN(T)

1 Initialize a Huffman tree according to the frequencies within T
2 **for** $k \leftarrow 1$ **to** $n - 1$ **do**
3   output the codeword for leaf(x_k)
4      $p \leftarrow$ leaf(x_k)
5   **while** $p \neq root$ **do**
6          $q \leftarrow$ lowest numbered node with same weight as p
7   **if** $q \neq p$ **then**
8   interchange p with q
9   decrement p's weight by 1
10         $p \leftarrow$ parent of p
11     $p \leftarrow$ leaf(x_k)
12   **if** p's weight is 0 **then**
13   replace p's parent node by p's sibling
14   delete the leaf p

Lines 11–14 deal with the case where the weight of leaf(x_k) has been reduced to 0, which means that the last occurrence of the character x_k is encountered. Note that in this case, the leaf is the lowest numbered one and must be the left child of its parent node q. It can thus be eliminated from the tree, by replacing q by its right child, which is the leaf's left sibling. Whenever the structure of the tree is altered (in lines 8 or 13), we assume that the numbering of the nodes, referred to in the sibling property, is updated as well.

For example, consider the text $T = $ Abrahamasantaclaragasse (the name of a street in Vienna) over the alphabet {A, a, b, c, e, g, h, l, m, n, r, s, t} with corresponding weights {1, 8, 1, 1, 1, 1, 1, 1, 1, 1, 2, 3, 1}. We started with one of the possible Huffman trees, shown in Fig. 1(a), in which the leaves are assigned in lexicographical order. The other trees in the figure are those obtained after the processing, by Algorithm 1, of the first three characters of T, A, b and r, respectively.

When A is processed, all the nodes on the emphasized path from the leaf (A,1) to the root are the lowest indexed of their corresponding weights, so no interchanges in lines 6–8 are needed; the weight of each node on this path is then decremented, the parent of (A,1) is overwritten by the sibling of (A,1), which is the leaf (b,1), and the leaf (A,1) is erased from the tree. The resulting tree is shown in Fig. 1(b) in which the leaf that corresponds to the only codeword that has been changed, appears in gray. The shape of the tree has changed, therefore, a new numbering is necessary.

When the following input character b is processed, the corresponding leaf, numbered 9, is not the node with the lowest index among those with weight 1. Nodes 9 and 1 are therefore interchanged. Climbing up the tree from the new position of the leaf (b,1), we get to node 10, which is the lowest numbered node of weight 2. However, its parent node 16 has weight 3, as has also node 15.

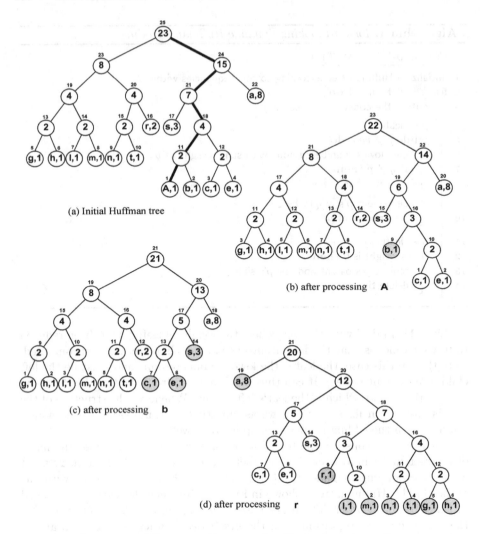

Fig. 1. Illustration of FORWARD-HUFFMAN for T=Abrahamasantaclaragasse.

We therefore interchange the leaf (s,3) with the subtree rooted at node 16. The process continues to nodes 19, 22 and 23, whose corresponding weights are the smallest indexed ones for the weights 6, 14 and 22, respectively, so no further updates are needed. The resulting tree is given in Fig. 1(c), which again displays the changed codewords in gray.

The processing of character r starts by swapping the nodes 12 and 9 in Fig. 1(c), and then continues to node 15, which is the lowest indexed of weight 4. Its parent node 19, is interchanged with leaf 18; the updates continue with nodes 20 and 21. Finally, the weights of the nodes on the new path from the root to leaf (r,2) are decremented, resulting in the tree given in Fig. 1(d).

4   Analysis

Dynamic Huffman coding repeatedly changes the shape of the tree, but there is a delay between the occurrence of a change and when such a change starts to influence the encoding. For encoding the current character we use the tree built in the previous stage, and the changes implied by the processed character do only affect the encoding in the subsequent stages, if at all. This behavior is demonstrated in the following extreme example, comparing the performances of the two dynamic variants. The example shows that the file constructed by traditional dynamic Huffman may be about twice as large as that produced by the FORWARD-HUFFMAN algorithm.

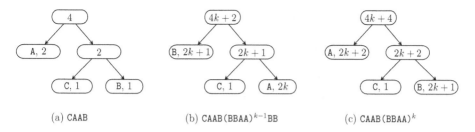

(a) CAAB   (b) CAAB(BBAA)$^{k-1}$BB   (c) CAAB(BBAA)^k

Fig. 2. Example for which classical dynamic Huffman coding produces a file twice the size of that constructed by FORWARD-HUFFMAN.

Let $T = $CAAB$\{$BBAA$\}^k$ for some positive integer k. We initialize our Huffman tree with $\Sigma = \{$A, B, C$\}$ as shown in Fig. 2(a) for the prefix CAAB of T. Consider first the standard dynamic algorithm. When the two Bs of the first quadruple BBAA are processed, only the second B causes a change in the structure of the Huffman tree, but this happens *after* the two Bs have already been encoded using 2 bits for each. The Huffman tree after reading the prefix CAABBB of T, is depicted in Fig. 2(b). When the following two As of the first quadruple BBAA are processed, again the positions of the A and B nodes are swapped only *after* the frequency of A exceeds that of B, so each of the As is also encoded by 2 bits. The resulting Huffman tree after processing CAABBBAA is presented in Fig. 2(c), and this is in fact the same Huffman tree as that in Fig. 2(a). This alternation between two different structures of the Huffman tree proceeds for each of the BBAA quadruples, and every character of T (except the first two As) uses 2 bits, for a total of $8k + 6$.

When compressing T using the proposed forward looking coding, the Huffman tree may start with the static Huffman tree identical to that of Fig. 2(a). Since the weight of the first read character C is 1, its node is deleted, and the Huffman tree is reduced to only two leaves, one for A and the other for B. All the codewords are then of length 1 and the size of the compressed file is exactly $4k + 5$, roughly half the size of the compressed file constructed by traditional dynamic Huffman methods.

If we were to use static Huffman coding, the tree would be the one of Fig. 2(c) and the size of the encoded file would be $6k + 6$.

Note that this example also shows that the standard dynamic Huffman coding may produce an encoding which is worse than that of static Huffman. The new forward looking dynamic algorithm, on the other hand, is at least as good as static Huffman, not only on this example, but in general, as proved in the following Theorem.

Theorem: *For a given distribution of frequencies, the average codeword length of* FORWARD-HUFFMAN *is at least as good as the average codeword length of static Huffman coding.*

Proof: Suppose the file has n characters and let us inspect the situation after t characters have already been encoded according to the static Huffman code. If $t = n$, the encoding is identical to static Huffman coding. For $t < n$, if we knew also the distribution of the characters in the first t characters, we should have built a Huffman code just for these t characters, but lacking this knowledge, we took the global frequencies and encoded accordingly. However, for the following $n - t$ characters, we know the true distribution, which might be different from the global one, so continuing with the static global distribution, one can only be worse (or at least not better) than changing to another Huffman code according to the frequencies in the last $n - t$ characters, which is exactly what is done in FORWARD-HUFFMAN. The overall sum of the codeword lengths for the static encoding is thus larger or equal to the sum of the codeword lengths of the first t characters, plus the sum of the codeword lengths of the static Huffman encoding based only on the frequencies of the last $n - t$ characters. But the same argument applies on this static encoding of the last $n - t$ characters, thus repeating the argument and choosing $t = 1$ gives that the expected codeword length of a static Huffman code is larger than or equal to the expected codeword length of the FORWARD-HUFFMAN presented in Algorithm 1. □

5   Empirical Results

To get empirical evidence how the three algorithms behave in practice, we considered texts of different languages and alphabet sizes: *ftxt* is the French version of the European Union's JOC corpus, a collection of pairs of questions and answers on various topics used in the ARCADE evaluation project [10]; *sources* is formed by C/Java source codes obtained by concatenating all the .c, .h and .java files of the linux-2.6.11.6 distributions; *English* is the concatenation of English text files selected from etext02 to etext05 collections of the Gutenberg Project, from which the headers related to the project were deleted so as to leave just the real text; *exe* is the executable file of the static Huffman source code we used; and *bible* is the Bible (King James version) in basic English, Textfile 980302. The alphabet consisted of individual ASCII characters for all test files, except the last one, for which the different words have been used.

Table 1. Compression performance.

File	Full size MB	m	Compression ratio		
			STATIC	FORWARD	DYNAMIC
ftxt	7.6	127	0.583	0.578	0.577
sources	200	208	0.699	0.686	0.647
English	50	217	0.580	0.571	0.566
exe	0.02	256	0.665	0.631	0.634
bible	4.3	22,180	0.236	0.234	0.245

Our goal was to compare the compression performance of the three methods: static Huffman, the proposed FORWARD-HUFFMAN algorithm, and the traditional dynamic Huffman. The results are presented in Table 1. The second column gives the original file sizes in MB. The third column gives the size of the encoded alphabet, m. The following three columns, entitled STATIC, FORWARD and DYNAMIC, show the compression ratios achieved by the compared algorithms. The compression ratio is defined as the size of the compressed file divided by the size of the original file. As mentioned, we included the overhead of the description of the model in the size of the compressed file.

As can be seen, our method is consistently better than static Huffman, as expected, and the traditional dynamic Huffman achieves the best results in most cases. However, FORWARD was better for the executable files we tested, as well as when a large alphabet consisting of words has been used.

6   Conclusion

The contribution of this paper is twofold, theoretical as well as practical. The theoretical result is that the standard static Huffman coding, well-known for its optimality, may in fact be improved. In practice, this is often achieved by traditional dynamic Huffman coding, but the traditional dynamic version can also be worse. The new forward looking Huffman coding is *provably* better than static Huffman *on all files*, though may be outperformed on certain files by the classical dynamic coding. For executables and for large alphabets, when the precise number of occurrences is already known to the decoder such as in Information Retrieval applications, the FORWARD algorithm was also better than DYNAMIC on our tests.

References

1. Faller, N.: An adaptive system for data compression. In: Record of the 7-th Asilomar Conference on Circuits, Systems and Computers, pp. 593–597 (1973)
2. Ferguson, T.J., Rabinowitz, J.H.: Self-synchronizing Huffman codes. IEEE Trans. Inf. Theory **30**(4), 687–693 (1984)

3. Gallager, R.: Variations on a theme by Huffman. IEEE Trans. Inf. Theory **24**(6), 668–674 (1978)
4. Huffman, D.: A method for the construction of minimum redundancy codes. Proc. IRE **40**, 1098–1101 (1952)
5. Klein, S.T., Shapira, D.: A new compression method for compressed matching. In: Data Compression Conference, DCC 2000, Snowbird, Utah, USA, March 28–30, 2000, pp. 400–409 (2000)
6. Knuth, D.E.: Dynamic Huffman coding. J. Algorithms **6**(2), 163–180 (1985)
7. Moffat, A.: Word-based text compression. Softw. Pract. Exper. **19**(2), 185–198 (1989)
8. Schwartz, E.S., Kallick, B.: Generating a canonical prefix encoding. Commun. ACM **7**, 166–169 (1964)
9. Storer, J.A., Szymanski, T.G.: Data compression via textural substitution. J. ACM **29**(4), 928–951 (1982)
10. Véronis, J., Langlais, P.: Evaluation of parallel text alignment systems: the arcade project. In: Véronis, J. (ed.) Parallel Text Processing, pp. 369–388. Kluwer Academic Publishers, Dordrecht (2000)
11. Vitter, J.S.: Design and analysis of dynamic Huffman codes. J. ACM **34**(4), 825–845 (1987)
12. Witten, I.H., Neal, R.M., Cleary, J.G.: Arithmetic coding for data compression. Commun. ACM **30**(6), 520–540 (1987)
13. Zobel, J., Moffat, A.: Inverted files for text search engines. ACM Comput. Surv. **38**(2), 6 (2006)

Computational Complexity of Real Powering and Improved Solving Linear Differential Equations

Ivan Koswara, Svetlana Selivanova$^{(\boxtimes)}$, and Martin Ziegler

School of Computing, KAIST, Daejeon, Republic of Korea
sseliv@kaist.ac.kr

Abstract. We re-consider the problem of solving systems of differential equations approximately up to guaranteed absolute error $1/2^n$ from the rigorous perspective of sequential and parallel time (i.e. Boolean circuit depth, equivalently: Turing machine space) complexity. While solutions to general smooth ODEs are known "PSPACE-complete" [Kawamura'10], we show that (i) The Cauchy problem for linear ODEs can be solved in NC^2, that is, within polylogarithmic parallel time $\mathcal{O}(\log^2 n)$ by Boolean circuits of polynomial size. (ii) The Cauchy problem for linear analytic PDEs, having a unique solution by the Cauchy-Kovalevskaya theorem, can be also solved in polylogarithmic parallel time, thus generalizing the case of analytic ODEs [Bournez/Graça/Pouly'11]. (iii) Well-posed Cauchy and boundary-value problems for linear PDEs in classes of continuously differentiable functions are solvable in the counting complexity class $\#P^{\#P}$: improving over common numerical approaches yielding exponential sequential time or parallel polynomial time. Our results build on efficient algorithms and their analyses for real polynomial, matrix and operator powering which do not occur in the discrete case and may be of independent interest.

1 Introduction and Brief Summary of Main Results

Recursive Analysis provides a rigorous algorithmic foundation to Numerics, that is, to operations on continuous data by means of approximations [18]. It has a long history of thorough investigations regarding ordinary (ODEs) and partial differential equations (PDEs) with respect to computability and real complexity. For example, it has been established that (a) ODEs with a polynomial/analytic right-hand side can be solved in (appropriately parameterized) sequential polynomial time [3]; while (b) general non-linear ODEs are optimally solved by Euler's Method using an amount of memory polynomial in the output precision

---

Supported by the National Research Foundation of Korea (grant NRF-2017R1E1A1A 03071032) and by the Korean Ministry of Science and ICT (grant NRF-2016K1A3A 7A03950702). Dedicated to the memory of Ker-I Ko. We thank Gleb Pogudin for seminal discussions that led to Theorem 7.

R. van Bevern and G. Kucherov (Eds.): CSR 2019, LNCS 11532, pp. 215–227, 2019.
https://doi.org/10.1007/978-3-030-19955-5_19

parameter n [9], equivalently: in polynomial parallel time [2]; and (c) solving Poisson's linear PDEs corresponds to the complexity class $\#P$ [11].

The main definitions and notions of real complexity theory (the framework in which we work) are summarized in Sect. 2. The discrete complexity hierarchy[1] [17, Corollary 2.34] translates to the real setting (see Definition 4) as

$$\mathbb{RNC}^1 \subseteq \mathbb{RSPACE}(\log n) \subseteq \mathbb{RNC}^2 \subseteq \mathbb{RPTIMESPACE}(\log^2 n) \subseteq \mathbb{RNC}^4$$
$$\text{and more generally } \mathbb{RNC}^i \subseteq \mathbb{RPTIMESPACE}(\log^i n) \subseteq \mathbb{RNC}^{2i} \ldots \subseteq \quad (1)$$
$$\mathbb{RPTIME} \subseteq \mathbb{R\#P} \subseteq \mathbb{R\#P}^{\#P} \subseteq \mathbb{RPSPACE} = \mathbb{RPAR} \subseteq \mathbb{REXP} \ .$$

The present work continues the above complexity investigations for linear ODEs and PDEs with the following contributions (see rigorous formulations later in this section and ideas of proofs in Sect. 4):

(a) Theorem 1 establishes linear systems of ODEs computable in polylogarithmic parallel time (=depth), specifically in $\mathbb{RNC}^2$, the real counterpart to NC^2.
(b) Also linear *analytic* PDEs can be solved in polylogarithmic parallel time according to Theorem 3, generalizing the case of analytic ODEs.
(c) By Theorem 2, a large class of continuously differentiable linear PDEs with boundary conditions, classically treated with numerical difference schemes, can be solved with computational complexity $\mathbb{RPSPACE}$;
(d) For many cases with periodic boundary conditions this can be improved to $\mathbb{R\#P}^{\#P}$ (probably even $\mathbb{R\#P}$, which will be a subject of future investigation).

In all these cases, the output consists of (approximations up to error 2^{-n} to) the real value $u(t)$ or $u(t,x)$ of the solution at given time/space t and x. Our results are obtained by applying matrix/operator exponentials, known but rarely used in classical Numerics. They in turn rely on efficient recursive algorithms for powering polynomials, matrices and operators developed in this paper (see Sect. 3), which may be of independent interest:

(e) Given $A \in [-1; 1]^{\text{poly}(n) \times \text{poly}(n)}$ with bounded powers $\|A^k\| \le 1$, its power $A^{\text{poly}(n)}$ can be computed in polylogarithmic depth $\mathbb{RNC}$; see Proposition 6(c). This result is used to prove Theorem 1.
(f) If the entries of $A \in [-1; 1]^{2^n \times 2^n}$ are computable in polynomial time, then the entries of A^{2^n} are computable in $\mathbb{RPSPACE}$, and this is optimal in general; see Proposition 6(g). This result is used to prove Theorem 2, general case of a difference scheme approximating the considered system of PDEs.
(g) For circulant matrices A of constant bandwidth with polynomial-time computable entries and bounded powers this result can be improved: the entries of the matrix power A^{2^n} are computable in $\mathbb{R\#P}$, see Theorem 7, where also powering of polynomials as an important auxiliary tool is considered. These results are used to prove Theorem 2, particular case of a difference scheme.

---

[1] Rigorously speaking, $\#P$ is a class of integer functions rather than decision problems and should here be read as $PTIME^{\#P}$.

(h) Theorem 8 and Example 9 establish complexity of powering and exponentiation of (e.g. differential) operators in Banach spaces. These results are used to prove Theorem 3.

Note that the hypothesis on uniformly bounded powers of the matrices in (e) and (g) makes the problems trivial over integers[2], yet interesting and new in the real case under consideration. The matrices corresponding to convergent difference schemes for PDEs meet this hypothesis because of the stability property required for a difference scheme to be convergent. Our work can be regarded as instances and confirmation of [1].

When translating the classical discrete parallel/depth complexity to the real setting, we restrict to 2^ℓ-Lipschitz functions f. A family C_n of Boolean circuits is required to approximate $y = f(x)$ up to error 2^{-n} for every $x \in \text{dom}(f)$, the latter given by approximations up to error $2^{-\ell-n}$. Space complexity of f is formalized as the number of working tape cells used by a Turing machine, not counting input nor output tape, to similarly approximate $y = f(x)$ up to error 2^{-n}: see Sect. 2 for details. We thus follow the general paradigm of measuring computational cost (depth=parallel/sequential time, memory) over the reals, other than in the discrete setting with finite inputs, in dependence on n reflecting roughly the number of correct bits of the output approximation attained in the worst-case over all (uncountably many) continuous arguments. 'Binary' error bounds 2^{-n}, rather than 'unary' $1/n$, capture roughly n correct binary digits and yield stronger statements.

We consider Cauchy (i.e. initial-value) problems for autonomous linear evolutionary differential equations in a general form

$$\frac{\partial}{\partial t}\vec{u}(t) = A\vec{u}(t), \quad t \in [0;1], \quad \vec{u}(0) = \vec{\varphi}. \tag{2}$$

Here A is a matrix in the case of ODEs and a more general operator (including partial differentiation for PDEs) for other cases. Similarly, $\vec{\varphi}$ is an initial real-valued vector for ODEs and an initial function in a more general setting. Let us now formally state our main contributions regarding differential equations:

Theorem 1. *Given $A \in [-1;1]^{d \times d}$ and $\vec{v} \in [-1;1]^d$ and $t \in [0;1]$, the solution*

$$\vec{u}(t) = \exp(tA)\vec{v} := \sum_k \frac{t^k}{k!}A^k\vec{v} \tag{3}$$

to the system of linear ordinary differential Eq. (2) is computable by Boolean circuits of depth $\mathcal{O}\big((\log d + \log n)^2\big)$ and size polynomial in $d + n$.

Recall that circuit depth is synonymous for parallel time. Theorem 1 thus formally captures the intuition that solving ODEs in the linear case is easier than in the analytic case [3], and much easier than in the general C^1 smooth case proven "PSPACE-complete" [9]. It is no loss of generality to impose unit bounds on $A, \vec{v}, t$: the general case is covered by rescaling.

---

[2] Cmp. Geoff Robinson's answer on https://math.stackexchange.com/q/59693.

Our next result is concerned with finitely-often differentiable solutions to Eq. (2) with $A = \sum_{|\alpha| \leq k} b_\alpha(x) D_x^\alpha$ and $\vec{\varphi} = \vec{\varphi}(x)$. They can be reduced [4], by adding extra variables, to first-order linear systems of PDEs with

$$A = \sum_{j=1}^{m} B_j(x) \frac{\partial}{\partial x_j}, \tag{4}$$

where $B_j(x)$ are matrices of a suitable dimension.

Theorem 2. *For $m \in \mathbb{N}$ and a convex open bounded $\Omega \subseteq \mathbb{R}^m$ consider the initial-value problem (IVP) (2) with the operator (4), and the boundary-value problem (BVP) with additionally given linear boundary conditions $\mathcal{L}u(t,x)|_{\partial\Omega} = 0, (t,x) \in [0,1] \times \partial\Omega$. Suppose the given IVP and BVP be well posed in that the classical solution $\vec{u} : [0;1] \times \overline{\Omega} \to \mathbb{R}$ (i) exists, (ii) is unique, and (iii) depends continuously on the data $\varphi(x)$ and $B_i(x)$ (for the BVP also on coefficients of $\mathcal{L}$). More precisely we assume that $u(t,x) \in C^2$ and its C^2-norm is bounded linearly by C^2-norms of the data (in functional spaces guaranteeing all the required properties). Moreover suppose that the given IVP and BVP admit a (iv) stable and (v) approximating with at least the first order of accuracy difference scheme (A_n) in the sense of Definition 10.*

(a) *If the difference scheme (meaning its matrix) A_n and the initial condition φ are (vi) computable in depth $s(n) \geq \log(n)$ in the sense of Definition 5, then evaluating the solutions $u : [0;1] \times \overline{\Omega} \ni (t,x) \mapsto u(t,x)$ of both IVP and BVP is feasible in depth $\mathcal{O}\big(s(2^n) + n \cdot \log n\big)$.*

(b) *If A_n and φ are (vi') computable in polynomial sequential time and A_n is additionally circulant of constant bandwidth, then the solution function u belongs to the real complexity class $\mathbb{R}\#\mathsf{P}^{\#\mathsf{P}}$.*

This second result establishes polynomial parallel time (equivalently: polynomial space or depth) complexity for the considered PDEs in the binary output precision parameter n and further down to the second level of the Counting Hierarchy. As a main tool we modify the classical difference schemes approach: standard step-by-step iterations would yield only exponential sequential time; we replace them with efficient matrix powering according to Proposition 6 and Theorem 7 below. It complements work like [14] measuring bit-cost in dependence on $N = 2^{\mathcal{O}(n)}$, the size of the grid under consideration, and implicitly supposing the difference scheme and initial data computable in logarithmic depth $s(n) = \mathcal{O}(\log n)$: where we consider the output precision parameter n, and allow for more involved difference schemes with $s(n) = \mathcal{O}(\log^2 n)$, say. Theorem 2 also complements rigorous cost analyses considering approximations up to *output* error $1/n$, or up to fixed error and in dependence on the length of the (algebraic) *input* [15,16]; and it vastly generalizes previous works on the computational complexity of Poisson's PDE [11].

Between linear ODEs and finitely-often continuously differentiable PDEs are analytic PDEs, captured by the Kovalevskaya Theorem; and their computational complexity also turns out to lie between the aforementioned two:

Theorem 3 (Polynomial-Time/Polylogarithmic-Space Kovalevskaya).
Let $f_1, \ldots, f_e : [-1; 1]^d \to \mathbb{R}^{e \times e}$ *and* $v : [-1; 1]^d \to \mathbb{R}^e$ *denote real functions analytic on some open neighborhood of* $[-1; 1]^d$ *and consider the system of linear partial differential equations*

$$\partial_t \vec{u}(\vec{x}, t) = f_1(\vec{x}) \partial_1 \vec{u} + \cdots + f_e(\vec{x}) \partial_e \vec{u} \qquad \vec{u}(\vec{x}, 0) \equiv v . \tag{5}$$

If $f_1, \ldots f_e$ *are computable in sequential polynomial time, then the unique real analytic local solution* $\vec{u} : [-\varepsilon; +\varepsilon]^{d+1} \ni (\vec{x}, t) \mapsto \vec{u}(\vec{x}, t) \in \mathbb{R}^e$ *to Eq. (5) is again computable in sequential polynomial time.*
If $f_1, \ldots f_e$ *are computable in polylogarithmic depth, then so is* $\vec{u}$.

We emphasize that the constructive proof of Kovalevskaya's Theorem [4, §4.6.3] expresses the solution's j-th coefficient as a multivariate integer polynomial p_j in the initial condition's and right-hand side's coefficients; however as p_j's total degree and number of variables grows with j, its number of terms explodes exponentially. Symbolic-numerical approaches employ Janet, Riquier or Gröbner bases whose worst-case complexity however is also exponential [13].

2   Real Complexity Theory

We consider the computational worst-case cost of computing continuous real functions on a compact domain, formalized as follows:

Definition 4. *Equip* $\mathbb{R}^d$ *with the maximum norm and fix* 2^ℓ*-Lipschitz* $f : \Omega \subseteq [-2^k; 2^k]^d \to [-2^m; 2^m]$, $k, \ell, m \in \mathbb{N}$.

(a) *Consider a Turing Machine* $\mathcal{M}$ *with* read-only *input tape, one-way output tape, and working tape(s).* $\mathcal{M}$ *is said to compute* f *(w.r.t.* μ) *if, given* $0^n 1 \operatorname{bin}(\vec{a}) \in \{0, 1\}^{n+1+d \cdot \mathcal{O}(k+\ell+n)}$ *for* $\vec{a} \in \{-2^{k+\ell+n}, \ldots, 0, \ldots + 2^{k+\ell+n}\}^d$ *with* $|\vec{x} - \vec{a}/2^{\ell+n}| \leq 2^{-\ell-n}$ *for some* $\vec{x} \in \Omega$, $\mathcal{M}$ *outputs* $\operatorname{bin}(b)$ *for some* $b \in \{-2^{m+n}, \ldots, 0, \ldots, 2^{m+n}\}$ *with* $|f(\vec{x}) - b/2^n| \leq 2^{-n}$ *and stops. Here* $\operatorname{bin}(\vec{a}) \in \{0, 1\}^*$ *denotes some binary encoding of integer vectors.*
(b) *Fix* $s, t : \mathbb{N} \to \mathbb{N}$ *with* $t(n) \geq n$ *and* $s(n) \geq \log_2(n)$. *The computation from (a) runs in* time t *and* space s *if* $\mathcal{M}$ *stops after at most* $t(n + m + d \cdot \mathcal{O}(n + k + \ell))$ *steps and uses at most* $s(n + m + d \cdot \mathcal{O}(n + k + \ell))$ *cells on its work tape, not counting input nor output tape usage and regardless of* $\vec{a}$ *as above. In this case write* $f \in \mathbb{R}\mathsf{TIME}(t) \cap \mathbb{R}\mathsf{SPACE}(s)$. *Polynomial time is abbreviated* $\mathbb{R}\mathsf{PTIME} = \bigcup_i \mathbb{R}\mathsf{TIME}(\mathcal{O}(n^i))$, *polynomial space means* $\mathbb{R}\mathsf{PSPACE} = \bigcup_i \mathbb{R}\mathsf{SPACE}(\mathcal{O}(n^i))$, *and* $\mathbb{R}\mathsf{PTIMESPACE}(\log^i n) := \mathbb{R}\mathsf{PTIME} \cap \mathbb{R}\mathsf{SPACE}(\log^i n)$.
(c) *Consider a Boolean circuit* C_n *having* $\mathcal{O}(n + m)$ *binary outputs and* $d \cdot \mathcal{O}(k + \ell + n)$ *binary inputs. Such a sequence* (C_n) *computes* f *if* C_n, *on every (possibly padded) input* $\operatorname{bin}(\vec{a})$ *with* $|\vec{x} - \vec{a}/2^{n+\ell}| \leq 2^{-k-\ell}$ *for some* $\vec{x} \in \Omega$, *it outputs some* $\operatorname{bin}(b)$ *such that* $|f(\vec{x}) - b/2^n| \leq 2^{-n}$.

(d) *We say that f is computable in* depth t, *written $f \in \mathrm{RDEPTH}(t)$, if there exists a sequence (C_n) of Boolean circuits over basis binary NAND, say, of depth at most $t(n + m + d \cdot \mathcal{O}(n + k + \ell))$ computing f. RNC^i abbreviates $\mathrm{RDEPTH}(\mathcal{O}(\log^i n))$ with the additional requirements that (i) the circuits be logspace uniform and (ii) their size (#gates) grows at most polynomially in $n + m + d \cdot \mathcal{O}(n + k + \ell)$. Similarly, RPAR abbreviates $\bigcup_i \mathrm{RDEPTH}(\mathcal{O}(n^i))$ with the additional requirement that the (possibly exponentially large) circuits C_n be polynomial-time uniform in that a polynomial-time Turing machine can, given $n \in \mathbb{N}$ in unary and $I < J \in \mathbb{N}$ in binary with respect to some fixed topological order, report whether in C_n the output of gate #I is connected to gate #J.*

(e) *We say f belongs to $\mathbb{R}\#\mathrm{P}$ if it can be computed by a Turing machine $\mathcal{M}^{\varphi}$ in polynomial time (b) given oracle access to some counting problem φ in the discrete complexity class $\#\mathrm{P}$. Similarly for $\mathbb{R}\#\mathrm{P}^{\#\mathrm{P}}$.*

We follow the classical conception of real numbers as 'streams' of approximations, both for input and output [6]: the alternative approach based on oracles [8,12] involves a *stack* of query tapes to ensure closure under composition.

Common numerical difference schemes are matrices A_m whose dimension D_m grows exponentially with the precision parameter m and therefore cannot be output entirely within polynomial time; instead Definition 5 requires any desired entry $(A_m)_{I,J}$ to admit efficient approximations, where indices I, J are given in binary such that their length remains polynomial in m:

Definition 5. (a) *Computing a vector $\vec{v} \in \mathbb{R}^D$ means to output, given $J \in \{0, \ldots D - 1\}$ in binary and $n \in \mathbb{N}$ in unary, some $a \in \mathbb{Z}$ in binary such that $|v_J - a/2^n| \leq 2^{-n}$; similarly for matrices $A \in \mathbb{R}^{D \times D}$, considered as vectors in $\mathbb{R}^{\mathcal{O}(D \cdot D)}$ via the pairing function $(I, J) \mapsto (I+J) \cdot (I+J+1)/2 + J$.*

(b) *For in both arguments monotonically non-decreasing $t(n, m)$ and $s(n, m) \leq t(n, m)$, a sequence $\vec{v}_m \in \mathbb{R}^{D_m}$ of vectors is computed in sequential time t and space s if its entries have binary length at most polynomial in its dimension $\sup_J |v_{m,J}| \leq 2^{\mathrm{poly}(D_m)}$ and it takes a Turing machine at most $t(n, m)$ steps and $s(n, m)$ tape cells to output, given $J \in \{0, \ldots D_m - 1\}$ in binary and $m \in \mathbb{N}$ in unary, some $a \in \mathbb{Z}$ in binary with $|v_{m,J} - a/2^n| \leq 2^{-n}$; similarly for sequences of matrices $A_m \in \mathbb{R}^{D_m \times D_m}$.*

(c) *Polynomial sequential time/space and poly/logarithmic space means polynomial and poly/logarithmic in $n + m$, respectively. $\mathbb{R}\#\mathrm{P}$ for (sequences of) vectors consists of those $\vec{v}_m \in \mathbb{R}^{D_m}$ computable in polynomial time by a Turing machine with a $\#\mathrm{P}$ oracle; similarly for (sequences of) matrices.*

(d) *A sequence $\vec{v}_m \in \mathbb{R}^{D_m}$ of vectors is computed in depth $s(n, m)$ if its entries have binary length at most polynomial in D_m: $\sup_J |v_{m,J}| \leq 2^{\mathrm{poly}(D_m)}$, and a family of Boolean circuits $C_{n,m}$ of depth $s(n, m)$ can output, given $J \in \{0, \ldots D_m - 1\}$ in binary, some $a \in \mathbb{Z}$ in binary with $|v_{m,J} - a/2^n| \leq 2^{-n}$.*

(e) *Polylogarithmic depth means Boolean circuits $C_{n,m}$ in (d) of depth polylogarithmic in $n + m$; RNC^i abbreviates $\mathrm{RDEPTH}(\mathcal{O}(\log^i n))$ with additional requirements (i) and (ii) as in Definition 4(f); similarly for RPAR.*

Note that already reading J in (b) takes time of order $\log D_m \le t(n,m)$, and space of order $s(n,m) \ge \log\log D_m$. Similarly, a circuit of depth $s(n,m)$ can access at most $2^{s(n,m)} \ge \log D_m$ input gates.

3   Efficient Real Polynomial/Matrix/Operator Powering

A major ingredient to our contributions are new efficient algorithms for real polynomial/matrix/operator powering, analyzed in dependence on various parameters that do not occur/make sense in the classical discrete setting:

Proposition 6. *(a) For $D, K \in \mathbb{N}$ and $A, A', B, B' \in [-2^L; 2^L]^{D \times D}$, it holds $|A^K|_\infty \le 2^{KL} \cdot D^{K-1}$ and $|A \cdot B - A' \cdot B'|_\infty \le D \cdot 2^{L+1} \cdot \max\big\{|A - A'|_\infty, |B - B'|_\infty\big\}$, where $|B|_\infty := \max_{I,J} |B_{I,J}|$.*

(b) For $D, K \in \mathbb{N}$, matrix powering $[-2^L; 2^L]^{D \times D} \ni A \mapsto A^K$ is computable by circuits C_n of depth $\mathcal{O}\big(\log(K) \cdot (\log n + \log D + \log K + \log L)\big)$ and size polynomial in $n + D + K + L$.

(c) Refining (b), if $|A^k|_\infty \le 2^L$ holds for all $k \le K$, the depth of circuits computing $\mathbb{R}^{D \times D} \ni A \mapsto A^K$ can be reduced to $\mathcal{O}\big((\log n + \log D + \mathbf{loglog}\, K + \log L) \cdot \log K\big)$.

(d) Suppose sequences $\vec{u}_m, \vec{v}_m \in \big[-2^{L_m}; +2^{L_m}\big]^{D_m}$ are computable in depth $s(n,m) \ge \log(n) + \log(m)$ and have inner product $\vec{u}_m^\perp \cdot \vec{v}_m \in \big[-2^{L_m}; +2^{L_m}\big]$. Then said inner product is computable in depth $\mathcal{O}\big(\log(D_m) + s(n + L_m + \log D_m), m\big)$.

(e) Suppose sequences $\vec{v}_m, \vec{w}_m$ are computable in polynomial sequential time. Then their inner product $\vec{u}_m^\perp \cdot \vec{v}_m \in \mathbb{R}$ is in $\mathbb{R}\#\mathsf{P}$; and this is optimal in general.

(f) Suppose $A_m \in \mathbb{R}^{D_m \times D_m}$ are computable in depth $s(n,m) \ge \log(n) + \log(m)$ and satisfy $A_m^k \in \big[-2^{L_m}; 2^{L_m}\big]^{D_m \times D_m}$ for all $k \le K_m$. Then the powers $A_m^{K_m} \in \mathbb{R}^{D_m \times D_m}$ are computable in depth

$$\mathcal{O}\Big(\log(K_m) \cdot (\log n + \log L_m + \log D_m + \mathbf{loglog}\, K_m)$$
$$+ \, s\big(n + (L_m + \log D_m) \cdot \log K_m, m\big)\Big)$$

(g) If A_m is computable in polynomial time and same for $\mathbb{N} \ni K_m < 2^{\mathrm{poly}(m)}$, then the powers $A_m^{K_m}$ are computable in $\mathbb{R}\mathsf{PSPACE}$; and this is optimal in general.

Proof (Sketch). Claims (b), (c) and (f) are based on repeated squaring, each matrix multiplication being D^2 inner products, realized as prefix sums (carry look-ahead) and known logarithmic-time integer multiplication (d). Regarding (g), encode PSPACE-complete reachability as matrix powering.   □

Theorem 7. *For a uni-variate polynomial $p = \sum_{j=0}^d p_j X^j \in \mathbb{R}[X]$, let $|p| := \sum_j |p_j|$ denote its norm. Fix $d \in \mathbb{N}$.*

(a) *Given $a, b \in \mathbb{R}$ with $|a| + |b| \leq 1$ and $J \leq K \in \mathbb{N}$, $\binom{K}{J} \cdot a^J \cdot b^{K-J}$ can be approximated to 2^{-n} in time polynomial in n and the binary length of K.*

(b) *Given $a_1, \ldots, a_d \in \mathbb{R}$ with $\sum_j |a_j| \leq 1$ and given $J_1, \ldots J_d, K \in \mathbb{N}$ with $\sum_j J_j = K$, $\binom{K}{J_1, \ldots J_d} \cdot a_1^{J_1} \cdots a_d^{J_d}$ can be approximated up to error 2^{-n} in time polynomial in n and the binary length of K. Here $\binom{K}{J_1, \ldots J_d} = \frac{K!}{J_1! \cdots J_d!}$ denotes the multinomial coefficient.*

(c) *For $p = \sum_{j=0}^{d} p_j X^j \in \mathbb{R}[X]$ with $|p| \leq 1$ and polynomial-time computable coefficients, the coefficient vector*

$$\sum_{\substack{J_0 + J_1 + \ldots + J_d = 2^n \\ J_1 + 2J_2 + \cdots + dJ_d = J}} p_0^{J_0} \cdots p_d^{J_d} \cdot \binom{2^n}{J_0, \ldots J_d}, \qquad J \leq d \cdot 2^n \qquad (6)$$

of p^{2^n} belongs to $\mathbb{R}\#\mathsf{P}$.

(d) *Let $C_m \in \mathbb{R}^{2^m \times 2^m}$ denote a circulant matrix of bandwidth d with polynomial-time computable entries $c_{-d}, \ldots c_{+d} \in \mathbb{R}$ such that $|p_{-d}| + \ldots + |p_{+d}| \leq 1$. Then the matrix power $C_m^{2^m}$ belongs to $\mathbb{R}\#\mathsf{P}$.*

We omit the proof of Theorem 7 because of space constraints. Item (d) is proved by means of Items (a)–(c). Items (a) and (c) use the Gaussian Distribution as approximation. Note that, again, Items (c) and (d) are trivial over integers; considering the real setting makes them meaningful and crucial for cost analyses of difference schemes.

Our next tool is about efficient operator powering. It generalizes Proposition 6(b) to compact subsets of some infinite-dimensional vector space.

Theorem 8. *Fix a Banach space $\mathcal{B}$ with norm $\|\cdot\|$ and linear map $A : \mathcal{B} \to \mathcal{B}$. And fix an increasing sequence $\mathcal{V}_d \subseteq \mathcal{V}_{d+1} \subseteq \mathcal{B}$ of non-empty compact convex symmetric subsets such that*

(i) *$A^K : \mathcal{V}_d \to \mathcal{V}_{\mathrm{poly}(d+K)}$ is well-defined for all $d, K \in \mathbb{N}$ and*

(ii) *satisfies $\|A^K v\| \leq \mathcal{O}(1)^d \cdot d^K \cdot K!$ for all $v \in \mathcal{V}_d$*

(iii) *and is computable in sequential time polynomial in $n + d + K$*

(iii') *or is computable in polylogarithmic depth $\mathrm{poly}(\log n + \log d + \log K)$.*

$$u(t) := \exp(tA)v = \sum_K t^K \cdot A^K v / K! \in \mathcal{B}$$

is well-defined for all $\vec{v} \in \mathcal{V}_d$ and all $|t| < 1/d$ and satisfies $u_t = Au$. Moreover $\mathcal{V}_d \times [0; 1/2d] \ni (v, t) \mapsto u(t)$ is computable in sequential time polynomial in $n + d$ (iii) or (iii') in depth $\mathrm{poly}(\log n + \log d)$.

Proof (Sketch). Under hypothesis (ii), the series $u(t) = \sum_K t^K \cdot A^K v / K!$ permits differentiation under the sum for $|t| \leq 1/2d$ and hence solves $u_t = Au$. Moreover the first n terms of the series approximate u up to error 2^{-n}. $\square$

Notice that a naïve hypothesis (i') $A : \mathcal{V}_d \to \mathcal{V}_{2d}$ would only imply $A^K : \mathcal{V}_d \to \mathcal{V}_{2^K d}$: blowing up exponentially in K. The stronger hypothesis (i), as well as (ii) to (iii'), are for instance satisfied for $\mathcal{V}_d := [-2^d; 2^d]^D$ by every $A \in [-2^d; 2^d]^{D \times D}$; recall Proposition 6(a). A more involved case revolves around univariate analytic functions.

Example 9. Consider the space $\mathcal{B}$ of complex-valued functions $v : [-1; 1] \to \mathbb{C}$ infinitely often continuously differentiable on the real interval. Write $v^{(j)}$ for its j-th iterated derivative, and abbreviate $|v|_\infty := \sup_x |v(x)|$. For each $d \in \mathbb{N}$ let

$$\mathcal{V}_d = \{v : [-1; 1] \to \mathbb{C}, \; \forall j \in \mathbb{N} \, |v^{(j)}|_\infty \le 2^d \cdot j! \cdot d^j\}, \quad \|v\| := \sum_j |v^{(j)}|_\infty / (j!)^2$$

Represent each $v \in \mathcal{V}_d$ by the $(2d + 1)$-tuple of its germs around positions 1, $(d - 1)/d, \ldots, (-d + 1)/d, 1 \in [-1; 1]$; and represent each germ by its power series coefficient sequence.

  (i) $\partial^K : \mathcal{V}_d \ni v \mapsto v^{(K)} \in \mathcal{V}_{3d + \lceil K \log d \rceil + \lceil K \log K \rceil}$ is well-defined.
  (ii) satisfies $\|\partial^K v\| \le (2e^2)^d \cdot d^K \cdot K!$ for all $v \in \mathcal{V}_d$
  (iii) $\partial^K|_{\mathcal{V}_d}$ is computable in sequential time polynomial in $n + d + K$
  (iii') and in polylogarithmic depth $\text{poly}(\log n + \log d + \log K)$.
  (iv) $\mathcal{V}_d$ is compact.

Proof (Sketch). For (i) and (iii) see the proof of Theorem 16(d) in [7, §3.2]. The underlying algorithm basically shifts and scales the coefficient sequences to symbolically take the derivative of the power series: easy to parallelize [19]. It also needs to add new germs/points of expansion in $[-1; 1]$: which again can be performed in parallel, thus establishing (iii'). □

4 Complexity of Differential Equations

In this section we apply results on matrix/operator powering from the previous section to prove Theorems 1–3.

Proof (Theorem 1, sketch). Recall from Proposition 6(a) that A^k has entries bounded by d^{k-1}. Hence the tail $\sum_{k>K} \frac{t^k}{k!} A^k \vec{v}$ is bounded, according to Stirling formula, by

$$\sum_{k>K} 1/k! \cdot 2^{k \cdot \log d} \le \sum_{k>K} \mathcal{O}\big(2^{-k \cdot (\log k - \log d)}\big) \le \sum_{k>K} 2^{-k} = 2^{-K}$$

for $K \ge 2d$. Thus we can calculate the first $K := \max\{2d, n\}$ terms of the power series in Eq. (3) simultaneously within depth $\mathcal{O}\big(\log(\max\{n, 2d\}) \cdot (\log n + \log d + \log \log \max\{n, 2d\})\big) = \mathcal{O}\big((\log n + \log d)^2\big)$ (by Proposition 6(c)); and then add them, incurring additional depth of the same magnitude, which completes the proof of the theorem. □

Proof (Theorem 3, Sketch). Example 9 generalizes to functions of several variables [10]. The statements of the theorem thus follow from Theorem 8 with $A := f_1 \partial_1 + \cdots + f_e \partial_e$. □

Before proceeding to the proof of Theorem 2 let us briefly recall basic definitions and facts about difference schemes. Theorem 2 implicitly assumes the domain Ω be "good enough"; w.l.o.g. we will consider uniform grids G_h on $\bar{\Omega}$ and G_h^τ on $[0,1] \times \bar{\Omega}$ for BVPs (resp. on the intersection of the domain of existence and uniqueness with $[0,1] \times \bar{\Omega}$ for IVPs). Here h, τ are respectively the space and time steps. Unlikely choosing them heuristically as it is usually done in Numerics, we will compute them from the output precision as $h = C_h/2^n$, $\tau = C_\tau/2^n$, giving precise estimates on C_h, C_τ. That's the reason why we denote the matrix of the difference scheme A_n (depending on n) in Definition 10 below.

Definition 10. *(a) For the IVP or BVP for a system of PDEs considered in Theorem 2, an explicit difference scheme is a system of algebraic equations*

$$u^{(h,(l+1)\tau)} = \mathbf{A_n} u^{(h,l\tau)}, \quad u^{(h,0)} = \varphi^{(h)} \tag{7}$$

Here A_n is a difference operator, which is in our case linear, i.e. a matrix of dimension $\mathcal{O}(2^n)$ (for BVPs it also includes the boundary conditions); $u^{(h,l\tau)}$ and $\varphi^{(h)}$ are grid functions, i.e. vectors of dimension $\mathcal{O}(2^n)$, approximating the corresponding continuously differentiable functions.

(b) The scheme (7) is said to approximate *the given differential problem with order of accuracy p (where p is a positive integer) on a solution $u(t,x)$ of the considered boundary-value problem if $\left|(u_t - Au)|_{G_h^\tau} - L_h u^{(h)}\right| \leq C_1 h^p$, and $\left|\varphi|_{G_h} - \varphi^{(h)}\right| \leq C_1 h^p$ for some constant C_1 not depending on h and τ. Here $u^{(h)} = \{u^{(h,l\tau)}\}_{l=1}^M$, L_h is the linear difference operator corresponding to (7) rewritten in the form $L_h u^{(h)} = 0$, M is the number of time steps.*

(c) The difference scheme (7) is called stable *if its solution $\mathbf{u}^{(h)}$ satisfies $|\mathbf{u}^{(h)}| \leq c_2|\varphi^{(h)}|$ for a constant c_2 independent of h, τ and $\varphi^{(h)}$.*
  Here $|\cdot|$ is the sup-norm, i.e. the maximal value over all grid cells.

(d) We will call the complexity *of the difference scheme (7) be the complexity of the corresponding matrix A_n in the in sense of Definition 5(b).*

Fact 11 *(Well-known facts, see e.g. [5]): (a) Let the difference scheme be stable and approximate (1) on the solution u with order p. Then the solution $u^{(h)}$ of the recursively defined linear algebraic systems (7) uniformly converges to the solution u in the sense that*

$$|u|_{G_h^\tau} - u^{(h)}| \leq Ch^p \tag{8}$$

for C not depending on h and τ (but possibly depending on the inputs; this dependence will occur later in the proofs).

(b) The difference scheme (7) is stable iff there is a constant C_2 uniformly bounding all powers of A_n:

$$|A_n^q| \leq C_2, \quad q = 1, 2, \ldots, M. \tag{9}$$

The stability property implies $\tau \leq \nu h$; ν is called the Courant number.

(c) The convergence constant in (8) is $C = C_1 \cdot C_2$, where C_2 is from (9); C_1 is from the approximation property.

Proof (Theorem 2, sketch). To evaluate the solution u at a fixed point (t, x) with the prescribed precision 2^{-n} and estimate the bit-cost of the computation, consider the following computation steps.

1. Choose binary-rational grid steps $h = 2^{-N}$ (where $N = \mathcal{O}(n)$) and $\tau \leq \nu h$ (where ν is the Courant number existing due to stability property): τ just any binary-rational meeting this inequality; h defined by the inequality (12) below.
2. For a grid point (t, x) put $l = \frac{t}{\tau}$ (note that $l \leq M = \lceil \frac{1}{\tau} \rceil = \mathcal{O}(2^n)$) and calculate the matrix powers and vector products

$$u^{(h, l\tau)} = A_n^l \varphi^{(h)}. \tag{10}$$

Note that (10) uses matrix powering instead of step-by-step iterations initially suggested by the difference scheme (7).

3. For non-grid points take (e.g.) a multilinear interpolation $\widetilde{u^{(h)}}$ of $u^{(h)}$ computed from the (constant number of) "neighbor" grid points.

Due to well-known properties of multilinear interpolations,

$$\sup_{t,x} |\widetilde{u^{(h)}}(t, x)| \leq \tilde{C} \sup_{G_h^\tau} |u^{(h)}|; \quad \sup_{t,x} |\widetilde{u \mid_{G_h^\tau}}(t, x)| \leq \bar{C} \sup_{t,x} |D^2 u(t, x)| \cdot h^2, \tag{11}$$

where $\tilde{C}$ and $\bar{C}$ are absolute constants. Based on (11) and on the continuous dependence property, as well as on linearity of the interpolation operator, infer

$$\sup_{t,x} |u(t, x) - \widetilde{u^{(h)}}(t, x)| \leq \sup_{t,x} \left(|u(t, x) - \widetilde{u \mid_{G_h^\tau}}(t, x)| + |\widetilde{u \mid_{G_h^\tau}}(t, x) - \widetilde{u^{(h)}}(t, x)| \right)$$

$$\leq \tilde{C} C_0 \sup_x |D^2 \varphi(x)| h^2 + \bar{C} C_1 C_2 \cdot h \leq 2^{-n}.$$

Thus choosing a grid step $h = 2^{-N}$ such that

$$h \leq C_h \cdot 2^{-n}, \quad C_h = \tilde{C} C_0 \sup_x |D^2 \varphi(x)| + \bar{C} C_1 C_2, \tag{12}$$

will guarantee the computed function $\widetilde{u^{(h)}}$ approximate the solution u with the prescribed precision 2^{-n} (here C_h depends only on the fixed $s(n)$ space computable functions φ, B_i and therefore is a fixed constant).

According to (10), item (a) of Theorem 2 follows from items (f) and (d) of Proposition 6; item (b) of Theorem 2 follows from item (d) of Theorem 7 combined with item (d) of Proposition 6. □

Conditions of Theorem 2 hold for large classes of IVPs and BVPs for linear PDEs. E.g. for certain BVPs for symmetric hyperbolic systems $u_t + \sum_{i=1}^m B_i u_{x_i} = 0$ with constant matrices $B_i = B_i^*$ (to which also the wave equation $p_{tt} - a^2 \sum_{i=1}^m p_{x_i x_i} = 0$ can be reduced), as well as for the heat equation

$p_t - a^2 \sum\limits_{i=1}^{m} p_{x_i x_i} = 0$, difference schemes with circulant constant bandwidth matrices can be constructed. For equations with non-constant coefficients the matrices are more complicated, thus the corresponding problems might have higher complexity. Deriving optimal complexity bounds for the considered (and possibly broader) classes of equations is one of the directions of future work.

References

1. Allender, E., Bürgisser, P., Kjeldgaard-Pedersen, J., Miltersen, P.B.: On the complexity of numerical analysis. SIAM J. Comput. **38**(5), 1987–2006 (2009)
2. Borodin, A.: Relating time and space to size and depth. SIAM J. Comput. **6**(4), 733–744 (1977). https://doi.org/10.1137/0206054
3. Bournez, O., Graça, D.S., Pouly, A.: Solving analytic differential equations in polynomial time over unbounded domains. In: Murlak, F., Sankowski, P. (eds.) MFCS 2011. LNCS, vol. 6907, pp. 170–181. Springer, Heidelberg (2011). https://doi.org/10.1007/978-3-642-22993-0_18
4. Evans, L.: Partial Differential Equations, Graduate Studies in Mathematics, vol. 19. American Mathematical Society, Providence (1998)
5. Godunov, S., Ryaben'kii, V.: Difference Schemes: An Introduction to the Underlying Theory. Elsevier Science Ltd., New York (1987)
6. Hoover, H.J.: Feasible real functions and arithmetic circuits. SIAM J. Comput. **19**(1), 182–204 (1990). https://doi.org/10.1137/0219012
7. Kawamura, A., Müller, N.T., Rösnick, C., Ziegler, M.: Computational benefit of smoothness: parameterized bit-complexity of numerical operators on analytic functions and Gevrey's hierarchy. J. Complex. **31**(5), 689–714 (2015)
8. Kawamura, A., Ota, H.: Small complexity classes for computable analysis. In: Csuhaj-Varjú, E., Dietzfelbinger, M., Ésik, Z. (eds.) MFCS 2014. LNCS, vol. 8635, pp. 432–444. Springer, Heidelberg (2014). https://doi.org/10.1007/978-3-662-44465-8_37
9. Kawamura, A., Ota, H., Rösnick, C., Ziegler, M.: Computational complexity of smooth differential equations. Logical Methods in Computer Science **10**, 1:6,15 (2014). https://doi.org/10.2168/LMCS-10(1:6)2014
10. Kawamura, A., Steinberg, F., Thies, H.: Parameterized complexity for uniform operators on multidimensional analytic functions and ODE solving. In: Moss, L.S., de Queiroz, R., Martinez, M. (eds.) WoLLIC 2018. LNCS, vol. 10944, pp. 223–236. Springer, Heidelberg (2018). https://doi.org/10.1007/978-3-662-57669-4_13
11. Kawamura, A., Steinberg, F., Ziegler, M.: On the computational complexity of the Dirichlet problem for Poisson's equation. Math. Struct. Comput. Sci. **27**(8), 1437–1465 (2017). https://doi.org/10.1017/S096012951600013X
12. Ko, K.I.: Complexity Theory of Real Functions. Progress in Theoretical Computer Science, Birkhäuser, Boston (1991)
13. Mayr, E., Meyer, A.: The complexity of the word problem for commutative semigroups and polynomial ideals. Adv. Math. **46**, 305–329 (1982)
14. Pan, V., Reif, J.: The bit-complexity of discrete solutions of partial differential equations: compact multigrid. Comput. Math. Appl. **20**, 9–16 (1990)
15. Selivanova, S., Selivanov, V.L.: Computing solution operators of boundary-value problems for some linear hyperbolic systems of PDEs. Logical Meth. Comput. Sci. **13**(4) (2017). https://doi.org/10.23638/LMCS-13(4:13)2017

16. Selivanova, S.V., Selivanov, V.L.: Bit complexity of computing solutions for symmetric hyperbolic systems of PDEs (extended abstract). Proc. CiE **2018**, 376–385 (2018). https://doi.org/10.1007/978-3-319-94418-0_38
17. Vollmer, H.: Introduction to Circuit Complexity. EATCS Text in Theoretical Computer Science. Springer, Heidelberg (1999)
18. Weihrauch, K.: Computable Analysis. Springer, Berlin (2000)
19. Yu, F., Ko, K.I.: On parallel complexity of analytic functions. Theoret. Comput. Sci. **489**(490), 48–57 (2013). https://doi.org/10.1016/j.tcs.2013.04.008

On the Quantum and Classical Complexity of Solving Subtraction Games

Dmitry Kravchenko[1(✉)], Kamil Khadiev[2,3], and Danil Serov[3]

[1] Center for Quantum Computer Science, Faculty of Computing,
University of Latvia, Riga, Latvia
kravchenko@gmail.com
[2] Smart Quantum Technologies Ltd., Kazan, Russia
kamilhadi@gmail.com
[3] Kazan Federal University, Kazan, Russia
serovdanilru@gmail.com

Abstract. We study algorithms for solving Subtraction games, which are sometimes referred as one-heap Nim games.

We describe a quantum algorithm which is applicable to any game on DAG, and show that its query complexity for solving an arbitrary Subtraction game of n stones is $O\left(n^{3/2}\log n\right)$.

The best known deterministic algorithms for solving such games are based on the dynamic programming approach [8]. We show that this approach is asymptotically optimal and that classical query complexity for solving a Subtraction game $\Theta\left(n^2\right)$ in general.

Of course, this difference between classical and quantum algorithms is far from the best known examples, but, up to our knowledge, this paper is the first constructive "quantum" contribution to the algorithmic game theory.

Keywords: Quantum computation · Quantum models ·
Quantum algorithm · Query model · Game theory · Nim ·
Subtraction game

1 Introduction

Quantum computing [5,20] is one of the hot topics in computer science of the last decades. There are many problems where quantum algorithms outperform the best known classical algorithms for different computational models [1–4,9,15–18].

On the other hand, *quantum game theory* [6,10,11] traditionally is being studied in the context of nonlocal properties of quantum particles, and usually stays apart from quantum computing.

We aim to bring these two topics together, and show that quantum computers will help also in computational game theory.

© Springer Nature Switzerland AG 2019
R. van Bevern and G. Kucherov (Eds.): CSR 2019, LNCS 11532, pp. 228–236, 2019.
https://doi.org/10.1007/978-3-030-19955-5_20

In this paper, we consider so called Subtraction games. A Subtraction game is similar to a canonical Nim game [12]. The difference is that, in the former game the players deal with just one heap of stones but with certain limitations imposed on the number of stones they can take from the heap. Nim is a very notable game in game theory, mainly because it serves as a "base case" for Sprague-Grundy theorem [14,21]. Many games are known to be reducible to some Nim games, and also many games on graphs can be reduced to some Subtraction games.

The most common limitation for Subtraction games is defining maximum for the number of stones to be taken away, and this kind of games has nice combinatorial solutions. Here we study a much more general class of such limitations and thus a broader class of Subtraction games.

We investigate an algorithm for *solving* the game, that is, determining whether the first or the second player has a winning strategy in the game. In this algorithm, we exploit quantum dynamic programming approach for DAGs (Directed Acyclic graphs) described in [19]. The proposed quantum algorithm for solving a Subtraction game has time complexity $O\left(\sqrt{n \cdot |E|} \log n\right)$, where n is the number of stones in the heap and $|E|$ is the number of edges in the graph of legal moves. At the same time, we suggest a huge subclass of Subtraction games that requires $\Theta(|E|)$ queries and thus has time complexity $\Theta(|E|)$. In the case of $|E| = O(n^2)$, we have complexity $O\left(n^{1.5} \log n\right)$ in the quantum case and complexity $O(n^2)$ in the classical case.

The paper is organized in the following way. Section 2 contains some basic definitions. In Sect. 3, we estimate the classical complexity of a Subtraction game. The quantum algorithm is described in Sect. 4. Finally, Sect. 5 contains generalization of the problem for two different graphs for two players.

2 Definitions

2.1 Subtraction Games

We use the following definition of a Subtraction game throughout this paper: It is a two-player game in which the players alternately remove some positive amounts of stones from a heap. Let n be the initial number of stones in the heap, and Γ be a (triangular) binary matrix of size $n \times n$. A player can remove $j - i$ stones ($0 \leq i < j \leq n$) from the heap with exactly j stones left iff $\Gamma_{j,i} = 1$. One who cannot make a legal move loses, and their opponent wins the game. Thus a player wins a game if he takes all the remaining stones, or leaves such number of stones j that no allowed moves remain: $\sum_{i=0}^{j-1} \Gamma_{j,i} = 0$.

Obviously, the rules of a Subtraction game are fully determined by such matrix Γ, so hereafter we sometimes use letter Γ to denote a corresponding game. We also reserve name n to denote the initial size of stones, which also corresponds to the dimension of the matrix Γ. Note that there are only $n(n+1)/2$ meaningful bits in the matrix Γ, as a player cannot take more stones than there remain in the heap: $\Gamma_{j,i} = 0$ for all $j < i$.

2.2 Winning Function

We define a Boolean function WIN: WIN $(\Gamma) = 1$ iff the first player has a winning strategy in game Γ.

We also extend its domain to all possible positions in the game: WIN $(\Gamma, j) = 1$ iff a player having j stones in game Γ has a winning strategy. In particular:

- WIN $(\Gamma, n) =$ WIN (Γ);
- WIN $(\Gamma, 0) = 0$;
- WIN $(\Gamma, j) \iff \exists i : \Gamma_{j,i} \wedge \neg$WIN (Γ, i).

Hereinafter we alternately use Boolean and integer forms of these and other values: TRUE = 1 and FALSE = 0.

2.3 Properties of Subtraction Games

We call a game Γ *k-balanced* if the number of winning positions differs from the number of losing positions by at most k:

$$\left| n/2 - \sum_{j=1}^{n} \text{WIN}\,(\Gamma, j) \right| \le k/2 \tag{1}$$

We call a game Γ *sensitive* if in each winning position a player has a unique winning move. Or, equivalently, if in each position a player can have at most one winning move:

$$\forall j : \left| \{i : \Gamma_{j,i} \wedge \neg \text{WIN}\,(\Gamma, i)\} \right| \le 1 \tag{2}$$

In the next section, we study sensitive k-balanced games for small k-s. Hereafter one can assume $k = 0$ for simplicity, but all our considerations also hold for any $k = o(n)$. The exact value of k only affects the size of the considered subset of games.

Finally, we call a game Γ *losing* if $\neg$WIN (Γ).

In the next section, we derive a lower bound for the classical complexity of solving a losing Subtraction game with these two properties. Hereafter we call this set LSB ("Losing, Sensitive and Balanced").

3 Classical Query Complexity

Lemma 1. *Let Γ be a losing sensitive $o(n)$-balanced Subtraction game Γ picked uniformly at random from LSB:* (1) $\wedge$ (2) $\wedge \neg$WIN (Γ). *Let game Γ' differ from Γ in exactly one random bit of their binary representations:* HAMMINGDISTANCE $(\Gamma, \Gamma') = 1$. *Then* $\mathbb{E}[\text{WIN}\,(\Gamma')] \ge 1/24$.

Proof. We first make three assumptions for $\forall j, i \, (0 \leq i < j \leq n)$:

- $\mathbb{E}\left[\text{WIN}\left(\Gamma, j\right)\right] = 1/2$, which does not fully correspond to the uniform distribution for the considered subset of games, but is asymptotically equivalent to it (i.e. expected difference between the probability mass functions of the two distributions is neglectable for large n-s). We leave this fact without formal proof, because even non-uniform distribution is sufficient for our purposes. We also neglect the addend $\pm k/n = \pm o\left(1\right)$ in this estimation.
- $\text{WIN}\left(\Gamma, i\right) \implies \mathbb{E}\left[\Gamma_{j,i}\right] = 1/2$. Informally: a winning position i is achievable from any preceding position j with probability $1/2$. This assumption is valid since possibility or impossibility to make a losing move in any position leaves a game strategically unaffected. The formal proof is simple: if $\text{WIN}\left(\Gamma, i\right)$ and Γ' is equal to Γ with the only exception $\Gamma'_{j,i} = \neg\Gamma_{j,i}$, then $\Gamma' \in \text{LSB} \iff \Gamma \in \text{LSB}$.
- $\text{WIN}\left(\Gamma, j\right) \wedge \neg\text{WIN}\left(\Gamma, i\right) \implies \mathbb{E}\left[\Gamma_{j,i}\right] = \dfrac{1}{\sum\limits_{i'=0}^{j-1} \neg\text{WIN}(\Gamma,i')}$. Informally: from a

winning position j, all subsequent losing positions i' are achievable equiprobably (and these probabilities sum up to 1). This assumption is also valid due to the definition of the considered subset of games.

Now let $\Gamma'_{j,i} \neq \Gamma_{j,i}$ for some pair of indices j, i picked at random. Then:

- $\Pr\left[\neg\text{WIN}\left(\Gamma, j\right)\right] = 0.5$
- $\forall j' (j < j' < n)$:

$$\Pr\left[\text{WIN}\left(\Gamma, j'\right) \wedge \Gamma_{j',j} \mid \neg\text{WIN}\left(\Gamma, j\right)\right] = \frac{0.5}{\mathbb{E}\left[\sum\limits_{i'=0}^{j'-1} \neg\text{WIN}\left(\Gamma, i'\right)\right]} = \frac{1}{j'}$$

Informally: each preceding j' gives some small chance of $1/j'$ for a losing position j to be accessible from a winning position j'.

- $\Pr\left[\nexists j' : \text{WIN}\left(\Gamma, j'\right) \wedge \Gamma_{j',j} \mid \neg\text{WIN}\left(\Gamma, j\right)\right] = \prod\limits_{j'=j+1}^{n-1} \left(1 - \frac{1}{j'}\right) = \frac{j}{n-1}$

Informally: even though each previously estimated chance was small, altogether they result in some significant probability for a losing position j to be accessible from at least one preceding winning position (of totally $n - j - 1$ preceding positions).

We follow that once j is a losing position, with probability $1 - \frac{j}{n-1}$ it is achievable from some winning position j'. We shall consider the smallest j' if there happen to occur several.

On the other side, $\Pr\left[\neg\text{WIN}\left(\Gamma, i\right)\right] = 0.5$, so once j is a losing position, then with probability 0.5 position i is also losing, and thus $\Gamma_{j,i} = 0$ (since one losing position cannot be achievable from another losing position). And altogether: with probability

$$\left(1 - \frac{j}{n-1}\right) \times 0.5 \times 0.5 \tag{3}$$

we have the case where $\left(\text{WIN}\left(\Gamma, j'\right) \wedge \Gamma_{j',j}\right) \wedge \neg\text{WIN}\left(\Gamma, j\right) \wedge \neg\text{WIN}\left(\Gamma, i\right)$.

Now let us observe what happens with Γ' in this (not so improbable) case: we have $\text{WIN}\,(\Gamma', i) = \text{WIN}\,(\Gamma, i) = 0$ and $\Gamma'_{j,i} = \neg\Gamma_{j,i} = 1$. It implies $\text{WIN}\,(\Gamma', j)$ since Γ' allows access from j to the losing position i. And then the only winning move in position j' becomes obsolete: $\text{WIN}\,(\Gamma', j') = 0$ (some other winning positions of game Γ may become also losing in Γ', but it definitely happens with the "smallest" former winning position j').

Finally, we note that, with probability $\Pr[\Gamma_{n,j'} = 1] = 0.5$, this difference also implies $\text{WIN}\,(\Gamma', n) = 1 \neq \text{WIN}\,(\Gamma, n)$. This probability together with (3) lead to the following lower bound:

$$\Pr\big[\text{WIN}\,(\Gamma', n) \neq \text{WIN}\,(\Gamma, n)\big] \geq \mathbb{E}\big[0.5 \times \big(1 - \tfrac{j}{n-1}\big) \times 0.5 \times 0.5\big]$$
$$= \big(1 - \tfrac{\mathbb{E}[j]}{n-1}\big)\big/8 \approx \tfrac{1}{24} \qquad\qquad\qquad \square$$

Theorem 1. *There is no deterministic or randomized algorithm for solving function* WIN *faster than in* $\Omega\,(n^2)$ *steps.*

We prove this impossibility result by analyzing the performance of the best possible classical algorithm on the set LSB of losing sensitive $o\,(n)$-balanced Subtraction games. But before that, we make a couple of remarks on the eligibility of such proof:

- One may argue that, LSB is a too small set, and that we are going to prove the lower bound for some negligible number of games. However, actually the number of such games is roughly $2^{n^2/4}$ while the total number of Subtraction games is roughly $2^{n^2/2}$. That is, a game from LSB and an arbitrary Subtraction game are representable by asymptotically similar numbers of bits. Although just one example would be sufficient to make a statement about the "worst case" complexity, we here demonstrate that a *significant* part of all Subtraction games are hard to solve. We believe that, similar bound should also hold for the whole set of Subtraction games.
- One may argue that, since LSB contains only losing games ($\text{WIN}\,(\Gamma) = 0$), one could design an algorithm which, one way or another, recognizes that a game belongs to this set and returns answer 0. Actually we can easily refute this criticism by extending this set with sufficient number of Γ'-s which often are winning and always are hardly-distinguishable from the losing games of LSB. Formally, we can consider set $\text{LSB}' = \bigcup_{\Gamma \in \text{LSB}} \text{HAMMINGBALL}\,(\Gamma, 1)$.

Proof. Suppose a classical algorithm, given $\Gamma \in \text{LSB}$, reports an answer after querying on average less than $n\,(n+1)\,/(2 \cdot 24)$ bits of Γ.

Obviously, such algorithm cannot be correct with probability more than 0.5, since Lemma 1 implies that there are on average $n\,(n+1)\,/(2 \cdot 24)$ such crucial bits $\Gamma_{j,i}$ that inverting the bit also inverts the value of the game $\text{WIN}\,(\Gamma)$. Leaving any such bit unchecked means failing to guess $\text{WIN}\,(\Gamma)$ with adequate probability. $\qquad\qquad \square$

4   Quantum Algorithm

In this section we suggest a quantum algorithm for solving an arbitrary Subtraction game (not only that from LSB). We assume the reader to be familiar with the basics of quantum computing and in particular with Grover Search algorithm [7,13]. We refer to the most remarkable textbook on quantum computing [20] for the details.

We consider directed acycling graph (DAG) $G = (V, E)$ with adjacency matrix Γ: set V corresponds to $n + 1$ positions of the game Γ, and set E corresponds to all legal moves.

The algorithm applies dynamic programming approach [8,19]: it solves a problem using precomputed solutions of smaller parts of the same problem. For analyzing DAGs, it typically means usage of some modification of Depth-first search algorithm (DFS) as a subroutine [8].

We exploit the same idea, but for solving WIN (Γ, j) (for each vertex j, starting from $j = 1$), we use Grover's Search algorithm [7,13] to find a losing vertex i' among directly accessible vertices $\text{ADJ}[j] \overset{\text{def}}{=} \{i : \Gamma_{j,i}\}$, s.t. WIN $(\Gamma, i') = 0$. This algorithm has two important properties:

- its time complexity is $O\left(\sqrt{\deg j}\right)$, where $\deg j \overset{\text{def}}{=} |\text{ADJ}[j]|$ is the number of vertices directly accessible from the vertex j;
- it returns some losing vertex i' with a constant probability (say 0.5) if there exist one or more losing vertices.

In Algorithm 1, we use Grover's Search in form of function GROVER_ISZEROAMONG which returns 0 or 1 equiprobably if there is zero among its input bits, and returns 0 if all inputs are ones. We store the search results in the array w and reuse them in all the subsequent searches.

Algorithm 1. Quantum Algorithm for solving WIN (Γ)

$w_0 \leftarrow 0$
  for $j = 1 \dots n$ **do**   ▷ $O(n)$
    $w_j \leftarrow 0$
    for $z = 1 \dots 2 \cdot \log_2 n$ **do**   ▷ $O(\log n)$
      if GROVER_ISZEROAMONG $\{w_i \mid i \in \text{ADJ}[j]\}$ **then**   ▷ $O\left(\sqrt{\deg j}\right)$
        $w_j \leftarrow 1$
      end if
    end for
  end for
  return w_n

Theorem 2. *Algorithm 1 computes* WIN (Γ) *in time* $O\left(\sqrt{n|E|}\log n\right)$ *and with error probability* $\epsilon \lesssim 1/n$.

Proof. The correctness of the algorithm is obvious: each of the variables w_j (for j running from 0 to n) is assigned a value WIN (Γ, j) according to the definition of the function WIN (Γ, j).

The time complexity follows from Cauchy-Bunyakovsky-Schwarz inequality:

$$\sum_{j=1}^{n} \sqrt{\deg j} \leq \sum_{j=1}^{n} \sqrt{\mathbb{E}_j \left[\deg j\right]} = \sum_{j=1}^{n} \sqrt{|E|/n} = \sqrt{n|E|}.$$

The probability of error in evaluating one particular w_j is $2^{-2\log_2 n} = 1/n^2$, so the probability of no error at all among evaluations of $w_1, \ldots, w_n$ is $\left(1 - 1/n^2\right)^n \gtrsim 1 - 1/n$. ☐

We note that, for a random Subtraction game, the expected number of edges $\mathbb{E}\left[|E|\right] = \Theta\left(n^2\right)$, and then we conclude that, while the best classical algorithms require time $\Theta\left(n^2\right)$ to solve a Subtraction game, there exists a polynomially faster quantum algorithm which runs in time $O\left(n^{3/2}\log n\right)$.

5   Generalization of the Subtraction Games

In this section, we consider the generalized version of a Subtraction game. Let both players have different matrices of legal moves in a Subtraction game: Γ^1 for Player 1 and Γ^2 for Player 2. We denote such game as (Γ^1, Γ^2).

All the considerations mostly repeat the ideas given in Sects. 3 and 4, but they illustrate that the proposed approach potentially covers some broader class of problems. Hence we only formulate the results and briefly outline the proofs.

Lemma 2. *Let (Γ^1, Γ^2) be a losing sensitive $o(n)$-balanced Subtraction game (Γ^1, Γ^2) picked uniformly at random from games with properties:*

- *(1) for Γ^1 and Γ^2;*
- *analog of the property (2) for Player 2:*

$$\forall j : \left|\{i : \Gamma_{j,i}^2 \wedge \neg Win((\Gamma^1, \Gamma^2), i)\}\right| \leq 1;$$

- *The game is a losing game for Player 1, i.e. $\neg$WIN $\left((\Gamma^1, \Gamma^2)\right)$.*

Let game $(\hat{\Gamma}^1, \hat{\Gamma}^2)$ differ from (Γ^1, Γ^2) in exactly one random bit of the binary representation: HAMMINGDISTANCE $\left((\hat{\Gamma}^1, \hat{\Gamma}^2), (\Gamma^1, \Gamma^2)\right) = 1$. *Then*

$$\mathbb{E}\left[\text{WIN}\left((\hat{\Gamma}^1, \hat{\Gamma}^2)\right)\right] \geq 1/48.$$

Proof. The proof is similar to the proof of Lemma 1. The only difference is considering Player 1 and Player 2 separately. ☐

Theorem 3. *There is no deterministic or randomized algorithm for solving function* WIN $\left((\Gamma^1, \Gamma^2)\right)$ *faster than in $\Theta\left(n^2\right)$ steps.*

Proof. The proof is similar to the proof of Theorem 1.   □

The quantum algorithm for Subtraction Game for two different matrices is similar to Algorithm 1. Let $\mathrm{ADJ}^p[j] \stackrel{\text{def}}{=} \{i : \Gamma^p_{j,i}\}$, for $p \in \{1,2\}$. Let $w^1_i = \mathrm{WIN}\left((\Gamma^1, \Gamma^2), i\right)$ be 1 if Player 1 starts and wins given heap of i stones. Let $w^2_i = \mathrm{WIN}\left((\Gamma^2, \Gamma^1), i\right)$ be 1 if Player 2 starts and wins given heap of i stones. Then Algorithm 2 solves the Subtraction game defined above in asymptotically the same number of steps as Algorithm 1.

Algorithm 2. Quantum Algorithm for solving $\mathrm{WIN}\left((\Gamma^1, \Gamma^2)\right)$

$w^1_0 \leftarrow 0$
$w^2_0 \leftarrow 0$
for $j = 1 \ldots n$ **do**   ▷ $O(n)$
   $w^1_j \leftarrow 0$
   $w^2_j \leftarrow 0$
   for $z = 1 \ldots 2 \cdot \log_2 n$ **do**   ▷ $O(\log n)$
      if GROVER_ISZEROAMONG $\{w^2_i \mid i \in \mathrm{ADJ}^1[j]\}$ **then**   ▷ $O\left(\sqrt{\deg^1 j}\right)$
         $w^1_j \leftarrow 1$
      end if
      if GROVER_ISZEROAMONG $\{w^1_i \mid i \in \mathrm{ADJ}^2[j]\}$ **then**   ▷ $O\left(\sqrt{\deg^2 j}\right)$
         $w^2_j \leftarrow 1$
      end if
   end for
end for
return w^1_n

Theorem 4. *Algorithm 2 computes* $\mathrm{WIN}\left((\Gamma^1, \Gamma^2)\right)$ *in time* $O\left(\sqrt{n\,|E|}\log n\right)$ *and with error probability* $\epsilon \lesssim 1/n$.

Acknowledgement. The research is supported by PostDoc Latvia Program, and by the ERDF within the project 1.1.1.2/VIAA/1/16/099 "Optimal quantum-entangled behavior under unknown circumstances".

The reported study was funded by RFBR according to the research project No. 19-37-80008.

We would like to thank the anonymous reviewers for their detailed and helpful comments. We also thank Dr. Abuzer Yakaryılmaz for the proofreading.

References

1. Ablayev, F., Ablayev, M., Khadiev, K., Vasiliev, A.: Classical and quantum computations with restricted memory. In: Böckenhauer, H.-J., Komm, D., Unger, W. (eds.) Adventures Between Lower Bounds and Higher Altitudes. LNCS, vol. 11011, pp. 129–155. Springer, Cham (2018). https://doi.org/10.1007/978-3-319-98355-4_9

2. Ablayev, F., Ambainis, A., Khadiev, K., Khadieva, A.: Lower bounds and hierarchies for quantum memoryless communication protocols and quantum ordered binary decision diagrams with repeated test. In: Tjoa, A.M., Bellatreche, L., Biffl, S., van Leeuwen, J., Wiedermann, J. (eds.) SOFSEM 2018. LNCS, vol. 10706, pp. 197–211. Springer, Cham (2018). https://doi.org/10.1007/978-3-319-73117-9_14

3. Ablayev, F., Gainutdinova, A., Khadiev, K., Yakaryılmaz, A.: Very narrow quantum OBDDs and width hierarchies for classical OBDDs. Lobachevskii J. Math. **37**(6), 670–682 (2016)

4. Ablayev, F., Gainutdinova, A., Khadiev, K., Yakaryılmaz, A.: Very narrow quantum OBDDs and width hierarchies for classical OBDDs. In: Jürgensen, H., Karhumäki, J., Okhotin, A. (eds.) DCFS 2014. LNCS, vol. 8614, pp. 53–64. Springer, Cham (2014). https://doi.org/10.1007/978-3-319-09704-6_6

5. Ambainis, A.: Understanding quantum algorithms via query complexity. arXiv preprint arXiv:1712.06349 (2017)

6. Benjamin, S.C., Hayden, P.M.: Multiplayer quantum games. Phys. Rev. A **64**(3), 030301 (2001)

7. Boyer, M., Brassard, G., Høyer, P., Tapp, A.: Tight bounds on quantum searching. Fortschritte der Physik **46**(4–5), 493–505 (1998)

8. Cormen, T.H., Leiserson, C.E., Rivest, R.L., Stein, C.: Introduction to Algorithms, 2nd edn. McGraw-Hill, New York (2001)

9. De Wolf, R.: Quantum computing and communication complexity. Ph.D. thesis (2001)

10. Eisert, J., Wilkens, M.: Quantum games. J. Mod. Opt. **47**(14–15), 2543–2556 (2000)

11. Eisert, J., Wilkens, M., Lewenstein, M.: Quantum games and quantum strategies. Phys. Rev. Lett. **83**(15), 3077 (1999)

12. Ferguson, T.S.: Game theory class notes for math 167, fall 2000 (2000). https://www.cs.cmu.edu/afs/cs/academic/class/15859-f01/www/notes/comb.pdf

13. Grover, L.K.: A fast quantum mechanical algorithm for database search. In: Proceedings of the Twenty-Eighth Annual ACM Symposium on Theory of Computing, pp. 212–219. ACM (1996)

14. Grundy, P.M.: Mathematics and games. Eureka **2**, 6–8 (1939)

15. Ibrahimov, R., Khadiev, K., Prūsis, K., Yakaryılmaz, A.: Error-free affine, unitary, and probabilistic OBDDs. In: Konstantinidis, S., Pighizzini, G. (eds.) DCFS 2018. LNCS, vol. 10952, pp. 175–187. Springer, Cham (2018). https://doi.org/10.1007/978-3-319-94631-3_15

16. Jordan, S.: Bounded error quantum algorithms zoo. https://math.nist.gov/quantum/zoo

17. Khadiev, K., Khadieva, A.: Reordering method and hierarchies for quantum and classical ordered binary decision diagrams. In: Weil, P. (ed.) CSR 2017. LNCS, vol. 10304, pp. 162–175. Springer, Cham (2017). https://doi.org/10.1007/978-3-319-58747-9_16

18. Khadiev, K., Khadieva, A., Mannapov, I.: Quantum online algorithms with respect to space and advice complexity. Lobachevskii J. Math. **39**(9), 1210–1220 (2018)

19. Khadiev, K., Safina, L.: Quantum algorithm for dynamic programming approach for dags. Applications for zhegalkin polynomial evaluation and some problems on dags. In: Proceedings of Unconventional Computation and Natural Computation 2019. LNCS, vol. 11493 (2019). https://doi.org/10.1007/978-3-030-19311-9_13

20. Nielsen, M.A., Chuang, I.L.: Quantum Computation and Quantum Information. Cambridge University Press, New York (2010)

21. Sprague, R.P.: Über mathematische kampfspiele. Tohoku Math. J. **41**, 438–444 (1935)

Derandomization for Sliding Window Algorithms with Strict Correctness

Moses Ganardi, Danny Hucke, and Markus Lohrey$^{(\boxtimes)}$

Universität Siegen, Siegen, Germany
lohrey@eti.uni-siegen.de

Abstract. In the sliding window streaming model the goal is to compute an output value that only depends on the last n symbols from the data stream. Thereby, only space sublinear in the window size n should be used. Quite often randomization is used in order to achieve this goal. In the literature, one finds two different correctness criteria for randomized sliding window algorithms: (i) one can require that for every data stream and every time instant t, the algorithm computes a correct output value with high probability, or (ii) one can require that for every data stream the probability that the algorithm computes at every time instant a correct output value is high. Condition (ii) is stronger than (i) and is called "strict correctness" in this paper. The main result of this paper states that every strictly correct randomized sliding window algorithm can be derandomized without increasing the worst-case space consumption.

1 Introduction

Sliding window streaming algorithms process an input sequence $a_1 a_2 \cdots a_m$ from left to right and receive at time t the symbol a_t as input. Such algorithms are required to compute at each time instant t a value $f(a_{t-n+1} \cdots a_t)$ that depends on the n last symbols (we should assume $t \geq n$ here). The value n is called the *window size* and the sequence $a_{t-n+1} \cdots a_t$ is called the *window content* at time t. In many applications, data items in a stream are outdated after a certain time, and the sliding window model is a simple way to model this. A typical application is the analysis of a time series as it may arise in medical monitoring, web tracking, or financial monitoring.

A general goal in the area of sliding window algorithms is to avoid the explicit storage of the window content, and, instead, to work in considerably smaller space, e.g. space polylogarithmic in the window size. In the seminal paper of Datar, Gionis, Indyk and Motwani [10], where the sliding window model was introduced, the authors prove that the number of 1's in a 0/1-sliding window of size n can be maintained in space $O(\frac{1}{\varepsilon} \cdot \log^2 n)$ if one allows a multiplicative error of $1 \pm \varepsilon$. Other algorithmic problems that were addressed in the extensive literature on sliding window streams include the computation of statistical data (e.g. computation of the variance and k-median [3], and quantiles [2]), optimal

© Springer Nature Switzerland AG 2019
R. van Bevern and G. Kucherov (Eds.): CSR 2019, LNCS 11532, pp. 237–249, 2019.
https://doi.org/10.1007/978-3-030-19955-5_21

sampling from sliding windows [7], membership problems for formal languages [11–14], computation of edit distances [8], database querying (e.g. processing of join queries over sliding windows [15]) and graph problems (e.g. checking for connectivity and computation of matchings, spanners, and minimum spanning trees [9]). The reader can find further references in [1, Chapter 8] and [6].

Many of the above mentioned papers deal with sliding window algorithms that only compute a good enough approximation of the exact value of interest. In fact, even for very simple sliding window problems it is unavoidable to store the whole window content. Examples are the exact computation of the number of 1's [10] or the computation of the first symbol of the sliding window for a 0/1-data stream [12]. In this paper, we consider a general model for sliding window *approximation problems*, where a (possibly infinite set) of *admissible output values* is fixed for each word. To be more accurate, a specific approximation problem is described by a function $\mathcal{A}\colon \Sigma^* \to 2^\Omega$ which associates to words over a finite alphabet Σ (the set of data values in the stream) admissible output values from a possibly infinite set Ω. A sliding window algorithm for such a problem is then required to compute at each time instant an admissible output value for the current window content. This model covers exact algorithms (where $\mathcal{A}$ is a function $\mathcal{A}\colon \Sigma^* \to \Omega$) as well as a wide range of approximation algorithms. For example the computation of the number of 1's in a 0/1-sliding window with an allowed multiplicative error of $1\pm\varepsilon$ is covered by our model, since for a word with k occurrences of 1, the admissible output values are the integers from $\lfloor (1-\varepsilon)k \rfloor$ to $\lceil (1+\varepsilon)k \rceil$.

A second ingredient of many sliding window algorithms is randomization. Following our recent work [11–13] we model a randomized sliding window algorithm as a family $\mathcal{R} = (R_n)_{n\geq 0}$ of *probabilistic automata* R_n over a finite alphabet Σ, where R_n is the algorithm for window size n. Probabilistic automata were introduced by Rabin [17] and can seen as a common generalization of deterministic finite automata and Markov chains. The basic idea is that for every state q and every input symbol a, the next state is chosen according to some probability distribution. In addition to the classical model of Rabin, we require that for every probabilistic automaton R_n, (i) states are encoded by bit strings (the memory contents of the algorithm – this allows to define the space consumption of R_n on a certain input) and (ii) every state is associated with an output value from the set Ω. The second point allows to associate with every input word $w \in \Sigma^*$ and every output value $\omega \in \Omega$ the probability that the automaton outputs ω on input w. In order to solve a specific approximation problem $\mathcal{A}\colon \Sigma^* \to 2^\Omega$ one should require that for every window size n, the probabilistic automaton R_n should have a small error probability ϵ (say $\epsilon = 1/3$) on every input stream. But what does the latter exactly mean? Two different definitions can be found in the literature:

– For every input stream $w = a_1 \cdots a_m$ and every window size n ($n \leq m$), the probability that R_n outputs on input w a value $\omega \notin \mathcal{A}(a_{m-n+1} \cdots a_m)$ is at most ϵ. In this case, we say that $\mathcal{R}$ is ϵ-*correct* for $\mathcal{A}$.

– For every input stream $w = a_1 \cdots a_m$ and every window size n, the probability that R_n outputs at some time instant t $(n \leq t \leq m)$ a value $w \notin A(a_{t-n+1} \cdots a_t)$ is at most ϵ. In this case, we say that R is *strictly ϵ-correct* for A.

One can rephrase the difference between strict ϵ-correctness and ϵ-correctness as follows: ϵ-correctness means that while the randomized sliding window algorithm runs on an input stream it returns at each time instant an admissible output value with probability at least $1 - \epsilon$. In contrast, strict ϵ-correctness means that while the randomized sliding window algorithm reads an input stream, the probability that the algorithm returns an admissible output value at every time instant is at least $1 - \epsilon$. Obviously this makes a difference: imagine that $\Omega = \{1, 2, 3, 4, 5, 6\}$ and that for every input word $w \in \Sigma^*$ the admissible output values are $2, 3, 4, 5, 6$, then the algorithm that returns at every time instant the output of a fair dice throw is $1/6$-correct. But the probability that this algorithm returns an admissible output value at every time instant is only $(5/6)^m$ for an input stream of length m and hence converges to 0 for $m \to \infty$. Of course, in general, the situation is more complex since successive output values of a randomized sliding window algorithm are not independent.

In the following discussion, let us fix the error probability $\epsilon = 1/3$ (using probability amplification, one can reduce ϵ to any constant > 0). In our recent paper [13] we studied the space complexity of the membership problem for regular languages with respect to ϵ-correct randomized sliding window algorithms. It turned out that in this setting, one can gain from randomization. Consider for instance the regular language ab^* over the alphabet $\{a, b\}$. Thus, the sliding window algorithm for window size n should output "yes", if the current window content is ab^{n-1} and "no" otherwise. From our results in [11,12], it follows that the optimal space complexity of a *deterministic* sliding window algorithm for the membership problem for ab^* is $\Theta(\log n)$. On the other hand, it is shown in [13] that there is an ϵ-correct randomized sliding window algorithm for ab^* with (worst-case) space complexity $O(\log \log n)$ (this is also optimal). In fact, we proved in [13] that for every regular language L, the space optimal ϵ-correct randomized sliding window algorithm for L has either constant, doubly logarithmic, logarithmic, or linear space complexity, and the corresponding four space classes can be characterized in terms of simple syntactic properties.

Strict ϵ-correctness is used (without explicit mentioning) for instance in [5,10].[1] In these papers, the lower bounds shown for deterministic sliding-window algorithms are extended with the help of Yao's minimax principle [18] to strictly ϵ-correct randomized sliding-window algorithms. The main result of this paper states that this is a general phenomenon: we show that every strictly ϵ-correct sliding window algorithm for an approximation problem A can be derandomized without increasing the worst-case space complexity (Theorem 4). To the best of

---

[1] For instance, Ben-Basat et al. write "We say that algorithm A is ϵ-correct on a input instance S if it is able to approximate the number of 1's in the last W bits, at every time instant while reading S, to within an additive error of $W\epsilon$".

our knowledge, this is the first investigation on the general power of randomization on the space consumption of sliding window algorithms. We emphasize that our proof does not utilize Yao's minimax principle, which would require the choice of a "hard" distribution of input streams specific to the problem. It remains open, whether such a hard distribution exists for every approximation problem.

We remark that the proof of Theorem 4 uses exponentially long input streams in the size of the sliding window. In fact, we show that for a certain problem a restriction to polynomially long input streams yields an advantage of strictly correct randomized algorithms over deterministic ones, see Propositions 8 and 9.

It is possible to extend Theorem 4 to average space complexity with the cost of an additional constant factor. More precisely, for every randomized strictly ϵ-correct sliding window algorithm $\mathcal{R}$ for $\mathcal{A}$ there exists a deterministic sliding window algorithm for $\mathcal{A}$ whose space complexity is only a constant factor larger than the average space complexity of $\mathcal{R}$. This is a direct corollary of Theorem 4 and Lemma 3, which allows to go from average space to worst-case space with the cost of a constant blow-up.

Let us add further remarks on our model. First of all, it is crucial for our proofs that the input alphabet (i.e., the set of data values in the input stream) is finite. This is for instance the case when counting the number of 1's in a 0/1-sliding window. On the other hand, the problem of computing the sum of all data values in a sliding window of arbitrary numbers (a problem that is considered in [10] as well) is not covered by our setting, unless one puts a bound on the size of the numbers in the input stream.

As a second remark, note that our sliding window model is non-uniform in the sense that for every window size we may have a different streaming algorithm. In other words: it is not required that there exists a single streaming algorithm that gets the window size as a parameter. Clearly, lower bounds get stronger when shown for the non-uniform model. Moreover, all proofs of lower bounds in the sliding window setting, we are aware of, hold for the non-uniform model.

2   Preliminaries

With $[0, 1]$ we denote the real interval $\{p \in \mathbb{R} : 0 \leq p \leq 1\}$. The set of all words over a finite alphabet Σ is denoted by Σ^*. The empty word is denoted by λ. The length of a word $w \in \Sigma^*$ is denoted with $|w|$. The sets of words over Σ of length exactly, at most and at least n are denoted by Σ^n, $\Sigma^{\leq n}$ and $\Sigma^{\geq n}$, respectively.

2.1   Approximation Problems

An *approximation problem* is a mapping $\mathcal{A} \colon \Sigma^* \to 2^{\Omega}$ where Σ is a finite alphabet and Ω is a (possibly infinite) set of output values. For a given input word $w \in \Sigma^*$ the set $\mathcal{A}(w)$ is the set of *admissible outputs* for w. Typical examples include:

- exact computation problems $\mathcal{A} \colon \Sigma^* \to \Omega$ (here we identify an element $\omega \in \Omega$ with the singleton subset $\{\omega\}$). A typical example is the mapping

$c_1 \colon \{0,1\}^* \to \mathbb{N}$ where $c_1(w)$ is the number of 1's in w. Another exact problem is given by the characteristic function $\chi_L \colon \Sigma^* \to \{0,1\}$ of a language $L \subseteq \Sigma^*$.

- approximation of some numerical value for the data stream, which can be modeled by a function $\mathcal{A} \colon \Sigma^* \to 2^{\mathbb{N}}$. A typical example would be the mapping $w \mapsto \{k \in \mathbb{N} \colon (1-\epsilon) \cdot c_1(w) \le k \le (1+\epsilon) \cdot c_1(w)\}$ for some $0 < \epsilon < 1$.

2.2   Probabilistic Automata with Output

In the following we will introduce probabilistic automata [16,17] as a model of randomized streaming algorithms which produce an output after each input symbol. A *probabilistic automaton* $R = (Q, \Sigma, \iota, \rho, \omega)$ consists of a (possibly infinite) set of states Q, a finite alphabet Σ, an initial state distribution $\iota \colon Q \to [0,1]$, a transition probability function $\rho \colon Q \times \Sigma \times Q \to [0,1]$ and an output function $\omega \colon Q \to \Omega$ such that

- $\sum_{q \in Q} \iota(q) = 1$,
- $\sum_{q \in Q} \rho(p, a, q) = 1$ for all $p \in Q$, $a \in \Sigma$.

If ι and ρ map into $\{0,1\}$, then R is a *deterministic automaton*; in this case we write R as $R = (Q, \Sigma, q_0, \delta, \omega)$, where $q_0 \in Q$ is the initial state and $\delta \colon Q \times \Sigma \to Q$ is the transition function. A *run* on a word $a_1 \cdots a_m \in \Sigma^*$ in R is a sequence $\pi = (q_0, a_1, q_1, a_2, \ldots, a_m, q_m)$ where $q_0, \ldots, q_m \in Q$ and $\rho(q_{i-1}, a_i, q_i) > 0$ for all $1 \le i \le m$. If $m = 0$ we obtain the empty run (q_0) starting and ending in q_0. We write runs in the usual way

$$\pi \colon q_0 \xrightarrow{a_1} q_1 \xrightarrow{a_2} \cdots \xrightarrow{a_m} q_m$$

or also omit the intermediate states: $\pi \colon q_0 \xrightarrow{a_1 \cdots a_m} q_m$. We extend ρ to runs in the natural way: if $\pi \colon q_0 \xrightarrow{a_1} q_1 \xrightarrow{a_2} \cdots \xrightarrow{a_m} q_m$ is a run in R then $\rho(\pi) = \prod_{i=1}^{m} \rho(q_{i-1}, a_i, q_i)$. Furthermore we define $\rho_\iota(\pi) = \iota(q_0) \cdot \rho(\pi)$. We denote by $\mathrm{Runs}(R, w)$ the set of all runs on w in R and denote by $\mathrm{Runs}(R, q, w)$ those runs on w that start in $q \in Q$. If R is clear from the context, we simply write $\mathrm{Runs}(w)$ and $\mathrm{Runs}(q, w)$. Notice that for each $w \in \Sigma^*$ the function ρ_ι is a probability distribution on $\mathrm{Runs}(R, w)$ and for each $q \in Q$ the restriction of ρ to $\mathrm{Runs}(R, q, w)$ is a probability distribution on $\mathrm{Runs}(R, q, w)$. If Π is a set of runs (which will often be defined by a certain property of runs), then $\mathrm{Pr}_{\pi \in \mathrm{Runs}(w)}[\pi \in \Pi]$ denotes the probability $\sum_{\pi \in \mathrm{Runs}(w) \cap \Pi} \rho_\iota(\pi)$ and $\mathrm{Pr}_{\pi \in \mathrm{Runs}(q,w)}[\pi \in \Pi]$ denotes $\sum_{\pi \in \mathrm{Runs}(q,w) \cap \Pi} \rho(\pi)$.

3   Randomized Streaming and Sliding Window Algorithms

We define a *randomized streaming algorithm* as a pair (R, enc) consisting of a probabilistic automaton $R = (Q, \Sigma, \iota, \rho, \omega)$ as above and an injective function $\mathrm{enc} \colon Q \to \{0,1\}^*$. Usually, we will only refer to the underlying automaton R. If R

is deterministic, we speak of a *deterministic streaming algorithm*. The maximum number of bits stored in a run $\pi : q_0 \xrightarrow{a_1} q_1 \xrightarrow{a_2} \cdots \xrightarrow{a_m} q_m$ is denoted by space(R, π), i.e.,

$$\text{space}(R, \pi) = \max\{|\text{enc}(q_i)| : 0 \le i \le m\}.$$

We are interested in two space measures for an input stream w:

- worst case space:

$$\text{space}(R, w) = \max\{\text{space}(R, \pi) : \pi \in \text{Runs}(R, w), \rho_\iota(\pi) > 0\}.$$

- expected space:

$$\text{space}_\varnothing(R, w) = \sum_{\pi \in \text{Runs}(R, w)} \rho_\iota(\pi) \cdot \text{space}(R, \pi)$$

Let $R = (Q, \Sigma, \iota, \rho, \omega)$ be a randomized streaming algorithm, let $\mathcal{A} : \Sigma^* \to 2^\Omega$ be an approximation problem and let $w = a_1 a_2 \cdots a_m \in \Sigma^*$ be an input stream.

- A run $\pi : q_0 \xrightarrow{w} q_m$ is *correct for* $\mathcal{A}$ if $\omega(q_m) \in \mathcal{A}(w)$. The *error probability* of R on w for $\mathcal{A}$ is

$$\epsilon(R, w, \mathcal{A}) = \sum_{\pi \in N} \rho_\iota(\pi),$$

where $N = \{\pi \in \text{Runs}(R, w) : \pi \text{ is not correct for } \mathcal{A}\}$.
- A run $\pi : q_0 \xrightarrow{a_1} q_1 \xrightarrow{a_2} \cdots q_{m-1} \xrightarrow{a_m} q_m$ is *strictly correct for* $\mathcal{A}$ if $\omega(q_t) \in \mathcal{A}(a_1 \cdots a_t)$ for all $0 \le t \le m$. The *strict error probability* of R on w for $\mathcal{A}$ is

$$\epsilon_*(R, w, \mathcal{A}) = \sum_{\pi \in N_*} \rho_\iota(\pi),$$

where $N_* = \{\pi \in \text{Runs}(R, w) : \pi \text{ is not strictly correct for } \mathcal{A}\}$.

3.1 Sliding Window Algorithms

For a window length $n \ge 0$ and a stream $w \in \Sigma^*$ we define $\text{last}_n(w)$ to be the suffix of $\square^n w$ of length n where $\square \in \Sigma$ is a fixed alphabet symbol. The word $\text{last}_n(\lambda) = \square^n$ is also called the *initial window*. Given an approximation problem $\mathcal{A} : \Sigma^* \to 2^\Omega$ and a window length $n \ge 0$ we define the sliding window problem $\mathcal{A}_n : \Sigma^* \to 2^\Omega$ as

$$\mathcal{A}_n(w) = \mathcal{A}(\text{last}_n(w))$$

for $w \in \Sigma^*$. Since we can identify a language $L \subseteq \Sigma^*$ with its characteristic function $\chi_L : \Sigma^* \to \{0, 1\}$, the definition of $\mathcal{A}_n$ specializes to

$$L_n = \{w \in \Sigma^* : \text{last}_n(w) \in L\}.$$

A *randomized sliding window algorithm* (*randomized SWA* for short) is a sequence $\mathcal{R} = (R_n)_{n \geq 0}$ of randomized streaming algorithms R_n over the same alphabet Σ and over the same set of output values Ω. If every R_n is deterministic, we speak of a *deterministic SWA*. Note that this is a non-uniform model in the sense that for every window length n we have a separate algorithm R_n. The *space complexity* of the randomized SWA $\mathcal{R} = (R_n)_{n \geq 0}$ is the function

$$f(\mathcal{R}, n) = \sup\{\text{space}(R_n, w) : w \in \Sigma^*\}$$

and its *expected space complexity* is the function

$$f_\varnothing(\mathcal{R}, n) = \sup\{\text{space}_\varnothing(R_n, w) : w \in \Sigma^*\}.$$

Clearly, if R_n is finite, then one can always find a state encoding such that $f(\mathcal{R}, n) = \lfloor \log_2 |R_n| \rfloor$ ($|R_n|$ denotes the number of states of R_n).

Definition 1. *Let $0 \leq \epsilon \leq 1$ be an error bound, let $\mathcal{R} = (R_n)_{n \geq 0}$ be a randomized SWA, and let $\mathcal{A}$ be an approximation problem.*

- $\mathcal{R}$ *is ϵ-correct for $\mathcal{A}$ if $\epsilon(R_n, w, \mathcal{A}_n) \leq \epsilon$ for all $n \geq 0$ and $w \in \Sigma^*$.*
- $\mathcal{R}$ *is strictly ϵ-correct for $\mathcal{A}$ if $\epsilon_*(R_n, w, \mathcal{A}_n) \leq \epsilon$ for all $n \geq 0$ and $w \in \Sigma^*$.*

A deterministic SWA for $\mathcal{A}$ is a deterministic SWA which is 0-correct (and hence strictly 0-correct) for $\mathcal{A}$.

Remark 2. Since one can store for window size n the window content with $\lceil \log_2 |\Sigma| \rceil \cdot n$ bits, every approximation problem has a deterministic SWA $\mathcal{D} = (D_n)_{n \geq 0}$ such that $f(\mathcal{D}, n) \leq \lceil \log_2 |\Sigma| \rceil \cdot n$. In particular, for every (strictly) ϵ-correct randomized SWA $\mathcal{R}$ for $\mathcal{A}$, there exists a (strictly) ϵ-correct randomized SWA $\mathcal{R}'$ for $\mathcal{A}$ such that $f(\mathcal{R}', n) \leq \min\{f(\mathcal{R}, n), \lceil \log_2 |\Sigma| \rceil \cdot n\}$.

It is not clear, whether the statement in Remark 2 also holds for average space complexity. On the other hand, the following statement holds:

Lemma 3. *Let $\mathcal{R}$ be a randomized SWA which is (strictly) ϵ-correct for $\mathcal{A}$ and let $\mu \geq 1$. Then there exists a randomized SWA $\mathcal{R}'$ which is (strictly) $(\epsilon + \frac{1}{\mu})$-correct for $\mathcal{A}$ such that $f(\mathcal{R}', n) \leq \mu \cdot f_\varnothing(\mathcal{R}, n)$.*

Proof. Fix $n \geq 0$ and let $s = f_\varnothing(\mathcal{R}, n)$ be the expected space complexity on window length n. If $s = \infty$ then the statement is trivial (take $R_n' = R_n$). So, let us assume that s is finite. Let $Q_\geq$ be the set of states in R_n with encoding length $\geq \mu \cdot s$. If $Q_\geq = \emptyset$, then $f(\mathcal{R}, n) \leq \mu \cdot f_\varnothing(\mathcal{R}, n)$ already holds. If $Q_\geq$ is nonempty, let R_n' be the algorithm obtained from R_n by merging all states $q \in Q_\geq$ into a single state $q_\perp$ encoded using an unused bit string of minimal length, which is at most $\mu \cdot s$. On an input stream $w \in \Sigma^*$, the probability that $q_\perp$ is reached is

$$\Pr_{\pi \in \text{Runs}(R_n', w)}[\pi \text{ contains } q_\perp] = \Pr_{\pi \in \text{Runs}(R_n, w)}[\text{space}(R_n, \pi) \geq \mu \cdot s] \leq \frac{s}{\mu \cdot s} = \frac{1}{\mu}$$

by Markov's inequality. Note that if an R'_n-run on w is not (strictly) correct, then (i) it must contain $q_\perp$ or (ii) it must be an R_n-run on w which is not (strictly) correct. Hence, an union bound yields

$$\epsilon(R'_n, w, \mathcal{A}) \le \frac{1}{\mu} + \epsilon(R_n, w, \mathcal{A}) \le \frac{1}{\mu} + \epsilon, \text{ respectively}$$

$$\epsilon_*(R'_n, w, \mathcal{A}) \le \frac{1}{\mu} + \epsilon_*(R_n, w, \mathcal{A}) \le \frac{1}{\mu} + \epsilon,$$

which shows the lemma. $\square$

4   Derandomization of Strictly Correct Algorithms

In this section we prove the main result of this paper, which states that strictly correct randomized SWAs can be completely derandomized:

Theorem 4. *Let $\mathcal{A}: \Sigma^* \to 2^\Omega$ be an approximation problem and let $\mathcal{R}$ be a randomized SWA which is strictly ϵ-correct for $\mathcal{A}$, where $0 \le \epsilon < 1$. There exists a deterministic SWA $\mathcal{D}$ for $\mathcal{A}$ such that $f(\mathcal{D}, n) \le f(\mathcal{R}, n)$ for all $n \ge 0$.*

Theorem 4 talks about the worst-case space complexity $f(\mathcal{R}, n)$. On the other hand, if we choose a constant $\mu \ge 1$ with $\epsilon + 1/\mu < 1$ then with Lemma 3 we obtain from a strictly ϵ-correct randomized SWA $\mathcal{R}$ for $\mathcal{A}$ a deterministic SWA $\mathcal{D}$ for $\mathcal{A}$ with $f(\mathcal{D}, n) \le \mu \cdot f_\varnothing(\mathcal{R}, n)$.

Let $\mathcal{A}: \Sigma^* \to 2^\Omega$, $\mathcal{R} = (R_n)_{n \ge 0}$, and $0 \le \epsilon < 1$ as in Theorem 4. By Remark 2, we can assume that every R_n has a finite state set. Fix a window size $n \ge 0$ and let $R_n = (Q, \Sigma, \iota, \rho, \omega)$. Consider a run

$$\pi : q_0 \xrightarrow{a_1} q_1 \xrightarrow{a_2} \cdots \xrightarrow{a_m} q_m$$

in R_n. The run π is *simple* if $q_i \ne q_j$ for $0 \le i < j \le m$. A *subrun* of π is a run

$$q_i \xrightarrow{a_{i+1}} q_{i+1} \xrightarrow{a_{i+2}} \cdots q_{j-1} \xrightarrow{a_j} q_j$$

for some $0 \le i \le j \le m$. Consider a nonempty subset $S \subseteq Q$ and a function $\delta: Q \times \Sigma \to Q$ such that S is closed under δ, i.e., $\delta(S \times \Sigma) \subseteq S$. We say that the run π is *δ-conform* if $\delta(q_{i-1}, a_i) = q_i$ for all $1 \le i \le m$. We say that π is *(S, δ)-universal* if for all $q \in S$ and $x \in \Sigma^n$ there exists a δ-conform subrun $\pi' : q \xrightarrow{x} q'$ of π. Finally, π is *δ-universal* if it is (S, δ)-universal for some nonempty subset $S \subseteq Q$ which is closed under δ.

Lemma 5. *Let π be a strictly correct run in R_n for $\mathcal{A}$, let $S \subseteq Q$ be a nonempty subset and let $\delta: Q \times \Sigma \to Q$ be a function such that S is closed under δ. If π is (S, δ)-universal, then there exists $q_0 \in S$ such that $D_n = (Q, \Sigma, q_0, \delta, \omega)$ is a deterministic streaming algorithm for $\mathcal{A}_n$.*

Proof. Let $q_0 = \delta(p, \square^n) \in S$ for some arbitrary state $p \in S$ and define $D_n = (Q, \Sigma, q_0, \delta, \omega)$. Let $w \in \Sigma^*$ and consider the run $\sigma : p \xrightarrow{\square^n} q_0 \xrightarrow{w} q$ in D_n of length $\geq n$. We have to show that $(\text{last}_n(w), \omega(q)) \in \mathcal{A}$. We can write $\square^n w = x \, \text{last}_n(w)$ for some $x \in \Sigma^*$. Thus, we can rewrite the run σ as $\sigma : p \xrightarrow{x} q' \xrightarrow{\text{last}_n(w)} q$. We know that $q' \in S$ because S is closed under δ. Since π is (S, δ)-universal, it contains a subrun $q' \xrightarrow{\text{last}_n(w)} q$. Strict correctness of π implies $(\text{last}_n(w), \omega(q)) \in \mathcal{A}$. □

For the rest of this section we fix an arbitrary function $\delta : Q \times \Sigma \to Q$ such that for all $q \in Q$, $a \in \Sigma$,

$$\rho(q, a, \delta(q, a)) = \max\{\rho(q, a, p) : p \in Q\}.$$

Note that

$$\rho(q, a, \delta(q, a)) \geq \frac{1}{|Q|}. \tag{1}$$

for all $q \in Q$, $a \in \Sigma$. Furthermore, let $D_n = (Q, \Sigma, q_0, \delta, \omega)$ where the initial state q_0 will be defined later. We define for each $i \geq 1$ a state p_i, a word $w_i \in \Sigma^*$, a corresponding run $\pi_i^* \in \text{Runs}(D_n, p_i, w_i)$ in D_n and a set $S_i \subseteq Q$. For $m \geq 0$, we use the abbreviation

$$\Pi_m = \text{Runs}(R_n, w_1 \cdots w_m).$$

Note that $\Pi_0 = \text{Runs}(R_n, \lambda)$. For $1 \leq i \leq m$ let H_i denote the event that for a random run $\pi = \pi_1 \cdots \pi_m \in \Pi_m$, where each π_j is a run on w_j, the subrun π_i is (S_i, δ)-universal. Notice that H_i is independent of $m \geq i$.

First, we choose for p_i a state that maximizes the probability

$$\Pr_{\pi \in \Pi_{i-1}} [\pi \text{ ends in } p_i \mid \forall j \leq i - 1 : \overline{H_j}],$$

which is at least $1/|Q|$. Note that p_1 is a state such that $\iota(p_1)$ is maximal, since Π_0 only consists of empty runs (q). For S_i we take any maximal strongly connected component of D_n (viewed as a directed graph) which is reachable from p_i. Here, maximality means that for every $q \in S_i$ and every $a \in \Sigma$, also $\delta(q, a)$ belongs to S_i. Finally, we define the run π_i^* and the word w_i. The run π_i^* starts in p_i. Then, for each pair $(q, x) \in S_i \times \Sigma^n$ the run π_i^* leads from the current state to state q via a simple run and then reads the word x from q. The order in which we go over all pairs $(q, x) \in S_i \times \Sigma^n$ is not important. Since S_i is a maximal strongly connected component of D_n such a run π_i^* exists. Hence, π_i^* is a run on a word

$$w_i = \prod_{q \in S_i} \prod_{x \in \Sigma^n} y_{q,x} \, x,$$

where $y_{q,x}$ is the word that leads from the current state via a simple run to state q. Since we choose the runs on the words $y_{q,x}$ to be simple, we have $|y_{q,x}| \leq |Q|$ and thus $|w_i| \leq |Q| \cdot |\Sigma|^n \cdot (|Q| + n)$. Let us define

$$\mu = \frac{1}{|Q|^{|Q| \cdot |\Sigma|^n \cdot (|Q| + n) + 1}}.$$

Note that by construction, the run π_i^* is (S_i, δ)-universal. Inequality (1) yields

$$\Pr_{\pi \in \mathrm{Runs}(p_i, w_i)}[\pi = \pi_i^*] \geq \frac{1}{|Q|^{|w_i|}} \geq \mu \cdot |Q|. \qquad (2)$$

Lemma 6. *For all $m \geq 0$ we have $\Pr_{\pi \in \Pi_m}[H_m \mid \forall i \leq m - 1 : \overline{H_i}] \geq \mu$.*

Proof. In the following, let π be a random run from Π_m and let π_i be the subrun on w_i. Under the assumption that the event $[\pi_{m-1}$ ends in $p_m]$ holds, the events $[\pi_m = \pi_m^*]$ and $[\forall i \leq m - 1 : \overline{H_i}]$ are conditionally independent.[2] Thus, we have

$$\Pr_{\pi \in \Pi_m}[\pi_m = \pi_m^* \mid \pi_{m-1} \text{ ends in } p_m \wedge \forall i \leq m - 1 : \overline{H_i}]$$
$$= \Pr_{\pi \in \Pi_m}[\pi_m = \pi_m^* \mid \pi_{m-1} \text{ ends in } p_m].$$

Since the event $[\pi_m = \pi_m^*]$ implies the event $[\pi_{m-1}$ ends in $p_m]$, we obtain:

$$\Pr_{\pi \in \Pi_m}[H_m \mid \forall i \leq m - 1 : \overline{H_i}]$$
$$\geq \Pr_{\pi \in \Pi_m}[\pi_m = \pi_m^* \mid \forall i \leq m - 1 : \overline{H_i}]$$
$$= \Pr_{\pi \in \Pi_m}[\pi_m = \pi_m^* \wedge \pi_{m-1} \text{ ends in } p_m \mid \forall i \leq m - 1 : \overline{H_i}]$$
$$= \Pr_{\pi \in \Pi_m}[\pi_m = \pi_m^* \mid \pi_{m-1} \text{ ends in } p_m \wedge \forall i \leq m - 1 : \overline{H_i}] \cdot$$
$$\Pr_{\pi \in \Pi_m}[\pi_{m-1} \text{ ends in } p_m \mid \forall i \leq m - 1 : \overline{H_i}]$$
$$= \Pr_{\pi \in \Pi_m}[\pi_m = \pi_m^* \mid \pi_{m-1} \text{ ends in } p_m] \cdot$$
$$\Pr_{\pi \in \Pi_m}[\pi_{m-1} \text{ ends in } p_m \mid \forall i \leq m - 1 : \overline{H_i}]$$
$$\geq \Pr_{\pi_m \in \mathrm{Runs}(p_m, w_m)}[\pi_m = \pi_m^*] \cdot \frac{1}{|Q|}$$
$$\geq \mu,$$

where the last inequality follows from (2). This proves the lemma.   □

Lemma 7. $\Pr_{\pi \in \Pi_m}[\pi \text{ is } \delta\text{-universal}] \geq \Pr_{\pi \in \Pi_m}[\exists i \leq m : H_i] \geq 1 - (1 - \mu)^m$.

Proof. The first inequality follows from the definition of the event H_i. Moreover, with Lemma 6 we get

$$\Pr_{\pi \in \Pi_m}[\exists i \leq m : H_i] = \Pr_{\pi \in \Pi_m}[\exists i \leq m - 1 : H_i] +$$
$$\Pr_{\pi \in \Pi_m}[H_m \mid \forall i \leq m - 1 : \overline{H_i}] \cdot \Pr_{\pi \in \Pi_m}[\forall i \leq m - 1 : \overline{H_i}]$$
$$= \Pr_{\pi \in \Pi_{m-1}}[\exists i \leq m - 1 : H_i] +$$
$$\Pr_{\pi \in \Pi_m}[H_m \mid \forall i \leq m - 1 : \overline{H_i}] \cdot \Pr_{\pi \in \Pi_{m-1}}[\forall i \leq m - 1 : \overline{H_i}]$$
$$\geq \Pr_{\pi \in \Pi_{m-1}}[\exists i \leq m - 1 : H_i] + \mu \cdot \Pr_{\pi \in \Pi_{m-1}}[\forall i \leq m - 1 : \overline{H_i}].$$

---

[2] Two events A and B are conditionally independent assuming event C if $\Pr[A \wedge B \mid C] = \Pr[A \mid C] \cdot \Pr[B \mid C]$, which is equivalent to $\Pr[A \mid B \wedge C] = \Pr[A \mid C]$.

Thus, $r_m := \Pr_{\pi \in \Pi_m}[\exists i \leq m : H_i]$ satisfies $r_m \geq r_{m-1} + \mu \cdot (1 - r_{m-1}) = (1 - \mu) \cdot r_{m-1} + \mu$. Since $r_0 = 0$, we get $r_m \geq 1 - (1 - \mu)^m$ by induction.   □

Proof of Theorem 4. We use the probabilistic method in order to show that there exists $q_0 \in Q$ such that $D_n = (Q, \Sigma, q_0, \delta, \omega)$ is a deterministic streaming algorithm for $\mathcal{A}_n$. With Lemma 7 we get

$$\Pr_{\pi \in \Pi_m} [\pi \text{ is strictly correct for } \mathcal{A} \text{ and } \delta\text{-universal}]$$

$$= 1 - \Pr_{\pi \in \Pi_m} [\pi \text{ is not strictly correct for } \mathcal{A} \text{ or is not } \delta\text{-universal}]$$

$$\geq 1 - \Pr_{\pi \in \Pi_m} [\pi \text{ is not strictly correct for } \mathcal{A}] - \Pr_{\pi \in \Pi_m} [\pi \text{ is not } \delta\text{-universal}]$$

$$\geq \Pr_{\pi \in \Pi_m} [\pi \text{ is } \delta\text{-universal}] - \epsilon$$

$$\geq 1 - (1 - \mu)^m - \epsilon.$$

We have $1 - (1 - \mu)^m - \epsilon > 0$ for $m > \log(1 - \epsilon)/\log(1 - \mu)$ (note that $\epsilon < 1$ and $0 < \mu < 1$ since we can assume that $|Q| \geq 2$). Hence there are $m \geq 0$ and a strictly correct δ-universal run $\pi \in \Pi_m$. We can conclude with Lemma 5.   □

The word $w_1 w_2 \cdots w_m$ (with $m > \log(1 - \epsilon)/\log(1 - \mu)$), for which there exists a strictly correct and δ-universal run has a length that is exponential in the window size n. In other words: We need words of length exponential in n in order to transform a strictly ϵ-correct randomized SWA into an equivalent deterministic SWA. We now show that this is unavoidable: if we restrict to inputs of length poly(n) then strictly ϵ-correct SWAs can yield a proper space improvement over deterministic SWAs.

For a word $w = a_1 \cdots a_n$ let $w^R = a_n \cdots a_1$ denote the reversed word. Take the language $K_{pal} = \{ww^R : w \in \{a, b\}^*\}$ of all palindromes of even length, which belongs to the class **DLIN** of deterministic linear context-free languages [4], and let $L = \$K_{pal}$.

Proposition 8. *If $\mathcal{D}$ is a deterministic SWA for L, then $f(\mathcal{D}, 2n + 1) = \Omega(n)$.*

Proof. Let $\mathcal{D} = (D_n)_{n \geq 0}$. Take two distinct words $\$x$ and $\$y$ where $x, y \in \{a, b\}^n$. Since D_{2n+1} accepts $\$xx^R$ and rejects $\$yx^R$, the automaton D_{2n+1} reaches two different states on the inputs $\$x$ and $\$y$. Therefore, D_{2n+1} must have at least $|\{a, b\}^n| = 2^n$ states and hence $\Omega(n)$ space.   □

Proposition 9. *Fix a polynomial $p(n)$. There is a randomized SWA $\mathcal{R} = (R_n)_{n \geq 0}$ such that (i) $f(\mathcal{R}, n) \in \mathcal{O}(\log n)$ and (ii) $\epsilon_*(R_n, w, L_n) \leq 1/e$ (e denotes Euler's number) for all input words $w \in \Sigma^*$ with $|w| \leq p(n)$.*

Proof. Babu et al. [4] have shown that for every language $K \in$ **DLIN** there exists a randomized streaming algorithm using space $\mathcal{O}(\log n)$ which, given an input w of length n,

- accepts with probability 1 if $w \in K$,
- and rejects with probability at least $1 - 1/n$ if $w \notin K$.

We remark that the algorithm needs to know the length of w in advance. To stay consistent with our definition, we view the algorithm above as a family $(S_n)_{n \geq 0}$ of randomized streaming algorithms S_n. Furthermore, the error probability $1/n$ can be further reduced to $1/n^d$ where $p(n) \leq n^d$ for sufficiently large n (by picking random primes of size $\Theta(n^{d+1})$ in the proof from [4]).

Now we prove our claim for $L = \$K_{\text{pal}}$. The streaming algorithm R_n for window size n works as follows: After reading a \$-symbol, the algorithm S_{n-1} from above is simulated on the longest factor from $\{a, b\}^*$ that follows (i.e. S_{n-1} is simulated until the next \$ arrives). Simultaneously we maintain the length ℓ of the maximal suffix over $\{a, b\}$, up to n, using $\mathcal{O}(\log n)$ bits. If ℓ reaches $n - 1$, then R_n accepts if and only if S_{n-1} accepts. Notice that R_n only errs if the stored length is $n - 1$ (with probability $1/n^d$), which happens at most once in every n steps. Therefore the number of time instants where R_n errs on w is at most $|w|/n \leq n^d/n = n^{d-1}$. The union bound yields $\epsilon_*(R_n, w, L_n) \leq n^{d-1}/n^d = \frac{1}{n}$ for every stream $w \in \{\$, a, b\}^{\leq p(n)}$. This concludes the proof.   □

Acknowledgment. The first author has been supported by the DFG research project LO 748/13-1.

References

1. Aggarwal, C.C.: Data Streams - Models and Algorithms. Springer, New York (2007)
2. Arasu, A., Manku, G.S.: Approximate counts and quantiles over sliding windows. In: Proceedings of PODS 2004, pp. 286–296. ACM (2004)
3. Babcock, B., Datar, M., Motwani, R., O'Callaghan, L.: Maintaining variance and k-medians over data stream windows. In: Proceedings of PODS 2003, pp. 234–243. ACM (2003)
4. Babu, A., Limaye, N., Radhakrishnan, J., Varma, G.: Streaming algorithms for language recognition problems. Theor. Comput. Sci. **494**, 13–23 (2013)
5. Ben-Basat, R., Einziger, G., Friedman, R., Kassner, Y.: Efficient summing over sliding windows. In: Proceedings of SWAT 2016. LIPIcs, vol. 53, pp. 11:1–11:14. Schloss Dagstuhl - Leibniz-Zentrum für Informatik (2016)
6. Braverman, V.: Sliding window algorithms. In: Kao, M.-Y. (ed.) Encyclopedia of Algorithms, pp. 2006–2011. Springer, New York (2016)
7. Braverman, V., Ostrovsky, R., Zaniolo, C.: Optimal sampling from sliding windows. J. Comput. Syst. Sci. **78**(1), 260–272 (2012)
8. Chan, H.-L., Lam, T.-W., Lee, L.-K., Pan, J., Ting, H.-F., Zhang, Q.: Edit distance to monotonicity in sliding windows. In: Asano, T., Nakano, S., Okamoto, Y., Watanabe, O. (eds.) ISAAC 2011. LNCS, vol. 7074, pp. 564–573. Springer, Heidelberg (2011). https://doi.org/10.1007/978-3-642-25591-5_58
9. Crouch, M.S., McGregor, A., Stubbs, D.: Dynamic graphs in the sliding-window model. In: Bodlaender, H.L., Italiano, G.F. (eds.) ESA 2013. LNCS, vol. 8125, pp. 337–348. Springer, Heidelberg (2013). https://doi.org/10.1007/978-3-642-40450-4_29
10. Datar, M., Gionis, A., Indyk, P., Motwani, R.: Maintaining stream statistics over sliding windows. SIAM J. Comput. **31**(6), 1794–1813 (2002)

11. Ganardi, M., Hucke, D., König, D., Lohrey, M., Mamouras, K.: Automata theory on sliding windows. In: Proceedings of STACS 2018. LIPIcs, vol. 96, pages 31:1–31:14. Schloss Dagstuhl - Leibniz-Zentrum für Informatik, (2018, to appear)
12. Ganardi, M., Hucke, D., Lohrey, M.: Querying regular languages over sliding windows. In: Proceedings of FSTTCS 2016. LIPIcs, vol. 65, pp. 18:1–18:14. Schloss Dagstuhl - Leibniz-Zentrum für Informatik (2016)
13. Ganardi, M., Hucke, D., Lohrey, M.: Randomized sliding window algorithms for regular languages. In: Proceedings of ICALP 2018. LIPIcs, vol. 107, pp. 127:1–127:13. Schloss Dagstuhl - Leibniz-Zentrum für Informatik (2018)
14. Ganardi, M., Jez, A., Lohrey, M.: Sliding windows over context-free languages. In: Proceedings of MFCS 2018. LIPIcs, vol. 117, pp. 15:1–15:15. Schloss Dagstuhl - Leibniz-Zentrum für Informatik (2018)
15. Golab, L., Özsu, M.T.: Processing sliding window multi-joins in continuous queries over data streams. In: Proceedings of VLDB 2003, pp. 500–511. Morgan Kaufmann (2003)
16. Paz, A.: Introduction to Probabilistic Automata. Academic Press, River Edge (1971)
17. Rabin, M.O.: Probabilistic automata. Inf. Control 6(3), 230–245 (1963)
18. Yao, A.C.: Probabilistic computations: toward a unified measure of complexity. In: Proceedings of FOCS 1977, pp. 222–227. IEEE Computer Society (1977)

On the Complexity of Restarting

Jan-Hendrik Lorenz[(✉)]

Institute of Theoretical Computer Science, Ulm University, 89069 Ulm, Germany
jan-hendrik.lorenz@uni-ulm.de

Abstract. Restarting is a technique used by many randomized local search and systematic search algorithms. If the algorithm has not been successful for some time, the algorithm is reset and reinitialized with a new random seed. However, for some algorithms and some problem instances, restarts are not beneficial. Luby et al. [12] showed that if restarts are useful, then there is a restart time t^* such that the so-called fixed-cutoff strategy is the best possible strategy in expectation.

In this work, we show that deciding whether restarts are useful is NP-complete. Furthermore, we show that there is no feasible approximation algorithm for the optimal restart time t^*. Lastly, we show that calculating the expected runtime for a known probability distribution and a given restart time is #P-complete.

Keywords: Restarts · Fixed-cutoff strategy ·
Probability distribution · Computational complexity · NP · #P ·
Inapproximability

1 Introduction

Some (randomized) algorithms employ an algorithmic paradigm called restarting: If a solution is not found after a certain number of steps, then the algorithm is reset and the search starts over with a new random seed. Restarts are especially prevalent in stochastic local search (e.g. [13,17]) and randomized systematic search algorithms (e.g. [3,9]). In practice, restarts help to improve the performance of some algorithms by orders of magnitude, and are presently employed in most state-of-the-art SAT solvers [4].

For Las Vegas algorithms the runtime behavior is often modeled with probability distributions. For example, Arbelaez et al. [1] use lognormal and exponential distributions to describe the runtime behavior of two SAT solvers. Frost et al. [8] model the runtime behavior of solvable CSP instances with Weibull distributions and the behavior of unsolvable CSP instances with lognormal distributions. Both papers use empirical observations to fit these distributions to the observed runtimes. Hence, it also seems natural for theoretical purposes to model the runtime behavior with cumulative distribution functions (cdfs) or probability mass functions (pmfs). It allows to succinctly represent the runtime behavior on a large (possibly infinite) support.

© Springer Nature Switzerland AG 2019
R. van Bevern and G. Kucherov (Eds.): CSR 2019, LNCS 11532, pp. 250–261, 2019.
https://doi.org/10.1007/978-3-030-19955-5_22

From the theoretical perspective, Luby et al. [12] showed that the so-called fixed-cutoff strategy is an optimal strategy w.r.t. the expected runtime for some restart time t^*. The fixed-cutoff strategy allows an unbounded number of restarts and the algorithm always restarts after t^* steps. It is, however, necessary to have nearly complete knowledge of the runtime behavior to compute the optimal restart time t^*.

A good restart strategy should improve the performance of the algorithm in question. For the case when the distribution is known, Lorenz [11] derived formulas to evaluate whether there is any restart strategy which is beneficial w.r.t. the expected runtime. Furthermore, the formulas can be used to calculate an optimal restart time.

Another crucial property for restart strategies is that the calculation of restart times should be efficient. In other words, the time spent choosing a restart strategy should be significantly less than the time dedicated to the actual algorithm that uses the restart strategy. So far, the complexity of choosing a restart strategy has not been considered.

Our Contribution: We study the computational complexity of problems directly related to restarts. It is assumed that the probability distribution is known and provided in symbolic form, i.e., the formulas of either the cdf or pmf of the distribution are known. Section 3 considers decision problems. It is shown that deciding whether there is a restart time such that the expected runtime is less than a threshold is NP-hard (Theorem 2) while a relaxation of the problem is still NP-complete (Theorem 1). Furthermore, Theorem 3 and Corollary 2 show that there is no feasible approximation algorithm for the restart time w.r.t. the expected runtime.

In Sect. 4, we investigate underlying reasons why the decision problems mentioned above are hard. Moorsel and Wolter [14] showed that restarts are beneficial if and only if $E[X] < E[X - t \mid X > t]$ for some t, where X is a random variable describing the runtime. Here, we show that calculating both sides of the inequality is #P-complete (Theorems 4 and 5).

Related Work: Initiated by the work of Batu et al. [2], *property testing* gained traction in recent years. In this field, the general objective is determining whether a probability distribution fulfills a specific desired property. For example, an algorithm should decide whether the pmf of a distribution is unimodal [7]. For an in-depth overview of the topic, refer to [6]. The difference to our model is that in the framework of property testing the probability distribution is not explicitly known. Instead, it is assumed that sample access to the distribution is available. In other words, there is an oracle which returns an element x with probability $p(x)$. Therefore, the pmf p cannot be observed directly, while in our model p is known. To the best of our knowledge, complexity properties of probability distributions provided in symbolic form have not been studied before.

2   Preliminaries

We presume that the reader is familiar with the basic notions of complexity theory. For a detailed description of the topic, we refer to a standard book on the topic like [16]. We assume that the runtime distribution is known and provided as input in symbolic form. First, the structure of the functions is specified: In this work, it is only required that a function F is well-defined on a bounded interval $I = \{0, \ldots, a\}$ with $0 < a < \infty$. In this work, we consider discrete cumulative distribution functions and probability mass functions.

Definition 1. *Let X be an integer-valued random variable. Then, the **cumulative distribution function (cdf)** F_X of X is defined by*

$$F_X(i) = \Pr(X \leq i).$$

*The **probability mass function (pmf)** p_X of X is defined by*

$$p_X(i) = \Pr(X = i).$$

In the following, the subscripts are often omitted. Cdfs and pmfs are bounded from below by zero and bounded from above by one. Moreover, cdfs are monotone on I. Let p be a pmf then its corresponding cumulative function $F(k) = \sum_{i=0}^{k} p(i)$ is required to be a cdf.

Each function F is defined with binary encoded input, i.e., $F : \{0,1\}^n \mapsto [0,1]$ where n is a suitable integer. The function F is given in symbolic form or in other words: as a formula. We use uniform families of arithmetic circuits to describe the formulas. In an arithmetic circuit, the gates are arithmetic operations, like addition or multiplication. An overview of arithmetic circuits can be found in [19]. We use binary encoded integers as input. Also, there is a large number of non-standard functions which could be used to achieve a succinct representation of an arbitrary function. Therefore, we only allow the use of addition, subtraction, multiplication, exponentiation, and division gates. Another work using a model with the same gates is, for example, described in [5]. Another significant property which is required for the following results is that $F(x)$ can be evaluated in polynomial time for all x. In the following, when we refer to formulas, we implicitly mean formulas which are given as arithmetic circuits.

In this work, it is often necessary to encode a conjunctive normal form (CNF) formula as a function G using only multiplication and subtraction such that the SAT formula is satisfiable iff $G(a) = 1$ for some input a. It is known that negation $\neg x$ can be expressed by $1 - x$, a logical **and** $x \wedge y$ is given by xy, and a logical **or** $x \vee y$ is $1 - (1-x)(1-y)$.

Here, we formally define the fixed-cutoff strategy.

Definition 2 ([12]). *Let $\mathcal{A}(x)$ an algorithm $\mathcal{A}$ on input x. Let t be a positive integer. A modified algorithm $\mathcal{A}_t$ is obtained by introducing a counter $T = 0$. Then, $\mathcal{A}(x)$ is allowed to run. Increment T after each step of $\mathcal{A}(x)$. If T exceeds t reset and reinitialize $\mathcal{A}(x)$ with a new random seed and set T to zero. If at* *any point $\mathcal{A}(x)$ finds a solution, then $\mathcal{A}_t(x)$ returns this solution. Repeat these steps until $\mathcal{A}(x)$ finds a solution. The integer t is called* **restart time**. *This algorithmic approach is called* **fixed-cutoff strategy**.

In this work, we only address fixed-cutoff strategies. Hence, whenever we refer to restart strategies or restarts, we implicitly mean a fixed-cutoff strategy.

Let X be a discrete random variable with cdf F and let $t \in \mathbb{N}$ be an arbitrary restart time. The expected value with restarts after t steps is denoted by $E[X_t]$. Luby et al. [12] showed that $E[X_t]$ is given by

$$E[X_t] = \frac{1 - F(t)}{F(t)} t + E[X \mid X \le t] = \frac{1}{F(t)} \left(t - \sum_{x < t} F(x) \right). \tag{1}$$

They also introduced some useful bounds in their work:

Lemma 1 ([12]). *Let $l^* = \inf_t E[X_t]$ be the optimal expected value under restart, then*

$$l^* \le \min_t \frac{t}{F(t)} \le 4l^* \tag{2}$$

holds.

For the rest of this work, the quotient $t/F(t)$ is called the upper bound (of the expected value with restarts).

3   Hardness and Inapproximability

We first focus on the upper bound $t/F(t)$ and show that this version of the problem is computationally hard.

RESTART-UPPER-BOUND-CDF

Input: A cdf $F : \mathbb{N} \mapsto [0, 1]$ and an integer k. The cdf F is given as an arithmetic circuit and $F(t)$ can be evaluated in polynomial time for every $t < k$. The symbolic form of F must be composed of summation, subtraction, multiplication and division operations.

Question: Is there a restart time t such that $\frac{t}{F(t)} < k$?

Theorem 1. RESTART-UPPER-BOUND-CDF *is NP-complete.*

Proof. We first show that RESTART-UPPER-BOUND-CDF is in NP. It is easy to see that the inequality $\frac{t}{F(t)} < k$ only holds if $t < k$. Then, the inequality $\frac{t}{F(t)} < k$ can be verified in polynomial time since by hypothesis F can be evaluated in polynomial time for every fixed $t < k$. Thus, RESTART-UPPER-BOUND-CDF is in NP.

Secondly, let G be any SAT formula with n variables. Any $0 \le t \le 2^n - 1$ can be interpreted as an assignment of the variables in G; this assignment is denoted by α_t. In the following, $G\alpha$ denotes the numeric value of G evaluated with the assignment α. We define the function $F_G : \{0, 1\}^{n+1} \mapsto [0, 1]$:

$$F_G(t) = \begin{cases} \frac{t+G\alpha_t}{2^n}, & 0 \le t \le 2^n - 1 \\ 1, & t \ge 2^n \end{cases} \qquad (3)$$

The function $F_G(t)$ can be evaluated in polynomial time for all $t \le 2^n$ because the division by 2^n can be expressed in polynomial length with the standard binary representation. Clearly, F_G is bounded from below by zero and from above by one. The formula F_G is monotonically increasing which can be derived from the fact that $G\alpha_t \le 1 + G\alpha_{t+1}$ holds.

$$G\alpha_t \le 1 + G\alpha_{t+1}$$
$$\Leftrightarrow \frac{t + G\alpha_t}{2^n} \le \frac{t + 1 + G\alpha_{t+1}}{2^n}$$

What remains to be shown is that F_G can be expressed with summation, subtraction, multiplication and division. Each of the two cases can clearly be expressed with these operations, since a CNF formula can be expressed as a function as described above. Only the case distinction has to expressed with the allowed operations. This can be achieved with the following representation:

$$F_G(t) = \frac{t + G\alpha_t}{2^n} \cdot (1 - x_{n+1}) + x_{n+1}.$$

Where x_{n+1} is the $(n+1)$-st bit of the binary representation of t. Hence, F_G is a cumulative distribution function which can be evaluated in polynomial time for each $t < 2^n + 1$. Let U be an unsatisfiable formula, then $F_U(t) = \frac{t}{2^n}$ for all t with $0 \le t \le 2^n$. Therefore,

$$\frac{t}{F_U(t)} = 2^n$$

holds for all $t \le 2^n$. On the other hand, let S be any satisfiable formula and let x be any integer such that α_x is a satisfying assignment of S. Then,

$$\frac{x}{F_S(x)} = \frac{x}{x+1} 2^n < 2^n.$$

A SAT instance G is satisfiable if and only if there is a t with $t/F_G(t) < 2^n$. Therefore, RESTART-UPPER-BOUND-CDF is NP-complete. $\qquad \square$

The version shown above is for the computation of an upper bound of the expected value under restart. This raises the question whether the exact version is also computationally hard. The next section addresses this problem. For this, the used function is also allowed to use the exponentiation operation.

RESTART-EXACT-CDF

Input: A cdf $F : \mathbb{N} \mapsto [0, 1]$ and an integer k. The arithmetic circuit of F must be composed of summation, subtraction, multiplication, division and exponentiation operations.

Question: Is there a restart time t such that $E[X_t] < k$?

Theorem 2. *The problem* RESTART-EXACT-CDF *is NP-hard.*

Proof. Let G and α_t be defined as in the proof of Theorem 1 and define the function F_G:

$$F_G(t) = \begin{cases} 0, & t < 1 \\ 1 - 2^{-t - G\alpha_{t-1}}, & 1 \le t \le 2^n \\ 1 - 2^{-t}, & t > 2^n \end{cases} \tag{4}$$

The value of $F(t)$ in standard binary representation is $0.11\ldots1$ with $t + G\alpha_{t-1}$ ones in the decimal places. This can be exponential in the length of the input. However, the value of $F(t)$ can be encoded unambiguously by just saving the number of ones in the decimal places. Thus, a representation of $F(t)$ can be computed in polynomial time. The function F is a cdf which can be verified in the same way as in Theorem 1. We define a random variable X such that F is the cdf of X. If U is an unsatisfiable SAT instance, then F_U is a geometric distribution with success probability 0.5. For this geometric distribution, it is known that $E[X_t] = 2$ for every $t \in \mathbb{N}$. If S is a satisfiable SAT instance and t is an integer such that $G\alpha_{t-1} = 1$, then the expected value $E[X_t]$ is at most

$$E[X_t] \le 2 \cdot \frac{1 - 2^{-t}}{1 - 2^{-t-1}} < 2.$$

Hence, a SAT instance is satisfiable if and only if there is a t with $E[X_t] < 2$. $\square$

Instead of the cdf, it is also possible to use the pmf for both problems. Let RESTART-UPPER-BOUND-PMF and RESTART-EXACT-PMF denote those problems. I.e., instead of a cdf the input consists of an arithmetic circuit calculating a pmf.

Corollary 1. RESTART-UPPER-BOUND-PMF *and* RESTART-EXACT-PMF *are NP-hard.*

Proof. Let F be a cdf which can be evaluated in polynomial time and which is given as an arithmetic circuit. Then, $p(t) = F(t) - F(t - 1)$ is a pmf which can be expressed as an arithmetic circuit. This can be used as a reduction from RESTART-UPPER-BOUND-CDF to RESTART-UPPER-BOUND-PMF and respectively from RESTART-EXACT-CDF to RESTART-EXACT-PMF. $\square$

The results shown so far are for the decision whether a restart strategy should be applied. Another important problem concerning restarts is the choice of a good restart time. In the following, we show that restart times cannot be approximated in polynomial time up to an arbitrary constant c unless P = NP. To make the notion more precise: Let X be a discrete random variable and let t^* be an optimal restart time such that the optimal expected value under restart is $E[X_{t^*}]$. Then, there is no polynomial-time algorithm which finds a restart time t such that $E[X_t] \le c \cdot E[X_{t^*}]$. First, we show that the upper bound $t/F(t)$ is also inapproximable.

Theorem 3. *Let X denote a discrete random variable, p the pmf of X and $\frac{t^*}{F(t^*)}$ its optimal upper bound for the expected value under restart. Let $c > 1$ be any arbitrary constant. Unless $P = NP$, there is no polynomial-time algorithm $\mathcal{A}(p)$ such that $\mathcal{A}(p)$ finds a restart time t such that the upper bound $\frac{t}{F(t)}$ is at most $c\frac{t^*}{F(t^*)}$ with p as input.*

Proof. Let G and α_t be defined as in the proof of Theorem 1. We define the function p_G:

$$p_G(t) = \begin{cases} \frac{G\alpha_{t-1}}{2^n+1}, & 1 \le t \le 2^n \\ \frac{1}{2^n+1}, & t = c \cdot 2^n \\ 0, & \text{else} \end{cases} \tag{5}$$

The function p_G is a pmf since it is non-negative for every t and sums up to at most one. Let U be an unsatisfiable SAT instance and let $F_U(t) = \sum_{i=0}^{t} p_U(i)$ be the cdf which corresponds to p_U. Let l_U denote the minimum of the upper bound $t/F_U(t)$. The value of $F_U(t)$ is greater than zero iff $t \ge c2^n$. Thus, the upper bound is minimal at $t = c2^n$:

$$l_U = \min_t \frac{t}{F_U(t)} = c(2^n + 1)2^n.$$

On the other hand, let S be a satisfiable SAT instance and, again, let F_S be the cdf which corresponds to p_S. Let l_S denote the minimum of the upper bound $t/F_S(t)$. The upper bound is minimal for t such that there is no $x < t$ with $S\alpha_{x-1} = 1$. Then, $(2^n + 1)2^n$ bounds l_S from above:

$$l_S = \min_t \frac{t}{F(t)} \le \frac{2^n}{F(2^n)} \le (2^n + 1)2^n.$$

Hence, a SAT instance G is satisfiable if and only if there is a restart time t with $t/F_G(t) < c(2^n + 1)2^n$.

Let $\mathcal{A}$ be a polynomial-time algorithm which takes a pmf as input such that $\mathcal{A}(p)$ finds a restart time x with

$$\frac{x}{F(x)} \le (c - \varepsilon) \min_t \frac{t}{F(t)},$$

where ε is an arbitrary constant with $0 < \varepsilon < c$. Consider the properties of $\mathcal{A}(p_S)$: The algorithm finds a restart time x such that

$$\frac{x}{F(x)} \le (c - \varepsilon)l_S \le (c - \varepsilon)(2^n + 1)2^n < c(2^n + 1)2^n = l_U.$$

As the constant c can be chosen arbitrarily, the existence of any polynomial-time approximation algorithm implies that checking whether the predicted restart time is less than $c \cdot 2^n$ decides SAT in polynomial time. $\square$

Corollary 2. *Let X denote a discrete random variable, p the pmf of X and $E[X_{t^*}]$ its optimal expected value under restart. Let $c > 1$ be any arbitrary constant. Unless $P = NP$, there is no polynomial-time algorithm $\mathcal{A}(p)$ which always finds a restart time t such that the expected value under restart $E[X_t]$ is at most $cE[X_{t^*}]$ with p as input.*

The proof is analogous to the proof of Theorem 3 and is therefore omitted.

4   The Hardness of the Mean

Valiant [18] introduced the complexity class #P. The following, equivalent definition is from [20].

Definition 3 ([18], [20]). *Let M be any non-deterministic Turing machine and let $acc_M(x)$ denote the number of accepting computations of M on input x. Then #P is given by*

$$\#P = \{f \mid f : \Sigma^* \mapsto \mathbb{N} \text{ such that } f = acc_M \text{ for some polynomial time}$$
$$non\text{-}deterministic \ Turing \ machine \ M\}.$$

In the following, we use *metric reductions* as defined by Krentel.

Definition 4 ([10]). *Let f, h be functions. The function f is **metrically reducible** to h if and only if there are two functions g_1, g_2 which are computable in polynomial time such that $f(x) = g_2(x, h(g_1(x)))$.*

The formal definition of #P-hardness and completeness requires Cook reductions. However, polynomial-time many-one reductions and metric reductions both imply the existence of Cook reductions (compare [15]). Thus, the formal definition of Cook reductions and #P-hardness is omitted. We refer the reader to [18] for both definitions. A canonical #P-complete problem is finding the number of satisfying assignments for a given boolean formula [18]. For the rest of this work, this problem is called #SAT.

> #SAT
> **Input:**   A boolean formula F in conjunctive normal form.
> **Question:** What is the number of satisfying assignments for F?

In the following, #SAT(G) denotes the number of satisfying assignments of a SAT formula G in CNF. Let $\mathcal{A}$ be an algorithm on some input x and let X be a random variable describing the runtime distribution of $\mathcal{A}(x)$. Moorsel and Wolter [14] showed that $\mathcal{A}(x)$ benefits from restarts if and only if there is a $t > 0$ with $E[X] < E[X - t \mid X > t]$. Moreover, the expected runtime $E[X_t]$ with restarts requires the conditional mean $E[X \mid X \leq t]$ (compare Eq. 1). Hence, computing the expected value $E[X]$ and the conditional expected values $E[X \mid X > t]$ and $E[X \mid X \leq t]$ is an important task in the context of restarts. A relevant step in computing the conditional expected value is calculating partial moments: $\mu_X(N) = \sum_{i=0}^{N} i \Pr(X = i)$. This is because $E[X \mid X \leq N] = \frac{\mu_X(N)}{\Pr(X \leq N)}$ and

$E[X \mid X > N] = \frac{E[X] - \mu_X(N)}{\Pr(X > N)}$. More generally, the k-th partial moment $\mu_X^{(k)}(N)$ is defined as:

$$\mu_X^{(k)}(N) = \sum_{i=0}^{N} i^k \Pr(X = i).$$

Here, we address the computational complexity of calculating partial moments. First, we study the complexity of computing the partial expectation $\mu_X^{(1)}(N)$. To achieve a better fit to standard computational models we restrict the allowed functions such that each cdf can easily be mapped to integers. More precisely, only functions of the form $F(i) = \frac{c(i)}{M}$ for some natural number M and some function $c : \mathbb{N} \mapsto \mathbb{N}$ are considered. Then, $ME[X]$ is computed instead of the expected value itself.

PARTIAL-EXPECTATION-CDF

Input: An integer N, a polynomial-time function $c : \{0, \ldots, N\} \mapsto \mathbb{N}$ and an integer M, such that $F(i) = \frac{c(i)}{M}$ is a (partial) cdf.

Question: What is the partial expectation $M\mu_X^{(1)}(N)$, where F defines X.

Here, a partial cdf denotes a function which is monotonically increasing and does not exceed 1.

Theorem 4. PARTIAL-EXPECTATION-CDF *is #P-complete with respect to metric reductions.*

Proof. First, we show that PARTIAL-EXPECTATION-CDF is in #P. Note that F has bounded support due to its definition. The partial expectation is then given by

$$M\mu_X^{(1)}(N) = M \sum_{i=1}^{N}(1 - F(i)) = \sum_{i=1}^{N}(M - c(i)).$$

We design a non-deterministic polynomial time Turing machine T such that the number of accepting computation paths of T is equal to $M\mu_X^{(1)}(N)$. The Turing machine T takes c, N, M as input. A positive integer $i \in \{0, \ldots, N\}$ is chosen non-deterministically and $c(i)$ is evaluated. Then, add $M - c(i)$ accepting paths to T which are identified by additional non-deterministic bits. For this, observe that the difference $M - c(i)$ is a positive number since F is a (partial) cdf. The total number of accepting paths of T is then $\sum_{i=1}^{N}(M - c(i)) = M\mu_X^{(1)}(N)$ which shows that PARTIAL-EXPECTATION-CDF is in #P.

Next, each $c(i)$ is bounded by M because otherwise, F would not be a cdf. Therefore, the computation of $M\mu_X^{(1)}(N)$ is in #P. In the following, we reduce #SAT to PARTIAL-EXPECTATION-CDF to show the #P-hardness. Let G and α_t be defined as in the proof of Theorem 1. The function $F_G(t)$ is given by:

$$F_G(t) = \begin{cases} \frac{(t-1) + G\alpha_{t-1}}{2^n}, & 1 \le t \le 2^n \\ 1, & t > 2^n \end{cases} \tag{6}$$

Let X be the random variable defined by F_G. Here, M is equal to 2^n and N is (at most) 2^n, the function c is given by $c(t) = (t - 1) + G\alpha_{t-1}$ for $t \leq 2^n$ and $c(t) = 2^n$ for $t > 2^n$. Note that $\mu_X^{(1)}(N)$ is equal to the expected value $E[X]$. Observe the value of $2^n E[X]$:

$$2^n E[X] = \sum_{i=1}^{2^n} 2^n - c(i) = 2^{2n} - \sum_{i=1}^{2^n} (i-1) + G\alpha_{i-1}$$

$$= 2^{2n} - 2^{n-1}(2^n - 1) - \sum_{i=0}^{2^n - 1} G\alpha_i$$

$$= 2^{2n} - 2^{n-1}(2^n - 1) - \#SAT(G).$$

Therefore, $\#SAT$ can be metrically reduced to PARTIAL-EXPECTATION-CDF by encoding the SAT instance as a cdf as in F_G for g_1 and setting $g_2(G, k)$ to $-\left(k - 2^{2n} + 2^{n-1}(2^n - 1)\right)$. $\qquad \square$

The problem can also be defined with probability mass functions as input.

PARTIAL-MOMENT-PMF

Input: An integer N, a polynomial-time function $c : \{0, \ldots, N\} \mapsto \mathbb{N}$, an integer k and an integer M, such that $p(i) = \frac{c(i)}{M}$ is a (partial) pmf and such that $i^k \cdot c(i)$ is a natural number for all i.

Question: What is the partial moment $M\mu_X^{(k)}(N)$, where F defines X.

Theorem 5. PARTIAL-MOMENT-PMF   is $\#P$-complete   with   respect   to polynomial-time many-one reductions.

Proof. We show that PARTIAL-MOMENT-PMF is in $\#P$. The partial moment $M\mu_X^{(k)}(N)$ is given by $M \sum_{i=0}^N i^k p(i)$. By definition, $M i^k p(i)$ is a natural number which can be computed in polynomial time. Therefore, PARTIAL-MOMENT-PMF is in $\#P$.

Let G and α_t be defined as in the proof of Theorem 1.

$$p_G(t) = \begin{cases} \frac{G\alpha_{t-1}}{2^n \cdot t^k}, & 1 \leq t \leq 2^n \\ 0, & \text{else} \end{cases} \qquad (7)$$

Here, M and N are 2^n and the function $c(t)$ is given by $\frac{G\alpha_{t-1}}{t^k}$. The function $p_G(t)$ is a partial pmf since the sum $\sum_{t=1}^{2^n} \frac{G\alpha_{t-1}}{2^n \cdot t^k}$ does not exceed 1 for all positive k. Clearly, the product $t^k c(t) = G\alpha_{t-1}$ is a natural number for all t with $1 \leq t \leq 2^n$. The partial expectation $2^n \mu_X^{(k)}(2^n)$ is given by

$$2^n \mu_X^{(k)}(2^n) = \sum_{i=1}^{2^n} i^k \cdot c(i) = \sum_{i=1}^{2^n} G\alpha_{i-1} = \#SAT(G).$$

Therefore, $\#SAT$ is polynomial-time many-one reducible to PARTIAL-MOMENT-PMF which completes the proof. $\qquad \square$

We conclude that the computations required to decide $E[X] < E[X - t \mid X > t]$ and to calculate $E[X_t]$ are $\#P$-complete.

5   Conclusion and Outlook

There are three major questions related to restarts in randomized algorithms:

1. Should the algorithm use a restart strategy?
2. If yes, when should it restart?
3. What is the expected runtime of the restarted process?

In this work, we evaluate the computational complexity of all three problems. We assume that the formula of the underlying probability distribution is known. I.e., the formula of either the probability mass function (pmf) or the cumulative distribution function (cdf) is used as input.

Deciding whether a restart strategy should be used is NP-hard (Theorems 1, 2 and Corollary 1). Finding a good restart time is addressed for the case when the pmf is known.

Theorem 3 and Corollary 2 show that there is no polynomial-time algorithm which only uses the pmf and has the following properties: The algorithm computes a restart time such that its corresponding expected runtime is worse by only a constant factor compared to the best restart strategy. Lastly, an essential step for computing the expected runtime with restarts is the computation of the so-called partial expectation. We show that the computation of the partial expectation is #P-complete (Theorems 4 and 5).

There are some loose ends in this work: We showed that there is no feasible approximation algorithm for the restart time if the pmf is known. However, the question whether there is an approximation scheme in the case when the cdf is known remains unanswered. Furthermore, while the general problems presented in this work are hard, there might be subclasses of problems which can be solved in polynomial time. On the other hand, we showed that RESTART-EXACT-CDF is NP-hard. It would be interesting to find out more about its computational complexity. We believe the problem could be undecidable if the support of the cdf is unbounded.

Moreover, we assume that the formulas provided as input describe cdfs or pmfs. Therefore, the properties proven in this work can be viewed in the context of promise problems. In fact, it is a non-trivial task to check whether the input formula is indeed either a cdf or a pmf.

Furthermore, to the best of our knowledge, the properties of probability distributions have not been studied before in the setting of computational complexity if the distribution is known as a formula. There are other attributes which could be analyzed, e.g., the hazard rate or the computation of quantiles.

References

1. Arbelaez, A., Truchet, C., O'Sullivan, B.: Learning sequential and parallel runtime distributions for randomized algorithms. In: ICTAI 2016, San Jose, California, USA, pp. 655–662. IEEE (2016)

2. Batu, T., Fortnow, L., Rubinfeld, R., Smith, W., White, P.: Testing that distributions are close. In: Proceedings of the Foundations of Computer Science, pp. 259–269. IEEE (2000)
3. Biere, A., Fröhlich, A.: Evaluating CDCL restart schemes. In: Proceedings of the International Workshop on Pragmatics of SAT (POS 2015) (2015)
4. Biere, A., Heule, M., van Maaren, H.: Handbook of Satisfiability. IOS Press, Amsterdam (2009)
5. Bshouty, D., Bshouty, N.H.: On interpolating arithmetic read-once formulas with exponentiation. J. Comput. Syst. Sci. **56**(1), 112–124 (1998)
6. Canonne, C.: A survey on distribution testing: your data is big. But is it blue? In: Electronic Colloquium on Computational Complexity (ECCC), vol. 22, no. 63 (2015)
7. Canonne, C., Diakonikolas, I., Gouleakis, T., Rubinfeld, R.: Testing shape restrictions of discrete distributions. Theor. Comput. Syst. **62**(1), 4–62 (2018)
8. Frost, D., Rish, I., Vila, L.: Summarizing CSP hardness with continuous probability distributions. In: Proceedings of the AAAI 1997/IAAI 1997, pp. 327–333. AAAI Press (1997)
9. Gomes, C., Selman, B., Kautz, H.: Boosting combinatorial search through randomization, pp. 431–437. AAAI Press (1998)
10. Krentel, M.: The complexity of optimization problems. J. Comput. Syst. Sci. **36**(3), 490–509 (1988)
11. Lorenz, J.-H.: Runtime distributions and criteria for restarts. In: Tjoa, A.M., Bellatreche, L., Biffl, S., van Leeuwen, J., Wiedermann, J. (eds.) SOFSEM 2018. LNCS, vol. 10706, pp. 493–507. Springer, Cham (2018). https://doi.org/10.1007/978-3-319-73117-9_35
12. Luby, M., Sinclair, A., Zuckerman, D.: Optimal speedup of Las Vegas algorithms. Inf. Process. Lett. **47**(4), 173–180 (1993)
13. Mengshoel, O., Wilkins, D., Roth, D.: Initialization and restart in stochastic local search: computing a most probable explanation in bayesian networks. IEEE Trans. Knowl. Data Eng. **23**(2), 235–247 (2011)
14. Moorsel, A., Wolter, K.: Analysis and algorithms for restart. In: Proceedings of the First International Conference on the Quantitative Evaluation of Systems, pp. 195–204 (2004)
15. Pagourtzis, A., Zachos, S.: The complexity of counting functions with easy decision version. In: Královič, R., Urzyczyn, P. (eds.) MFCS 2006. LNCS, vol. 4162, pp. 741–752. Springer, Heidelberg (2006). https://doi.org/10.1007/11821069_64
16. Papadimitriou, C.: Computational Complexity. Addison Wesley Pub. Co., Boston (1994)
17. Schöning, U.: A probabilistic algorithm for k-SAT and constraint satisfaction problems. In: 40th Annual Symposium on Foundations of Computer Science, pp. 410–414 (1999)
18. Valiant, L.: The complexity of computing the permanent. Theoret. Comput. Sci. **8**(2), 189–201 (1979)
19. Vollmer, H.: Introduction to Circuit Complexity: A Uniform Approach. Texts in Theoretical Computer Science, An EATCS Series. Springer, Heidelberg (1999). https://doi.org/10.1007/978-3-662-03927-4
20. Welsh, D.: Complexity: Knots, Colourings and Countings. Cambridge University Press, New York (1993)

On the Complexity
of MIXED DOMINATING SET

Jayakrishnan Madathil[1], Fahad Panolan[2], Abhishek Sahu[1(✉)],
and Saket Saurabh[1,2]

[1] The Institute of Mathematical Sciences, HBNI, Chennai, India
{jayakrishnanm,asahu,saket}@imsc.res.in
[2] Department of Informatics, University of Bergen, Bergen, Norway
fahad.panolan@uib.no

Abstract. A mixed dominating set (mds) of a graph G is a set $S \subseteq V(G) \cup E(G)$ such that every element $x \in (V(G) \cup E(G)) \setminus S$ is either adjacent to or incident with an element of S. In the MIXED DOMINATING SET (MDS) problem, we are given an n-vertex graph G and a positive integer k, and the objective is to decide whether G has an mds of size at most k. On general graphs, MDS parameterized by k is fixed-parameter tractable, but has no polynomial kernel unless coNP $\subseteq$ NP/Poly. In this paper, we study the restriction of MDS to several graph classes and establish the following results.
- On proper interval graphs, MDS is polynomial time solvable.
- On graphs that exclude $K_{d,d}$ as a subgraph, MDS admits a kernel of size $\mathcal{O}(k^d)$.
- On split graphs, MDS does not admit a polynomial kernel unless coNP $\subseteq$ NP/Poly.

In addition, we show that on general graphs, MDS admits an exact algorithm with running time $2^n n^{\mathcal{O}(1)}$.

1 Introduction

MIXED DOMINATING SET (MDS) bears a close resemblance to two rather well-studied problems in the area of graph algorithms—VERTEX COVER and DOMINATING SET—and yet diverges significantly from both these problems. (See Table 1.) In this article, we try to further advance a study of MDS that had previously been undertaken by three of the present authors along with Jain [14]. While self-contained, this article is meant to be a follow-up on [14], and answers several questions left open by [14]. The reader is encouraged to peruse [14] for a comprehensive understanding of MDS.

We first define the concept of *domination* in a graph, that is, what a vertex or an edge can dominate. A *vertex dominates* itself, all its neighbors and all the edges incident with it. An *edge dominates* its two endpoints as well as all the

This work is supported by the European Research Council (ERC) via grant LOPPRE, reference 819416 and Norwegian Research Council via project MULTIVAL.

R. van Bevern and G. Kucherov (Eds.): CSR 2019, LNCS 11532, pp. 262–274, 2019.
https://doi.org/10.1007/978-3-030-19955-5_23

Table 1. Complexity of VERTEX COVER, DOMINATING SET and MIXED DOMINATING SET on various graph classes. Results proved in this paper are highlighted.

Graph class	VERTEX COVER	DOMINATING SET	MIXED DOMINATING SET
General graphs	NP-hard FPT $\mathcal{O}(k)$ kernel	NP-hard W[2]-hard	NP-hard FPT No poly kernel
Bipartite graphs	P	NP-hard W[2]-hard	NP-hard FPT No poly kernel
Split graphs	P	NP-hard W[2]-hard	NP-hard FPT No poly kernel
Interval graphs	P	P	open
Proper interval graphs	P	P	P
$K_{d,d}$-free graphs	NP-hard FPT $\mathcal{O}(k)$ kernel	NP-hard FPT $\mathcal{O}(k^d)$ kernel	NP-hard FPT $\mathcal{O}(k^d)$ kernel

edges incident with either of its endpoints. A mixed dominating set (mds) of a graph G is a set $S \subseteq V(G) \cup E(G)$ that dominates all vertices and edges of G, i.e., for every $x \in (V(G) \cup E(G)) \setminus S$, x is either adjacent to or incident with an element of S. This definition raises the obvious computational question of testing whether it is possible to dominate the entire graph with a given number of vertices and edges. We study this problem in the frameworks of classical and parameterized complexity. (See, for instance, [4] for the fundamentals of parameterized algorithms.) Accordingly, the problem MIXED DOMINATING SET is defined as follows.

MIXED DOMINATING SET (MDS) **Parameter:** k
Input: An undirected graph G and a non-negative integer k.
Question: Is there a mixed dominating set of size at most k in G?

Previous Work. The concept of mixed dominating set was first introduced by Alavi et al. [1] as a generalization of matching and covering, and since then it has been studied extensively in graph theory [2,9,12,18,19]. On the algorithms and complexity side, Majumdar [16] showed that MIXED DOMINATING SET is NP-complete on general graphs and admits a linear-time algorithm on trees. Hedetniemi et al. [13] showed that MDS remains NP-complete on bipartite and chordal graphs. Manlove [17] and Zhao et al. [21] showed that MDS remains NP-complete on planar bipartite graphs of maximum degree 4 and split graphs, respectively. Lan and Chang [15] gave a linear time algorithm for MDS on cacti.

(A cactus is an undirected graph in which any two cycles have at most one vertex in common.) Hatami [11] gave a factor 2 approximation algorithm for MDS on general graphs.

Even as these results were emerging, MDS received scant attention from the parameterized complexity community until Rajaati et al. [20] studied MDS parameterized by the treewidth of the input graph and designed an algorithm with running time $\mathcal{O}^\star(3^{\mathsf{tw}(G)^2})$.[1] This was soon followed by Jain et al. [14], who studied MDS with respect to two different parameters—the solution size k and the treewidth of the input graph tw. They showed that MDS parameterized by k can be solved in time $\mathcal{O}^\star(7.465^k)$, and hence is fixed-parameter tractable, but admits no polynomial kernel unless coNP $\subseteq$ NP/poly. They also showed that MDS parameterized by tw can be solved in time $\mathcal{O}^\star(6^{\mathsf{tw}})$.

We would like to mention that MDS is equivalent to DISTANCE 2-DOMINATION on the incidence graph of the input graph. As a result, MDS when parameterized by the solution size admits (i) a linear bi-kernel on graphs of bounded genus by the meta kernelization result of Bodlaender et al. [3], (ii) a linear bi-kernel on any graph class of bounded expansion [7], and (iii) a pseudo-linear bi-kernel of size $f(r, \epsilon) \cdot k^{1+\epsilon}$ on every nowhere dense class, for some function f and any $\epsilon > 0$ [8]. But, since DISTANCE 2-DOMINATION is $W[2]$-hard on d-degenerate graphs (and hence on $K_{d,d}$-free graphs) [10], this equivalence does not imply anything meaningful about the kernelization complexity of MDS on $K_{d,d}$-free graphs.

Our Results and Methods. We study MDS (parameterized by k) on various graph classes and prove the following results.

- On proper interval graphs, MDS is polynomial time solvable. We utilize the property of proper interval graphs' admitting a *clique-partition*—a partition of the vertex set into an ordered sequence such that the graph induced by each part is a clique, and each edge of the graph is contained within one clique or between consecutive cliques. Then we do dynamic programming over the clique-partition to show polynomial time solvability of MDS.
- On graphs that do not contain $K_{d,d}$ as a subgraph (biclique-free graphs), MDS admits a kernel of size $\mathcal{O}(k^d)$. Biclique-free graphs contain well known sparse graph classes such as graphs of bounded expansion and nowhere dense graphs. Our kernel relies on a crucial relationship between vertex cover and mixed dominating set, namely, if a graph G has an mds of size at most k, then G has a vertex cover of size at most $2k$.
- On split graphs, MDS does not admit a polynomial kernel unless coNP $\subseteq$ NP/Poly. The proof is by a reduction from the RED-BLUE DOMINATING SET problem, parameterized by the number of red vertices.
- We use the standard branching technique to design an exact algorithm with running time $2^n n^{\mathcal{O}(1)}$ for MDS on general graphs.

---

[1] $\mathcal{O}^\star$ notation suppresses the polynomial factor. That is, $\mathcal{O}(f(k)n^{\mathcal{O}(1)}) = \mathcal{O}^\star(f(k))$.

2   Preliminaries

For a positive integer q, $[q]$ denotes the set $\{1, 2, \ldots, q\}$. All graphs in this paper are simple and undirected. For a graph G, we let $V(G)$ denote the vertex set of G and $E(G)$ the edge set of G. For sets $X, Y \subseteq V(G)$, we denote by $E(X, Y)$ the set of edges in G that have one endpoint in X and the other in Y, and $E(X) = E(X, X)$. For a set $E' \subseteq E(G)$, we denote by $V(E')$ the set of vertices that are the endpoints of the edges in E'. For a set $S \subseteq V(G) \cup E(G)$, we define $V(S) = (S \cap V(G)) \cup V(S \cap E(G))$. For $v \in V(G)$, $N_G(v)$ denotes the set of neighbors of v, and $N_G[v] = N_G(v) \cup \{v\}$. Similarly, for a subset $V' \subseteq V(G)$, $N_G(V') = (\bigcup_{v \in V'} N_G(v)) \setminus V'$ and $N_G[V'] = N_G(V') \cup V'$. We write $N(v)$ for $N_G(v)$ when the underlying graph G is clear from the context. Also, for $V' \subseteq V(G)$, we denote by $G[V']$, the subgraph of G induced by V'. We direct the reader to the book by Diestel [5] for any graph terminology or notations that are not explicitly defined here.

The following observation follows directly from the definition of a mixed dominating set, and we shall use it throughout the paper.

Observation 1. *Let G be a graph and $V' \cup E'$ be an mds of size k, where $V' \subseteq V(G)$ and $E' \subseteq E(G)$. Let $E_1 \subseteq E'$. Then, (i) $V' \cup V(E')$ is a vertex cover of G of size at most $2k$, (ii) any vertex v of degree at least $2k + 1$ belongs to $V' \cup V(E')$, (iii) $V' \cup (E' \setminus E_1) \cup E_2$ is an mds of G for any $E_2 \subseteq E(G)$ with $V(E_2) = V(E_1)$, and (iv) if G is a complete graph, then $V' \cup V(E')$ must contain all but one vertices of G.*

Now we define a *special mixed dominating set* and prove that there is in fact an optimum mds which is also a special mds.

Definition 1. *A mixed dominating set $V' \cup E'$ of a graph G, where $V' \subseteq V(G)$ and $E' \subseteq E(G)$ is said to be a special mixed dominating set if E' is a matching and $V' \cap V(E') = \emptyset$.*

Observation 2. *A graph G has an mds of size k if and only if G has a special mds of size k.*

Proof. Among all the k-sized mixed dominating sets of G, let S be an mds such that $|E(G) \cap S|$ is minimum. We claim that S is a special mds. If S contains two edges uv and uw, then $(S \setminus \{uv\}) \cup \{v\}$ is also an mds of size k with less number of edges than that in S. If S contains a vertex v and an edge uv, then $(S \setminus \{uv\}) \cup \{u\}$ is also an mds of size k with less number of edges than that in S. In either case, we have a contradiction to the assumption. $\square$

3   MDS **Restricted to Proper Interval Graphs Is in** P

In this section, we briefly outline our strategy for proving the following theorem. The complete proof can be found in the full version.

Theorem 1. MDS *on proper interval graphs can be solved in time* $\mathcal{O}(n^{13})$, *where n is the number of vertices of the input graph.*

Consider a proper interval graph G. It is known that $V(G)$ can be partitioned into an ordered sequence $\mathcal{Q} = (Q_1, Q_2, \ldots, Q_q)$ such that (i) for each $i \in [q]$, $G[Q_i]$ is a clique, and (ii) for each edge uv of G, there exists $i \in [q]$ such that either $u, v \in Q_i$ or $u \in Q_i, v \in Q_{i+1}$. Such an ordered sequence $\mathcal{Q}$ is called a *clique-partition* of G. And given a proper interval graph G, a clique-partition G can be computed in polynomial time. From now on, we assume that we have found such a clique-partition $\mathcal{Q}$ of our input graph G.

Consider G and Q. For $i \in [q]$, let $Q_{\leq i}$ denote $Q_1 \cup Q_2 \cup \cdots \cup Q_i$. For an mds S of G, and for $i \in [q]$, let $S_i = V_i \cup E_i$ be the set of all vertices and edges in S that have at least one endpoint in $Q_{\leq i}$. We can show that G has a minimum-sized special mixed dominating set $S = V' \cup E'$ with the following properties.

(PI) For every $i \in [q-1]$, $|S \cap E(Q_i, Q_{i+1})| \leq 1$, i.e., S contains at most one edge with one endpoint in Q_i and the other in Q_{i+1}.

(PII) For every $i \in [q]$, $|S \cap Q_i| \leq 3$, i.e., S contains at most 3 vertices from Q_i.

(PIII) For every $i \in [q]$, $|Q_{\leq i} \setminus \{N[V_i] \cup V[E_i]\}| \leq 1$, i.e., at most one vertex in $Q_{\leq i}$ is not dominated by S_i.

(PIV) For every $i \in [q]$, $|Q_i \setminus V(S_i)| \leq 1$, i.e., Q_i contains at most one vertex, say v, such that S_i does not contain v or any edge incident on v.

Our goal is to find such an mds S. Now consider set of edges of $\hat{E} = E(Q_i) \cap S$, and the set of endpoints of $\hat{E}$, $V(\hat{E})$. First, since S is a special mds, $\hat{E}$ is a matching. Second, consider $G[V(\hat{E})]$, the clique induced on $V(\hat{E})$. Let $\hat{P}$ be any perfect matching in $G[V(\hat{E})]$. Then $\hat{S} = (S \setminus \hat{E}) \cup \hat{P}$ is also a minimum sized special mds of G. That is, in S, if we replace $\hat{E}$ with any other perfect matching in $G[V(\hat{E})]$ we will still have a minimum sized special mds of G. Finally, properties (PI)-(PIV) imply that $Q_i \setminus V(\hat{E})$ contains at most 6 vertices.

Based on these observations, we do dynamic programming as follows. Let S, as described above be the optimal solution that we are seeking. At stage i, we consider the subgraph of G induced on $Q_{\leq i}$. As mentioned above, $|Q_i \setminus V(\hat{E})| \leq 6$. We "guess" these six vertices. And for each such guess, we compute a partial solution, one of which will eventually lead to S.

Remark. As an aside, we record that the complexity of MDS on interval graphs is still open. While the vertex set of an interval graph admits an ordered partition so that each part induces a clique, we cannot ensure that the edges are within one clique or between consecutive cliques. For this reason, an approach similar to the one in the algorithm above may not work.

4   Polynomial Kernel for MDS on $K_{d,d}$-free Graphs

In this section, we show that MIXED DOMINATING SET admits a kernel with $\mathcal{O}(k^d)$ vertices on $K_{d,d}$-free graphs. A graph G is said to be $K_{d,d}$-free if G does not contain $K_{d,d}$ as a subgraph (not necessarily induced). Before focusing exclusively on $K_{d,d}$-free graphs, we explore some structural properties of graphs that have mixed dominating sets of size at most k, for a non-negative integer k. For a graph G and a non-negative integer k, we define the k-*induced partition* of G as follows.

Definition 2. *Let G be a graph without isolated vertices and k a non-negative integer. The k-induced partition of G is a triplet (H, I, R), where $H = \{v \in V(G) \mid d_G(v) \geq 2k + 1\}$, $I = \{v \in V(G) \setminus H \mid N_G(v) \subseteq H\}$ and $R = V(G) \setminus (H \cup I)$.*

Notice the following immediate consequences of the above definition. First, the k-induced partition is unique, and that $H \cup I \cup R$ is indeed a partition of $V(G)$. Also, by definition, I is an independent set (i.e., $E(I) = \emptyset$) and there is no edge in G with one endpoint in I and the other in R (i.e., $E(I, R) = \emptyset$). Finally, it is easy to see that H must be contained in every vertex cover of G of size at most $2k$, if such a vertex cover exists. We now define what we call a k-*induced mixed dominating set*.

Definition 3. *Let G be a graph without isolated vertices and k be a non-negative integer. Let (H, I, R) be the k-induced partition of G. A mixed dominating set S of G is called a k-induced mixed dominating set if $V(S) \cap I = \emptyset$.*

The following lemma shows that if G has an mds of size at most k, then G has a k-induced mds.

Lemma 1. *Let G be a graph without isolated vertices and k a non-negative integer. Suppose (G, k) is a yes-instance of MDS. Then G has a k-induced mds.*

Proof. Let S' be a minimum-sized mds of G. Since (G, k) is a yes-instance, by Observation 1, we have that $H \subseteq V(S')$. We construct S from S' as follows. Add all vertices of $(H \cup R) \cap S'$ to S. Add all edges of $E(G[H \cup R]) \cap S'$ to S. For every edge $xy \in E(H, I) \cap S'$, where $x \in H$ and $y \in I$, add x to S. For every vertex $y \in S' \cap I$, add a neighbor of y to S. Then, $|S| \leq |S'|$ and $V(S) \cap I = \emptyset$. We claim that S is a mixed dominating set of G. Consider $x \in R \cup H$. Note that our construction of S ensures that if $x \in V(S')$, then $x \in V(S)$ as well. Thus S dominates $H \cup R$ and all edges incident with $H \cup R$. Therefore, in order to show that S is an mds of G, we only need to show that S dominates I. Consider $y' \in I$. Then, since S' dominates y', either $x'y' \in S'$ for some $x' \in H$, in which case $x' \in S$, and hence S dominates y'; or $y' \in S'$, in which case we added a neighbor of y' to S, and hence S dominates y; or S' contains a neighbor of y', say $z' \in H$, in which case S contains z' too, and hence S dominates y'. Thus S is a minimum-sized mds of G and $V(S) \cap I = \emptyset$. Hence, S is a k-induced mds of G. This completes the proof of the lemma.   □

In what follows, let (G, k) be an instance of MDS where G is a $K_{d,d}$-free graph. Note that all isolated vertices in G must belong to any mixed dominating set. Consequently, we apply the following reduction rule.

Reduction Rule 1: If $v \in V(G)$ is an isolated vertex, then delete v from G and decrease k by one.

From now on, we assume that G has no isolated vertices. Let (H, I, R) be the k-induced partition of G. We now separately bound the sizes of H, R and I in the event that (G, k) is a yes-instance. The proof of the following lemma, which bounds $|H|$, follows from Observation 1.

Lemma 2. *If* $|H| > 2k$, *then* (G, k) *is a no-instance.*

Lemma 3. *If* $|R| > 8k^2$, *then* (G, k) *is a no-instance.*

Proof. Assume that (G, k) is a yes-instance. Let S be an mds of G of size at most k. We shall show that $|R| \leq 8k^2$. Note that $V(S)$ is a vertex cover of G. Now, consider the graph $G[R]$, the subgraph of G induced by R. Then, $V(S) \cap R$ is a vertex cover of $G[R]$. Since every vertex in R has degree at most $2k$, $V(S) \cap R$ can cover at most $2k|V(S) \cap R|$ edges of $G[R]$. Since $V(S) \cap R$ is a vertex cover of $G[R]$, $V(S) \cap R$ covers all edges of $G[R]$. We thus get that $|E(G[R])| \leq 2k|V(S) \cap R| \leq 2k|V(S)| \leq 2k \times 2k = 4k^2$.

Recall that Rule 1 removed all isolated vertices from G. Observe that $G[R]$ has no isolated vertices either. If $v \in R$ were isolated in $G[R]$, then $N_G(v) \subseteq H$, which then would imply that $v \in I$, a contradiction. Since $G[R]$ has no isolated vertices and $|E(G[R])| \leq 4k^2$, we have $|V(G[R])| \leq 2|E(G[R])| \leq 8k^2$.   □

As Lemmas 2 and 3 show, if either $|H| > 2k$ or $|R| > 8k^2$, then we can immediately conclude that (G, k) is a no-instance. This leads to the following reduction rule.

Reduction Rule 2: Let (G, k) be an instance of MDS and let (H, I, R) be the k-induced partition of $V(G)$. If $|H| > 2k$ or $|R| > 8k^2$, then return a trivial no-instance.

Hence from now on, we assume that $|H| \leq 2k$ and $|R| \leq 8k^2$. In order to obtain a kernel, we now need to upper bound $|I|$. Towards that end, we introduce the following reduction rule.

Reduction Rule 3: If there exist $2k + 2$ distinct vertices $u_0, u_1, \ldots, u_{2k+1}$ in I such that $N_G(u_0) = N_G(u_1) = \cdots = N_G(u_{2k+1})$, then delete u_0 from G.

Lemma 4. *Reduction Rule 3 is safe.*

Proof. Consider $u_0, u_1, \ldots, u_{2k+1} \in I$ with $N_G(u_0) = N_G(u_1) = \cdots = N_G(u_{2k+1})$. We shall show that (G, k) is a yes-instance if and only if $(G - u_0, k)$ is a yes-instance. Now, if (G, k) is a yes-instance, then by Lemma 1, G has a k-induced mds, say S, of size at most k. Then $u_0 \notin V(S)$, and hence S is an mds of $G - u_0$ as well.

To see the reverse direction, assume that $(G - u_0, k)$ is a yes-instance. Then, since $d_{G-u_0}(x) \geq 2k + 1$ for every $x \in H$, $(H, I \setminus \{u_0\}, R)$ is the k-induced partition of $G - u_0$. By Lemma 1, $G - u_0$ has a k-induced mds, say S', of size at most k. That is, $V(S') \cap (I \setminus \{u_0\}) = \emptyset$. We claim that S' is a mixed dominating set of G as well. Note that we only need to show that S' dominates u_0 and all edges incident with u_0. By Observation 1, we have $H \subseteq V(S')$. In particular, $N_G(u_0) \subseteq H \subseteq V(S')$, and hence S' dominates all edges incident with u_0. Now, since S' dominates u_1, and since S' is k-induced, S' must contain a neighbor of u_1, say v. But we have $N_G(u_0) = N_G(u_1)$. Thus $v \in N_G(u_0) \cap S'$, and hence S' dominates u_0. $\qquad\square$

It is easy to see that all the Reduction Rules 1–3 can be applied in polynomial time. Assume that Reduction Rules 1–3 have been applied exhaustively. Let (G', k') be the reduced instance with the k'-induced partition (H, I, R). Notice that $k' \leq k$. We partition I into two parts as follows. Let $I_{<d}$ be the set of vertices in I that have at most $(d-1)$ neighbors and $I_{\geq d}$ be the set of vertices in I that have at least d neighbors. That is, $I_{<d} = \{v \in I \mid |N_{G'}(v)| \leq d - 1\}$ and $I_{\geq d} = I \setminus I_{<d} = \{v \in I \mid |N_{G'}(v)| \geq d\}$. We shall separately bound the sizes of $I_{<d}$ and $I_{\geq d}$. Let $\mathcal{H}_{<d}$ be the family of subsets of H of size at most $(d-1)$, i.e., $\mathcal{H}_{<d} = \{X \subset H : |H| \leq d - 1\}$. Let $\mathcal{H}_{=d}$ be the family of subsets of H of size exactly d, i.e., $\mathcal{H}_{=d} = \{X \subset H : |H| = d\}$.

Observation 3. $|\mathcal{H}_{<d}| \leq d(2k')^{d-1}$ and $|\mathcal{H}_{=d}| \leq (2k')^d$.

Lemma 5. $|I_{<d}| \leq d(2k' + 1)(2k')^{d-1}$ and $|I_{\geq d}| \leq (d-1)(2k')^d$.

Proof. Since Reduction Rule 3 has been applied exhaustively, for every $X \in \mathcal{H}_{<d}$ there exist at most $2k'+1$ vertices $x \in I_{<d}$ such that $N_{G'}(x) = X$. Hence, $|I_{<d}| \leq (2k' + 1)|\mathcal{H}_{<d}|$, and then using Observation 3, we get $|I_{<d}| \leq (2k' + 1)d(2k')^{d-1}$.

Note that for every $x \in I_{\geq d}$, there exists $X \in \mathcal{H}_{=d}$ such that $N_{G'}(x) \supseteq X$; and call x and X *partners* of each other. Now, given $Y \in \mathcal{H}_{=d}$, note that Y can have at most $(d-1)$ partners in $I_{\geq d}$. For otherwise, if Y has at least d partners, then the graph induced on Y and all its partners contains $K_{d,d}$ as a subgraph. But this is not possible as G is $K_{d,d}$-free. Thus, every $x \in I_{\geq d}$ has a partner in $\mathcal{H}_{=d}$, and every $Y \in \mathcal{H}_{=d}$ has at most $(d-1)$ partners. We thus have $|I_{\geq d}| \leq (d-1)|\mathcal{H}_{=d}| \leq (d-1)(2k')^d$. $\qquad\square$

Lemma 5 shows that $|I| = |I_{<d} \cup I_{\geq d}| \leq d(2k' + 1)(2k')^{d-1} + (d-1)(2k')^d = (2k')^{d-1}(4dk' - 2k' + d)$. Thus, as $k' \leq k$, we get the following theorem.

Theorem 2. MIXED DOMINATING SET *on $K_{d,d}$-free graphs admits a kernel with $(2k)^{d-1}(4dk - 2k + d) + 8k^2 + 2k = \mathcal{O}(k^d)$ vertices.*

5   MDS on Split Graphs has No Polynomial Kernel

In this section, we show that MDS restricted to split graphs does not admit a polynomial kernel unless $\text{coNP} \subseteq \text{NP/poly}$. The proof is by a polynomial parameter transformation from RED-BLUE DOMINATING SET (RBDS), parameterized by the number of red vertices.

In the RBDS problem, the input is a bipartite graph G with bipartition $R \cup B$ and a positive integer ℓ. The objective is to test whether there exists a set $X \subseteq R$ of size at most ℓ that dominates the set B, i.e., $N(X) = B$. Such a set X is called a red-blue dominating set (rbds, for short) of G. This problem when parameterized by $|R|$ is the same as SMALL UNIVERSE HITTING SET, which does not have a polynomial kernel unless coNP $\subseteq$ NP/poly (see [6]). Thus we get the following result from [6].

Lemma 6 ([6]). RBDS *parameterized by* $|R|$ *has no polynomial kernel unless* coNP $\subseteq$ NP/poly.

We first state an auxiliary lemma, (the proof of which follows from Observation 1) and then prove the main theorem of the section.

Lemma 7. *Let G be a graph and let $X \subseteq V(G)$ be such that $G[X]$ is a complete graph. Let $S \subseteq V(G) \cup E(G)$ be such that S dominates all edges of $G[X]$. Then, $|S| \geq (|X| - 1)/2$.*

Theorem 3. MDS *on split graphs does not admit a polynomial kernel, unless* coNP $\subseteq$ NP/poly.

Proof. The proof is by a polynomial parameter transformation from the RED-BLUE DOMINATING SET problem, parameterized by the number of red vertices. Consider an instance (G, ℓ) of RBDS, where $V(G) = R \cup B$. We assume that the instance (G, ℓ) has the following properties: (i) the set R contains an isolated vertex, and (ii) $|R|$ is odd and ℓ is even. It is easy to verify that these assumptions are safe.

We construct an equivalent instance (G', ℓ') of MDS as follows. Let G' be the graph obtained from G by turning $G[R]$ into a clique. That is, $V(G') = V(G) = R \cup B$ and $E(G') = E(G) \cup \{r_1 r_2 \mid r_1, r_2 \in R\}$. Note that G' is a split graph. We set $\ell' = (|R| + \ell - 1)/2$. We claim that G has an rbds of size at most ℓ if and only if G' has an mds of size at most $\ell' = (|R| + \ell - 1)/2$.

Assume that (G, ℓ) is a yes-instance of RBDS and let $X \subseteq R$ be an rbds of G of size ℓ. Let $v \in R$ be an isolated vertex in G. If $v \in X$, then replace v with any other vertex in $R \setminus X$, which is still an rbds of G. So from now on, we assume that $v \notin X$. Now note that $|R \setminus (X \cup \{v\})| = |R| - (\ell + 1)$ is even and $G'[R \setminus (X \cup \{v\})]$ is a complete graph and hence $G'[R \setminus (X \cup \{v\})]$ admits a perfect matching M. Then, $|M| = (|R| - \ell - 1)/2$ and $|X \cup M| = |X| + |M| = \ell + (|R| - \ell - 1)/2 = (|R| + \ell - 1)/2$. We claim that $X \cup M$ is an mds of G'. Note that X dominates all vertices in B, as X is an rbds of G. Since $G'[R]$ is a clique, X dominates all vertices of R. (In fact, any one vertex of X dominates all of R.) The set X also dominates all edges of G' that are incident with X. Thus, the only elements of $V(G') \cup E(G')$ that are not dominated by X are the edges incident with $R \setminus X$, but not incident with X. Now note that, since M is a perfect matching of $G'[R \setminus (X \cup \{v\})]$, we have $V(M) = R \setminus (X \cup \{v\})$, and hence M dominates all edges incident with $R \setminus (X \cup \{v\})$. Notice that all the edges incident with v and X are dominated by X. This implies that $M \cup X$ dominates all edges incident with $R \setminus X$. Thus, $X \cup M$ is an mds of G' of size $(|R| + \ell - 1)/2 = \ell'$.

Conversely, assume that G does not have an rbds of size at most ℓ. Let $S = V' \cup E'$ be a minimum-sized mds of G', where $V' \subseteq V(G')$ and $E' \subseteq E(G')$. We shall show that $|S| > \ell'$. Let $S' \subseteq S$ be a minimum-sized subset of S that dominates all vertices in B. Thus, S' consists of some vertices in $R \cup B$ and some edges in $E(R, B)$. Construct a set $S'' \subseteq R$ as follows. Add all vertices in $S' \cap R$ to S''. For every $xy \in S' \cap E(R, B)$, where $x \in R$ and $y \in B$, add x to S''. For every $y \in S' \cap B$, add a neighbor of y to S''. Then, $|S''| \leq |S'|$ and $S'' \supseteq V(S') \cap R$.

We claim that S'' is an rbds of G. To see this, consider $w \in B$. We shall show that $w \in N_G(S'')$. Since S is an mds of G', (i) either $w \in V'$, in which case we added a neighbor of w to S'', and hence $w \in N_G(S'')$, or (ii) S' contains an edge incident with w, say uw, in which case we added u to S'', and hence $w \in N_G(S'')$, or (iii) S' contains a neighbor (in G') of w, say u, and then $u \in S' \cap R \subseteq S''$, in which case also $w \in N_G(S'')$. Thus S'' is an rbds of G. Because of our assumption that G has no mds of size at most ℓ, we have $\ell < |S''| \leq |S'|$.

Now consider the subgraph $G'[R \setminus S'']$ of G'. Note that $G'[R \setminus S'']$ is a clique and none of its edges are dominated by S', because $S'' \supseteq V(S') \cap R$. Equivalently, (since S is an mds of G',) all edges of the clique $G'[R \setminus S'']$ are dominated by $S \setminus S'$. Therefore, by Lemma 7, $|S \setminus S'| \geq (|R \setminus S''| - 1)/2 = (|R| - |S''| - 1)/2$. Hence,

$$
\begin{aligned}
|S| &= |S'| + (|R| - |S''| - 1)/2 \\
&\geq |S'| + (|R| - |S'| - 1)/2 \quad \text{(because } |S''| \leq |S'|) \\
&= (|R| + |S'| - 1)/2 \\
&> (|R| + \ell - 1)/2 \quad\quad \text{(because } |S'| \geq |S''| > \ell) \\
&= \ell'.
\end{aligned}
$$

That is $|S| > \ell'$. This completes the proof of the lemma. $\square$

6   Exact Exponential Time Algorithm for MDS

In this section, we design an exponential time algorithm that computes the size of a minimum mixed dominating set of an n-vertex graph in time $2^n n^{\mathcal{O}(1)}$. The problem is formally defined as follows.

MINIMUM MIXED DOMINATING SET (MIN MDS)
Input: An undirected graph G.
Output: The size of a minimum mixed dominating set of G.

Let G be an n-vertex graph. In light of Observation 2, in order to find a minimum-sized mds of G, we can restrict the search space to the collection of all special mixed dominating sets of G. Moreover, by Observation 1 (iii) if $S = V' \cup E'$ is a special mixed dominating set of G, then for any arbitrary perfect matching E'' of $G[V(E')]$, $V' \cup E''$ is a special mixed dominating set of size $|S|$. Hence, to test whether G has an mds of size ℓ, it is enough to check the existence of a partition (V_1, V_2, V_3) of $V(G)$ such that $|V_1| + \frac{|V_2|}{2} = \ell$ and $V_1 \cup E_2$ is an mds of G, where E_2 is an arbitrary perfect matching in $G[V_2]$. Thus, our search space here is the collection of all partitions of $V(G)$ into at most three parts, which is upper bounded by 3^n. Now by checking if each partition corresponds to a special mds of G, (which can be done in polynomial time because a maximum matching can be found in polynomial time), and by finding the minimum-sized one among the special mixed dominating sets, we get the required result.

Our goal here is to design a faster exponential time algorithm. We now state the following two lemmas that form the basis of our algorithm. Their proofs can be found in the full version.

Lemma 8. *Let $S = V' \cup E'$ be a minimum-sized special mixed dominating set of a graph G, where $V' \subseteq V(G)$ and $E' \subseteq E(G)$. Let G' be the subgraph of G induced by $V(G) \setminus V(E')$. Let V'' be the set of isolated vertices in G'. Then $V'' \subseteq V'$ and $V' \setminus V''$ is a minimum-sized vertex cover of G'. Moreover, for any vertex cover U of G', $U \cup V'' \cup E''$ is a mixed dominating set of G, where E'' is any perfect matching in $G[V(E')]$.*

Lemma 9. *There is an algorithm that, given an n-vertex graph G, runs in time $2^n n^{\mathcal{O}(1)}$ and outputs a minimum vertex cover of $G[U]$ for every $U \subseteq V(G)$. That is, the algorithm outputs 2^n minimum vertex covers, one for each $G[U]$, where $U \subseteq V(G)$.*

We are now ready to describe our algorithm for MIN MDS.

Algorithm for MIN MDS

Step 1. Run the algorithm in Lemma 9 on G and let $M[X]$ be the value returned by the algorithm for $X \subseteq V(G)$.

Step 2. For each $X \subseteq V(G)$, do the following.
- Let $G_X = G - X$, and let I_X be the set of isolated vertices in G_X.
- If $G[X]$ has a perfect matching, then set $val(X) := |X|/2 + |I_X| + M[X]$; otherwise, set $val(X) := \infty$.

Step 3. Return $\min_{X \subseteq V(G)} val(X)$.

The correctness of the algorithm follows from Lemma 8. As for the running time of the algorithm, note that each of step 1 and step 2 takes time $2^n n^{\mathcal{O}(1)}$. Hence our algorithm runs in time $2^n n^{\mathcal{O}(1)}$. We thus have the following result.

Theorem 4. MIN MDS *on a n-vertex graph can be solved in time $2^n n^{\mathcal{O}(1)}$.*

7   Conclusion

In this paper, we studied the complexity of MIXED DOMINATING SET on restricted graph classes and answered several questions left open by previous works. While we showed that MDS is polynomial time solvable on proper interval graphs, the complexity status of MDS on interval graphs is still unknown, and is worth investigating. Another open question is to improve the size of the kernel for MDS on $K_{d,d}$-free graphs or prove a matching lower bound.

References

1. Alavi, Y., Behzad, M., Lesniak-Foster, L.M., Nordhaus, E.A.: Total matchings and total coverings of graphs. J. Graph Theor. **1**(2), 135–140 (1977)
2. Alavi, Y., Liu, J., Wang, J., Zhang, Z.: On total covers of graphs. Discrete Math. **100**(1–3), 229–233 (1992)
3. Bodlaender, H.L., Fomin, F.V., Lokshtanov, D., Penninkx, E., Saurabh, S., Thilikos, D.M.: (meta) kernelization. J. ACM **63**(5), 44:1–44:69 (2016)
4. Cygan, M., Fomin, F.V., Kowalik, L., Lokshtanov, D., Marx, D., Pilipczuk, M., Pilipczuk, M., Saurabh, S.: Parameterized Algorithms. Springer, Cham (2015)
5. Diestel, R.: Graph Theory. Graduate Texts in Mathematics, vol. 173. Springer, Heidelberg (2012)
6. Dom, M., Lokshtanov, D., Saurabh, S.: Kernelization lower bounds through colors and IDs. ACM Trans. Algorithms **11**(2), 13:1–13:20 (2014)
7. Drange, P.G., et al.: Kernelization and sparseness: the case of dominating set. In: 33rd Symposium on Theoretical Aspects of Computer Science, STACS 2016, Orléans, France, 17–20 February 2016, pp. 31:1–31:14 (2016)
8. Eickmeyer, K., et al.: Neighborhood complexity and kernelization for nowhere dense classes of graphs. In: 44th International Colloquium on Automata, Languages, and Programming, ICALP 2017, Warsaw, Poland, 10–14 July 2017, pp. 63:1–63:14 (2017). https://doi.org/10.4230/LIPIcs.ICALP.2017.63
9. Erdös, P., Meir, A.: On total matching numbers and total covering numbers of complementary graphs. Discrete Math. **19**(3), 229–233 (1977)
10. Golovach, P.A., Villanger, Y.: Parameterized complexity for domination problems on degenerate graphs. In: Broersma, H., Erlebach, T., Friedetzky, T., Paulusma, D. (eds.) WG 2008. LNCS, vol. 5344, pp. 195–205. Springer, Heidelberg (2008). https://doi.org/10.1007/978-3-540-92248-3_18
11. Hatami, P.: An approximation algorithm for the total covering problem. Discussiones Mathematicae Graph Theory **27**(3), 553–558 (2007)
12. Haynes, T.W., Hedetniemi, S., Slater, P.: Fundamentals of Domination in Graphs. CRC Press, Boca Raton (1998)
13. Hedetniemi, S.M., Hedetniemi, S.T., Laskar, R., McRae, A., Majumdar, A.: Domination, independence and irredundance in total graphs: a brief survey. In: Proceedings of the 7th Quadrennial International Conference on the Theory and Applications of Graphs. vol. 2, pp. 671–683 (1995)
14. Jain, P., Jayakrishnan, M., Panolan, F., Sahu, A.: MIXED DOMINATING SET: a parameterized perspective. In: Bodlaender, H.L., Woeginger, G.J. (eds.) WG 2017. LNCS, vol. 10520, pp. 330–343. Springer, Cham (2017). https://doi.org/10.1007/978-3-319-68705-6_25
15. Lan, J.K., Chang, G.J.: On the mixed domination problem in graphs. Theor. Comput. Sci. **476**, 84–93 (2013)
16. Majumdar, A.: Neighborhood hypergraphs: a framework for covering and packing parameters in graphs. Ph.D. thesis, Clemson University (1992)
17. Manlove, D.: On the algorithmic complexity of twelve covering and independence parameters of graphs. Discrete Appl. Math. **91**(1–3), 155–175 (1999)
18. Meir, A.: On total covering and matching of graphs. J. Comb. Theory, Ser. B **24**(2), 164–168 (1978)
19. Peled, U.N., Sun, F.: Total matchings and total coverings of threshold graphs. Discrete Appl. Math. **49**(1–3), 325–330 (1994)

20. Rajaati, M., Hooshmandasl, M.R., Dinneen, M.J., Shakiba, A.: On fixed-parameter tractability of the mixed domination problem for graphs with bounded tree-width. CoRR abs/1612.08234 (2016)
21. Zhao, Y., Kang, L., Sohn, M.Y.: The algorithmic complexity of mixed domination in graphs. Theor. Comput. Sci. **412**(22), 2387–2392 (2011)

Uniform CSP Parameterized by Solution Size is in W[1]

Ruhollah Majdoddin$^{(\boxtimes)}$ (iD)

Institut für Informatik, Humboldt-Universität zu Berlin, Berlin, Germany
r.majdodin@gmail.com

Abstract. We show that the uniform Constraint Satisfaction Problem (CSP) parameterized by the size of the solution is in W[1] (the problem is W[1]-hard and it is easy to place it in W[3]). Given a single "free" element of the domain, denoted by **0**, we define the *size* of an assignment as the number of variables that are mapped to a value other than **0**. Named by Kolaitis and Vardi (2000), *uniform* CSP means that the input contains the domain and the list of tuples of each relation in the instance. Uniform CSP is polynomial time equivalent to homomorphism problem and also to evaluation of conjunctive queries on relational databases. It also has applications in artificial intelligence.

We do not restrict the problem to any (finite or infinite) family of relations. Marx and Bulatov (2014) showed that Uniform CSP restricted to some finite family of relations (thus with a bound on the arity of relations) and over any finite domain is either W[1]-complete or fixed parameter tractable.

We then prove that the parameterized subset sum problem with weights bounded by n^k is in W[1]. Abboud et al. (2014) have already proved it, but our proof is much shorter and arguably more intuitive.

Lastly, we study the weighted CSP over the Boolean domain, where each variable is assigned a weight, and given a target value, it should be decided if there is a satisfying assignment of size k (the parameter) such that the weight of its 1-variables adds up to the target value. We prove that if the weights are bounded by n^k, then the problem is in W[1].

Our proofs give a nondeterministic RAM program with special properties deciding the problem. First defined by Chen et al. (2005), such programs characterize W[1].

1 Introduction

The Constraint Satisfaction Problem (CSP) is a fundamentally important problem in computer science, that can express a large number of problems in artificial intelligence and operational research [16]. An instance I of CSP is specified by a finite domain D, a set of relations over domain D, a set V of variables, and a set of constraints C of the form $R(x_1, \ldots, x_r)$, where R is one of the relations with arity $r \geq 1$ and $x_1, \ldots, x_r \in V$. An *assignment* to a set of variables $S \subseteq V$ is a mapping from S to D. An assignment to the set of variables of a constraint *satisfies* the constraint if evaluating the tuple of variables of the constraint according

© Springer Nature Switzerland AG 2019
R. van Bevern and G. Kucherov (Eds.): CSR 2019, LNCS 11532, pp. 275–285, 2019.
https://doi.org/10.1007/978-3-030-19955-5_24

to the assignment, gives a tuple in the corresponding relation. An assignment to V is a *satisfying assignment* of I, if it satisfies all the constraints in I. When seen as a decision problem, the question in CSP is whether the given instance has a satisfying assignment.

Kolaitis and Vardi [10] made a distinction between *nonuniform* CSP, where the domain and the family of relations are fixed, and *uniform* CSP, where the input contains the domain and the list of tuples of each relation in the instance. They showed that uniform CSP is polynomial time equivalent to evaluation of conjunctive queries on relational databases. Feder and Vardi [8] observed that uniform CSP (which had already applications in artificial intelligence) and the *homomorphism problem* are polynomial time equivalent.

We study uniform CSP in the settings of parameterized complexity. Given a single "free" element of the domain, denoted by **0**, we define the *size* of an assignment as the number of variables that are mapped to a value other than **0**. The *Parameterized Size* CSP is defined as follows:

---

p-SIZE-CSP
   Instance: A domain D including **0**, a set of variables V, a set of
             constraints, list of tuples of each relation, and $k \geq 0$.
   Parameter: k.
   Problem: Decide whether there is a satisfying assignment of size k.

Many parameterized problems ask, given a structure $\mathcal{A}$ (on universe A), if there is a set $S \subset A$ of a given cardinality (the parameter) such that the substructure induced by S has a special property. Many of these problems can be readily reduced to p-SIZE-CSP, such that the size parameter in the resulting p-SIZE-CSP instance has the same value as the cardinality of the set looked for. A good example is p-CLIQUE, which, given an instance (G, k), asks if Graph G has a clique of size k. Another example is p-Vertex-Cover. These problems, however, can be expressed with p-SIZE-CSP restricted to a finite family of relations. To capture the full expressiveness of p-SIZE-CSP on Boolean domain, we introduce the following problem which is fixed parameter equivalent to p-SIZE-CSP:

---

p-W-HYPERGRAPHS-HITTING-SET
   Instance: A set W, hypergraphs $(V_1, E_1), \ldots, (V_m, E_m)$ where $V_i \subseteq W$,
          and $k \geq 0$.
   Parameter: k.
   Problem: Decide whether there is a set $S \subset W$ of cardinality k such that
        $S \cap V_i \in E_i$ for $1 \leq i \leq m$.

Our main contribution is the following containment theorem:

Theorem 1. p-SIZE-CSP $\in$ W[1].

Corollary 1. p-W-HYPERGRAPHS-HITTING-SET $\in$ W[1].

We prove the theorem by giving a *tail-nondeterministic κ-restricted NRAM program* (explained in the next section) deciding the problem. The significance of our containment result is that it is for the general problem, without restricting it to any (finite or infinite) family of relations.

Our work builds upon the work of Cesati [3], which, answering a longstanding open problem, proved that p-PERFECT-CODE is in W[1]. Downey and Fellows [7] had already shown that this problem is W[1]-hard and had conjectured that it either represents a natural degree intermediate between W[1] and $W[2]$, or is complete for $W[2]$. There is a natural reduction from p-PERFECT-CODE to p-EXACT-WSAT(CNF$^+$), and the proof of [3] can be readily adapted to decide the latter problem. This problem is to decide, given a CNF without negation symbols and a natural number k, whether there is an assignment of size k, such that exactly one variable in each clause is mapped to 1. This can be seen as p-SIZE-CSP restricted to a specific (infinite) family of Boolean relations, where a tuple is in a relation, if and only if the tuple has exactly one 1 (this implies that p-SIZE-CSP is W[1]-hard). Notice that because we do not restrict the problem to any family of relations, our result generalizes that of Cesati in at least three ways: Size of the tuples in the relations are not bounded, the (Boolean) relations do not need to be *symmetric* (symmetric means that a tuple being in the relation depends only on the number of 1s in the tuple), and an instance can have any finite domain.

In fact, p-EXACT-WSAT(CNF$^+$) is an example of an interesting special case of our containment result: p-SIZE-CSP restricted to any (infinite) family of symmetric Boolean relations, provided that there is a bound on the size of the tuples of any relation in the family. Notice that the bound implies that the number of tuples of each relation is bounded by a polynomial in the arity of the relation. Thus, listing all the tuples in the input makes the size of input at most polynomially bigger, and uniform and nonuniform CSP in this case have the same complexity.

It is not hard to see that p-SIZE-CSP over the Boolean domain is in W[3], by reducing it to the *parameterized weighted satisfiability problem* for a class of circuits with bounded depth and *weft* 3: one for the conjunction of all constraints in the instance, one for the disjunction of all satisfying assignments of each constraint, and one to specify each satisfying assignment of a constraint. So what is the significance of placing a problem from (at least) W[3] down to W[1]? First, although it is a fundamental conjecture that W[1]-complete problems are not fixed-parameter tractable, many of them can still be solved substantially faster than exhaustive search over all $\binom{n}{k}$ subsets. For example, [13] gives an $O(n^{.793k})$ time algorithm for p-CLIQUE. In contrast, the W[2]-complete problem p-DOMINATING-SET, was shown by [15] not to have such algorithms, unless SAT(CNF) has an $O(2^{\delta n})$ time algorithm for some $\delta < 1$, which is an important open problem. Second, we can express the problems in W[1] by a logic that is (conjectured to be) a proper subclass of any logic that can express the problems in W[3] (see [9]). This means that putting a problem in W[1] decreases the descriptive complexity of the problem.

It is easy to see that p-SIZE-CSP restricted to some finite family of relations (implying the arity of relations is bounded) is in W[1]. Notice that listing the tuples of all relations in the input adds just a constant to the size of input. Thus, uniform and nonuniform CSP in this case have the same complexity. These problems are studied by Marx [12], where he provides a dichotomy: If the family of relations has a property that he calls *weak-separability*, then the problem is fixed-parameter tractable (like p-Vertex-Cover), otherwise it is W[1]-complete (like p-CLIQUE). This result is extended by Bulatov and Marx [2] to any finite domain.

There is a variant of CSP on the Boolean domain where the variables are weighted. That is, each instance comes with a weight function over the variables and a target value. It should be decided if there is an assignment that satisfies the constraints and the weights of its 1-variables add up to the target value. Special cases of this variant are studied in the literature. For example, [5,6,18] study the parameterized problem of finding a clique in a weighted graph, where the parameter is the target value.

In studying this kind of problems, one is in some way dealing with SUBSET-SUM. For every computable function $f : \mathbb{N} \to \mathbb{N}$ we let:

---

p-SUBSET-SUM(f)
>   *Instance:* $k \geq 0$, n integers $x_1, \ldots, x_n \in [0, n^{f(k)}]$, $t \in [0, n^{f(k)}]$.
>   *Parameter:* k.
>   *Problem:* Decide whether there exists a subset $B \subset [n]$ of size
> $|B| = k$ such that $\sum_{i \in B} x_i = t$.

Theorem 2. p-SUBSET-SUM$(f) \in$ W[1], *for all computable functions f.*

Our proof gives a *tail-nondeterministic κ-restricted NRAM program* deciding the problem. Abboud, Lewi and Williams [1] have also proved this theorem. Given an instance of the problem, they generate $g(k) \cdot n^{o(1)}$ instances of p-CLIQUE on n node graphs, such that one of these graphs contains a k-clique if and only if the p-SUBSET-SUM instance has a solution. The proof follows because p-CLIQUE is W[1]-complete. Our proof is considerably shorter and arguably more intuitive. They also prove that the weighted variant of p-CLIQUE is in W[1]. Notice that here the parameter is the size of the clique, thus it is a substantially weaker parameterization than that of [5,6,18]. Generalizing this problem in the language of CSP, for every computable function $f : \mathbb{N} \to \mathbb{N}$, we introduce the *Parameterized Weighted* CSP:

---

p-WCSP(f)
>   *Instance:* A set of variables V, the domain $\{0, 1\}$, a set of constraints,
>   list of tuples of each relation, $k \geq 0$,
> $w : V \to [0, n^{f(k)}]$, $t \in [0, n^{f(k)}]$.
>   *Parameter:* k.
>   *Problem:* Decide whether there is a satisfying assignment B of size
> $|B| = k$, such that $\sum_{(v,1) \in B} w(v) = t$.

---

where n is the size of the input.

Theorem 3. p-WCSP$(f) \in$ W[1], *for all computable functions* $f : \mathbb{N} \to \mathbb{N}$.

The proof employs our proofs of Theorems 1 and 2.

For the basic concepts, definitions and notation of parameterized complexity theory, we refer the reader to [9].

Notation. For integers n, m with $n \leq m$, we let $[n, m] := \{n, n+1, \ldots, m\}$ and $[n] := [1, n]$.

2   A Machine Characterization of W[1]

We use a nondeterministic random access machine model. It is based on a standard deterministic random access machine (RAM) model. Registers store nonnegative integers. Register 0 is the *accumulator*. The arithmetic operations are addition, subtraction (cut off at 0), and division by two (rounded off), and we use a uniform cost measure. For more details see [9]. We define a *nondeterministic* RAM, or NRAM, to be a RAM with an additional instruction "GUESS" whose semantics is:

> *Guess a natural number less than or equal to the number stored in the accumulator and store it in the accumulator.*

Acceptance of an input by an NRAM program is defined as usually for nondeterministic machines. Steps of a computation of an NRAM that execute a GUESS instruction are called *nondeterministic steps*.

Definition 1. *Let* $\kappa : \Sigma^* \to \mathbb{N}$ *be a parameterization. An NRAM program* P *is* κ-*restricted if there are computable functions* f *and* g *and a polynomial* $p(X)$ *such that on every run with input* $x \in \Sigma^*$ *the program* P

- *performs at most* $f(k) \cdot p(n)$ *steps, at most* $g(k)$ *of them being nondeterministic;*
- *uses at most the first* $f(k) \cdot p(n)$ *registers;*
- *contains numbers* $\leq f(k) \cdot p(n)$ *in any register at any time.*

Here $n := |x|$, *and* $k := \kappa(x)$.

Definition 2. *A* κ-*restricted NRAM program* P *is tail-nondeterministic if there is a computable function* q *such that for every run of* P *on any input* x *all nondeterministic steps are among the last* $q(\kappa(x))$ *steps of the computation.*

The machine characterization of W[1] reads as follows:

Theorem 4 ([4]). *Let* (Q, κ) *be a parameterized problem. Then* $(Q, \kappa) \in$ W[1] *if and only if there is a tail-nondeterministic* κ-*restricted NRAM program deciding* (Q, κ).

3   Partially Ordered Sets

The *Möbius function* of a poset $(P, \leq)$ is a function $\mu : P \times P \to \mathbb{Z}$ defined recursively as follows.

$$\mu(x, y) = \begin{cases} 1 & x=y, \\ -\sum_{x \leq z < y} \mu(x, z) & x < y, \\ 0 & x > y. \end{cases}$$

Theorem 5 (Möbius inversion formula). *Let $(P, \leq)$ be a finite partially ordered set with a minimum element. For functions $f, g : P \to \mathbb{Z}$, suppose that*

$$g(x) = \sum_{y \leq x} f(y).$$

Then

$$f(x) = \sum_{y \leq x} g(y)\mu(y, x). \tag{1}$$

Furthermore, there is a computable function $h : \mathbb{N} \to \mathbb{N}$ such that

$$|f(x)| \leq h(t) \max_x |g(x)| \quad \text{for all } x, \tag{2}$$

where t is the maximum size of any interval in P.

Proof. See [17] for a proof of the first claim. For the bound, it is easy to see that there is a computable function $h' : \mathbb{N} \to \mathbb{N}$ such that

$$\max_{x,y} |\mu(x, y)| \leq h'(t). \tag{3}$$

Thus, $f(x) \leq t \; h'(t) \; \max_x |g(x)|$ for all x. Set $h(t) := t \; h'(t)$ and the claim follows.

Let $(P, \leq)$ be a poset, and $x, y \in P$. We say y covers x if $x < y$ and there is no element $z \in P$ such that $x < z < y$. Let $Q \subseteq P$. We denote by $\tilde{Q}$ (with respect to P) the set of all $y \in P$, such that $y \notin Q$ and y covers some $x \in Q$.

Lemma 6. *Let $(P, \leq)$ be a poset and suppose $Q \subseteq P$ and $y \in P$. Then $y \notin Q$ if and only if all maximal elements of $\{w \in Q \cup \tilde{Q} | w \leq y\}$ are in $\tilde{Q}$.*

Proof. Suppose $y \notin Q$. If $y \in Q \cup \tilde{Q}$, then clearly $y \in \tilde{Q}$ and the claim follows. Otherwise, let $x < y$ be a maximal element of $\{w \in Q \cup \tilde{Q} | w \leq y\}$. There is at least one $z \in P$ such that $z \leq y$ and z covers x. Now, if $x \in Q$, then either $z \in Q$ or $z \in \tilde{Q}$, thus $z \in \{w \in Q \cup \tilde{Q} | w \leq y\}$, contradicting the maximality of x. Therefore, $x \in \tilde{Q}$.

The other direction is implied by the trivial fact that if $y \in Q$, then y is the maximum element of $\{w \in Q \cup \tilde{Q} | w \leq y\}$.

4 p-SIZE-CSP is in W[1]

In this section we present a tail-nondeterministic κ-restricted NRAM program Q that decides p-SIZE-CSP. Given an instance I_0 of the problem with the set of variables V and parameter value k, our program first constructs a second instance I with the same set of variables and the same parameter value, such that I_0 and I have the same set of satisfying assignments of size $\leq k$, and I has the following properties. Each variable appears in each constraint at most once, and for each subset $S \subseteq V$, there is at most one constraint with this set of variables, thus each constraint is characterized by its set of variables. This construction, invented by Papadimitriou and Yannakakis [14], is as follows.

Henceforth, we characterize an assignment A with the set of all (v, d), such that v is mapped to d and $d \neq 0$.

Fix an order on V. For each subset $S \subseteq V$, if I_0 has a constraint such that the set of variables of the constraint is exactly S (possibly with repetitions), then I has Relation R_S of arity $|S|$ and Constraint $\mathcal{C}_S$ defined as follows. An assignment A of S satisfies $\mathcal{C}_S$ if and only if $|A| \leq k$ and A satisfies every constraint $\mathcal{C}$ in I_0, such that the set of variables of $\mathcal{C}$ is exactly S (possibly with repetitions). The order of variables in $\mathcal{C}_S$ is determined by the order on V. Relation R_S is defined accordingly. Notice that there is a natural bijective mapping of the tuples in R_S to the satisfying assignments of $\mathcal{C}_S$.

Program Q, in its nondeterministic part, guesses an assignment and checks if it satisfies all the constraints of I. But because Q is tail-nondeterministic κ-restricted, we need some method, explained below, instead of trivially going over all constraints.

For a constraint $\mathcal{C}_S$ in I, let $\tilde{C}_S$ be the set of all satisfying assignments of $\mathcal{C}_S$, and P_S be the set of all possible assignments of S. Clearly, $\tilde{C}_S \subseteq P_S$. Thus, we define $\tilde{C}_S$ with respect to Poset $(P_S, \subseteq)$.
Define

$$\tilde{C} := \bigcup_{\mathcal{C}_S \in I} \tilde{C}_S,$$

and for each $T \in \tilde{C}$ define

$$I_T := \{\mathcal{C}_S \in I | T \in \tilde{C}_S\}.$$

For each T in $\tilde{C}$ and each $\mathcal{C}_S \in I_T$, let $H_{T,S}$ be the set of proper supersets of T in $\mathcal{C}_S$:

$$H_{T,S} := \{U \in \mathcal{C}_S | T \subset U\}.$$

Lemma 7. *Let B be an assignment. If B does not satisfy I, then there is a $T_B \in \tilde{C}$ such that if B does not satisfy some $\mathcal{C}_S \in I_{T_B}$ (and there is at least one such constraint), then $H_{T_B,S} = \varnothing$.*

Proof. If B does not satisfy I, then set T_B to an element of maximal size in $\tilde{C}$, such that at least one constraint $\mathcal{C}_S \in I_{T_B}$ is not satisfied by B. By applying

Lemma 6 to Poset $(P_S, \subseteq)$, C_S and B (restricted to S), it follows that all the maximal elements (with respect to inclusion) of $\{U \in C_S \cup \tilde{C}_S | U \subseteq B\}$ are in $\tilde{C}_S$, and T_B is one of them, otherwise some $T \in \tilde{C}_S$, $T_B \subset T$ is a maximal element, contradicting that T_B has maximal size. This means that $H_{T_B, S} = \varnothing$.

Lemma 7 implies that Program Q has to just check for any $T \in \tilde{C}$, $T \subseteq B$, that for every $C_S \in I_T$, there is a $U \in C_S$ that $T \subset U \subseteq B$. But Q does not have enough *time* to go over all constraints in I_T, so the idea is to enumerate all $T \subset U \subseteq B$ and use some *machinery* that (implicitly) adds 1 for each constraint that passes the check and 0 for each constraint that fails, to get a number that equals $|I_T|$ iff the check is passed for T. This is not simple, because the same U can be a superset of T in many constraints, and T can have many supersets in one constraint. In an earlier work [11], we applied the inclusion-exclusion principle as the machinery, but with a limited success. In the current paper, we will use the Möbius inversion formula. In fact the Inclusion-Exclusion principal can be proved by Möbius inversion formula [17], thus can be seen as a special case of it.

We define the function $f_{T,S} : H_{T,S} \to \mathbb{Z}$ recursively, subject to

$$\sum_{W \subseteq U} f_{T,S}(W) = 1, \tag{4}$$

for all $U \in H_{T,S}$. Applying Theorem 5 to Poset $(H_{T,S}, \subseteq)$ and $f_{T,S}$, it follows that $f_{T,S}$ is evaluated by Möbius inversion formula (1).

Lemma 8. *Let B be an assignment. If B satisfies I then for each $T \in \tilde{C}$,*

$$\sum_{C_S \in I_T} \sum_{\substack{W \in H_{T,S} \\ W \subseteq B}} f_{T,S}(W) = |I_T|.$$

If B does not satisfy I, then

$$\sum_{C_S \in I_{T_B}} \sum_{\substack{W \in H_{T_B,S} \\ W \subseteq B}} f_{T_B,S}(W) < |I_{T_B}|. \tag{5}$$

Proof. If B satisfies a constraint C_S, then for all $T \in \tilde{C}_S$, $T \subset B$, it is clear that B is the maximum element of $\{W \in H_{T,S} | W \subseteq B\}$, with respect to inclusion. Thus, by (4), the inner summation equals 1.

Now, if B satisfies I, then clearly B satisfies all $C_S \in I_T$ for each $T \in \tilde{C}$, and the outer summation equals $|I_T|$. And if B does not satisfy I, then by Lemma 7, for any constraint $C_S \in I_{T_B}$ that is not satisfied by B, the inner summation is empty and equals 0. Thus the outer summation is $< |I_{T_B}|$.

Now, apply the Fubini's Principle to swap the summations:

$$\sum_{\substack{W \in \bigcup_{C_S \in I_T} H_{T,S} \\ W \subseteq B}} \sum_{C_S \in I_T} f_{T,S}(W).$$

Program Q evaluates the inner summation in its deterministic part and the outer summation in its nondeterministic part, for all $T \in \tilde{C}$.

Now we are ready for our main theorem:

Theorem 9. p-SIZE-CSP $\in$ W[1].

Proof. We give a tail-nondeterministic κ-restricted NRAM program Q deciding the problem. The result follows by Theorem 4. Let I_0 be the given instance with the set of variables V and the parameter value k. Program Q first constructs Instance I from Instance I_0 as described above. This can easily be done in polynomial time.

Next, Q calculates two tries:

- Trie 1 stores the values $d[T] := |I_T|$ for all $T \in \tilde{C}$.
- Trie 2 stores the values

$$l[T, W] := \sum_{C_S \in I_T} f_{T,S}(W), \tag{6}$$

for every $T \in \tilde{C}$, and every $W \in \bigcup_{C_S \in I_T} H_{T,S}$ (if $W \notin H_{T,S}$, then we extend $f_{T,S}$ to $f_{T,S}(W) := 0$).

For the queries with nonexistent keys, 0 is returned.

Now the nondeterministic part of the computation starts: Program Q guesses an assignment B of size k. Then Q iterates over subsets $T \subseteq B$, and if $d[T] > 0$, checks if

$$\sum_{T \subset W \subseteq B} l[T, W] = d[T]. \tag{7}$$

By our argument above, it should be clear that Q decides p-SIZE-CSP.

We claim that each Trie has at most polynomially many entries. This is because the problem is uniform (of size, say, n), and the input contains the list of tuples of each relation in I_0. Thus, we have $|C_S| \leq n$ and $|\tilde{C}_S| \leq |S||D|(|C_S| + 1) \leq n^3$.

We also claim that $f_{T,S}$ is an fpt-function. This is because by construction of I, for all constraints $C \in I$, for all $W \in C$, we have $|W| \leq k$. This implies that the size of an interval in $H_{T,S}$ is at most 2^k, and the claim follows by (4) and Theorem 5.

The above claims imply that the tries can easily be populated in fpt-time.

Lastly, we show that Program Q is tail-nondeterministic κ-restricted. This is because the tries are arranged such that for all assignments T, U of size $\leq k$, the query with key T or T, U is answered in $O(k)$ time (a general property of the trie data structure). Moreover, the summation in Check (7) has at most 2^{2k} summands. This completes the proof.

5 p-WCSP and p-SUBSET-SUM are in W[1]

In this section we prove that p-WCSP and p-SUBSET-SUM are in W[1].

Theorem 10. p-SUBSET-SUM$(f) \in W[1]$, *for all computable functions* f.

Proof. For an integer $a \in [0, n^{f(k)}]$, define numbers $\bar{a}^j \in [0, n-1]$ for $j \in [0, f(k)]$ as $a = \sum_{j \in [0, f(k)]} \bar{a}^j n^j$ (this is presentation of a in base n, thus $\bar{a}^j$ are unique).

For a fixed f, we present a tail-nondeterministic κ-restricted program P deciding p-SUBSET-SUM(f). On input $x_1, \ldots, x_n, t$, Program P first computes two tables. Table 1 stores the values $\bar{x}_i^j$ for $i \in [n]$ and $j \in [0, f(k)]$, and Table 2 stores the values $\bar{t}^j$ for $j \in [0, f(k)]$. The tables are arranged in such a way that the numbers can be accessed in constant time. The tables can be easily computed in polynomial time.

Now the nondeterministic part of the computation starts: Program P guesses k elements in $[n]$ and checks if they are distinct. Let B be the set of guessed elements. Then, for $j \in [0, f(k)]$, Program P divides $c_{j-1} + \sum_{i \in B} \bar{x}_i^j$ by n, sets c_j as the quotient ($c_{-1} := 0$) and checks if the remainder equals $\bar{t}^j$. Finally, P checks if $c_{f(k)} = 0$.

Notice that $c_j \leq k+1$ for $j \in [0, f(k)]$, and P can perform each division operations with $O(k)$ of its arithmetic operations. Thus, the number of steps in the nondeterministic part is $O(kf(k))$, and Program P is κ-restricted tail-nondeterministic.

Theorem 11. p-WCSP$(f) \in W[1]$, *for all computable functions* $f : \mathbb{N} \to \mathbb{N}$.

Proof. We present a tail-nondeterministic κ-restricted program H deciding the problem. Let Q be the program that decides p-SIZE-CSP, and P be the program that decides p-SUBSET-SUM(f), as described in the proofs of Theorems 9 and 10, respectively. H first performs the deterministic part of Q and then that of P. Then, H performs the nondeterministic part of P: it guesses an assignment B of size $|B| := k$ and checks if weights of variables in B add up to t. If no, then this nondeterministic branch rejects. If yes, then H performs the nondeterministic part of Q, omitting the guessing step, to check if B is a satisfying assignment. If yes, then H accepts. The number of steps in the nondeterministic part of H is bounded by the sum of that of Q and P. Thus, Program H is κ-restricted and tail-nondeterministic.

Acknowledgment. The author thanks Johannes Köbler, Frank Fuhlbrück, and Amir Abboud for helpful discussions.

References

1. Abboud, A., Lewi, K., Williams, R.: Losing weight by gaining edges. In: Schulz, A.S., Wagner, D. (eds.) ESA 2014. LNCS, vol. 8737, pp. 1–12. Springer, Heidelberg (2014). https://doi.org/10.1007/978-3-662-44777-2_1
2. Bulatov, A.A., Marx, D.: Constraint satisfaction parameterized by solution size. SIAM J. Comput. **43**(2), 573–616 (2014). https://doi.org/10.1137/120882160
3. Cesati, M.: Perfect code is W[1]-complete. Inf. Process. Lett. **81**(3), 163–168 (2002). https://doi.org/10.1016/S0020-0190(01)00207-1

4. Chen, Y., Flum, J., Grohe, M.: Machine-based methods in parameterized complexity theory. Theor. Comput. Sci. **339**(2–3), 167–199 (2005). https://doi.org/10.1016/j.tcs.2005.02.003

5. du Cray, H.P., Sau, I.: Improved FPT algorithms for weighted independent set in bull-free graphs. Discrete Math. **341**(2), 451–462 (2018). https://doi.org/10.1016/j.disc.2017.09.012

6. Dabrowski, K., Lozin, V., Müller, H., Rautenbach, D.: Parameterized complexity of the weighted independent set problem beyond graphs of bounded clique number. J. Discrete Algorithms **14**, 207–213 (2012). https://doi.org/10.1016/j.jda.2011.12.012

7. Downey, R.G., Fellows, M.R.: Fixed-parameter tractability and completeness II: on completeness for W[1]. Theor. Comput. Sci. **141**(1–2), 109–131 (1995). https://doi.org/10.1016/0304-3975(94)00097-3

8. Feder, T., Vardi, M.Y.: The computational structure of monotone monadic SNP and constraint satisfaction: a study through datalog and group theory. SIAM J. Comput. **28**(1), 57–104 (1998). https://doi.org/10.1137/S0097539794266766

9. Flum, J., Grohe, M.: Parameterized Complexity Theory. Texts in Theoretical Computer Science. An EATCS Series. Springer, Heidelberg (2006). https://doi.org/10.1007/3-540-29953-X

10. Kolaitis, P.G., Vardi, M.Y.: Conjunctive-query containment and constraint satisfaction. J. Comput. Syst. Sci. **61**(2), 302–332 (2000). https://doi.org/10.1006/jcss.2000.1713

11. Majdoddin, R.: Parameterized complexity of CSP for infinite constraint languages. CoRR abs/1706.10153 (2017). http://arxiv.org/abs/1706.10153

12. Marx, D.: Parameterized complexity of constraint satisfaction problems. Comput. Complex. **14**(2), 153–183 (2005). https://doi.org/10.1007/s00037-005-0195-9

13. Nešetřil, J., Poljak, S.: On the complexity of the subgraph problem. Commentationes Mathematicae Universitatis Carolinae **26**(2), 415–419 (1985)

14. Papadimitriou, C.H., Yannakakis, M.: On the complexity of database queries. J. Comput. Syst. Sci. **58**(3), 407–427 (1999). https://doi.org/10.1006/jcss.1999.1626

15. Patrascu, M., Williams, R.: On the possibility of faster SAT algorithms. In: Charikar, M. (ed.) Proceedings of the Twenty-First Annual ACM-SIAM Symposium on Discrete Algorithms, SODA 2010, Austin, Texas, USA, 17–19 January 2010, pp. 1065–1075. SIAM (2010). https://doi.org/10.1137/1.9781611973075.86

16. Rossi, F., van Beek, P., Walsh, T. (eds.): Handbook of Constraint Programming, Foundations of Artificial Intelligence, vol. 2. Elsevier (2006). http://www.sciencedirect.com/science/bookseries/15746526/2

17. Rota, G.C.: On the foundations of combinatorial theory i. Theory of möbius functions. Probab. Theory Relat. Fields **2**(4), 340–368 (1964). https://doi.org/10.1007/BF00531932

18. Thomassé, S., Trotignon, N., Vuskovic, K.: A polynomial turing-kernel for weighted independent set in bull-free graphs. Algorithmica **77**(3), 619–641 (2017). https://doi.org/10.1007/s00453-015-0083-x

On the Parameterized Complexity
of Edge-Linked Paths

Neeldhara Misra[1], Fahad Panolan[2(✉)], and Saket Saurabh[3]

[1] Indian Institute of Technology, Gandhinagar, Gandhinagar, India
`neeldhara.m@iitgn.ac.in`
[2] Department of Informatics, University of Bergen, Bergen, Norway
`fahad.panolan@ii.uib.no`
[3] The Institute of Mathematical Sciences, HBNI, Chennai, India
`saket@imsc.res.in`

Abstract. An edge Hamiltonian path of a graph is a permutation of its edge set where every pair of consecutive edges have a vertex in common. Unlike the seemingly related problem of finding an Eulerian walk, the edge Hamiltonian path is known to be a NP-hard problem, even on fairly restricted classes of graphs. We introduce a natural optimization variant of the notion of an edge Hamiltonian path, which seeks the longest sequence of distinct edges with the property that every consecutive pair of them has a vertex in common. We call such a sequence of edges an edge-linked path, and study the parameterized complexity of the problem of finding edge-linked paths with at least k edges. We show that the problem is FPT when parameterized by k, and unlikely to admit a polynomial kernel even on connected graphs.

On the other hand, we show that the problem admits a Turing kernel of polynomial size. To the best of our knowledge, this is the first problem on general graphs to admit Turing kernels with adaptive oracles (for which a non-adaptive kernel is not known). We also design a single-exponential parameterized algorithm for the problem when parameterized by the treewidth of the input graph.

Keywords: FPT · Turing kernelization · Edge Hamiltonian cycle

1 Introduction

An edge Hamiltonian path of a graph is a permutation of its edge set where every pair of consecutive edges have a vertex in common. The notion is a classical one, well-studied in the context of structural graph theory, and more recently, has received attention from the computational perspective as well. Unlike the seemingly related problem of finding an Eulerian walk, the edge Hamiltonian

---

This work is supported by the European Research Council (ERC) via grant LOPPRE, reference 819416 and Norwegian Research Council via project MULTIVAL.

R. van Bevern and G. Kucherov (Eds.): CSR 2019, LNCS 11532, pp. 286–298, 2019.
https://doi.org/10.1007/978-3-030-19955-5_25

path is known to be a NP-hard problem, even on fairly restricted classes of graphs such as bipartite graphs and graphs of maximum degree three [1,11,13].

We introduce a natural optimization variant of the notion of an edge Hamiltonian path, which is the following: what is the longest sequence of distinct edges with the property that every consecutive pair of them has a vertex in common? Note that this subsumes the question of finding an edge Hamiltonian path as a special case, and therefore the classical hardness of the problem follows immediately. For ease of discussion, we use the phrase "edge-linked paths" to refer to sequences of distinct edges that have the property of every consecutive pair of edges having a common vertex between them. We use the abbreviation LELP to refer to the problem of finding the longest edge-linked path.

It turns out that studying LELP from a parameterized perspective leads to interesting new algorithms for the problem. There are two natural parameters that emerge for LELP: the first is the standard parameter, which is the length of the edge sequence, and the second is the treewidth of the input graph. The treewidth parameter turns out to be useful also in the context of the original edge Hamiltonian path problem, see, for instance [12]. In this contribution, we establish the following results:

- LELP is FPT when parameterized by either the standard parameter or the treewidth. In particular, for the treewidth parameter, we demonstrate a single exponential algorithm for this problem.
- We argue that the problem is unlikely to admit a polynomial kernel with the standard parameter, even when the graph is connected. On the other hand, we show that the problem does admit a Turing kernel of polynomial size.

We remark here that our demonstration of a Turing kernel is particularly interesting, since our algorithm is based on oracle queries made in an adaptive fashion. Recall that a *Turing kernel* is a polynomial time algorithm that can solve the problem with access to an oracle for the problem, operating under the constraint that the oracle can only answer queries for small instances. The size of the largest instance on which we invoke the oracle is the "size" of the Turing kernel. Turing kernels are central to the study of problems that are unlikely to admit polynomial kernels, and yet, they have been demonstrated for only a small number of problems. Further, most Turing kernels work by producing polynomially many instances of bounded size without using the oracle at all[1]. Indeed, only a few Turing kernels are known that take full advantage of the oracle framework. These include the problems of finding long paths and cycles on restricted classes of graphs [9,10] and the weighted independent set problem on bull-free graphs [14]. Our contribution adds a natural problem to this limited list of problems known to admit Turing kernels using oracles in an adaptive fashion. To the best of our knowledge LELP is the first problem on *general*

---

[1] These instances typically have the property that the input instance is a Yes-instance if and only if one of these instances is a Yes-instance: therefore, the oracle can be applied to each instance in turn to solve the problem.

graphs to admit Turing kernels using adaptive oracles (for which a non-adaptive kernel is not known).

Apart from the standard parameter and treewidth, we also consider a natural "above-guarantee" parameter. Note that the maximum degree of a graph provides an easy lower bound on the length of a longest edge-linked path, since one has the sequence of edges incident on the vertex with the largest degree. Therefore, an interesting question to ask is if there is an edge-linked path of length at least $\Delta(G) + k'$, where $\Delta(G)$ denotes the maximum degree of G. This is an example of an *above-guarantee parameter*, where the parameter signifies the length of the path that is possible beyond what is structurally guaranteed to exist. We show that LELP is also FPT by this parameterization.

Methodology. It is quite straightforward to observe that LELP is FPT when parameterized by the standard parameter k. One method is to find a path of length k in the line graph of the input graph. Another approach, for instance, would be to observe that we can say YES if either the depth of a DFS traversal exceeds k, or if there is a vertex of degree at least k, since any usual path is also an edge-linked path, and a vertex of degree k naturally corresponds to an edge-linked path of length k. If both of these cases do not arise, then the size of the graph is easily seen to be bounded as a function of k. In fact it is known that if the depth of the DFS tree is at most k, then the treewidth of the graph is at most $k - 1$. We propose an efficient algorithm parameterized by treewidth, using techniques based on representative families. In particular, our main result in this context is the following.

Theorem 1. *Let G be an n-vertex graph given together with its tree decomposition of width* **tw**. *Then LELP can be solved in time* $\mathcal{O}\Big(\big(1 + 2^{(\omega+3)}\big)^{\mathbf{tw}}$ $\mathbf{tw}^{\mathcal{O}(1)} \cdot nm \Big)$ *where $m = |E(G)|$, and ω is the matrix-multiplication exponent.*

In the context of kernelization, we begin by observing that the problem is unlikely to admit a polynomial kernel with the standard parameter, by a standard application of the disjoint union construction. However, the graphs obtained by this construction are not connected, so we establish, using an explicit cross-composition, the hardness of kernelization for connected graphs. On the other hand, we also establish a polynomial-size Turing kernel for the problem. The algorithm we propose makes use of Tutte decompositions, which are tree decompositions that have additional properties—most notably that the torsos of the bags are 3-edge-connected. In this context, we are able to exploit the fact that such graphs are known to admit large Eulerian subgraphs [2], which imply the existence of long edge-linked paths. Therefore, if the Tutte decomposition of the given graph has a large bag, then we already have a Yes-instance on our hands. Otherwise, it turns out that we can use the decomposition to find a separation (A, B) of order at most two where one of A or B has bounded size.

From this point, the algorithm is based on careful invocations of an oracle that can find long edge-linked paths on the smaller side of the separation, either to determine that we have a Yes-instance or to discover a vertex that can be safely removed, thus making progress at every step. This involves a careful analysis of all possible ways in which a solution can split across the separation. The overall approach that we employ is inspired by the techniques used for obtaining Turing kernels for the problem of finding long paths and cycles on special classes of graphs [10].

Theorem 2. LELP, *when parameterized by* k, *does not admit a polynomial kernel on general graphs and connected graphs, unless* CoNP $\subseteq$ NP/poly. *On the other hand, the problem admits a Turing kernel with* $O(k^{3.656})$ *vertices and* $O(k^{4.656})$ *edges.*

Finally, for parameterized "above maximum degree", we show an FPT algorithm obtained by relying on bounding the treewidth of certain connected components in the graph.[2]

Related Work. The EDGE HAMILTONIAN PATH is known to be NP-complete even on bipartite graphs or graphs with maximum degree 3 [1,11,13]. Demaine et al. [6] presented an XP (i.e. running in time $n^{f(k)}$) algorithm for EDGE HAMILTONIAN PATH on bipartite graphs, where k is the size of the smaller part and asked whether it can be improved to an FPT algorithm. Lampis et al. [12] answered this question affirmatively by giving a cubic edge kernel for EDGE HAMILTONIAN PATH when parameterized by vertex cover. They also show that it is FPT on hypergraphs when parameterized by the size of a hitting set. They also studied the problem with parameters treewidth and clique-width of the input graph. The running times of their algorithms when parameterized by treewidth and clique width are $\mathbf{tw}^{\mathcal{O}(\mathbf{tw})} n^{\mathcal{O}(1)}$ and $\mathbf{cw}^{\mathcal{O}(\mathbf{cw}^2)} n^{\mathcal{O}(1)}$, respectively, where $\mathbf{tw}, \mathbf{cw}$ and n are the treewidth, clique-width and number of vertices of the input graph. To the best of our knowledge, this is the first study that uses the length of an edge-linked path as a parameter. We also note LELP can be solved also by using known algorithms for LONG PATH on the line graph of the input graph. However, in the context of the standard parameter, this yields no insight into the kernelization complexity. Also, when parameterized by treewidth, this approach does not give us FPT algorithms: note that the treewidth of the line graph can be arbitrarily larger than the treewidth of the input graph: for instance, note that the line graph of a star (treewidth one) is a clique (treewidth n).

2   Preliminaries

We refer the reader to [4, 7] for standard terminology and notions in graph theory and parameterized algorithms. Unless made explicit, we use standard notation throughout. We introduce and recall some important definitions below.

---

[2] Due to lack of space, the algorithm parameterized by treewidth and the arguments for the above-guarantee parameter are deferred to the full version of the paper.

For a graph $G, S \subseteq V(G)$, and $F \subseteq E(G)$, we use $E_{in}^G(S)$ to denote the set of edges incident to vertices in S, and $V(F)$ to denote the set of end-vertices of F. When the graph is clear from the context we use $E_{in}(S)$ instead of $E_{in}^G(S)$. For a graph G, a pair (A, B), where $A, B \subseteq V(G)$ and $A \cup B = V(G)$ is called a *separation* of G if there is no edge in G with one endpoint in $A \setminus B$ and the other in $B \setminus A$. The *order* of the separation (A, B) is $|A \cap B|$. A minimal separator in a connected graph G is an inclusion minimal vertex set $S \subseteq V(G)$ such that $G - S$ is disconnected. A vertex set of a disconnected graph is a minimal separator if it is a minimal separator for one of its connected components. For a vertex set U, define torso(G, U) as the graph obtained from $G[U]$ by adding an edge between each pair of vertices in U that are connected by a path in G whose internal vertices are not from U. Following simple observation is used many times in the paper.

Observation 1. *The number of odd degree vertices in a graph is even.*

Definition 1. *A tree-decomposition of a graph G is a pair $(\mathbb{T}, \mathcal{X} = \{X_t\}_{t \in V(\mathbb{T})})$, where $\mathbb{T}$ is a rooted tree, such that (i) $\bigcup_{t \in V(\mathbb{T})} X_t = V(G)$, (ii) for every edge $xy \in E(G)$ there is a $t \in V(\mathbb{T})$ with $\{x, y\} \subseteq X_t$, and (iii) for every vertex $v \in V(G)$ the subgraph of $\mathbb{T}$ induced by the set $\{t \mid v \in X_t\}$ is connected.*

The width *of a tree decomposition is* $\max_{t \in V(\mathbb{T})} |X_t| - 1$ *and the treewidth of G is the minimum width over all tree decompositions of G and is denoted by* tw(G)*. The* adhesion *of an edge $\{t, t'\} \in E(\mathbb{T})$ is $|X_t \cap X_{t'}|$. The* adhesion *of a tree decomposition is the maximum adhesion of an edge in $\mathbb{T}$.*

Proposition 1. *For every graph G, there is a tree decomposition $(\mathbb{T}, \mathcal{X} = \{X_t\}_{t \in V(\mathbb{T})})$ of adhesion at most two, called a* Tutte decomposition*, such that the following conditions hold.*

- *For each node $t \in V(\mathbb{T})$, the graph* torso(G, X_t) *is a 3-vertex-connected topological minor of G.*
- *For each edge $\{t, t'\} \in E(\mathbb{T})$, either $X_t \cap X_{t'} = \emptyset$ or $X_t \cap X_{t'}$ is a minimal separator in G.*

Proposition 2 (Hopcroft and Tarjan [8]). *There is a linear time algorithm to compute a Tutte decomposition of a given graph.*

Recall that an edge-linked path is a sequence of distinct edges $e_1, e_2, \ldots, e_k$ such that every consecutive pair of edges in the sequence have a vertex in common. The length of an edge-linked path is the number of edges that belong to the path. We introduce the following problem:

LONG EDGE-LINKED PATHS   Parameter: k/tw
  Input: A graph G and an integer k
  Question: Does G have an edge-linked path of length at least k?

We denote by LELP(k) the problem of finding an edge-linked path of length at least k, parameterized by k. We will need the following result from [2]. In the following statement, following the notation of [2], we use K_2^3 to denote the graph with two vertices joined by three parallel edges. Further, a graph is said to be *Eulerian* if it is connected and all its vertices have even degree.

Theorem 3 ([2]). *Let G be a 3-edge-connected graph, $e, f \in E(G)$, and assume $G \neq K_2^3$. Then G contains an Eulerian subgraph H such that $e, f \in E(H)$ and $|E(H)| \geq (|E(G)|/6)^\alpha + 2$, where $\alpha \approx 0.753$ is the real root of $4^{1/x} - 3^{1/x} = 2$.*

Parameterized Complexity. We refer to [4] for a detailed introduction to parameterized complexity. We provide some key definitions pertinent to our arguments below.

Definition 2 (polynomial equivalence relation [3]**).** *An equivalence relation $\mathcal{R}$ on Σ^*, where Σ is a finite alphabet, is called a* polynomial equivalence relation *if the following holds: (1) equivalence of any $x, y \in \Sigma^*$ can be checked in time polynomial in $|x| + |y|$, and (2) any finite set $S \subseteq \Sigma^*$ has at most $(\max_{x \in S} |x|)^{\mathcal{O}(1)}$ equivalence classes.*

Definition 3 (cross-composition [3]**).** *Let $L \subseteq \Sigma^*$ and $Q \subseteq \Sigma^* \times \mathbb{N}$ be a parameterized problem. We say that L* cross-composes *into Q if there is a polynomial equivalence relation $\mathcal{R}$ on Σ^* and an algorithm which given t strings $x_1, \ldots, x_t$ belonging to the same equivalence class of $\mathcal{R}$, computes an instance $(x^*, k^*) \in \Sigma^* \times \mathbb{N}$ in time polynomial in $\sum_{i=1}^{t} |x_i|$ such that: (i) $(x^*, k^*) \in Q \Leftrightarrow x_i \in L$ for some $1 \leq i \leq t$ and (ii) k^* is bounded by a polynomial in $(\max_{1 \leq i \leq t} |x_i| + \log t)$.*

The following theorem allows us to rule out the existence of a polynomial kernel for a parameterized problem.

Theorem 4 ([3]). *If an* NP-*hard problem $L \subseteq \Sigma^*$ has a cross-composition into the parameterized problem Q and Q has a polynomial kernel then* coNP $\subseteq$ NP/poly.

Definition 4 (Turing kernelization). *Let Q be a parameterized problem and let $f \colon \mathbb{N} \to \mathbb{N}$ be a computable function. A* Turing kernelization *for Q of size f is an algorithm that decides whether a given instance (x, k) is contained in Q in time polynomial in $|x| + k$, when given access to an oracle that decides membership in Q for any instance (x', k') with $|x'|, k' \leq f(k)$ in a single step. We call such an oracle as f-oracle for Q.*

3   Kernelization Complexity for the Standard Parameter

As noted earlier, the hardness of kernelization for LELP follows by a standard disjoint union argument (see [4] for a similar example). We now demonstrate a cross-composition algorithm for LELP(k) on the class of connected graphs.

We introduce an auxiliary problem, where we seek to find an edge-linked path of length at least k that starts at a specified start vertex s. We denote this variant by $\text{LELP}_s(k)$. We will show that this version is NP-complete by a reduction from $\text{LELP}(k)$. We will then demonstrate a cross-composition from $\text{LELP}_s(k)$ to $\text{LELP}(k)$ where the composed instance will turn out to be connected. We defer the details of all these arguments to the full version of the paper due to lack of space. However, to provide some intuition, we describe briefly the construction that we employ to show the NP-completeness of $\text{LELP}_s(k)$.

Let (G, k) be an instance of $\text{LELP}(k)$. We denote the vertices of G by $\{v_1, \ldots, v_n\}$. Assume, without loss of generality, that $|V(G)| = 2^h$ for some h (if this is not the case, we can obtain an equivalent instance by adding an appropriate number of isolated vertices and at most doubling the size of the instance). Our construction involves a complete binary tree T of height h, and therefore, n leaves. Let r denote the root of T, and let $\ell_1, \ell_2, \ldots, \ell_n$ denote the leaves of T. We make n copies of G, denoted by $G_1, \ldots, G_n$. We make the vertex v_i in the copy G_i adjacent to the leaf ℓ_i. Denote this graph by H. Our reduced instance is now given by $(H, r, k + 2h + 1)$.

We now turn to the Turing kernel. Our goal will be to demonstrate a Turing kernel with $\mathcal{O}(k^{3.656})$ vertices for $\text{LELP}(k)$. To explain Turing kernelization for the problem we first define a closely related problem, namely: LONG EDGE-LINKED CYCLE. If the first edge and the last edge in an edge-linked path $P = e_1, e_2, \ldots, e_k$ has a common vertex, then we call P an *edge-linked cycle*. We use LELC to refer to the problem of determining if a graph has an edge-linked cycle of length at least k, parameterized by k.

Lemma 1. *If there is a Turing kernel for* LELC *of size* f, *then there is a Turing kernel for* $\text{LELP}(k)$ *of size* f *for any computable function* f.

Proof Sketch. It is easy to see that if G has an edge-linked path of length k, then there exist $u, v \in V(G)$ such that $G + e$ has an edge-linked cycle of length $k + 1$, where $e = \{u, v\}$. Moreover, for any $u, v \in V(G)$, if $G + e'$ (where $e' = \{u, v\}$) has an edge-linked cycle of length at least $k + 1$, then G has an edge-linked path of length k. As a result to test whether a graph G has an edge linked-path of length k, it is enough to test whether $G + \{\{u, v\}\}$ has an edge-linked cycle of length at least $k + 1$ for some $u, v \in V(G)$. But there is caveat here; we require an oracle for the same problem. In fact it is not hard to show that an oracle for LELC can be obtained using an oracle for $\text{LELP}(k)$ and vice versa, where the query lengths are asymptotically the same.   □

Thus, in this section we focus on proving the following theorem, which when combined with the hardness result mentioned above, amounts to a proof of Theorem 2.

Theorem 5. *There is a Turing kernel for* LELC *with* $\mathcal{O}(k^{3.656})$ *vertices and* $O(k^{4.656})$ *edges.*

We remark that our overall methodology is inspired by the approaches used by [10] to obtain Turing kernels for long path and cycle problems on special classes of graphs. Towards the proof of Theorem 5 we prove Lemma 2. Before stating Lemma 2 we define the notion of an irrelevant vertex.

Definition 5. *Let (G, k) be an instance of* LELC. *A vertex $v \in V(G)$ is called an* irrelevant vertex *for (G, k) if the following holds: (G, k) is a Yes-instance of* LELC *if and only if $(G - v, k)$ is a Yes-instance of* LELC.

We would like to mention that the size bound of the f-oracle for LELC refers to the bound on the number of vertices in the graph. Moreover, if there is a vertex of degree at least k, then the input is a Yes-instance and thus the bound on the edges will be at most k times the number of vertices.

Lemma 2. *There is a constant c and a function $f : \mathbb{N} \to \mathbb{N}$ defined as $f(x) = cx^{3.656}$ for any $x \in \mathbb{N}$, such that there is a polynomial time algorithm that given an instance (G, k) of* LELC, *uses an f-oracle for* LELC *to:*

- *either correctly determine if (G, k) is a Yes-instance of* LELC, *or,*
- *output an irrelevant vertex $v \in V(G)$.*

Given Lemma 2, the proof of Theorem 5 is straight forward: we can either solve the instance or make progress by deleting the irrelevant vertex. The rest of the section is devoted to prove Lemma 2. To this end, we first introduce and prove some auxiliary lemmas.

Proposition 3. *Let $k \in \mathbb{N}$. Any 3-edge-connected graph (or 3-vertex-connected graph) G with at least $k^{1.328}$ vertices contains an edge-linked cycle of length $\geq k$.*

Proof. Let G be a 3-edge-connected graph, $n = |V(G)|$ and $m = |E(G)|$. Theorem 3 implies the following argument. If $k \leq m^{0.753}$, then G has an Eulerian subgraph with at least k edges, and hence G has an edge-linked cycle of length at least k. Any 3-edge-connected graph G has at least $\frac{3}{2}|V(G)|$ edges. Therefore if $k \leq \left(\frac{3}{2} \cdot n\right)^{0.753}$, then (G, k) is a Yes-instance of LELC. Otherwise $n \leq \frac{2}{3} \cdot k^{1.328}$. Any 3-vertex-connected graph is also a 3-edge-connected graph. □

Lemma 3 ($\star^3$). *There is a constant c' and a polynomial time algorithm which given an instance (G, k) of* LELC *such that G is connected and $|V(G)| > c'k^{3.656}$, either correctly concludes that (G, k) is a Yes-instance or outputs a separation (A, B) of G of order at most 2 such that $7k < |A| \leq c' \cdot k^{3.656}$.*

The proof of Lemma 3 uses Proposition 2. Next we define an equivalent characterization for edge-linked paths and cycles in terms of trails and tours. A trail is a walk in which no edges repeats and a tour is a closed walk in which no edges repeats. If a trail (tour) covers all the edges in a graph G, then it is a Eulerian path (Eulerian cycle). As a result we have the following simple observations.

---

[3] The proofs of results marked with $\star$ are deferred to the full version of the paper.

Observation 2. *Let G be a graph. Let C be a tour in G and P be a trail with endpoints u and v, where $u \neq v$. Then, (i) degree of each vertex in the subgraph $(V(C), E(C))$ is even, and (ii) degree of any vertex $w \in V(P) \setminus \{u, v\}$ in the graph $G' = (V(P), E(P))$ is even and degrees of u and v in G' are odd.*

Observation 3. *Let G be a graph. Let C be a connected subgraph of G such that $d_C(v)$ is even for all $v \in V(C)$. Let P be a connected subgraph of G such that there exist two distinct vertices $x, y \in V(P)$ with the following properties: $d_P(x)$ and $d_P(y)$ are odd numbers, and $d_P(v)$ is an even number for all $v \in V(P) \setminus \{x, y\}$. Then, C is a tour, and P is a trail from x to y in G.*

The following observation characterizes edge linked paths and cycles in terms of trails and tours, respectively.

Observation 4. *Let G be a graph. There is an edge linked path of length k in G if and only if there is a trail P in G such that $|E_{in}(V(P))| \geq k$. Also, there is an edge linked cycle of length at least k in G if and only if there is a tour C in G such that $|E_{in}(V(C))| \geq k$.*

Proof. Let $P = u_1 \ldots u_\ell$ be a trail in G such that $|E_{in}(V(P))| \geq k$. Let $e_i = \{u_i, u_{i+1}\}$ for all $i \in \{1, \ldots, \ell - 1\}$. Let $\{i_1, \ldots, i_{\ell'}\} \subseteq \{1, \ldots, \ell\}$ be the maximal subset such that u_{i_j} appear first in i_j th position in P for all $j \in \{1, \ldots, \ell'\}$. For each $j \in \{1, \ldots, \ell'\}$ let S_{i_j} be the set of edges incident to u_{i_j}, but not in $E(P)$ and not incident to any u_{i_r} for some $r < j$. For all $q \notin \{i_1, \ldots, i_{\ell'}\}$, $S_q = \emptyset$. Then the sequence of edges in $S_1 e_1 S_2 \ldots S_{\ell-1} e_{\ell-1} S_\ell$ forms an edge-linked path containing all the edges in $E_{in}(V(P))$. Similarly, we can prove that there is an edge-linked cycle containing all the edges in $E_{in}(V(C))$. □

Due to Observation 4, to solve LELC, it is enough to test for the existence of a tour C such that $|E_{in}(V(C))| \geq k$. In the following two lemmas we summarize the behaviour of a tour C across a separation of order at most 2.

Lemma 4. *Let G be a graph and (A, B) be a separation of G. Let C be a tour in G such that $|V(C) \cap (A \cap B)| = 1$. Let $x \in V(C) \cap (A \cap B)$. Then, one of the following is true.*

- *C is contained in $G[A]$ or $G[B]$.*
- *$E(C) \cap E(G[A])$ forms a tour C' in $G[A]$ and $x \in V(C')$.*

Proof. Suppose C is contained in $G[A]$ or $G[B]$, then we are done. Otherwise, let F be the subgraph induced on $E(C) \cap E(G[A])$. Clearly $x \in V(F)$. By Observation 2, for any $u \in V(F) \setminus \{x\}$, we know that $d_F(u)$ is even. Then by Observation 1, $d_F(x)$ is even. Since C is connected, F is also connected. Therefore, by Observation 3, F is a tour in $G[A]$. This completes the proof of the lemma. □

Lemma 5 ($\star$). *Let G be a graph and (A, B) is a separation of G of order 2. Let C be a tour in G and $A \cap B = \{x, y\} \subseteq V(C)$. Let $G_A = G[A]$ and $G_B = (B, E(G) \setminus E(G_A))$. Then, one of the following is true.*

1. C is contained in $G_{B'}$.
2. There exists a tour C' in G_A such that $E(C') = E(C) \cap E(G_A)$.
3. There exist two vertex disjoint tours C_1 and C_2 in G_A such that $x \in V(C_1), y \in V(C_2)$, and $E(C_1) \cup E(C_2) = E(C) \cap E(G_A)$.
4. There exists a trail P from x to y in G_A such that $E(P) = E(C) \cap E(G_A)$.

By Observation 4, we know that to test for the existence of an edge-linked cycle of length at least k it is enough to test for the existence of a tour such that at least k edges are incident on the vertices of the tour.

Lemma 6. *There exists two constant c_1 and c_2 (where $c_2 < c_1$ and c_2 is same as the constant in Lemma 3), and a function $f : \mathbb{N} \to \mathbb{N}$ defined as $f(x) = c_1 x^{3.656}$ for any $x \in \mathbb{N}$ such that the following holds. There is a polynomial time algorithm $\mathcal{A}$ which given an instance (G, k) of LELC and a separation (A, B) of G of order at most 2 such that $7 \cdot k < |A| \le c_2 \cdot k^{3.656}$ and max degree of G at most k, uses an f-oracle for LELC and outputs one of the following.*

- *Correctly concludes that (G, k) is a Yes-instance of LELC.*
- *Outputs an irrelevant vertex $v \in V(G)$.*

Proof. Let $G_A = G[A]$, $G_B = (B, E(G) \setminus E(G_A))$. Algorithm $\mathcal{A}$ first queries the f-oracle with input (G_A, k). If the oracle returns Yes, then $\mathcal{A}$ concludes that (G, k) is a Yes-instance of LELC. From now on we assume that (G_A, k) is a No-instance. We have the following three cases based on $|A \cap B|$.

Case 1 : $|A \cap B| = 0$. In this case any edge-linked cycle is either contained in G_A or contained in G_B. Since (G_A, k) is a No-instance, any vertex in A is an irrelevant vertex.

Case 2 : $|A \cap B| = 1$. Let $\{x\} = A \cap B$. First we prove the following claim.

Claim 1 ($\star$). *Let (G_A, k) be a No-instance of LELC. Let $k' < k$ be the largest integer such that there is a tour Q in G_A such that $x \in V(Q)$ and $|E_{in}^{G_A}(V(Q))| = k'$. If there is a tour Q' in $G_A - v$ such that $x \in V(Q')$ and $|E_{in}^{G_A-v}(V(Q'))| = k'$ for some $v \in A \setminus N[x]$, then v is an irrelevant vertex.*

We note that the integer k' can be identified using f-oracle as follows. We construct a new graph G' by adding a new cycle D of length $2k$ with vertices x and $2k - 2$ new vertices. Since (G_A, k) is a No-instance, any tour C with $|E_{G'}^{in}(C)| \ge 2k + k'$ contains D (i.e., $E(D) \subseteq E(C)$). Therefore, we can identify k' by calling f-oracle on $(G', 2k + k_1)$ for all $k_1 < k$. The largest k_1 for which $(G', 2k + k_1)$ is a Yes-instance is k'. We will set the value of c_1 to be at least $c_2 + 2$ so that we can use f-oracle. After identifying the value of k' we again use f-oracle to identify an irrelevant vertex v. Towards that we query f-oracle with $(G' - u, k' + 2k)$ for all $u \in A \setminus N[x]$ and there will be at least one vertex v for which f-oracle outputs Yes on input $(G' - v, k' + 2k)$. That vertex v is an irrelevant vertex because of Claim 1.

Case 3 : $|A \cap B| = 2$. Let $A \cap B = \{x, y\}$. First we prove the following claim.

Claim 2. Suppose there is a path between x and y in G_B and there is trail P from x to y in G_A such that $|E_{in}^{G_A}(V(P))| \geq k-1$. Then (G, k) is a Yes instance.

Proof. Let P' be a path from x to y in G_B. Then the edges of P and P' forms a tour C in G. Since $E(P')$ is disjoint from $E(G_A)$ and $|E_{in}^{G_A}(V(P))| \geq k - 1$, we have that $|E_{in}^{G}(V(C))| \geq k$. Then, by Observation 4, (G, k) is a Yes-instance. $\square$

Algorithm $\mathcal{A}$ will test whether there is a path between x and y in G_B and the existence of a trail P from x to y in G_A with $|E_{in}^{G_A}(V(P))| \geq k-1$ using f-oracle as follows. We construct a graph G' by adding a $2k$ length path L between x and y. Since (G_A, k) is a No-instance, and any tour C with $|E_{in}^{G'}(V(P))| \geq 3k - 1$ will contain the path L. Therefore $(G', 3k - 1)$ is a Yes-instance of LELC if and only if there is a trail P from x to y in G_A such that $|E_{in}^{G_A}(V(P))| \geq k-1$. Then by Claim 2, we conclude that (G, k) is a Yes instance.

Now on we assume that if G_B contains a path between x and y, then there is no trail P from x to y in G_A with $|E_{in}^{G_A}(V(P))| \geq k - 1$. We use f-oracle to output an irrelevant vertex as follows. We construct three graphs G_1, G_2, G_3, and G_4 as follows. The graph G_1 is created by adding a cycle L_1 of length $4k$ containing x and $4k - 1$ new vertices. The graph G_2 is created by adding a cycle L_2 of length $4k$ containing y and $4k - 1$ new vertices. The graph G_3 is created by adding two internally vertex disjoint paths Q_3 and Q_3' of length $2k$ each between x and y. The graph G_4 is created by adding a path Q_4 of length $4k$ between x and y. For each $i \in \{1, \ldots, 4\}$, we use f-oracle to compute the largest $1 \leq k_i < k$ such that $(G_i, 4k + k_i)$ is a Yes-instance. If no such k_i exists we set $k_i = -\infty$. Clearly this can be done by querying the f-oracle at most k times for each $i \in \{1, \ldots, 4\}$. For any $i \in \{1, \ldots, 4\}$, if $k_i \neq -\infty$, any tour C_i in G_i with $|E_{in}^{G_i}(V(C_i))| \geq 4k + k_i$ will contain the newly added $4k$ edges. Therefore all but k_i vertices in G_A are irrelevant for the instance $(G_i, 4k + k_i)$. These vertices can be identified using f-oracle. So Algorithm $\mathcal{A}$ identifies at most $4k$ vertices that are relevant for at least one of $(G_i, 4k + k_i)$. Since $|A| > 7k$, there is one vertex $v \in A \setminus (N[x] \cup N[y])$ which is irrelevant for $(G_i, 4k + k_i)$ for all $i \in \{1, \ldots, 4\}$. Algorithm $\mathcal{A}$ will output such a vertex v as an irrelevant vertex for the instance (G, k). One can show that vertex v is indeed an irrelevant vertex for (G, k). This concludes the proof of the Lemma.

$\square$

We are now ready to give the proof of Lemma 2.

Proof of Lemma 2. First we set constants c_1 and c_2 mentioned by Lemma 6. Then we set the constant c mentioned in the lemma to be c_1. Let (G, k) be the input instance. If there is a vertex $v \in V(G)$ with degree at least k, then we conclude that (G, k) is a Yes-instance. Let H be a connected component of G. If $|V(H)| \leq c_2 k^{3.656}$, then we use f-oracle to test whether (H, k) is a Yes-instance or not. If (H, k) is not a Yes-instance, then we output any vertex in $V(H)$ as an irrelevant vertex.

Now consider the case $|V(H)| > c_2 k^{3.656}$. We use Lemma 3 on (H, k). If Lemma 3 concludes that (H, k) is a Yes-instance, then (G, k) is indeed a Yes-instance. Otherwise Lemma 3 outputs a separation (A, B') of H of order at most 2 such that $7k < |A| \leq c_2 \cdot k^{3.656}$. Notice that $(A, B = B' \cup (V(G) \setminus V(H)))$ is a separation of G of order at most 2. Now we use Lemma 6 to either conclude that (G, k) is a Yes-instance or to output an irrelevant vertex.   □

4   Concluding Remarks

In this contribution, we introduced and studied a natural optimization variant of the notion of an edge Hamiltonian path. Specifically, we ask if a graph admits such an edge-linked path of length at least k. We show that the problem is FPT when parameterized by k, and admits a single-exponential algorithm when parameterized by treewidth. The latter generalizes and improves known algorithms for the special case of the Edge Hamiltonicity problem (where $k = m$) when parameterized by treewidth. While we used the technique of representative families, possibly, other techniques can also be employed to obtain similar running time dependency on the treewidth parameter (for example, using rank-based approaches [5]).

We also studied the kernelization complexity of the problem, showing that it is unlikely to admit a polynomial kernel even if the input graph is connected. Our main result in this context was a Turing kernel for the problem, which made use of Tutte decompositions and the existence of long edge-linked paths on 3-connected graphs. An interesting open problem here is to determine the kernelization complexity of the problem on two-connected graphs. We conjecture that the problem remains hard for this class of graphs as well.

References

1. Bertossi, A.A.: The edge Hamiltonian path problem is NP-complete. Inf. Process. Lett. **13**(4/5), 157–159 (1981)
2. Bilinski, M., Jackson, B., Ma, J., Yu, X.: Circumference of 3-connected claw-free graphs and large Eulerian subgraphs of 3-edge-connected graphs. J. Comb. Theory Ser. B **101**(4), 214–236 (2011)
3. Bodlaender, H.L., Jansen, B.M.P., Kratsch, S.: Kernelization lower bounds by cross-composition. SIAM J. Discrete Math. **28**(1), 277–305 (2014)
4. Cygan, M., et al.: Parameterized Algorithms. Springer, Cham (2015). https://doi.org/10.1007/978-3-319-21275-3
5. Cygan, M., Kratsch, S., Nederlof, J.: Fast hamiltonicity checking via bases of perfect matchings. J. ACM **65**(3), 12:1–12:46 (2018)
6. Demaine, E.D., Demaine, M.L., Harvey, N.J.A., Uehara, R., Uno, T., Uno, Y.: UNO is hard, even for a single player. Theor. Comput. Sci. **521**, 51–61 (2014)
7. Diestel, R.: Graph Theory. Graduate Texts in Mathematics, vol. 173, 4th edn. Springer, Heidelberg (2012)
8. Hopcroft, J., Tarjan, R.: Dividing a graph into triconnected components. SIAM J. Comput. **2**(3), 135–158 (1973)

9. Jansen, B.M.P., Pilipczuk, M., Wrochna, M.: Turing kernelization for finding long paths in graphs excluding a topological minor. In: IPEC, vol. 89. Schloss Dagstuhl - Leibniz-Zentrum fuer Informatik (2017)

10. Jansen, B.M.: Turing kernelization for finding long paths and cycles in restricted graph classes. J. Comput. Syst. Sci. **85**(C), 18–37 (2017)

11. Lai, T.H., Wei, S.S.: The edge Hamiltonian path problem is NP-complete for bipartite graphs. Inf. Process. Lett. **46**(1), 21–26 (1993)

12. Lampis, M., Makino, K., Mitsou, V., Uno, Y.: Parameterized edge hamiltonicity. Discrete Appl. Math. **248**, 68–78 (2018)

13. Ryjáček, Z., Woeginger, G.J., Xiong, L.: Hamiltonian index is NP-complete. Discrete Appl. Math. **159**(4), 246–250 (2011)

14. Thomassé, S., Trotignon, N., Vuskovic, K.: A polynomial turing-kernel for weighted independent set in bull-free graphs. Algorithmica **77**(3), 619–641 (2017)

The Parameterized Complexity
of Dominating Set and Friends Revisited
for Structured Graphs

Neeldhara Misra[(✉)] and Piyush Rathi

Indian Institute of Technology, Gandhinagar, Palaj, Gandhinagar 382355, India
{neeldhara.m,piyush.rathi}@iitgn.ac.in

Abstract. We consider variants and generalizations of the dominating set problem on special classes of graphs, specifically, graphs that are a small distance from a tractable class. Here, our focus is mainly on the problems of domination and efficient domination (a variant where we want every vertex to be dominated uniquely) and their respective generalizations to r-distance domination.

We consider graphs which are at most k vertices away from the following classes: edgless graphs, cluster graphs, split graphs, and complements of bipartite graphs. For the newly introduced parameter CBDS, we show that DOMINATING SET is W[2]-hard, while in contrast, r-DS is FPT for $r > 1$. For this parameter, EFFICIENT DOMINATING SET turns out to be FPT as well. We generalize known results for DOMINATING SET parameterized by CVD to r-DOMINATING SET parameterized by CVD, inspired by algorithms for r-DS parameterized by VC, which is in general a larger parameter. We are also able to do this for the EFFICIENT DOMINATING SET problem. Finally, for the problem of efficient domination parameterized by the distance to split graphs, we improve the complexity of the known algorithm from $\mathcal{O}^\star(1.732^k)$ to $\mathcal{O}^\star(1.616^k)$.

For the most part, our results generalize what is known in the current landscape, both in terms of obtaining algorithms for smaller parameters, as well as for the more general notion of domination itself, which is to dominate over larger distances.

Keywords: Dominating set · Efficient domination · FPT ·
Kernelization · W-hardness · Structural parameters

1 Introduction

A dominating set of a graph $G = (V, E)$ is a subset of vertices S such that every vertex is either a member of S or has a neighbor in S. The associated computational question, namely that of determining a dominating set of the smallest size, is a classical NP-hard problem, has numerous applications, and has been one of the most widely studied optimization problems. Since the dominating set problem is intractable in both the classical and parameterized settings, much work in

© Springer Nature Switzerland AG 2019
R. van Bevern and G. Kucherov (Eds.): CSR 2019, LNCS 11532, pp. 299–310, 2019.
https://doi.org/10.1007/978-3-030-19955-5_26

the literature is devoted to the study of approximation algorithms, algorithms for special classes of graphs, and parameterized algorithms.

We also consider the efficient dominating set problem, which requires that every vertex in the given graph be dominated *exactly* once. While a dominating set in a graph can be interpreted as a set cover—by thinking of the vertices as constituting the universe and treating the closed neighborhoods of vertices as the family—an efficient dominating set, within the same interpretation, can be thought of as a set packing. Note that since a vertex dominates itself, an efficient dominating set induces an independent set.

Finally, we also consider the problem of r-domination, which is a generalization where vertices can dominated not only their neighbors but all vertices that are distance at most r away from them. While the notion of a r-dominating set is clear already from this description, we state explicitly what we mean by a r-efficient dominating set. Informally, a r-efficient dominating set corresponds to a set packing by r-balls around the elements of the dominating set. More precisely, if S is a r-efficient dominating set, then every vertex of the graph is at distance at most r from exactly one vertex in S. Just as an efficient dominating set induces an independent set, we note that an r-efficient dominating set induces a $(r+1)$-scattered set (i.e, a set of vertices where the pairwise distances are at least $r+1$, note that an independent set is a 2-scattered set).

The Framework. We consider these problems in the context of special classes of graphs, specifically, graphs that are a small distance from a tractable class. In this setting, our typical framework is the following: we are given as input a graph $G = (V, E)$ along with a "modulator" X to a graph class $\mathcal{G}$, which is a subset of vertices such that the graph remaining after the removal of X belongs to the class $\mathcal{G}$. The goal is to determine if the problem considered is tractable with respect to the size of the modulator. For this reason, we restrict our attention to classes $\mathcal{G}$ on which the problems under consideration are already tractable. Indeed, note that if the problem is already intractable (say, for instance, NP-hard) on the class $\mathcal{G}$, then it is intractable for the special case of an empty modulator and therefore such cases are not interesting in this particular framework.

We consider graphs that are at most k vertices away from the following classes: edgless graphs, cluster graphs, split graphs, and co-bipartite graphs, which are complements of bipartite graphs (i.e, graphs whose vertex sets can be partitioned into two cliques). We refer the reader to Sect. 2 for the definitions of these graph classes. For each of these classes, we remark that finding a relevant modulator of size at most k is known to be fixed-parameter tractable in k, which justifies the promise versions that we are considering. In particular, we remark that for the first three classes, the problem of finding a modulator corresponds to the well-studied problems of Vertex Cover (VC), Cluster Vertex Deletion (CVD), and Split Vertex Deletion (SVD); while we can find a modulator to a co-bipartite graph (CBDS) by finding a modulator to a bipartite graph in the complement graph, which corresponds to the Odd Cycle Traversal problem. For further discussions about these parameters, we will usually refer to the problem considered by phrases such as "dominating set parameterized by CVD".

Our Contributions

For DOMINATING SET, a FPT algorithm with respect to CVD was given by [3]. This algorithm runs in time $\mathcal{O}^\star(3^k)$. For SVD, the problem is para-NP-hard. Note that DOMINATING SET can be solved immediately on co-bipartite graphs, so unlike SVD, it is not immediately para-NP-hard when parameterized by CBDS. However, we show that the problem is intractable in the parameterized setting.

Theorem 1 (Dominating Set by CBDS). DS *parameterized by* CBDS *is* W[2]-*hard*.

In extending the results for DS to r-DS, we show that r-DS is FPT when parameterized by CVD and CBDS, generalizing the FPT algorithm given by [4] for r-DS for the vertex cover parameter. The result for CVD is reasonably predictable, however it is worth noting that the running time does not grow exponentially with r. The result for CBDS is interesting in that there is a significant change in the complexity: while DOMINATING SET is already W[2]-hard for the parameter, r-DOMINATING SET turns out to be FPT for $r > 1$. Observe that every connected split graph has a two-dominating set of size one, therefore the two-dominating set problem on any split graph is in P, and the problem of parameterizing by SVD is now a reasonable question. A similar reasoning holds for the CBDS parameter.

Theorem 2 (r-Dominating Set by CVD and CBDS). r-DS *is fixed-parameter tractable parameterized by* CVD *and* CBDS *with running times* $\mathcal{O}^\star(7^k)$ *and* $\mathcal{O}^\star(11^k)$ *respectively*.

We now turn our attention to EFFICIENT DOMINATING SET. Like DOMINATING SET, a FPT algorithm with respect to CVD was given by [3], while for SVD, the authors demonstrated a branching algorithm with running time $\mathcal{O}^\star(1.732^k)$. We improve this running time to $\mathcal{O}^\star(1.616^k)$ and also show that the problem is FPT when parameterized by the CBDS parameter.

Theorem 3 (Efficient Dominating Set by SVD and CBDS). EDS *parameterized by* SVD *and* CBDS *admits* FPT *algorithms with running times* $\mathcal{O}^\star(1.616^k)$ *and* $\mathcal{O}^\star(5^k)$ *respectively*.

Finally, for the EFFICIENT r-DOMINATING SET problem, analogous to r-DOMINATING SET, we show that EFFICIENT r-DOMINATING SET is FPT when parameterized by CVD.

Theorem 4 (Efficient r-Dominating Set by CVD). r-EDS *is fixed-parameter tractable parameterized by* CVD *with running time* $\mathcal{O}^\star(7^k)$.

Table 1. Summary of new and known results. Results obtained in this contribution are shown in bold.

	VC	CVD	SVD	CBDS
DS	$\mathcal{O}^\star(3^k)$	$\mathcal{O}^\star(3^k)$	para NP-hard	W[2]-hard
	(due to CVD)	[3, Theorem 3]	(hardness on split graphs [3])	**Theorem 1**
r-DS	$\mathcal{O}^\star(5^k)$	$\mathcal{O}^\star(7^k)$	Open	$\mathcal{O}^\star(11^k)$ for $r > 7$
	[4, Theorem 19]	**Theorem 2**		**Theorem 2**
EDS	$\mathcal{O}^\star(1.61^k)$	$\mathcal{O}^\star(3^k)$	$\mathcal{O}^\star(1.616^k)$	$\mathcal{O}^\star(3^k)$
	Theorem 3	[3, Theorem 4]	**Theorem 3 (improves [3])**	**Theorem 3**
r-EDS	$\mathcal{O}^\star(5^k)$	$\mathcal{O}^\star(7^k)$	Open	Open
	(due to CVD)	**Theorem 4**		

Related Work. Our contribution most directly builds on the works of Goyal et al. [3] and Katsikarelis et al. [4], where several FPT algorithms for structural parameters are suggested alongside lower bounds based on ETH and SETH. The DOMINATING SET problem has enjoyed a long history of being studied on special classes of graphs especially from a parameterized perspective. For instance, the contributions of [5–7] study the dominating set problem on graphs that exclude short cycles and other related subgraphs, while, for instance, the work of [1] studies the r-dominating set problem over tree decompositions.

2   Preliminaries

For any two vertices u and v, we let $d(u,v)$ denote the length of the shortest path between the vertices. For $S \subseteq V$, $G[S]$ denotes the graph *induced* by S in G. The vertex set of $G[S]$ is S, and the edge set is $\{(u,v) \mid u \in S, v \in S \text{ and } (u,v) \in E\}$. The r-*neighborhood* of a vertex v is the set of all vertices that are at distance at most r from v. The r-neighborhood of a vertex is denoted by $N_r(v)$, and the *closed* r-neighborhood of a vertex, given by $N_r(v) \cup \{v\}$, is denoted by $N_r[v]$. The r-neighborhood of a subset of vertices S is $\cup_{v \in S} N_r(v)$, and is denoted by $N_r(S)$. Likewise, the closed r-neighborhood of a subset of vertices S is $N_r(S) \cup S$, and is denoted by $N_r[S]$. We say that a vertex v is *global* to a set S of vertices if v is adjacent to every vertex in S.

The DOMINATING SET and EFFICIENT DOMINATING SET problems are defined as follows:

DOMINATING SET   **Parameter:** k
Input: A graph $G = (V, E)$, and an integer k.
Question: Does G have a subset S of at most k vertices such that for every $v \in V$, $|N[v] \cap S| \geqslant 1$?

EFFICIENT DOMINATING SET   **Parameter:** k
Input: A graph $G = (V, E)$, and an integer k.
Question: Does G have a subset S of at most k vertices such that for every $v \in V$, $|N[v] \cap S| = 1$?

We defer the definitions of the r-domination analogs of these problems to the full version of the paper.

DOMINATING SET is a fundamental NP-complete and W[2]-complete problem. We also define COLORFUL RED-BLUE DOMINATING SET: here, we are given a bipartite graph $G = (V, E)$ with bipartition (R, B), a partition of R into k parts $R_1, \ldots, R_k$, and an integer k as input, and the question is if G has a subset S vertices such that for every $v \in B$, $|N(v) \cap S| \geqslant 1$ and $|S \cap R_i| = 1$ for all $i \in [k]$. This problem is also W[2]-hard parameterized by k.

When considering structural parameters, and in particular those that are based on modulators to graph classes, we will typically continue dealing with the decision versions of the questions as above, but the instance will additionally be equipped with a modulator—usually denoted X. The parameter for all such problems is $|X|$. We refer the reader to [2] for a comprehensive introduction to the framework of parameterized algorithms.

In the classical SET COVER problem, we are given as input a family $\mathcal{F}$ of sets over an universe $\mathcal{U}$, and a budget k. The question is if there is a subfamily of size at most k whose union is $\mathcal{U}$. The problem is W[2]-hard when parameterized by k, but FPT when parameterized by $|\mathcal{U}|$. The SET PARTITION problem is analogously defined, with the only difference being that we require the solution to be a partition, rather than a cover of the universe. We will also require the notion of a SET COVER WITH PARTITION, defined below, which was introduced by [3], where it was also shown to be FPT.

SET COVER WITH PARTITION   **Parameter:** $|\mathcal{U}|$
Input: A universe $\mathcal{U}$, a family of sets $\mathcal{F} = \{S_1, \ldots, S_m\}$, a partition $\mathcal{B} = (B_1, B_2, ..., B_q)$ of $\mathcal{F}$, a $(0, 1)$ vector $(f_1, f_2, \ldots, f_q)$ corresponding to each block in the partition $\mathcal{B} = (B_1, B_2, ..., B_q)$ and an integer ℓ.
Question: Does there exist a subset $\mathcal{H} \subseteq \mathcal{F}$ of size ℓ covering $\mathcal{U}$ and from each block B_i with flags $f_i = 1$ at least one set is picked?

The partition analog of SET COVER WITH PARTITION can also be seen to be FPT. We refer to this version as EXACT SET COVER WITH PARTITION.

We briefly define some of the graph classes that are relevant to our discussions. A *clique* is a collection of vertices that are mutually adjacent, while an *independent set* is a collection of vertices that have no edges between them. A *cluster graph* is a disjoint union of cliques. A *split graph* is a graph whose vertex set can be partitioned into a clique and an independent set. A *bipartite graph* is a graph whose vertex set can be partitioned into two independent sets, while a *co-bipartite graph* is a graph whose vertex set can be partitioned into two cliques.

The rest of this paper is organized as follows. Our main focus will be to sketch the details of the algorithm for r-DOMINATING SET parameterized by CVD, which outlines some of the main ideas that are relevant to all the other algorithms we propose. The algorithms for EFFICIENT DOMINATING SET require a notion of inseparable sets for an accurate set cover formulation. The details of all the other algorithms, including the branching algorithm for the SVD parameter, are given in the full version of this manuscript.

3   Dominating Set

We already know that DOMINATING SET admits a FPT algorithm when parameterized by CVD and is para-NP-hard when parameterized by SVD. In this section, we consider the CBDS parameter, with respect to which DS turns out to have an "intermediate" complexity.

Theorem 1 (Dominating Set by CBDS). DS *parameterized by* CBDS *is* W[2]-hard.

Proof. We establish this by a reduction from the COLORFUL RED-BLUE DOMINATING SET problem. Let $\mathcal{J} = ((G = (V, E); V = (R, B); R = R_1, \ldots, R_k), k)$ be an instance of COLORFUL RBDS. We let $\mathcal{H} = (H, \ell; X)$ denote the reduced instance, where X denotes a CBDS modulator, ℓ denotes the upper bound on the size of the dominating set sought, and the parameter is $|X|$. We first describe the graph H. To begin with, we make a copy of G and make cliques out of the vertex subsets R and B. Further, we add k vertices $r_1, \ldots, r_k$, and make the vertex r_i global to the subset R_i. Note that $X := \{r_1, \ldots, r_k\}$ is a CBDS modulator for the graph H described so far. We let $\ell = k$, and this completes the reduction. We now turn to the equivalence of the two instances.

In the forward direction, let $S \subseteq R$ be a colorful red-blue dominating set. We choose the vertices corresponding to S in H and note that every vertex in X is dominated since $|S \cap R_i| = 1$ for all $i \in [k]$; further, since R induces a clique, every vertex in R is dominated by all the vertices of S and finally, every vertex in B is dominated because S dominates them in G by definition and the edges between R and B are identical in H and G.

In the reverse direction, let $S \subseteq V(H)$ be a dominating set. Note that $N(r_i) = R_i$, therefore:

$$S \cap (R_i \cup \{r_i\}) \neq \emptyset \text{ for all } i \in [k] \text{ and } |S| \leqslant k \Rightarrow |S \cap (R_i \cup \{r_i\})| = 1 \text{ for all } i \in [k].$$

In particular, note that $S \cap B = \emptyset$. Now, define $S^\star \subseteq V(G)$ as follows: for $i \in [k]$, if $|S \cap R_i| = 1$, then we choose the corresponding vertex from R_i in $S^\star$, and if $r_i \in S$, then we choose an arbitrary vertex from R_i in S. It is easy to see that $S^\star$ is a colorful red-blue dominating set for G: indeed, $|S^\star \cap R_i| = 1$ for all $i \in [k]$ by construction, and if any vertex $v \in B$ is not dominated by $S^\star$, then the corresponding vertex is not dominated by S in H, which contradicts our assumption that $S \subseteq V(H)$ was a dominating set in H. This completes the argument for the equivalence of the instances. □

4 r-Dominating Set

In this section, we discuss our FPT algorithm for r-DOMINATING SET parameterized by CVD.

Let $\mathcal{J} = (G = (V, E), k; X)$ be an instance of r-DOMINATING SET, where X denotes the modulator under consideration. Here, we are considering the CVD parameter. To begin with, we make the simplifying assumption that $r > 3$, while noting that similar ideas with appropriate modifications can be used to handle the special cases $r = 2$ and $r = 3$. For CVD, recall that $G \setminus X$ is a disjoint union of cliques, which we will denote by $C_1, C_2, \ldots, C_t$. Assume, without loss of generality, that every C_i has a neighbor in X—indeed, note that any clique for which this is not true would be isolated and can be easily dealt with separately. We use S to denote an arbitrary but fixed (and unknown) r-DS of optimal size, and let $\ell := |S|$. We also recall that we will often use, for a vertex $v \in V$ and a subset of vertices $Z \subseteq V$, the phrase "distance of v from Z" to refer to the distance between v and the vertex in Z that is the closest to v.

We begin by "guessing" the intersection of an optimal solution with the modulator, namely $Y := S \cap X$. Based on Y, we obtain a classification of the cliques in $G \setminus X$ determined by the extent to they are r-dominated by the vertices of Y:

- A clique $C \in G \setminus X$ is *completely dominated* if all vertices in C are at distance at most r away from some vertex in Y. Note that this happens either if C has a vertex v whose distance from Y is at most $r - 1$ or every vertex in C is at distance exactly r from Y.
- A clique $C \in G \setminus X$ is *not dominated* if the distance of every vertex in C from Y is strictly greater than r.
- Any clique that does not fall into the categories above is said to be *partially dominated*. Such cliques have a non-trivial subset of vertices that are at distance exactly r from Y (which we refer to as *peripheral vertices*), while the distance of all other vertices from Y is greater than r (we refer to these as *internal vertices*).

We now guess a partition of $X \setminus Y$ into three parts:

$$D_{\leqslant r-2} \uplus D_{r-1} \uplus D_r$$

with the semantics that the distances of the vertices in D_r (respectively, D_{r-1}) from S is exactly r (respectively, $r - 1$) and the distance of any vertex in $D_{\leqslant r-2}$ from S is at most $r - 2$. This motivates the notion of a *good partition*.

Definition 1. *A partition $\mathcal{P}$ of X into* $Y \uplus D_{\leqslant r-2} \uplus D_{r-1} \uplus D_r$ *is called a good partition if there exists an optimal r-dominating set, say S, such that* $S \cap X = Y$, *and the distances of the vertices in* D_r *(respectively,* D_{r-1}*) from S is exactly r (respectively, $r-1$) and the distance of any vertex in* $D_{\leqslant r-2}$ *from S is at most $r-2$. We refer to S as a witness for $\mathcal{P}$.*

Now, we make some observations about the nature of this partition.

Lemma 1. *Let* $Y \uplus D_{\leqslant r-2} \uplus D_{r-1} \uplus D_r$ *be a good partition of X with a witness S. Further, let C be a clique that is not dominated by Y. Then, C either has a vertex v which has a neighbor in* $D_{\leqslant r-2}$ *or every vertex in C has a neighbor in* D_{r-1}.

Proof. First, assume that S contains no vertex from C. Since S is a r-dominating set, every vertex in C is at distance at most r from S. Suppose there is a vertex v in C whose distance from S is less than r. Let $u \in S$ be such that $d(u,v) = d < r$. Consider the shortest path P from u to v. Since $S \cap C = \emptyset$, and C has no neighbors in Y, note that $|V(P)| \geqslant 3$. Now, let w be the neighbor of v in P. If w belongs to X, then note that $w \in D_{\leqslant r-2}$, and we are done. Note that $w \notin Y$ because C is not dominated by Y (by assumption). If $w \in C$, then we can repeat this argument using w instead of v as our choice of a vertex whose distance from S is less than r.

If there is no vertex v whose distance from S is less than r, then the distance of every vertex in C from S is exactly r. Let v be any vertex in C, and let $u \in S$ be such that $d(u,v) = d < r$. Consider the shortest path P from u to v. Further, let w be the neighbor of v in P. Observe that $w \in D_{r-1}$. In particular, $w \notin C$—if $w \in C$, then the distance of w from S via P would be less than r, contradicting the case we are in. Also, $w \notin Y$ for the same reason as in the previous case. Since the choice of v was arbitrary, we have that every vertex in C indeed has a neighbor in D_{r-1} in this situation.

Now suppose S contains a vertex v from C. If v has a neighbor u in X, then note that $u \notin Y$ by the assumption that C is a clique that is not dominated by Y. Therefore, $u \in D_{\leqslant r-2}$. On the other hand, if v has no neighbors in X, then let $w \in C$ be such that w has a neighbor u in $X \setminus Y$. Note that such a w must exist because of our assumption that C has a neighbor in X and that C is not dominated by Y. However, note that $d(v,u) = 2$, and since $r \geqslant 4$ by assumption, we have that $r-2 \geqslant 2$, and therefore $u \in D_{\leqslant r-2}$. $\square$

It is easy to see that a similar statement holds also for the partially dominated cliques. For these cliques, the peripheral vertices are already "taken care of", but it turns out that we don't have to account for them separately. We state the following without proof since the arguments are identical to those used in Lemma 1.

Lemma 2. *Let* $Y \uplus D_{\leqslant r-2} \uplus D_{r-1} \uplus D_r$ *be a good partition of X with witness S. Further, let C be a clique that is partially dominated by Y. Then, C either has a vertex v which has a neighbor in* $D_{\leqslant r-2}$ *or every vertex in C has a neighbor in* D_{r-1}.

The following observation is easy to check and follows from the fact that $Y \subseteq S$.

Lemma 3. *Let* $Y \uplus D_{\leqslant r-2} \uplus D_{r-1} \uplus D_r$ *be a good partition of* X *with witness* S. *Then there is no vertex in* D_r *whose distance from* Y *is less than* r, *and there is no vertex in* D_{r-1} *whose distance from* Y *is less than* $r - 1$.

Recall that our algorithm so far has guessed a partition of X into $Y \uplus D_{\leqslant r-2} \uplus D_{r-1} \uplus D_r$. Our first step for any guess will be to implement the conditions in Lemmas 1, 2 and 3 to check if this is a potentially good partition. In other words, if any of the conditions are violated, then we discard the partition in question and move on to the next one. We refer to this as the *sanity check phase*.

For the potentially good partitions, our task is to find the remaining vertices of S from $G \setminus X$. We do this by framing an instance of the SET COVER problem based on G. Before describing this instance, however, we identify vertices in $G \setminus X$ which will not belong to any witness solution S. Indeed, the following is easy to see:

Lemma 4. *Let* $Y \uplus D_{\leqslant r-2} \uplus D_{r-1} \uplus D_r$ *be a good partition of* X *with witness* S. *If* $v \in G \setminus X$ *and the distance of* v *from* D_r *is less than* r, *then* $v \notin S$. *Similarly, if* $v \in G \setminus X$ *and the distance of* v *from* D_{r-1} *is less than* $r - 1$, *then* $v \notin S$.

We say that a vertex is *irrelevant* if it satisfies the conditions in Lemma 4, and we use Q to denote the set of all irrelevant vertices. Further, we say that a vertex in $X \setminus Y$ is *happy* if it is already at the desired distance from Y. Specifically, a vertex in $D_{\leqslant r-2}$ is happy if its distance from Y is at most $r - 2$, a vertex in D_{r-1} is happy if its distance from Y is exactly $r-1$, and a vertex in D_r is happy if its distance from Y is exactly r. Notice that in a potentially good partition, there is no vertex in D_{r-1} (respectively, D_r) whose distance from Y is less than $r - 1$ (respectively, r). Let $D^\star$ denote the set of happy vertices in $X \setminus Y$, and for $R \in \{D_{\leqslant r-2}, D_{r-1}, D_r\}$, let $R^\star$ denote $R \setminus D^\star$.

We are now ready to describe the instance of SET COVER which will help us find the remaining vertices of S. The universe $\mathcal{U}$ is defined as the set of all vertices in $X \setminus Y$ which are not happy. For every vertex $v \in G \setminus X$ that is not irrelevant, we let S_v denote the set of all vertices in $D^\star_{\leqslant r-2}$ that are at a distance of at most $r - 2$ from v, vertices in $D^\star_{r-1}$ that are at a distance of exactly $r - 1$ from v, and vertices in $D^\star_r$ that are at a distance of exactly r from v. Overall, our instance of set cover is the family of sets $\mathcal{F} := \{S_v \mid v \in G \setminus (X \cup Q)\}$ over the universe $\mathcal{U} := X \setminus (Y \cup D^\star)$, with $S_v = A_v \cup B_v \cup C_v$, where:

$$A_v := \{u \mid u \in D^\star_{\leqslant r-2} \cap N^{(r-2)}(v)\},$$

$$B_v := \{u \mid u \in D^\star_{r-1} \cap N^{(=r-1)}(v)\},$$

and

$$C_v := \{u \mid u \in D^\star_r \cap N^{(=r)}(v)\}.$$

Our main claim is the following:

Lemma 5. *Let* $\mathcal{P} := (Y \uplus D_{\leqslant r-2} \uplus D_{r-1} \uplus D_r)$ *be a partition of* X *and* $\mathcal{F}$ *be the defined as above. If* $\mathcal{H} \subseteq \mathcal{F}$ *is a set cover for* $\mathcal{F}$, *then the vertices corresponding to the sets in* $\mathcal{H}$, *along with the vertices in* Y, *form a* r*-dominating set of* G *that serves as a witness to* $\mathcal{P}$. *Conversely, sets corresponding to* $S \setminus Y$ *form a set cover for* $\mathcal{F}$ *for any* S *that is a witness to* $\mathcal{P}$.

Proof. We first establish the forward direction. Let $\mathcal{H}$ be a set cover for $\mathcal{F}$ and let Z be the vertices of $G \setminus X$ corresponding to $\mathcal{H}$. We claim that $S := Y \cup Z$ is a r-dominating set that witness the partition $\mathcal{P}$.

Let $v \in D_{\leqslant r-2}$. If v is happy, then v is, by definition, at a distance of at most $r-2$ from Y, and therefore S. If v is not happy, then v belongs to the universe of the set cover instance $(\mathcal{U}, \mathcal{F})$ and we let $S_w \in \mathcal{H}$ be the set that covers v. Then, the definition of S_w implies that $d(v, w) \leqslant r-2$. Since $w \in Z$, v is at a distance of at most $r-2$ from Z, and therefore S.

Let $v \in D_{r-1}$. Observe that there is no vertex in S whose distance from v is less than $r-1$. Indeed, vertices in $G \setminus X$ that are at a distance of less than $r-1$ are deemed irrelevant and there are no sets in $\mathcal{F}$ corresponding to such vertices. Similarly, if a vertex in Y was at a distance of less than $r-1$ from v, then the entire partition in question would have been discarded in the part of the sanity check phase based on Lemma 3. Now we argue that there is at least one vertex in S whose distance from v is exactly $r-1$. If v is a happy vertex, then, by definition, there exists a vertex in S whose distance from v is exactly r. If v is not happy, then v belongs to the universe of the set cover instance $(\mathcal{U}, \mathcal{F})$ and we let $S_w \in \mathcal{H}$ be the set that covers v. Then, the definition of S_w implies that $d(v, w) = r-1$. Since $w \in Z$, v is at a distance of exactly $r-1$ from Z, and therefore S. The argument for the case when $v \in D_r$ is analogous.

Finally, consider any vertex in a clique $C \in G \setminus X$. If C is fully dominated, then all vertices in C are r-dominated by vertices in Y. If C is not dominated by Y, then, by the part of the sanity check phase that is based on Lemma 1, we know that one of the following scenarios is true:

- C has a vertex v that has a neighbor in $D_{\leqslant r-2}$.
- Every vertex $v \in C$ has a neighbor in D_{r-1}.

Since we have already argued that every vertex in $D_{\leqslant r-2}$ is at a distance of at most $r-2$ from S and every vertex in D_{r-1} is at a distance of exactly $r-1$ from S, it is easy to see that every vertex in C is r-dominated by S in both the scenarios above. The argument for partially dominated cliques is analogous.

We now discuss the other direction. To this end, consider any r-dominating set S that is a witness for the partition $\mathcal{P}$. Let $\mathcal{H}$ be the naturally defined family consisting of sets corresponding to vertices in $Z := S \cap (V \setminus X)$. Recall that the universe consists of every element that is not happy in X. Such vertices must be r-dominated by vertices in Z—and since S is a witness for $\mathcal{P}$, the vertices in $D_{\leqslant r-2}$ that are not happy must be at a distance of at most $r-2$ from Z, and similarly, vertices in D_{r-1} (respectively, D_r) that are not happy must be at a distance of exactly $r - 1$ (respectively, r) from Z. The fact that these vertices, consequently, are covered by the corresponding sets follows from the definition of the sets. In particular, for instance, if a vertex $v \in D_{\leqslant r-2}$ is at a distance of at most $r - 2$ from $w \in Z$, then $v \in S_w$, and $S_w \in \mathcal{H}$, and so on. This concludes the argument in the reverse direction. □

We are now ready to summarize our algorithm. We begin by guessing a partition of X into $Y \uplus D_{\leqslant r-2} \uplus D_{r-1} \uplus D_r$, and retain only the potentially good partitions. We then form an appropriate instance of Set Cover based on vertices that are relevant to the semantics of the partition. By Lemma 5, a set cover corresponds to the desired r-dominating set, and we are done. We now turn to an analysis of the running time of the algorithm, which is given by:

$$\sum_{i=0}^{k} \binom{k}{i} 3^{k-i} \cdot 2^{k-i} = \sum_{i=0}^{k} \binom{k}{i} 6^{k-i} = 7^k$$

This concludes our discussion for parameterization by CVD.

5   (r-)Efficient Dominating Set

For the EFFICIENT DOMINATING SET problem parameterized by SVD, we propose a branching algorithm, while the FPT algorithm for the CBDS parameter is based on studying interactions with the modulator. For the r-DOMINATING SET problem parameterized by CVD, the first part of our FPT algorithm, especially pertaining to the overall approach and the sanity check phase, will be quite similar to our algorithm for the r-DS problem parameterized CVD. The key difference emerges in our set cover formulation.

6   Concluding Remarks

We generalized several known algorithms for the problems of DOMINATING SET and EFFICIENT DOMINATING SET when parameterized by the vertex-deletion distance to tractable graph classes. We introduced a new parameter in this context, which was the deletion to cobipartite graphs, which is smaller than the CVD parameter. The most immediate directions for future work is to understand the parameterized complexity for the cases that we have not resolved as stated in the summary table (c.f. Table 1). These include the complexity of r-DS and r-EDS parameterized by SVD as well as r-EDS parameterized by CBDS. It would also be interesting to understand if there is a meta property of $G \setminus X$ that makes domination problems tractable when parameterized by $|X|$.

Acknowledgements. The authors thank the reviewers for their detailed comments which have helped improve the presentation of the manuscript. This work is supported by the SERB project MTR/2017/001033.

References

1. Borradaile, G., Le, H.: Optimal dynamic program for r-domination problems over tree decompositions. In: 11th International Symposium on Parameterized and Exact Computation, IPEC 2016, August 24–26, 2016, Aarhus, Denmark, pp. 8:1–8:23 (2016)
2. Cygan, M., et al.: Parameterized Algorithms. Springer, Cham (2015). https://doi.org/10.1007/978-3-319-21275-3
3. Goyal, D., Jacob, A., Kumar, K., Majumdar, D., Raman, V.: Structural parameterizations of dominating set variants. In: Fomin, F.V., Podolskii, V.V. (eds.) CSR 2018. LNCS, vol. 10846, pp. 157–168. Springer, Cham (2018). https://doi.org/10.1007/978-3-319-90530-3_14
4. Katsikarelis, I., Lampis, M., Paschos, V.Th.: Structural parameters, tight bounds, and approximation for (k, r)-center. In: Proceedings of the 28th International Symposium on Algorithms and Computation, ISAAC, pp. 50:1–50:13 (2017)
5. Misra, N., Philip, G., Raman, V., Saurabh, S.: The effect of girth on the kernelization complexity of connected dominating set. In: IARCS Annual Conference on Foundations of Software Technology and Theoretical Computer Science, FSTTCS 2010, December 15–18, 2010, Chennai, India, pp. 96–107 (2010)
6. Philip, G., Raman, V., Sikdar, S.: Solving dominating set in larger classes of graphs: FPT algorithms and polynomial kernels. In: Fiat, A., Sanders, P. (eds.) ESA 2009. LNCS, vol. 5757, pp. 694–705. Springer, Heidelberg (2009). https://doi.org/10.1007/978-3-642-04128-0_62
7. Raman, V., Saurabh, S.: Short cycles make W-hard problems hard: FPT algorithms for W-hard problems in graphs with no short cycles. Algorithmica 52(2), 203–225 (2008)

Transition Property for Cube-Free Words

Elena A. Petrova and Arseny M. Shur[⊠]

Ural Federal University, Ekaterinburg, Russia
{elena.petrova,arseny.shur}@urfu.ru

Abstract. We study cube-free words over arbitrary non-unary finite alphabets and prove the following structural property: for every pair (u, v) of d-ary cube-free words, if u can be infinitely extended to the right and v can be infinitely extended to the left respecting the cube-freeness property, then there exists a "transition" word w over the same alphabet such that uwv is cube free. The crucial case is the case of the binary alphabet, analyzed in the central part of the paper.

The obtained "transition property", together with the developed technique, allowed us to solve cube-free versions of three old open problems by Restivo and Salemi. Besides, it has some further implications for combinatorics on words; e.g., it implies the existence of infinite cube-free words of very big subword (factor) complexity.

1 Introduction

The concept of power-freeness is in the center of combinatorics on words. This concept expresses the restriction on repeated blocks (factors) inside a word: an α-power-free word contains no block which consecutively occurs in it α or more times. For example, the block an in the word banana is considered as having $5/2$ consecutive occurrences; thus the word banana is 3-power-free (cube-free) but not $(5/2)$-power-free or 2-power-free (square-free); the block mag in the word magma occurs consecutively $5/3$ times, and so magma is square-free. Power-free words and languages are studied in lots of papers starting with the seminal works by Thue [19,20], who proved, in particular, the infiniteness of the sets of binary cube-free words and ternary square-free words. However, many phenomena related to power-freeness are still not understood. One group of problems about power-free words concerns their structure and extendability. In 1985, Restivo and Salemi presented [15] a list of five problems, originally considered only for ternary square-free words and binary overlap-free words, but equally important for every power-free language. Suppose that a finite alphabet Σ is fixed and we study α-power-free words over Σ. Here are the problems.

Problem 1. Given an α-power-free word u, decide whether there are infinitely many α-power-free words having (a) the prefix u; or (b) the suffix u; or

E. A. Petrova—Supported by the Russian Science Foundation, grant 18-71-00043.
A. M. Shur—Supported by the Russian Ministry of Education and Science, project 1.3253.2017.

R. van Bevern and G. Kucherov (Eds.): CSR 2019, LNCS 11532, pp. 311–324, 2019.
https://doi.org/10.1007/978-3-030-19955-5_27

(c) the form vuw, where v and w have equal length. (Such words u are called, respectively, right extendable, left extendable, and two-sided extendable.)

Problem 2. Given an α-power-free word u, construct explicitly an α-power-free infinite word having u as prefix, provided that u is right extendable.

Problem 3. Given an integer $k \geq 0$, does there exist an α-power-free word u such that (i) for some word v of length k the word uv is α-power free and (ii) for every word v' of bigger length, uv' is not α-power free.

Problem 4. Given α-power-free words u and v, decide whether there is a "transition" from u to v (i.e., does there exist a word w such that uwv is α-power free).

Problem 5. Given α-power-free words u and v, find explicitly a transition word w, if it exists.

These natural problems appear to be rather hard. Only for Problem 1a,b there is a sort of a general solution: a backtracking decision procedure exists for all k-power-free languages, where $k \geq 2$ is an integer [3,4]. In a number of cases, the parameters of backtracking were found by computer search, so it is not clear whether this technique can be extended for α-power-free words with rational α. The decision procedure also gives no clue to Problem 2.

There is a particular case of binary overlap-free words, for which all problems are solved in [1,15] (more efficient solutions were given in [2]). These words have a regular structure deeply related to the famous Thue-Morse word, and it seems that all natural algorithmic problems for them are solved. For example, the asymptotic order of growth for the binary overlap-free language is computed exactly [6,7], and even the word problem in the corresponding syntactic monoid has a linear-time solution [18]. Most of the results can be extended, with additional technicalities, to binary α-power-free words for any $\alpha \leq 7/3$, because the structure of these words is essentially the same as of overlap-free words (see, e.g., [8]). However, the situation changes completely if we go beyond the polynomial-size language of binary $(7/3)$-power-free words. In the exponential-size α-power-free languages[1] the diversity of words is much bigger, so it becomes harder to find a universal decision procedure. The only results on Problems 1–5 apart from those mentioned above are the positive answers to Problem 3 (including its two-sided analog) for the two classical test cases: for ternary square-free words [12] and for binary cube-free words [11].

In this paper, we study cube-free words over arbitrary alphabets. Still, the crucial case is the one of the binary alphabet; the central part of the paper is the proof of the following transition property of binary cube-free words.

Theorem 1. *For every pair (u, v) of binary cube-free words such that u is right extendable and v is left extendable, there exists a binary word w such that uwv is cube free.*

---

[1] For $\alpha > 7/3$, the language of binary α-power-free words has exponential size [8]. *The exponential conjecture* says that for $k \geq 3$ *all* infinite power-free languages over k letters have exponential size. This conjecture is proved for $k \leq 10$ [9,10] and odd k up to 101 [21].

After proving Theorem 1 in Sect. 3, we use it and its proof to derive further results. In Sect. 4 we prove the transition property for arbitrary alphabets, while in Sect. 5 we use this property to solve Restivo-Salemi Problems 2, 4, and 5. Thus, all Restivo–Salemi problems for binary cube-free words are solved; this is the first fully solved case since the original publication of the problems. For cube-free words over bigger alphabets, only Problem 3 is not yet solved.

We finish the introduction with two remarks. First, the result of Theorems 1 and 3 was conjectured, in a slightly weaker form, for all infinite power-free languages [17, Conj 1]. This conjecture is related to the properties of finite automata recognizing some approximations of power-free languages and was supported by extensive numerical studies. The transition words can be naturally interpreted as transitions in those automata and the transition property forces the automata to be strongly connected. Second, recently it was shown [16, Thm 39] that the transition property implies the existence of infinite α-power-free words of very big subword complexity. Namely, Theorems 1 and 3 imply that for every $d \geq 2$ there exists a d-ary cube-free infinite word which contains *all* two-sided extendable d-ary cube-free finite words as factors.

2   Preliminaries

Notation and Definitions. By default, we study words over finite alphabets Σ_d of cardinality $d \geq 2$, writing $\Sigma_d = \{a, b, c_1, \ldots, c_{d-2}\}$ (mostly we work with $\Sigma_2 = \{a, b\}$). Standard notions of factor, prefix, and suffix are used. The set of all finite (nonempty finite, infinite) words over an alphabet Σ is denoted by Σ^* (resp., Σ^+, Σ^∞). Elements of Σ^+ (Σ^∞) are treated as functions $w : \{1, \ldots, n\} \to \Sigma$ (resp., $\mathbf{w} : \mathbb{N} \to \Sigma$). We write $[i..j]$ for the range $i, i+1, \ldots, j$ of positive integers; the notation $w[i..j]$ stands for the factor of the word w occupying this range as well as for the particular occurrence of this factor in w at *position* i. Note that $w[i..i] = w[i]$ is just the ith letter of w. Let $w[i_1..j_1]$ and $w[i_2..j_2]$ be two factors of w. If the ranges $[i_1..j_1]$ and $[i_2..j_2]$ have a nonempty intersection, their intersection and union are also ranges; we refer to the factors of w, occupying these ranges, as the *intersection* and the *union* of $w[i_1..j_1]$ and $w[i_2..j_2]$. The word $\overleftarrow{w} = w[n] \ldots w[1]$ is called the *reversal* of the word w of length n.

We write λ for the empty word and $|w|$ for the length of a word w (infinite words have length ∞). A word w has *period* $p < |w|$ if $w[1..|w|-p] = w[p+1..|w|]$; the prefix $w[1..p]$ of w is the *root* of this period of w. One of the most useful properties of periodic words is the following.

Lemma 1 (Fine, Wilf [5]). *If a word u has periods p and q and $|u| \geq p + q - \gcd(p, q)$ then u has period $\gcd(p, q)$.*

A *cube* is a nonempty word of the form uuu, also written as u^3; we refer to $u, |u|$ as the root and the period of this cube. A word is *cube-free* (*overlap-free*) if it has no cubes as factors (resp., no factors of the form $cwcwc$, where c is a letter). There exist binary overlap-free (and thus cube-free) infinite words [20].

Let Σ_d be fixed. A word $w \in \Sigma_d^* \cup \Sigma_d^\infty$ is called a *right context* of a cube-free word $u \in \Sigma_d^*$ if uw is cube free; we call u *right extendable* if it has an infinite right context (or, equivalently, infinitely many finite right contexts). Left contexts and left extendability are defined in a symmetric way.

The *Thue–Morse morphism* θ is defined over Σ_2^+ by the rules $\theta(a) = ab$, $\theta(b) = ba$. The fixed points of θ are the infinite *Thue-Morse word*

$$\mathbf{T} = abbabaabbaababbabaababbaabbabaab\cdots$$

and its complement, obtained from $\mathbf{T}$ by exchanging a's and b's. We refer to the factors of $\mathbf{T}$ as *Thue-Morse factors*. The word $\mathbf{T}$, first introduced by Thue in [20] and rediscovered many times, possesses a huge number of nice properties; we need just a few. The Thue-Morse word is overlap free, *uniformly recurrent* (every Thue-Morse factor occurs in $\mathbf{T}$ infinitely many times with a bounded gap), and *closed under reversals* (u is a Thue-Morse factor iff $\overleftarrow{u}$ is).

Uniform words and markers. We call a word $w \in \Sigma_2^*$ *uniform* if $w = c\theta(u)d$ for some $c, d \in \{a, b, \lambda\}$, $u \in \Sigma_2^*$; a uniform word with $d = \lambda$ is *right aligned*. Similarly, a uniform infinite word has the form $c\theta(\mathbf{u})$ for $c \in \{a, b, \lambda\}$, $u \in \Sigma_2^\infty$. Such "almost" θ-images play a crucial role in further considerations. Note that all factors and suffixes of $\mathbf{T}$ are uniform. The following observations are straightforward.

Observation 1. *A word $u \in \Sigma_2$ is uniform iff all occurrences of factors of the form cc in u, where $c \in \Sigma_2$, are at positions of the same parity.*

Observation 2. *A cube-free word $u \in \Sigma_2$ is non-uniform iff it contains at least one of the factors aabaa, aababaa, bbabb, bbababb.*

All right (resp., left) contexts of the word $ababa$ begin (resp., end) with a, so $ababa$ occurs in a cube-free word only as a prefix/suffix or inside the non-uniform factor $aababaa$ (same argument applies to $babab$). This allows us to view binary cube-free words as sequences of uniform factors separated by *markers* $aabaa, ababa, babab$, and $bbabb$, which break uniformity. The importance of markers for the analysis of cube-free words is demonstrated by the following theorem, proved in Sect. 3.2.

Theorem 2. *Every right-extendable cube-free word $u \in \Sigma_2$ has an infinite right context with finitely many markers.*

3   Proof of the Transition Property for Binary Words

The proof of Theorem 1 consists of two stages. In the first stage we show that its result is implied by Theorem 2. In the second stage we prove Theorem 2. All words in this section are over Σ_2 if the converse is not stated explicitly. Proofs omitted due to space constraints can be found in the full version: arXiv:1812.11119.

3.1   Reduction to Theorem 2

Lemma 2. *Suppose that cube-free words u and v have right contexts which are Thue-Morse factors of length $2|u|$ and $2|v|$ respectively. Then there exists a word w such that $uw\overleftarrow{v}$ is cube free.*

Proof. Let u_1 and v_1 be the mentioned contexts of u and v respectively. Since $\mathbf{T}$ is recurrent and closed under reversals, there exists a Thue-Morse factor $w = u_1 w_1 \overleftarrow{v_1}$ for some $w_1 \neq \lambda$. Assume to the contrary that $uw\overleftarrow{v}$ contains a cube of period p. Since w is overlap free, such a cube cannot intersect u and $\overleftarrow{v}$ simultaneously by the length argument. W.l.o.g. assume that the cube intersects u. Then it must contain the whole u_1. Hence p is a period of u_1 and thus $p \geq |u_1|/2 \geq |u|$ because u_1 is overlap free. Further, the overlap-freeness of w means that u contains at least the whole period of the cube, implying $p \leq |u|$. So we have $|u| = p$ and the cube is a prefix of $uw\overleftarrow{v}$. But in this case $|u_1| \geq 2p$ and the cube is contained in the cube-free word uu_1, resulting in a contradiction. $\square$

We say that a cube-free word u is *T-extendable* if it has a right context of the form $w\mathbf{T}[n..\infty]$ for some $w \in \Sigma^*$, $n \geq 1$. By Lemma 2, if the words u and $\overleftarrow{v}$ are T-extendable, there is a word w such that uwv is cube-free. We analyze T-extendability in Lemmas 3–5.

Lemma 3. *If a uniform cube-free word u has a right context of length 3, or is right aligned and has a right context of length 2, then u is T-extendable.*

Here we give only the construction. There is nothing to prove if u is a factor of $\mathbf{T}$, so assume it is not. If $u = u'ab$ is right aligned then one of the words

$$\mathbf{v} = u\mathbf{T}[6..\infty] = u'ab\,aabbaababba\cdots, \quad \mathbf{v} = u\mathbf{T}[20..\infty] = u'ab\,babbaabbaba\cdots$$

is cube free. The case of non-aligned word ($u = u'aba$ or $u = u'baa$) is reduced to the aligned case for the words $u'ab$ and $u'baab$ respectively.

Next we extend T-extendability to the words with long uniform contexts.

Lemma 4. *If a cube-free word u has a uniform right context w such that $|w| \geq 2|u| + 3$ and w has no prefix ababa or babab, then u is T-extendable.*

Some right-extendable words have no long uniform right contexts, as Fig. 1 shows. However, a weaker property is enough for our purposes.

Fig. 1. A right-extendable word of length 76 having no long uniform right contexts: all its infinite right contexts begin with the marker $aabaa$.

Lemma 5. *Every cube-free word having an infinite right context with finitely many markers is T-extendable.*

Proof. Let u be the word and w be its context from the conditions of the lemma. The finiteness of the number of markers allows us to write $w = w_1v$, where v is uniform. Then uw_1 has an infinite uniform right context, and hence is T-extendable by Lemma 4. Then u is T-extendable as well.   □

Thus if Theorem 2 holds, then Lemmas 5 and 2 imply Theorem 1.

3.2   Proof of Theorem 2

We prove Theorem 2 by reductio ad absurdum; to obtain a contradiction, we use the following lemma on cube-free words over an arbitrary alphabet. (For Theorem 2, bounding k in Lemma 6 by *any* function of n would be sufficient; better bounds can be useful for algorithmic applications.)

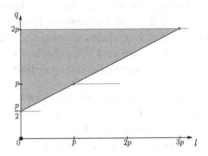

Fig. 2. The restrictions on periods of periodic factors in the word uw (Lemma 6).

Lemma 6. *Let u be a cube-free word of length n over an* arbitrary *fixed alphabet and let u have a length-k right context w with the following property: for each $i = 1,\ldots,k$, there is an integer $p_i \geq 2$ such that the suffix of length $3p_i - 2$ of the word $u{\cdot}w[1..i]$ has period p_i and, moreover, $p_i \neq p_{i+1}$. Then $k = O(\log n)$; more precisely, $k \leq \max\{1, 8.13 \log n - 15.64\}$.*

Proof. In the proof we can assume $k \geq 2$. Let $1 \leq i < j \leq k$, $p = p_i$, $q = p_j$, $l = j - i$, and let v be the intersection of the periodic suffixes of $u{\cdot}w[1..i]$ and $u{\cdot}w[1..j]$ (see Fig. 3a,b). If $|v| \geq p+q-\gcd(p,q)$, then v has the period $\gcd(p,q)$ by Lemma 1. If $p \neq q$, this means that the root of the longer periodic suffix is an integer power of a shorter word; thus uw contains a cube, which is impossible. If $p = q$, then we are in the situation shown in Fig. 3b, and the union of two suffixes has period p and the length $3p - 2 + l$. In this case, $l \geq 2$ by conditions of the lemma, so we again obtain a cube. Thus we conclude that

$$|v| \leq p + q - \gcd(p,q) - 1. \tag{1}$$

Fig. 3. The mutual location of periodic factors in the word uw (Lemma 6).

The case in Fig. 3b corresponds to $|v| = 3q - 2 - l$. Comparing this condition to (1), we get $q \leq \frac{p+l}{2}$ (but $q = p$ only if $l \geq 2p - 1$ and $q = p/2$ only if $l \geq p/2 - 1$). Similarly, the case in Fig. 3a corresponds to $|v| = 3p - 2$ and we get $q > 2p$ from (1). Thus, all possible values of q are outside the red area in Fig. 2.

Now we estimate how many elements of the sequence $\{p_1, \ldots, p_k\}$ can belong to the range $[p..2p]$ for some fixed $p \geq 2$. This is an analog of [13, Lemmas 4,5] and [14, Lemma 9]. Let $i_0 < i_1 < \cdots < i_s$ be the list of all positions such that the periodic suffix of $u \cdot w[1..i_j]$ has the period from the range $[p..2p]$; let $q_0, \ldots, q_s$ denote these periods. Then Fig. 2 gives us the lower bound for the distance $l_j = i_{j+1} - i_j$ between consecutive positions from the list:

$$l_j \geq 2q_{j+1} - q_j; \; l_j \geq 2q_{j+1} - 1 \text{ if } q_{j+1} = q_j; \; l_j \geq q_{j+1} - 1 \text{ if } q_{j+1} = q_j/2. \quad (2)$$

The densest packing of the numbers i_j, satisfying the restrictions (2), is achieved for $q_0 = 2p - 1$, $q_1 = q_3 = q_5 = \cdots = p$, $q_2 = q_4 = q_6 = \cdots = p + 1$: one can take $i_0 = 1$, $i_1 = 2$, and $i_{2j} = i_{2j-1} + p + 2$, $i_{2j+1} = i_{2j} + p - 1$ for all subsequent positions. Since $i_s \leq k$, we have $\lceil \frac{s-1}{2} \rceil \cdot (p+2) + \lfloor \frac{s-1}{2} \rfloor \cdot (p-1) + 2 \leq k$, implying the upper bound for the number of periods from the range $[p..2p]$:

$$s + 1 \leq \frac{2(k-2)}{2p+1} + 2. \quad (3)$$

Since $3p - 2 \leq |uw| = n + k$, the maximum possible value of p is $\lfloor \frac{n+k+2}{3} \rfloor$. We partition all possible periods into r ranges of the form $[p..2p]$:

$$[2..4], [5..10], [11..22], \ldots, \left[3 \cdot 2^{r-1} - 1 .. \left\lfloor \tfrac{n+k+2}{3} \right\rfloor\right].$$

The number of ranges thus satisfies $r \leq \log \frac{n+k+2}{9} + 1$. The sum of the upper bounds (3) for all ranges is at least k; observing that the number $2p+1$ in (3) is the first period from the range next to $[p..2p]$, we can write

$$k \leq 2(k-2) \cdot \sum_{i=1}^{r} \frac{1}{3 \cdot 2^i - 1} + 2r. \quad (4)$$

The sum in (4) is bounded by $\frac{1}{5} + \frac{1}{11} + \frac{1}{23} + \frac{1}{47} \cdot \sum_{i=0}^{\infty} \frac{1}{2^i} < 0.377$; substituting this value and the upper bound for r, we get

$$0.246(k - 2) \leq 2 \log \tfrac{n+k+2}{9}. \quad (5)$$

For $k \geq n - 1$, (5) implies $0.246(k - 2) \leq 2 \log \frac{2k+3}{9}$, but this inequality fails for $k \geq 2$. So $k \leq n - 2$ and we replace (5) with the inequality $0.246(k-2) \leq 2 \log \frac{2n}{9}$; it gives, after arithmetic transformations, the required bound on k. $\qquad \square$

Proof (Proof of Theorem 2). For the sake of contradiction, assume that all infinite right contexts of some right-extendable cube-free word u contain infinitely many markers. W.l.o.g. we can assume that u ends with a marker (if not, choose a prefix v of an infinite right context of u such that uv ends with a marker, and replace u with the word uv having the same property of right contexts). Let z be the marker which is a suffix of u. For example, if u is the word written in the "trunk" of the tree in Fig. 1, then $z = bbabb$.

By our assumption, u has no infinite uniform right contexts. Thus u has finitely many uniform right contexts (in Fig. 1 such contexts are λ, a, aa, aab, $aaba$, and $aabb$). Other uniform words, being appended to u, produce cubes; in Fig. 1, appending a word beginning with ab (resp., $aabab$, $aabba$) gives the cube $(bab)^3$ (resp., $(babbaabab)^3$, $(babbaababbabbaababbabbaabba)^3$). All these cubes contain markers and illustrate three types of cubes with respect to the occurrences of markers (x below denotes the root of the cube):

- *mini*: x^2 contains no markers, x^3 contains a marker (example: $x = bab$);
- *midi*: x contains no markers, x^2 contains a marker (example: $x = babbaabab$);
- *maxi*: x contains markers (example: $x = babbaababbabbaababbabbaabba$);

Note that mini cubes are exactly those having the root $x \in \{ab, ba, aba, bab\}$. Further, the intersection of two markers in a cube-free word is either empty or one-letter; this fact implies that each midi cube contains exactly two markers.

We call w a *semi-context* of u if uw ends with a cube but $u \cdot w[1..|w|-1]$ is cube free. Next we show the following fact:

(∗) The word u has two distinct uniform semi-contexts w_1 and w_2 such that uw_1 and uw_2 end with midi or maxi cubes.

To prove (∗) we need a case analysis. W.l.o.g., z begins with a. Let $\mathbf{w}$ be an infinite right context of u.

Case 1: $z = aabaa$. We have $u = \cdots baabaa$, so $\mathbf{w}$ cannot begin with a or baa because $u\mathbf{w}$ is cube-free. So $\mathbf{w}$ begins with bba, $babba$, or $baba$. In the first case, some prefixes of words

$$u\mathbf{T}[2..\infty] = \cdots baabaa\,bba\,baababba \cdots , \tag{6}$$

$$u\mathbf{T}[22..\infty] = \cdots baabaa\,bba\,abbabaab \cdots \tag{7}$$

must end with cubes. The longest common prefix $ubba$ of these words is cube free as a prefix of $u\mathbf{w}$, so these cubes are different, contain the marker z, and are not mini. Hence we can take some prefixes of $\mathbf{T}[2..\infty]$ and $\mathbf{T}[22..\infty]$ as the semi-contexts required in (∗). If $\mathbf{w}$ begins with $babba$, the same result is obtained with prefixes of the words

$$u\mathbf{T}[12..\infty] = \cdots baabaa\,babba\,baabbaab \cdots , \tag{8}$$

$$u\mathbf{T}[20..\infty] = \cdots baabaa\,babba\,abbabaab \cdots \tag{9}$$

Finally, if $\mathbf{w}$ begins with $baba$, we can take one word from each pair (say, $u\mathbf{T}[2..\infty]$ and $u\mathbf{T}[12..\infty]$). Their longest common prefix is the cube-free word ub, so the cubes given by the corresponding semi-contexts are distinct.

Case 2: $z = ababa$. Here $\mathbf{w} = ab \cdots$. Taking the pair of words

$$u\mathbf{T}[7..\infty] = \cdots ababa\,ab\,baababba \cdots , \tag{10}$$

$$u\mathbf{T}[19..\infty] = \cdots ababa\,ab\,abbaabba \cdots , \tag{11}$$

we achieve the same result as in Case 1: some prefixes of these words end with midi or maxi cubes, and these cubes are distinct because the common prefix uab of the presented words is cube-free. Thus, $(*)$ is proved.

Now take the semi-contexts w_1, w_2 given by $(*)$ such that uw_1 and uw_2 end with cubes x_1^3 of period p_1 and x_2^3 of period p_2 respectively. Let w be the longest common prefix of w_1 and w_2; w.l.o.g., $w_1 = waw_1'$, $w_2 = wbw_2'$. In both x_1^3 and x_2^3, the suffix z of u is the rightmost marker and hence matches an earlier occurrence of the same marker in u. These occurrences are different, because z is followed by wa in x_1^3 and by wb in x_2^3. In particular, $p_1 \neq p_2$. W.l.o.g., $p_1 > p_2$; then x_1 contains z and so x_1^3 is maxi (see Fig. 4).

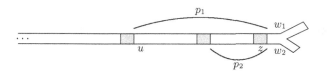

Fig. 4. Semi-contexts w_1 and w_2 of the word u: periods of cubes and corresponding markers. Three grey factors are markers equal to z, other markers are not shown.

Let $z_1 = z, z_2, \ldots, z_m$ be all markers in u, right to left. We factorize u as $u = y_m \cdots y_2 y_1$, where y_i begins with the first letter after z_{i+1} (y_m is a prefix of u) and ends with the last letter of z_i (even if z_{i+1} and z_i overlap); see Fig. 5 for the example. Assume that z_{j+1} matches z_1 in the maxi cube x_1^3 (note that $j \geq 2$, because a marker with a smaller number matches z_1 in x_2^3; $j = 3$ in Fig. 5). Then $z_{j-1}, \ldots, z_1$ are in the rightmost occurrence of x_1, and z_j is either also in this occurrence or on the border between the middle and the rightmost occurrences (in Fig. 5, the latter case is shown). Depending on this, x_1^3 contains either $3j$ or $3j-1$ markers. Further, we see that w_1 is a prefix of y_j, $y_1 = y_{j+1}, \ldots,$ $y_{2j-2} = y_{3j-2}$.

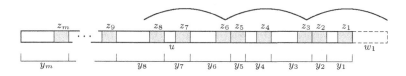

Fig. 5. Marker-based factorization of the word u. Markers are grey, arcs indicate the cube after appending w_1 to u.

Let us extend u to the right by a context y_0 such that uy_0 is right extendable, y_0 ends with a marker z_0, and all proper prefixes of y_0 are uniform. Applying all the above argument to uy_0 and its factorization $y_m \cdots y_1 y_0$, we get another maxi cube (say, x_0^3) and the corresponding set of equalities between y_i's.

Note that $y_0 \neq y_j$: as was mentioned in the previous paragraph, y_j has the prefix w_1, while y_0 cannot have this prefix because uw_1 contains a cube.

Let us iterate the procedure of appending a context k times, getting a right-extendable word $uy = y_m \cdots y_1 y_0 \cdots y_{1-k}$ as the result (according to our assumption on u, the number k can be arbitrarily big). Now consider the finite alphabet $\Gamma = \{y_m, \ldots, y_1, y_0, \ldots, y_{1-k}\}$ and let $U = y_m \cdots y_2 y_1$, $Y = y_0 \cdots y_{1-k}$ be words over Γ. They are cube free and Y is a length-k right context of U. Each word $U \cdot Y[1..i]$ ends with a suffix having some period p_i and length $3p_i - 2$ or $3p_i - 1$. In addition, $p_i \neq p_{i+1}$, because $y_{-i} \neq y_{p_i - i}$. So all conditions of Lemma 6 are satisfied, and we apply it to get an upper bound on k. The existence of this bound contradicts our assumption that all infinite right contexts of u have infinitely many markers. The theorem is proved.   □

4   Transition Property for Big Alphabets

Here we extend the results of the previous section to arbitrary finite alphabets.

Theorem 3. *For every $d \geq 3$ and every pair (u, v) of d-ary cube-free words such that u is right extendable and v is left extendable, there exists a d-ary word w such that uwv is cube free.*

As in the binary case, we use an auxiliary theorem about the existence of a context with finitely many markers (but the markers are different now).

Theorem 4. *Let $d \geq 3$. Every right-extendable cube-free word $u \in \Sigma_d^*$ has an infinite right context with finitely many occurrences of all letters except a and b.*

To prove this result, we follow the proof of Theorem 2, with the role of markers now played by the "c-letters" $c_1, \ldots, c_{d-2}$. The details are in arXiv:1812.11119.

Proof (of Theorem 3). By Theorem 4, the word $u \in \Sigma_d^*$ has an infinite right context with finitely many c-letters. First we note that we can choose such a context containing a c-letter (if a context $\mathbf{w}$ is over Σ_2, one can get another context of u replacing, say, the letter $\mathbf{w}[|u|]$ with c_1). So we can write this context $\mathbf{w}$ as $x\mathbf{u}_1$, where x ends with a c-letter and $\mathbf{u}_1 \in \Sigma_2^\infty$. Let u_1 be the prefix of $\mathbf{u}_1$ of length $\lceil |ux|/2 \rceil$. In the same way, we take a right context $\overleftarrow{y} \overleftarrow{\mathbf{v}_1}$ of $\overleftarrow{v}$ and the prefix $\overleftarrow{v_1} \in \{a, b\}^*$ of $\overleftarrow{\mathbf{v}_1}$ of length $\lceil |yv|/2 \rceil$. Then the binary words u_1, v_1 are cube free, u_1 is right extendable, and v_1 is left extendable. Applying Theorem 1, we take a binary transition word w_1 such that $u_1 w_1 v_1$ is cube free. Then $s = u x u_1 w_1 v_1 y v$ is cube free. Indeed, x ends with a c-letter, y begins with a c-letter, and these c-letters are separated by a cube-free word over Σ_2. Hence a cube in s, if any, must contain one of these c-letters. But the lower bounds on $|u_1|$ and $|v_1|$ imply that this c-letter cannot match another c-letter to produce a cube (recall that uxu_1 and v_1yv are cube free). Thus s is cube-free and we obtain a transition word $xu_1w_1v_1y$ for the pair (u, v).   □

5   Solving the Restivo–Salemi Problems and Future Work

To give the solutions to the Restivo–Salemi Problems 2, 4, and 5, recall the solution to Problem 1a [4]: a d-ary α-power-free word u is right extendable iff it has a right context of length $f_{\alpha,d}(|u|)$ for some computable function $f_{\alpha,d}$. Algorithm 1 below solves Problem 4.

Algorithm 1. Deciding the existence of a transition word for cube-free words $u, v \in \Sigma_d^*$

- For all $w \in \Sigma_d^*$ such that $|w| \leq f_{3,d}(|u|)$ check whether w is a right context of u
- If a context w with the suffix v is found, return "yes"
- If no context of length $f_{3,d}(|u|)$ is found, return "no"
- Else $\qquad\qquad\qquad\qquad\qquad\qquad\qquad\qquad\qquad\qquad$ ▷ u is right extendable
  - For all $w \in \Sigma_d^*$ such that $|w| \leq f_{3,d}(|v|)$ check whether w is a left context of v
  - If a context w with the prefix u is found, return "yes"
  - If no context of length $f_{3,d}(|v|)$ is found, return "no"
  - Else return "yes" $\qquad\qquad\qquad\qquad$ ▷ v is left extendable; apply Theorem 1 or 3

The natural next step is to find an *efficient* algorithm for Problem 4. The function $f_{3,d}(n)$ is sublinear, but the search space is still of size $2^{n^{\Omega(1)}}$. The possible way to a polynomial-time solution is to strengthen the connection with Lemma 6 to show that it is sufficient to process the contexts of length $O(\log n)$, where $n = \max\{|u|, |v|\}$.

For Problem 2, the first step is the reduction to the binary case. Let $u \in \Sigma_d$, $d \geq 3$, be a right-extendable cube-free word; we write $u = u'cu''$, where c is the rightmost c-letter in u. We check all cube-free words $w \in \Sigma_2^*$ such that $u''w$ is right extendable and $|u''w| = \lceil |u|/2 \rceil$ for being right contexts of u. If w is a right context of u, then any (binary) right context of $u''w$ is a right context of uw, so the problem is reduced to binary words. If no word w suits, we take the shortest right context of u of the form vc, where $v \in \Sigma_2^*$, $c \in \{c_1, \ldots, c_{d-2}\}$ such that the word $u_1 = uvc$ is right extendable; such a context can be found in finite time because $|v| < |u|/2$. Then we replace u by u_1 and repeat the search of long binary right contexts. By Theorem 4, we will succeed after a finite number of iterations, and Lemma 6 gives the upper bound on the maximum number k of iterations depending on $|u|$. Thus we end this step getting a word $y\hat{u}$ such that $uy\hat{u}$ is cube free, $\hat{u} \in \Sigma_2^*$, and all binary right contexts of $\hat{u}$ are right contexts of $uy\hat{u}$. If u is binary, we skip this step setting $\hat{u} = u$.

On the second step we further reduce the problem to uniform words. We act as in the first step, using Theorem 2 and Lemma 4. Namely, we check for uniform contexts and if $\hat{u}$ has no uniform context w of length $2|\hat{u}| + 3$ such that $\hat{u}w$ is right extendable and w has no prefix $ababa/babab$, we append the shortest context v ending with a marker, repeating the search for $\hat{u}_1 = \hat{u}v$. Theorem 2 guarantees that we will find the required uniform context in at most k iterations, where k is as in Lemma 6. Thus at this step we build a right context $\hat{y}\hat{w}$ of $\hat{u}$ such that $\hat{u}\hat{y}\hat{w}$ is right extendable, $|\hat{w}| \geq 2|\hat{u}\hat{y}| + 3$ and $\hat{w}$ is uniform and has no prefix $ababa/babab$.

Finally we choose, as described in Lemma 4, a suffix $\mathbf{T}[r..\infty]$ of $\mathbf{T}$ which is a right context of $\hat{w}$: if $\hat{w} = \mathbf{T}[i..j]$ for some i, j, then we take $r = j + 1$, otherwise the choice is performed according to Case 1 in the proof of Lemma 3. Now Lemma 4 guarantees that $\hat{u}\hat{y}\hat{w}\mathbf{T}[r..\infty]$ is cube free. Thus the infinite right context of the original word u is given by the finite word $Y = y\hat{u}\hat{y}\hat{w}$ and the number r. The above description is summarized below as Algorithm 2.

Algorithm 2. Finding an infinite right context of a right-extendable cube-free word $u \in \Sigma_d^*$

- $U \leftarrow u$, $Y \leftarrow \lambda$
- If U is not binary   ▷ 1st step
  - While U has no long binary right context
    - * Find the shortest right context vc ending with a c-letter
    - * $U \leftarrow Uvc$, $Y \leftarrow Yvc$
  - $\hat{u} \leftarrow$ long binary right context of U, $Y \leftarrow Y\hat{u}$
- Else $\hat{u} = u$
- While $\hat{u}$ has no long uniform right context without prefix $ababa/babab$ ▷ 2nd step
  - Find the shortest right context v such that $\hat{u}v$ ends with a marker
  - $\hat{u} \leftarrow \hat{u}v$, $Y \leftarrow Yv$
- $\hat{w} \leftarrow$ long uniform right context of $\hat{u}$ without prefix $ababa/babab$, $Y \leftarrow Y\hat{w}$
- Find r such that $\mathbf{T}[r..\infty]$ is a right context of $\hat{w}$ ▷ final step
- return Y, r

Again, the natural direction of the future work is to make Algorithm 2 efficient.

Finally we approach Problem 5. We first run Algorithm 1, which can provide us with an example of a transition word if u or $\overleftarrow{v}$ is not right extendable. If both $u, \overleftarrow{v}$ are right extendable, we run for each of them Algorithm 2, getting Y_1, Y_2, r_1, r_2 such that $uY_1\mathbf{T}[r_1..\infty]$ and $\overleftarrow{\mathbf{T}}[\infty..r_2]Y_2v$ are cube free. It remains to use Lemma 2: take big enough r_1', r_2' and find a word w such that $\mathbf{T}[r_1..r_1']w\overleftarrow{\mathbf{T}}[r_2'..r_2]$ is a factor of $\mathbf{T}$ and a transition word for the pair (uY_1, Y_2v); the uniform recurrence of $\mathbf{T}$ ensures that the word w can be found in finite time. Thus $Y_1\mathbf{T}[r_1..r_1']w\overleftarrow{\mathbf{T}}[r_2'..r_2]Y_2$ is the transition word for the pair (u, v), so Problem 5 is solved.

Once again, it is clear that some steps of the above solution can be sped up, so it would be nice to finally get a polynomial-time algorithm for Problem 5 (and thus for Problems 1, 2, 4 as well). From the experimental study we learned that if a length-n cube-free word is not right extendable, then likely not only all its right contexts have the length $O(\log n)$, but the number of such contexts is $O(\log n)$. The proof of this fact would lead to a linear-time solution of Problem 1.

Another obvious continuation of the current research is the study of the same problems for other power-free languages. One line is to use Thue-Morse words to solve Problems 2, 4, and 5 for other binary power-free languages. The other line is to obtain similar results for ternary square-free words, in the absence of such a strong tool as Thue-Morse words.

References

1. Carpi, A.: On the centers of the set of weakly square-free words on a two-letter alphabet. Inf. Process. Lett. **19**, 187–190 (1984)
2. Carpi, A.: Overlap-free words and finite automata. Theor. Comput. Sci. **115**, 243–260 (1993)
3. Currie, J.D.: On the structure and extendibility of k-power free words. Eur. J. Comb. **16**, 111–124 (1995)
4. Currie, J.D., Shelton, R.O.: The set of k-power free words over σ is empty or perfect. Eur. J. Comb. **24**, 573–580 (2003)
5. Fine, N.J., Wilf, H.S.: Uniqueness theorems for periodic functions. Proc. Am. Math. Soc. **16**, 109–114 (1965)
6. Guglielmi, N., Protasov, V.: Exact computation of joint spectral characteristics of linear operators. Found. Comput. Math. **13**, 37–97 (2013)
7. Jungers, R.M., Protasov, V.Y., Blondel, V.D.: Overlap-free words and spectra of matrices. Theor. Comput. Sci. **410**, 3670–3684 (2009)
8. Karhumäki, J., Shallit, J.: Polynomial versus exponential growth in repetition-free binary words. J. Combin. Theory Ser. A **104**, 335–347 (2004)
9. Kolpakov, R., Rao, M.: On the number of Dejean words over alphabets of 5, 6, 7, 8, 9 and 10 letters. Theoret. Comput. Sci. **412**, 6507–6516 (2011)
10. Ochem, P.: A generator of morphisms for infinite words. RAIRO Inf. Théor. App. **40**, 427–441 (2006)
11. Petrova, E.A., Shur, A.M.: Constructing premaximal binary cube-free words of any level. Internat. J. Found. Comp. Sci. **23**(8), 1595–1609 (2012)
12. Petrova, E.A., Shur, A.M.: Constructing premaximal ternary square-free words of any level. In: Rovan, B., Sassone, V., Widmayer, P. (eds.) MFCS 2012. LNCS, vol. 7464, pp. 752–763. Springer, Heidelberg (2012). https://doi.org/10.1007/978-3-642-32589-2_65
13. Petrova, E.A., Shur, A.M.: On the tree of ternary square-free words. In: Manea, F., Nowotka, D. (eds.) WORDS 2015. LNCS, vol. 9304, pp. 223–236. Springer, Cham (2015). https://doi.org/10.1007/978-3-319-23660-5_19
14. Petrova, E.A., Shur, A.M.: On the tree of binary cube-free words. In: Charlier, É., Leroy, J., Rigo, M. (eds.) DLT 2017. LNCS, vol. 10396, pp. 296–307. Springer, Cham (2017). https://doi.org/10.1007/978-3-319-62809-7_22
15. Restivo, A., Salemi, S.: Some decision results on non-repetitive words. In: Apostolico, A., Galil, Z. (eds.) Combinatorial Algorithms on Words. NATO ASI Series, vol. F12, pp. 289–295. Springer-Verlag, Heidelberg (1985)
16. Shallit, J., Shur, A.: Subword complexity and pattern avoidance. Theor. Comput. Sci. (2018). https://doi.org/10.1016/j.tcs.2018.09.010
17. Shur, A.M.: Two-sided bounds for the growth rates of power-free languages. In: Diekert, V., Nowotka, D. (eds.) DLT 2009. LNCS, vol. 5583, pp. 466–477. Springer, Heidelberg (2009). https://doi.org/10.1007/978-3-642-02737-6_38

18. Shur, A.M.: Deciding context equivalence of binary overlap-free words in linear time. Semigroup Forum **84**, 447–471 (2012)
19. Thue, A.: Über unendliche Zeichenreihen. Norske vid. Selsk. Skr. Mat. Nat. Kl. **7**, 1–22 (1906)
20. Thue, A.: Über die gegenseitige Lage gleicher Teile gewisser Zeichenreihen. Norske vid. Selsk. Skr. Mat. Nat. Kl. **1**, 1–67 (1912)
21. Tunev, I.N., Shur, A.M.: On two stronger versions of dejean's conjecture. In: Rovan, B., Sassone, V., Widmayer, P. (eds.) MFCS 2012. LNCS, vol. 7464, pp. 800–812. Springer, Heidelberg (2012). https://doi.org/10.1007/978-3-642-32589-2_69

A Polynomial Time Delta-Decomposition Algorithm for Positive DNFs

Denis Ponomaryov[(⊠)]

Ershov Institute of Informatics Systems,
Novosibirsk State University, Novosibirsk, Russia
ponom@iis.nsk.su

Abstract. We consider the problem of decomposing a positive DNF into a conjunction of DNFs, which may share a (possibly empty) given set of variables Δ. This problem has interesting connections with traditional applications of positive DNFs, e.g., in game theory, and with the broad topic of minimization of boolean functions. We show that the finest Δ-decomposition components of a positive DNF can be computed in polynomial time and provide a decomposition algorithm based on factorization of multilinear boolean polynomials.

1 Introduction

The interest in decomposition of positive DNFs stems from computationally hard problems in game theory, reliability theory, the theory of hypergraphs and set systems. A survey of relevant literature can be found in [1]. In the context of voting games, boolean variables are used to represent voters and the terms of a positive DNF correspond to the winning coalitions, i.e., the groups of voters, who, when simultaneously voting in favor of an issue, have the power to determine the outcome of the vote (i.e., in this case the DNF is evaluated as true). Dual to them are blocking coalitions, i.e., those that force the outcome of the vote to be negative, irrespective of the decisions made by the remaining voters. The problem to find blocking coalitions with a minimal number of voters is easily shown to be equivalent to the hitting set problem, which is NP-complete. Decomposing a DNF into components allows for reducing this problem to inputs having fewer variables.

In this paper, we consider decomposition of a positive DNF into a *conjunction of DNFs* sharing a given (possibly empty) subset of variables Δ, with the remaining subsets of variables being disjoint (in this case we say that a DNF is Δ-decomposable). In particular, each of the components has fewer variables than the original formula. Besides dimensionality reduction, decomposition of

This work was supported by the grant of Russian Foundation for Basic Research No. 17-51-45125 and by the Ministry of Science and Education of the Russian Federation under the 5-100 Excellence Program.

R. van Bevern and G. Kucherov (Eds.): CSR 2019, LNCS 11532, pp. 325–336, 2019.
https://doi.org/10.1007/978-3-030-19955-5_28

this kind facilitates finding a more compact representation of a positive boolean function. For example, the following DNF can be represented as a conjunction of two formulas:

$$xa \lor xb \lor ya \lor yb \equiv (x \lor y)(a \lor b) \tag{1}$$

i.e., it is $\varnothing$-decomposable into the components $x \lor y$ and $a \lor b$. The following DNF is not $\varnothing$-decomposable, but it is $\{d_1, d_2\}$-decomposable:

$$xad_1 \lor xbd_1d_2 \lor yad_1d_2 \lor ybd_2 \equiv (xd_1 \lor yd_2)(ad_1 \lor bd_2) \tag{2}$$

which can be easily verified by converting the expression into DNF.

In other words, Δ-decomposition allows for common Δ-variables between the components and partitions the remaining variables of a formula. Decomposition into variable disjoint components (i.e., Δ-decomposition for $\Delta = \varnothing$) is known as disjoint conjunctive decomposition, or simply as AND-decomposition. The notion of OR-decomposition is defined similarly.

The minimization of positive DNFs via decomposition in the sense above is related to open questions, not sufficiently addressed in the previous literature. For example, there is a fundamental work by Brayton et al. on the multilevel synthesis [2], which provides minimization methods with heuristics working well for arbitrary boolean functions. However, this contribution leaves space for research on minimization for special classes of functions, where the problem is potentially simpler. This is evidenced by the research in [3] and [4], for example.

It has been observed that the quality of multilevel decomposition (i.e., alternating AND/OR–decomposition) of DNFs strongly depends on the kind of decomposition used at the topmost level. As a rule, OR–decomposition has a priority over AND-decomposition in applications, since it is computationally trivial (while AND-decomposition is considered to be hard and no specialized algorithms for DNFs are known). However, choosing AND-decomposition at the topmost level may provide a more compact representation of a boolean function. For example, application of "AND–first" strategy gives a representation of the following positive DNF

$$absu \lor absv \lor absw \lor abtu \lor abtv \lor abtw \lor abxy \lor abxz \lor$$
$$acsu \lor acsv \lor acsw \lor actu \lor actv \lor actw \lor acxy \lor acxz \lor$$
$$desu \lor desv \lor desw \lor detu \lor detv \lor detw \lor dexy \lor dexz$$

in the form $(ab \lor ac \lor de)(su \lor sv \lor sw \lor tu \lor tv \lor tw \lor xy \lor xz)$. Further, OR–decomposition of the second component gives $su \lor sv \lor sw \lor tu \lor tv \lor tw$ and $xy \lor xz$ (the first component similarly OR-decomposes syntactically). Finally, AND-decomposition of the obtained formulas gives a representation

$$(a(b \lor c) \lor de)((s \lor t)(u \lor v \lor w) \lor x(y \lor z)),$$

which is a read-once formula of depth 4 having 13 occurrences of variables.

On the contrary, Espresso[1], which implements OR–decomposition at the top-most level, gives a longer expression:

$$x(a(c \vee b) \vee de)(z \vee y) \vee (a(c \vee b) \vee de)(t \vee s)(w \vee v \vee u)$$

which is a formula of depth 5 having 18 occurrences of variables and this formula is not read-once.

Bioch [1] studied (variable disjoint) decompositions of positive boolean functions in the form $\varphi \equiv F(G(X_A), X_B)$, where $\{X_A, X_B\}$ is a partition of the variables of φ and F, G are some (positive) boolean functions. The set X_A is called a modular set of φ in this case. By taking $\varphi = x_1 \vee \ldots \vee x_n$, one can see that the number of modular sets of φ is exponential in n (since any subset of the variables is modular). Bioch showed that one can compute a tree in time polynomial in the size of an input positive DNF, which succinctly represents all its modular sets. Given a modular set X_A, the corresponding component $G(X_A)$ can be also computed in polynomial time. These important results leave the question open however, which modular sets one should choose, when trying to find a compact representation of a boolean formula. For example, the modular tree for the DNF φ from Eq. (1) consists of the singleton variable subsets (plus the set of all the variables of φ being modular by definition), from which one can obtain representations of the form $\varphi \equiv x(a \vee b) \vee ya \vee yb$ (and similarly for y and a, b selected). On the other hand, $\varnothing$-decomposition of this formula gives the representation $\varphi \equiv (x \vee y)(a \vee b)$, which is more compact.

It has been shown in [5] that computing $\varnothing$-decomposition of a boolean function is coNP-hard in general, but for a positive DNF it can be computed in time polynomial in the size of the input formula (given as a string). In fact, it has been proved that $\varnothing$-decomposition reduces to factorization of a multilinear boolean polynomial efficiently obtained from the input positive DNF. For the latter problem the authors have provided a polynomial time algorithm based on the computation of formal derivatives. In this paper, we generalize the results from [5]. First, we provide an algorithm, which computes the finest $\varnothing$-decomposition components of a positive DNF. Since $\varnothing$-decomposable functions are expected to be rare, we consider the more general notion of Δ-decomposition. The problem of computing Δ-decomposition (for an arbitrary given Δ) can be reduced to $\varnothing$-decomposition: it suffices to test whether each of the (exponentially many) DNFs, obtained from the input one for (all possible) evaluations of Δ-variables, is $\varnothing$-decomposable with the same variable partition. The reduction holds for arbitrary boolean functions, not necessarily positive ones. We show however that for positive DNFs, it suffices to make $\varnothing$-decomposition tests only for a polynomial number of (positive) DNFs obtained from the input one. As a result, we obtain a polynomial time algorithm, which computes the finest Δ-decomposition components of a positive DNF for a subset of variables Δ.

---

[1] A well–known heuristic optimizer based on the work of Brayton et al., which is often used as a reference tool for optimization of boolean functions.

2 Preliminaries

A boolean expression is a combination of constants $0, 1$ and boolean variables using conjunction, disjunction, and negation. A boolean expression is a *DNF* if it is a disjunction of *terms* (where each term is a conjunction of literals and constants). We assume there are no double negations of variables in boolean expressions, no double occurrences of the same term in a DNF, or of the same literal in a term of a DNF. A DNF is *positive* if it does not contain negated variables and the constant 0. For a set of variables V, a V-*literal* is acfrom V. We use the notation $\mathsf{vars}(\varphi)$ for the set of variables of an expression φ. For boolean expressions φ and ψ, we write $\varphi \equiv \psi$ if they are logically equivalent.

Definition 1 (Δ-decomposability). *Let φ be a boolean expression and $\Delta \subseteq \mathsf{vars}(\varphi)$ a subset of variables. The expression φ is Δ -decomposable if it is equivalent to the conjunction of boolean expressions $\psi_1, \dots, \psi_n$, where $n \geqslant 2$, such that the following holds:*

a. $\bigcup_{i=1,\dots,n} \mathsf{vars}(\psi_i) \subseteq \mathsf{vars}(\varphi)$;
b. $\mathsf{vars}(\psi_i) \cap \mathsf{vars}(\psi_j) \subseteq \Delta$, *for all* $1 \leqslant i, j \leqslant n$, $i \neq j$;
c. $\mathsf{vars}(\psi_i) \setminus \Delta \neq \varnothing$, *for* $i = 1, \dots, n$.

The expressions $\psi_1, \dots, \psi_n$ are called (Δ-) decomposition components of φ. If $\Delta = \varnothing$ and the above holds then we call φ decomposable, for short.

Clearly, Δ-decomposition components can be subject to a more fine-grained decomposition, wrt the same or different delta's. It immediately follows from this definition that a boolean expression, which contains at most one non-Δ-variable, is not Δ-decomposable. Observe that conditions a, c in the definition are important: if any of them is omitted then every boolean expression turns out to be Δ–decomposable, for any (proper) subset of variables Δ.

Definition 2 (Finest Variable Partition wrt Δ). *Let φ be a boolean expression, $\Delta \subseteq \mathsf{vars}(\varphi)$ a subset, and $\pi = \{V_1, \dots, V_{|\pi|}\}$ a partition of $\mathsf{vars}(\varphi) \setminus \Delta$. The expression φ is said to be Δ-decomposable with partition π if it has Δ-decomposition components ψ_i, for $i = 1, \dots, |\pi|$, such that $\mathsf{vars}(\psi_i) = V_i \cup \Delta$.*

The finest variable partition of φ (wrt Δ) is $\{\mathsf{vars}(\varphi)\}$ if φ is not Δ-decomposable. Otherwise it is the partition π, which corresponds to the non-Δ-decomposable components ψ_i of φ.

It will be clear from the results of this paper that for any $\Delta \subseteq \mathsf{vars}(\varphi)$, the finest variable partition of a positive DNF φ is unique. In the general case, (e.g., for arbitrary boolean expressions), this property follows from the result proved in [6] for a broad class of logical calculi including propositional logic. As will be shown below, once the finest variable partition of a DNF φ is computed, the corresponding (non-decomposable) components of φ are easily obtained.

Throughout the text, we use the term *assignment* as a synonym for a consistent set of literals. Given a set of variables $V = \{x_1, \dots, x_n\}$, where $n \geqslant 1$, a V-*assignment* is a set of literals $\{l_1, \dots, l_n\}$, where l_i is a literal over variable x_i, for $i = 1, \ldots, n$. Let φ be a DNF, $V \subseteq \mathtt{vars}(\varphi)$ a subset, and X a V-assignment such that there is a term in φ, whose set of V-literals is contained in X. Then the *substitution* of φ with X, denoted by $\varphi[X]$, is a DNF defined as follows:

- if there is a term t in φ such that every literal from t is contained in X then $\varphi[X] = 1$ (and X is called *satisfying assignment* for φ, notation: $X \models \varphi$);
- otherwise $\varphi[X]$ is the DNF obtained from φ by removing terms, whose set of V-literals is not contained in X, and by removing V-literals in the remaining terms.

For example, for the positive DNF φ from (2) we have $\varphi[\{d_1, \neg d_2\}] = xa$.

For DNFs φ and ψ, we say that φ *implies* ψ if for any assignment X, it holds $X \not\models \varphi$ or $X \models \psi$. For a set of variables V and a DNF φ, a *projection* of φ onto V, denoted as $\varphi|_V$, is the DNF obtained from φ as follows. If there is a term in φ which does not contain a variable from V then $\varphi|_V = 1$. Otherwise $\varphi|_V$ is the DNF obtained from φ by removing literals with a variable not from V in the terms of φ. It should be clear from this definition that φ implies $\varphi|_V$ (in the literature, projection is also known as a *uniform interpolant* or the *strongest consequence* of φ wrt V). For instance, for the DNF φ from (2) we have $\varphi|_{\{x,y,d_1,d_2\}} = xd_1 \vee xd_1d_2 \vee yd_1d_2 \vee yd_2$.

Lemma 1 (Decomposition Components as Projections). *Let φ be a DNF, which is Δ-decomposable with a variable partition $\pi = \{V_1, \ldots, V_n\}$. Then $\varphi|_{U_1}, \ldots, \varphi|_{U_n}$, where $U_i = V_i \cup \Delta$, $i = 1, \ldots, n$, are Δ-decomposition components of φ.*

Proof. Since φ implies $\varphi|_{U_i}$, for $i = 1, \ldots, n$, it suffices to demonstrate that $\bigwedge_{i=1,\ldots,n} \varphi|_{U_i}$ implies φ. Assume $\varphi \equiv \varphi_1 \wedge \ldots \wedge \varphi_n$, where $\varphi_1, \ldots, \varphi_n$ are Δ-decomposition components of φ, with $\mathtt{vars}(\varphi_i) = U_i$, for $i = 1, \ldots, n$. We show that $\varphi|_{U_i}$ implies φ_i, for all $i = 1, \ldots, n$. For suppose there is a U_i-assignment X such that $X \models \varphi|_{U_i}$ and $X \not\models \varphi_i$, for some $i \in \{1, \ldots, n\}$. Then by the definition of the projection $\varphi|_{U_i}$ there exists an assignment $X' \supseteq X$ such that $X' \models \varphi$ and $X' \not\models \varphi_i$, which is a contradiction, since φ implies φ_i, for $i = 1, \ldots, n$. As $\varphi_1 \wedge \ldots \wedge \varphi_n$ implies φ, we conclude that $\bigwedge_{i=1,\ldots,n} \varphi|_{U_i}$ implies φ. $\square$

Let V be a set of variables, $\Delta \subseteq V$ a subset, and A a set of V-assignments. By $\mathsf{A}|_\Delta$ we denote the set of all Δ-assignments d, for which there is an assignment $X \in \mathsf{A}$ such that $d \subseteq X$. For a Δ-assignment d, the notation $\mathsf{A}\langle d \rangle$ stands for the set of $(V \setminus \Delta)$-assignments X such that $X \cup d \in \mathsf{A}$. Let V_1, V_2 be disjoint sets of variables and for $i = 1, 2$, let A_i be a set of V_i-assignments. Then the notation $\mathsf{A}_1 \bowtie \mathsf{A}_2$ stands for the set of all assignments $X_1 \cup X_2$ such that $X_1 \in \mathsf{A}_1$ and $X_2 \in \mathsf{A}_2$. The intuitive relationship between the cartesian combinations of assignments is illustrated by the following remark and is put formally in the subsequent Lemma 2.

Remark 1 (Conjunction of DNFs is Similar to Cartesian Product)
Taking the conjunction of DNFs $\xi_1 \vee \ldots \vee \xi_m$ and $\zeta_1 \vee \ldots \vee \zeta_n$ gives a DNF, which has the form $\bigvee(\xi_i \wedge \zeta_j)$, for all pairs i, j, with $1 \leqslant i \leqslant m$, $1 \leqslant j \leqslant n$.

Lemma 2 (Decomposability Criterion). *Let φ be a DNF and $\mathtt{A}$ the set of satisfying assignments for φ. Then φ is Δ-decomposable with a partition $\pi = \{V_1, \ldots, V_{|\pi|}\}$ iff for all $d \in \mathtt{A}|_\Delta$ it holds that $\mathtt{A}\langle d \rangle = \mathtt{A}\langle d \rangle|_{V_1} \bowtie \ldots \bowtie \mathtt{A}\langle d \rangle|_{V_{|\pi|}}$.*

In this paper, we are concerned with the problem of finding the finest variable partition of a positive DNF φ wrt a subset Δ of its variables. By Lemma 1, decomposition components of φ can be easily obtained from the finest variable partition. We assume that boolean expressions are given as strings and thus, the size of an expression φ is the length of the string, which represents φ.

For the sake of completeness, first we describe a factorization algorithm for multilinear boolean polynomials, which is based on the results from [5,7]. Then we provide a $\varnothing$-decomposition algorithm for a positive DNF φ based on factorization of a boolean polynomial, which is obtained from φ by a simple syntactic transformation. Finally, we demonstrate that Δ-decomposition of a positive DNF reduces to $\varnothing$-decomposition of (a polynomial number of) positive DNFs obtained from the input one and devise the corresponding polynomial time Δ-decomposition algorithm.

3   Factorization of Boolean Polynomials

In [8], Shpilka and Volkovich established the prominent result on the equivalence of polynomial factorization and identity testing. It follows from their result that a multilinear boolean polynomial can be factored in time cubic in the size of the polynomial given as a string. This result has been rediscovered in [5,7], where the authors have provided a factorization algorithm based on the computation of derivatives of multilinear boolean polynomials, which allows for deeper optimizations. Without going into implementation details, we employ this result here to formulate an algorithm, which computes the finest $\varnothing$-decomposition components of a positive DNF φ. Hereafter we assume that polynomials do not contain double occurrences of the same monomial.

Definition 3. *A polynomial F is factorable if $F = G_1 \cdot \ldots \cdot G_n$, where $n \geqslant 2$ and $G_1, \ldots, G_n$ are some non-constant polynomials. Otherwise F is irreducible. The polynomials $G_1, \ldots, G_n$ are called* factors *of F. For a polynomial F, the* finest variable partition *of F is $\{\mathtt{vars}(F)\}$ if F is irreducible and otherwise consists of the sets of the variables of the irreducible factors of F.*

It is important to stress that we consider here multilinear polynomials (every variable can occur only in the power of $\leqslant 1$) and thus, the factors are polynomials *over disjoint sets of variables*. Note that the finest variable partition of a multilinear boolean polynomial is unique, since the ring of these polynomials is a unique factorization domain. We now formulate the first important observation, which is a strengthening of Theorem 5 from [5].

Theorem 1 (Computing Finest Variable Partition for Polynomial). *For a multilinear boolean polynomial F, the (unique) finest partition of the variables of F can be found in time polynomial in the size of F.*

It is proved in [5] that testing whether F is factorable and computing its factors can be done in time polynomial in the size of F given as a string. By applying the factorization procedure to the obtained factors recursively, one obtains a partition of the variables of F, which corresponds to the irreducible factors of F. This is implemented in FindPartition procedure given below, which is a modification of the factorization algorithm from [5]. It is also shown in [5,7] that once a partition of variables, which corresponds to the factors of F is computed, the factors can be easily obtained as projections of F onto the components of the partition (see the notion of projection below).

The FindPartition procedure takes a boolean polynomial F as an input and outputs the finest partition of $\mathrm{vars}(F)$ in time polynomial in the size of F. A few notations are required. For a polynomial F, we denote by $\mathrm{vars}(F)$ the set of the variables of F. For a variable $x \in \mathrm{vars}(F)$ and a value $a \in \{0,1\}$, we denote by $F_{x=a}$ the polynomial obtained from F by substituting x with a. Given a set of variables V and a monomial m, the *projection* of m onto V (denoted as $m|_V$) is 1 if m does not contain any variable from V, or is equal to the monomial obtained from m by removing all the variables not contained in V, otherwise. The *projection* of a polynomial F onto V, denoted as $F|_V$, is the polynomial obtained by projecting the monomials of F onto V and by removing duplicate monomials.

Lines 2–4 of FindPartition is a test for a simple sufficient condition for irreducibility: if a polynomial is a constant then it cannot be factorable. Lines 5–15 implement a test for trivial factors: if some variable z is present in every monomial of F, then z is an irreducible factor. In the recursive part of the procedure, the remaining sets from the finest variable partition of F are computed as the values of the variable Σ and are added to FinestPartition.

1: **procedure** FINDPARTITION(F)
2:     **if** $F == 0$ or $F == 1$ **then**
3:         **return** vars(F)
4:     **end if**
5:     **for** z a variable occurring in every monomial of F **do**
6:         FinestPartition.add($\{z\}$)
7:         $F \leftarrow F_{z=1}$
8:     **end for**
9:     **if** F does not contain any variables **then**
10:         **return** FinestPartition
11:     **end if**
12:     **if** F contains a single variable, e.g., x **then**
13:         FinestPartition.add($\{x\}$)
14:         **return** FinestPartition
15:     **end if**
16:     $V \leftarrow$ variables of F
17:     **repeat**
18:         $\Sigma \leftarrow \varnothing$; $F \leftarrow F|_V$
19:         pick a variable x from V
20:         Σ.add(x); $V \leftarrow V \setminus \{x\}$
21:         $G \leftarrow F_{x=0} \cdot \frac{\partial F}{\partial x}$
22:         **for** a variable y from V **do**
23:             **if** $\frac{\partial G}{\partial y} \neq 0$ **then**
24:                 Σ.add(y)
25:             **end if**
26:         **end for**
27:         FinestPartition.add(Σ)
28:         $V \leftarrow V \setminus \Sigma$
29:     **until** $V = \varnothing$
30:     **return** FinestPartition
31: **end procedure**

4 ∅-Decomposition of Positive DNFs

A term t of a DNF φ is called *redundant* in φ if there exists another term t' of φ such that every literal of t' is present in t (i.e., $t' \subseteq t$). For example, the term xy is redundant in $xy \vee x$. It is easy to see that removing redundant terms gives a logically equivalent DNF.

Let us note the following simple fact:

Lemma 3 (Existence of Positive Components). *Let φ be a positive boolean expression and $\Delta \subseteq \mathtt{vars}(\varphi)$ a subset of variables. If φ is Δ-decomposable then it has decomposition components, which are positive expressions.*

Proof. It is known (e.g., see Theorem 1.21 in [9]) that a boolean expression ψ is equivalent to a positive one in a variable x iff for the set of satisfying assignments A for φ the following property holds: if $\{l_1, \ldots, l_n, \neg x\} \in \mathsf{A}$, where $l_1, \ldots, l_n$ are literals, then $\{l_1, \ldots, l_n, x\} \in \mathsf{A}$. Clearly, this property is preserved under decomposition: if a set of assignments A satisfies the property and it holds that $\mathsf{A} = \mathsf{A}_1 \bowtie \ldots \bowtie \mathsf{A}_n$, then so do the sets A_i, for $i = 1, \ldots, n$. Thus, the claim follows directly from Lemma 2. $\square$

The next important observation is a strengthening of the result from [5], which established the complexity of ∅-decomposition for positive DNFs.

Theorem 2 (Computing the Finest Variable Partition wrt $\Delta = \varnothing$). *The finest variable partition of a positive DNF φ can be computed in time polynomial in the size of φ.*

Let P be a 1-1 mapping, which for a positive DNF φ gives a multilinear boolean polynomial $\mathsf{P}(\varphi)$ over $\mathtt{vars}(\varphi)$ obtained by replacing the conjunction and disjunction with $\cdot$ and $+$, respectively. The theorem is proved by showing that decomposition components of a positive DNF φ can be recovered from factors of a polynomial $\mathsf{P}(\psi)$ constructed for a DNF ψ, which is obtained from φ by removing redundant terms. The idea is illustrated in ∅Decompose procedure below, which for a given positive DNF φ computes the finest variable partition of φ. It relies on the factorization procedure from Sect. 3 and is employed as a subroutine in Δ-decomposition algorithm in Sect. 5. The procedure uses a simple preprocessing, which removes redundant terms. The preprocessing also allows for detecting those variables (line 5 of the procedure) that φ does not depend on. By the definition of decomposability, these variables are decomposition components of φ, so they are added as singleton sets into the resulting finest variable partition (at line 6).

1: **procedure** $\varnothing\text{DECOMPOSE}(\varphi)$
2:     FinestPartition $\leftarrow \varnothing$
3:     $\psi \leftarrow \text{REMOVEREDUNDTERMS}(\varphi)$
4:     FinestPartition $\leftarrow$
    $\text{FINDPARTITION}(\mathbf{P}(\psi))$ ▷ see Sect. 3
5:     **for all** $x \in \text{vars}(\varphi) \setminus \text{vars}(\psi)$ **do**
6:         FinestPartition.add($\{x\}$)
7:     **end for**
8:     **return** FinestPartition
9: **end procedure**

1: **procedure** REMOVEREDUNDTERMS(φ)
2:     **for all** terms t in φ **do**
3:         **if** there exists a term t' in φ s.t.
    $t' \subseteq t$ **then**
4:             remove t from φ
5:         **end if**
6:     **end for**
7:     **return** φ
8: **end procedure**

5 Δ-decomposition of Positive DNFs

Definition 4 (Δ-atom). *For a positive DNF φ and a subset $\Delta \subseteq \text{vars}(\varphi)$, the set of Δ-variables of a term of φ is called Δ-atom of φ.*

Note that by definition a Δ-atom can also be the empty set. Let U be the set of unions of Δ-atoms of φ. Given a set $X \in U$, we introduce the notation $\varphi\langle X \rangle$ as a shortcut for the DNF $\varphi[X \cup \bar{X}]$, where $\bar{X} = \{\neg x \mid x \in \Delta \setminus X\}$.

Let π be a partition of $\text{vars}(\varphi) \setminus \Delta$. We say that a boolean expression ψ *supports* π if every set from the finest variable partition of ψ wrt Δ is contained in some set from π. It is easy to see that if φ is Δ-decomposable with π, then $\varphi[X]$ supports π, for any set of literals X such that $\varphi[X]$ is defined.

We formulate two lemmas that are the key to the main result, Theorem 3, in this section.

Lemma 4 (Δ-Decomposability Criterion for Positive DNF). *Let φ be a positive DNF, $\Delta \subseteq \text{vars}(\varphi)$ a subset, and U the set of unions of Δ-atoms of φ. Then φ is Δ-decomposable with a variable partition π iff $\varphi\langle X \rangle$ supports π, for all $X \in U$.*

Proof. ($\Rightarrow$): Take $X \in U$. Since φ is positive, X is a consistent set of literals, $\varphi\langle X \rangle$ is defined, and clearly, supports π.

($\Leftarrow$): Let $\pi = \{V_1, \ldots, V_{|\pi|}\}$, A be the set of satisfying assignments for φ, and $d \in A|_\Delta$ a Δ-assignment. Then there is $X \in U$ such that $X \subseteq d$, since φ is a DNF. Let X be the maximal set from U with this property. Then we have $\varphi[d] = \varphi\langle X \rangle$, so $\varphi[d]$ supports π. This yields $A\langle d \rangle = A\langle d \rangle|_{V_1} \bowtie \ldots \bowtie A\langle d \rangle|_{V_{|\pi|}}$ and since d was arbitrarily chosen, it follows from Lemma 2 that φ is Δ-decomposable with π. $\square$

Lemma 5 (Decomposition Lemma). *Let $\varphi_1, \ldots, \varphi_n$, where $n \geqslant 1$, be DNFs with the following property: for all $1 \leqslant j, k \leqslant n$ there is a subset $I \subseteq \{1, \ldots, n\}$, with $j, k \in I$, such that $\bigvee_{i \in I} \varphi_i$ is decomposable with π. Then so is $\varphi_1 \vee \ldots \vee \varphi_n$.*

Proof. Let $\pi = \{X, Y\}$ and denote $\varphi = \varphi_1 \vee \ldots \vee \varphi_n$. By Lemma 1 we need to show that $\varphi \equiv \varphi|_X \wedge \varphi|_Y$, which is equivalent to:

$$\varphi \equiv (\varphi_1|_X \vee \ldots \vee \varphi_n|_X) \wedge (\varphi_1|_Y \vee \ldots \vee \varphi_n|_Y) \tag{3}$$

Observe that the right-hand side of this equation can be written as the expression $D = \bigvee_{1 \leqslant j,k \leqslant n} \varphi_j|_X \ \varphi_k|_Y$. Take any $j, k \in \{1, \ldots, n\}$. By the condition of the lemma there is a subset $I \subseteq \{1, \ldots, n\}$, with $j, k \in I$, such that $\bigvee_{a,b \in I} \varphi_a|_X \ \varphi_b|_Y \equiv \bigvee_{i \in I} \varphi_i$. That is, a disjunction of formulas from D containing both, $\varphi_j|_X \ \varphi_k|_Y$ and $\varphi_k|_X \ \varphi_j|_Y$ is a equivalent to a disjunction of formulas from φ. Since the choice of j, k was arbitrary, we conclude that (3) holds and thus, the lemma is proved. $\square$

Theorem 3 (Computing the Finest Variable Partition wrt Δ). *Given a positive DNF φ and a subset $\Delta \subseteq \mathtt{vars}(\varphi)$, the finest variable partition of φ wrt Δ can be computed in time polynomial in the size of φ.*

Proof. Let A be the set of Δ-atoms of φ and U consist of all unions of sets from A. Note that $|A|$ is bounded by the size of φ, while $|U|$ is exponential. By Lemma 4, φ is Δ-decomposable with a partition π iff $\varphi\langle X \rangle$ supports π, for all $X \in U$.

For any $X \in U$, we have $\mathtt{vars}(\varphi\langle X \rangle) \subseteq \mathtt{vars}(\varphi)$. Observe that $\varphi\langle X \rangle$ is equivalent to the DNF $\psi = \varphi\langle X \rangle \vee t$, where t is a term redundant in ψ, $\mathtt{vars}(t) = \mathtt{vars}(\varphi) \setminus \mathtt{vars}(\varphi\langle X \rangle)$ (in case $\mathtt{vars}(t) = \varnothing$ we assume that $\psi = \varphi\langle X \rangle$) and it holds $\mathtt{vars}(\psi) = \mathtt{vars}(\varphi)$. Therefore, $\varphi\langle X \rangle$ supports π iff ψ is decomposable with π.

For any $X \in U$, $\varphi\langle X \rangle$ is equivalent to $\varphi_1 \vee \ldots \vee \varphi_n$ where $n \geqslant 1$ and for $i = 1, \ldots, n$, $\varphi_i = \varphi\langle a_i \rangle$, where $a_i \subseteq X$, a Δ-atom of φ. Notice further that $\varphi\langle X \rangle$ is equivalent to $\varphi'_1 \vee \ldots \vee \varphi'_n$, where $\varphi'_i = \varphi_i \vee t_i$ and t_i is a redundant term as introduced above. By Lemma 4, if φ is Δ-decomposable with a partition π then $\varphi\langle a_1 \cup a_2 \rangle$ supports π, for any $a_1, a_2 \in A$. For the other direction, if $\varphi\langle a_1 \cup a_2 \rangle$ supports π, for any $a_1, a_2 \in A$ then the condition of Lemma 5 holds for $\varphi'_1 \vee \ldots \vee \varphi'_n$. It follows that $\varphi\langle X \rangle$ supports π, for any $X \in U$ and hence by Lemma 4, φ is Δ-decomposable with π.

By Theorem 2, a variable partition σ, which corresponds to the finest decomposition of $\varphi\langle a_1 \cup a_2 \rangle$, can be found in time polynomial in the size of $\varphi\langle a_1 \cup a_2 \rangle$ (and hence, in the size of φ, as well). For any variables $x, y \in \mathtt{vars}(\varphi)$ and a set $S \in \sigma$, if $x, y \in S$ then x and y cannot belong to different decomposition components of $\varphi\langle a_1 \cup a_2 \rangle$. Let $\sim$ be an equivalence relation on $\mathtt{vars}(\varphi)$ such that $x \sim y$ iff there are $a_1, a_2 \in A$ such that x and y belong to the same component of the finest variable partition of $\varphi\langle a_1 \cup a_2 \rangle$. Since $|A|$ is bounded by the size of φ, one can readily verify that the equivalence classes wrt $\sim$ can be computed in time polynomial in the size of φ and are equal to its finest variable partition. $\square$

We conclude the paper with a description of $\Delta \mathtt{Decompose}$ procedure, which for a positive DNF φ and a (possibly empty) subset $\Delta \subseteq \mathtt{vars}(\varphi)$ computes the finest variable partition of φ wrt Δ and outputs Δ-decomposition components, which correspond to the partition.

In Lines 8–10 of the procedure, a set of Δ-atoms of φ is computed, while skipping those ones, which subsume some term of φ. Clearly, if there is a term t of φ, which consists only of d-variables for some subset $d \subseteq \Delta$, then it holds $\varphi[d] = 1$, which implies that $\varphi\langle d \rangle$ supports any partition π of $\mathrm{vars}(\varphi) \setminus \Delta$ (at this point φ necessarily contains at least 2 non-Δ-variables due to the test in line 4). Therefore, these atoms are irrelevant in computing decomposition and they can be omitted (similarly, the unions of Δ-atoms in line 13).

Lines 11–17 implement a call for computing the finest variable partition wrt the empty Δ for each DNF $\varphi\langle L \rangle$ obtained from φ for a union L of relevant Δ-atoms. The result is a family of partitions, which are further aligned by computing equivalence classes on the variables of φ. This is implemented in AlignPartitions procedure by computing connected components of a graph, in which vertices correspond to the variables of φ.

Finally, in lines 22–25 the decomposition components of φ are computed as projections onto the sets of variables corresponding to the finest partition. The components are cleaned up by removing redundant terms and are sent to the output.

```
 1: procedure ΔDECOMPOSE(φ, Δ)
 2:     FinestPartition ← ∅
 3:     Components ← ∅
 4:     if φ contains at most one non-Δ-
        variable then
 5:         return {φ}              ▷ φ is not
            Δ-decomposable
 6:     end if

 7:     ΔAtoms ← ∅
 8:     for every term t of φ, which contains
        at least one non-Δ-variable do
 9:         ΔAtoms.add(the set of Δ-variables of
            t)
10:     end for
11:     for all a₁, a₂ from ΔAtoms do
12:         L ← a₁ ∪ a₂
13:         if there is no term t in φ, whose
            every variable is from L then
14:         PartitionForL ← ∅DECOMPOSE(φ⟨L⟩) ▷
            see Sect. 4
15:         PartitionFamily.add(PartitionForL)
16:         end if
17:     end for
18:     FinestPartition ←
        ALIGNPARTITIONS(PartitionFamily)
19:     if FinestPartition.isSingleton then
20:         return {φ}              ▷ φ is not
            Δ-decomposable
```

```
21:     else
22:         for V ∈ FinestPartition do
23:             ψ ←REMOVEREDUNDTERMS(φ|V∪Δ)   ▷
                see Sect. 4
24:             Components.add(ψ)
25:         end for
26:         return Components
27:     end if

28: end procedure


 1: procedure ALIGNPARTITIONS(PFamily)
 2:     G← ∅       ▷ a graph with vertices being
        vars. of φ
 3:     for Partition ∈ PFamily do
 4:         for VarSet ∈ Partition do
 5:             G.add(a path involving all x ∈
                VarSet)
 6:         end for
 7:     end for

 8:     ResultPartition ← ∅
 9:     for C a connected component of G do
10:         ResultPartition.add(the set of vars
            from C)
11:     end for
12:     return ResultPartition
13: end procedure
```

References

1. Bioch, J.C.: Decomposition of Boolean functions. In: Crama, Y., Hammer, P.L. (eds.) Boolean Models and Methods in Mathematics, Computer Science, and Engineering. Encyclopedia of Mathematics and its Applications, vol. 134, pp. 39–78. Cambridge University Press, New York (2010)
2. Villa, T., Brayton, R.K., Sangiovanni-Vincentelli, A.: Synthesis of multilevel Boolean networks. In: Crama, Y., Hammer, P.L. (eds.) Boolean Models and Methods in Mathematics, Computer Science, and Engineering. Encyclopedia of Mathematics and its Applications, vol. 134, pp. 675–722. Cambridge University Press, New York, NY, USA (2010)
3. Boros, E.: Horn functions. In: Crama, Y., Hammer, P.L. (eds.) Boolean Functions: Theory, Algorithms, and Applications. Encyclopedia of Mathematics and its Applications, vol. 134, pp. 269–325. Cambridge University Press, New York (2011)
4. Gursky, S.: Special classes of Boolean functions with respect to the complexity of their minimization. Ph.D. thesis, Charles University in Prague (2014)
5. Emelyanov, P., Ponomaryov, D.: Algorithmic issues of conjunctive decomposition of Boolean formulas. Program. Comput. Softw. **41**(3), 162–169 (2015)
6. Ponomaryov, D.: On decomposability in logical calculi. Bull. Novosib. Comput. Cent. **28**, 111–120 (2008)
7. Emelyanov, P., Ponomaryov, D.: On tractability of disjoint AND-decomposition of Boolean formulas. In: Voronkov, A., Virbitskaite, I. (eds.) PSI 2014. LNCS, vol. 8974, pp. 92–101. Springer, Heidelberg (2015). https://doi.org/10.1007/978-3-662-46823-4_8
8. Shpilka, A., Volkovich, I.: On the relation between polynomial identity testing and finding variable disjoint factors. In: Abramsky, S., Gavoille, C., Kirchner, C., Meyer auf der Heide, F., Spirakis, P.G. (eds.) ICALP 2010. LNCS, vol. 6198, pp. 408–419. Springer, Heidelberg (2010). https://doi.org/10.1007/978-3-642-14165-2_35
9. Crama, Y., Hammer, P.L.: Boolean Functions - Theory, Algorithms, and Applications. Encyclopedia of Mathematics and Its Applications, vol. 142. Cambridge University Press, Cambridge (2011)

Unpopularity Factor in the Marriage and Roommates Problems

Suthee Ruangwises$^{(\boxtimes)}$ and Toshiya Itoh

Department of Mathematical and Computing Science,
Tokyo Institute of Technology, Tokyo, Japan
`ruangwises.s.aa@m.titech.ac.jp`, `titoh@c.titech.ac.jp`

Abstract. Given a set A of n people, we consider the *Roommates Problem* (RP) and *Marriage Problem* (MP) where each person has a list that ranks a subset of A as his/her acceptable partner in order of preference. Ties among two or more people are allowed in the lists. In RP there is no further restriction, while in MP only people with opposite genders can be matched. For a pair of matchings X and Y, we say a person prefers X to Y if he/she prefers the person matched by X to the person matched by Y, and let $\phi(X, Y)$ denote the number of people who prefer X to Y. Define an *unpopularity factor* $u(M)$ of a matching M to be the maximum ratio $\phi(M', M)/\phi(M, M')$ among all possible other matchings M'. In this paper, we develop an algorithm to efficiently compute the unpopularity factor of a given matching. The algorithm runs in $O(m\sqrt{n}\log^2 n)$ time for RP and in $O(m\sqrt{n}\log n)$ time for MP, where m is the total length of people's preference lists. We also generalize the notion of unpopularity factor to the weighted setting where people are given different voting weights, and show that our algorithm can be slightly modified to support that setting as well with the same runtime.

Keywords: Unpopularity factor · Popular matching ·
Perfect matching · Marriage Problem · Roommates Problem

1 Introduction

Stable Marriage Problem is one of the most classic and actively studied problems in theoretical computer science, with many applications in other areas such as economics [11,20]. In the setting called Marriage Problem (MP), we have a set of n men and n women, with each person having a list that ranks all people of opposite gender in strict order of preference. A matching M is called *stable* if there is no *blocking pair*: a man m and a woman w who are not matched to each other but prefer each other to their own partners in M. Gale and Shapley [7] showed that a stable matching always exists and can be found in $O(n^2)$ time. The same algorithm can also be adapted to the setting where each person's preference list may not contain all people of opposite gender, as he/she may regard some people as unacceptable. The algorithm runs in $O(m)$ time in this setting, where m is the total length of people's preference lists [11].

© Springer Nature Switzerland AG 2019
R. van Bevern and G. Kucherov (Eds.): CSR 2019, LNCS 11532, pp. 337–348, 2019.
https://doi.org/10.1007/978-3-030-19955-5_29

Stable Roommates Problem is a generalization of the original Stable Marriage Problem to a non-bipartite setting called Roommates Problem (RP), where each person can be matched with any other one regardless of gender. Unlike in MP, a stable matching in this case does not always exist. Irving [14] developed an $O(n^2)$ algorithm to find a stable matching or report that none exists in a given RP instance, where n is the number of people.

1.1 Popular Matchings

Apart from the well-known stability, a less restrictive property of a "good" matching is popularity. For a pair of matchings X and Y, we say a person prefers X to Y if he/she prefers a person matched by X to a person matched by Y, and let $\phi(X, Y)$ denote the number of people who prefer X to Y. A matching M is called *popular* if $\phi(M, M') \geq \phi(M', M)$ for any other matching M'. The concept of popularity of a matching was first introduced by Gardenfors [8] in the context of the Stable Marriage Problem. He also proved that every stable matching is also popular, but not vice-versa, hence a popular matching always exists in every MP instance.

In contrast to MP, in RP a popular matching does not always exist, and there is no fast algorithm to check whether it exists in a given instance. Biró et al. [3] proved that when ties among people in the preference lists are allowed, the problem of determining whether a popular matching exists in a given RP instance is NP-hard. Recently, Faenza et al. [6] and Gupta et al. [10] independently proved that this problem is still NP-hard even when people's preference lists are *strict* (containing no tie). Furthermore, in a complete graph RP instance where each person's preference list is strict and contains all other people, Cseh and Kavitha [4] showed that the problem of determining whether a popular matching exists can be solved in polynomial time for an odd n but is NP-hard for an even n.

Popular matching problems were also extensively studied in the setting of one-sided preference lists (matching each person with a unique item, where each person has a list that ranks items but each item does not have a list that ranks people) called the House Allocation Problem (HAP). Abraham et al. [1] introduced the first polynomial time algorithm to find a popular matching in a given HAP instance, or report that none exists. The algorithm runs in $O(m + n)$ time when people's preference lists are strict and in $O(m\sqrt{n})$ time when ties are allowed, where m is the total length of people's preference lists and n is the total number of people and items. Mestre [19] later generalized that algorithm to the case where people are given different voting weights, while Manlove and Sng [17] also generalized it to the case where each item is allowed to be matched with more than one person called the Capacitated House Allocation Problem (CHAP). Abraham and Kavitha [2] proved that starting at any matching in an instance, a popular matching can be achieved by at most two majority votes to force a change in assignments as long as at least one popular matching exists. Mahdian [16] investigated the existence of a popular matching when people's preference lists are strict, *complete* (containing all items), and randomly generated, and showed that a popular matching exists with high probability in a random HAP instance if the ratio of the number of items to the number of people is greater than a specific constant. Ruangwises and Itoh [21] later generalized Mahdian's study to the case where preference lists are not complete and found a similar behavior of the probability of existence of a popular matching. Kavitha et al. [15] introduced the concept of a *mixed matching*, which is a probability distribution over a set of matchings, and proved that a mixed matching that is "popular" always exists.

1.2 Unpopularity Measures

While a popular matching does not exist in some instances, several measures of "badness" of a matching that is not popular have been introduced. In the one-sided preference lists setting, McCutchen [18] introduced two such measures: the *unpopularity factor* and the *unpopularity margin*. The unpopularity factor $u(M)$ of a matching M is the maximum ratio $\phi(M',M)/\phi(M,M')$ among all possible other matchings M', while the unpopularity margin $g(M)$ is the maximum difference $\phi(M',M) - \phi(M,M')$ among all possible other matchings M'. Note that the two measures are not equivalent as $\phi(M',M)$ and $\phi(M,M')$ may not add up to n since some people may like M and M' equally, thus it is possible for a matching to have higher unpopularity factor but lower unpopularity margin than another matching. See Example 1.

McCutchen then developed an algorithm to compute $u(M)$ and $g(M)$ of a given matching M of an HAP instance in $O(m\sqrt{n_2})$ and $O((g+1)m\sqrt{n})$ time, respectively, where n_2 is the number of items and $g = g(M)$ is the unpopularity margin of M. He also proved that the problem of finding a matching that minimizes either measure is NP-hard. Huang et al. [13] later developed an algorithm to find a matching with bounded values of these measures in HAP instances with certain properties.

The notions of unpopularity factor and unpopularity margin also apply to the setting of two-sided preference lists. Huang and Kavitha [12] proved that an RP instance with strict preference lists always has a matching with unpopularity factor $O(\log n)$, and it is NP-hard to find a matching with lowest unpopularity factor, or even the one with less than $4/3$ times of the optimum.

Example 1. Consider the following RP instance. A set in a preference list means that all people in that set are ranked equally, e.g. a_2 equally prefers a_1 and a_4 as his first choices over a_3.

Preference Lists
a_1 : a_4, a_2, a_3	$M_0 = \{\{a_1, a_2\}, \{a_3, a_4\}\}$
a_2 : $\{a_1, a_4\}, a_3$	$M_1 = \{\{a_1, a_3\}, \{a_2, a_4\}\}$
a_3 : $\{a_1, a_4\}, a_2$	$M_2 = \{\{a_1, a_4\}, \{a_2, a_3\}\}$
a_4 : $\{a_2, a_3\}, a_1$	

In this example, $\phi(M_0, M_1) = 1$, $\phi(M_1, M_0) = 0$, $\phi(M_0, M_2) = 3$, $\phi(M_2, M_0) = 1$, $\phi(M_1, M_2) = 3$, and $\phi(M_2, M_1) = 1$. Therefore, M_0 is popular, while

$u(M_1) = \infty$, $g(M_1) = 1 - 0 = 1$, $u(M_2) = 3/1 = 3$, and $g(M_2) = 3 - 1 = 2$. Observe that M_1 has higher unpopularity factor but lower unpopularity margin than M_2.   □

1.3 Our Contribution

Biró et al. [3] developed an algorithm to determine whether a given matching M is popular in $O(m\sqrt{n\alpha(n,m)}\log^{3/2} n)$ time for RP (improved to $O(m\sqrt{n}\log n)$ time when running with the most recent algorithm to find a maximum weight perfect matching of Duan et al. [5]) and in $O(m\sqrt{n})$ time for MP, where α is the inverse Ackermann's function. Their algorithm also simultaneously computes the unpopularity margin of M during the run. However, there is currently no algorithm to efficiently compute an unpopularity factor of a given matching in MP and RP.

Using a similar idea to [3], in this paper we develop an algorithm to compute the unpopularity factor of a given matching. The algorithm runs in $O(m\sqrt{n}\log^2 n)$ time for RP and in $O(m\sqrt{n}\log n)$ time for MP. We also generalize the notion of unpopularity factor to the weighted setting where people are given different voting weights, and show that our algorithm can be slightly modified to support that setting as well with the same runtime.

2 Preliminaries

Let I be an RP or MP instance consisting of a set $A = \{a_1, ..., a_n\}$ of n people, with each person having a preference list that ranks a subset of A as his/her acceptable partner in order of preference. In RP there is no further restriction, while in MP people are classified into two genders, and each person's preference list can contain only people with opposite gender. Throughout this paper, we consider a more general setting where ties among two or more people are allowed in the preference lists. Also, let m be the total length of people's preference lists.

For a matching M and a person $a \in A$, let $M(a)$ be the person matched with a in M (for convenience, let $M(a) = null$ if a is unmatched in M). Also, let $r_a(b)$ be the rank of a person b in a's preference list, with the most preferred item(s) having rank 1, the second most preferred item(s) having rank 2, and so on (for convenience, let $r_a(null) = \infty$).

For any pair of matchings X and Y, we define $\phi(X, Y)$ to be the number of people who strictly prefer X to Y, i.e. $\phi(X, Y) = |\{a \in A | r_a(X(a)) < r_a(Y(a))\}|$. Also, let

$$\Delta(X, Y) = \begin{cases} \phi(Y, X)/\phi(X, Y), & \text{if } \phi(X, Y) > 0; \\ \infty, & \text{otherwise.} \end{cases}$$

Finally, define an unpopularity factor

$$u(M) = \max_{M' \in M - M_M} \Delta(M, M'),$$

where $\mathbb{M}$ is the set of all matchings of I and $\mathbb{M}_M$ is the set of all matchings M' with $\phi(M, M') = \phi(M', M) = 0$. Note that a matching M is popular if and only if $u(M) \leq 1$.

3 Main Results

Let I be an RP instance, M be a matching of I, and k be an arbitrary positive real number. Similarly to [3], we construct an undirected graph $H_{(M,k)}$ with vertices $A \cup A'$, where $A' = \{a'_1, ..., a'_n\}$ is a set of "copies" of people in A. An edge $\{a_i, a_j\}$ exists if and only if a_i is in a_j's preference list and a_j is in a_i's preference list; an edge $\{a'_i, a'_j\}$ exists if and only if $\{a_i, a_j\}$ exists; an edge $\{a_i, a'_j\}$ exists if and only if $i = j$.

However, we will assign weights to edges of $H_{(M,k)}$ differently from [3]. For each pair of i and j with an edge $\{a_i, a_j\}$, define $\delta_{i,j}$ as follows.

$$\delta_{i,j} = \begin{cases} 1, & \text{if } a_i \text{ is unmatched in } M \text{ or } a_i \text{ prefers } a_j \text{ to } M(a_i); \\ -k, & \text{if } a_i \text{ prefers } M(a_i) \text{ to } a_j; \\ 0, & \text{if } \{a_i, a_j\} \in M \text{ or } a_i \text{ likes } a_j \text{ and } M(a_i) \text{ equally.} \end{cases}$$

For each pair of i and j, we set the weights of both $\{a_i, a_j\}$ and $\{a'_i, a'_j\}$ to be $\delta_{i,j} + \delta_{j,i}$. Also, for each edge $\{a_i, a'_i\}$, we set its weight to be $-2k$ if a_i is matched in M, and 0 otherwise. See Example 2.

Example 2. Consider the following matching M in an RP instance.

Preference Lists
a_1 : a_2, a_3, a_4
a_2 : a_3, a_1
a_3 : a_1, a_2, a_4
a_4 : a_1, a_3

$M = \{(a_1, a_2), (a_3, a_4)\}$

$\delta_{i,j}$		j		
	1	2	3	4
1		0	-2	-2
2	0		1	
3	1	1		0
4	1		0	

(i labels the rows)

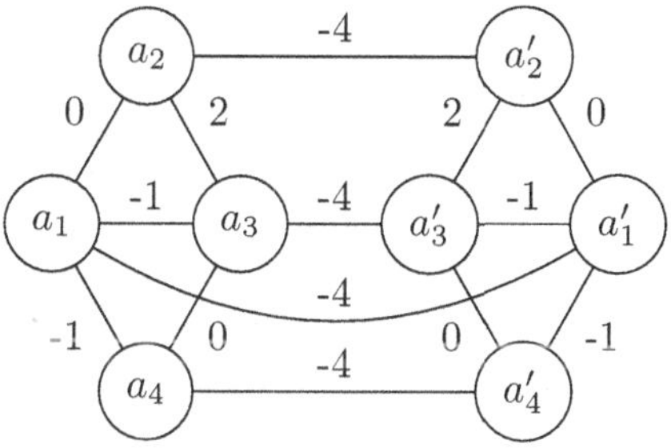

The values of all $\delta_{i,j}$ are shown in the above table, and the auxiliary graph $H_{(M,2)}$ is shown on the right. □

The intuition of this auxiliary graph is that we want to check whether $u(M) > k$, i.e. whether there exists another matching M' with the number of people who prefer M' to M more than k times the number of those who prefer M to M'. Each matching M' is represented by a perfect matching in $H_{(M,k)}$ consisting of the edges of M' in A as well as their "copies" in A', with each unmatched person a_i being matched with his own copy a'_i. The intuition of assigning weights to the edges is that we add 1 for each person who prefers M' to M and subtract k for each one who prefers M to M', and then check whether the sum is positive.

The relation between $u(M)$ and $H_{(M,k)}$ is formally shown in the following lemma.

Lemma 1. $u(M) > k$ if and only if $H_{(M,k)}$ contains a positive weight perfect matching.

Proof. For any matching M', define $A_1(M') = \{a_i \in A | a_i \text{ is matched in } M'\}$ and $A_2(M') = \{a_i \in A | a_i \text{ is unmatched in } M'\}$. Also, define

$A_1^+(M') = \{a_i \in A_1(M') | a_i \text{ is unmatched in } M \text{ or } a_i \text{ prefers } M'(a_i) \text{ to } M(a_i)\}$;

$A_1^-(M') = \{a_i \in A_1(M') | a_i \text{ prefers } M(a_i) \text{ to } M'(a_i)\}$;

$A_2^-(M') = \{a_i \in A_2(M') | a_i \text{ is matched in } M\}$.

We have $\phi(M', M) = |A_1^+(M')|$ and $\phi(M, M') = |A_1^-(M')| + |A_2^-(M')|$.

Suppose that $u(M) > k$. From the definition of $u(M)$, there must be a matching M_0 such that $\phi(M_0, M) > k\phi(M, M_0)$. In the graph $H_{(M,k)}$, consider a perfect matching

$$S_0 = M_0 \cup \{\{a'_i, a'_j\} | \{a_i, a_j\} \in M_0\} \cup \{\{a_i, a'_i\} | a_i \text{ is unmatched in } M_0\}.$$

From the definition, the weight of S_0 is

$$\begin{aligned} W(S_0) &= 2\left(|A_1^+(M_0)| - k|A_1^-(M_0)|\right) - 2k|A_2^-(M_0)| \\ &= 2\left(|A_1^+(M_0)| - k\left(|A_1^-(M_0)| + |A_2^-(M_0)|\right)\right) \\ &= 2(\phi(M_0, M) - k\phi(M, M_0)) \\ &> 0, \end{aligned}$$

hence $H_{(M,k)}$ contains a positive weight perfect matching.

On the other hand, suppose there is a positive weight perfect matching S_1 of $H_{(M,k)}$ with weight $W(S_1)$. See Example 3. Let $M_1 = \{\{a_i, a_j\} \in S_1\}$ and $M_2 = \{\{a_i, a_j\} | \{a'_i, a'_j\} \in S_1\}$. Since S_1 is a perfect matching in $H_{(M,k)}$, we have $A_2(M_1) = A_2(M_2)$, and

$$\begin{aligned} 0 &< W(S_1) \\ &= \left(|A_1^+(M_1)| - k|A_1^-(M_1)|\right) + \left(|A_1^+(M_2)| - k|A_1^-(M_2)|\right) - 2k|A_2^-(M_1)| \\ &= \left(|A_1^+(M_1)| - k|A_1^-(M_1)|\right) + \left(|A_1^+(M_2)| - k|A_1^-(M_2)|\right) - k|A_2^-(M_1)| - k|A_2^-(M_2)| \\ &= (\phi(M_1, M) - k\phi(M, M_1)) + (\phi(M_2, M) - k\phi(M, M_2)). \end{aligned}$$

Thus, we must have $\phi(M_1, M) > k\phi(M, M_1)$ or $\phi(M_2, M) > k\phi(M, M_2)$, which means $u(M) > k$. □

Example 3. Consider the auxiliary graphs $H_{(M,2)}$ and $H_{(M,3)}$ constructed from a matching M in Example 2.

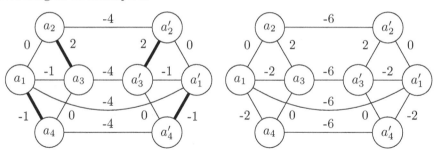

On the left, $H_{(M,2)}$ has a positive weight perfect matching consisting of the bold-faced edges, but on the right $H_{(M,3)}$ does not. This implies $2 < u(M) \leq 3$.

$\square$

For a given value of k, the problem of determining whether $u(M) > k$ is now reduced to detecting a positive weight perfect matching in $H_{(M,k)}$, which can be done by finding the maximum weight perfect matching of $H_{(M,k)}$.

Lemma 2. *Given an* RP *instance I, a matching M of I, and a number $k = x/y$, where $x \in [0, n-1]$ and $y \in [1, n]$ are integers, there is an algorithm to determine whether $u(M) > k$ in $O(m\sqrt{n}\log n)$ time.*

Proof. From Lemma 1, the problem is equivalent to determining whether $H_{(M,k)}$ has a positive weight perfect matching. Observe that $H_{(M,k)}$ has $O(n)$ vertices and $O(m)$ edges, and we can multiply the weights of all edges by y so that they are all integers with magnitude $O(n)$. Using the recently developed algorithm of Duan et al. [5], we can find a maximum weight perfect matching in a graph with integer weight edges of magnitude poly(n) in $O(m\sqrt{n}\log n)$ time, hence we can detect a positive weight perfect in $O(m\sqrt{n}\log n)$ time. $\square$

We can now efficiently compute $u(M)$ by performing a binary search on all possible values of it.

Theorem 1. *Given an* RP *instance I and a matching M of I, there is an algorithm to compute $u(M)$ in $O(m\sqrt{n}\log^2 n)$ time.*

Proof. Observe that if $u(M)$ is not ∞, it must be in the form of x/y, where $x \in [0, n-1]$ and $y \in [1, n]$ are integers, meaning that there are at most $O(n^2)$ possible values of $u(M)$. By performing a binary search on the value of $k = x/y$ (if $u(M) > n - 1$, then $u(M) = \infty$), we run the algorithm in Lemma 2 for $O(\log n^2) = O(\log n)$ times, hence the total running time is $O(m\sqrt{n}\log^2 n)$. $\square$

The running time of the above algorithm is for a general RP instance. However, we can do faster for an MP instance using the following approach. First, consider a matching

$$S = M \cup \{\{a_i', a_j'\}|\{a_i, a_j\} \in M\} \cup \{\{a_i, a_i'\}|a_i \text{ is unmatched in } M\}$$

of $H_{(M,k)}$. Since S is a perfect matching, every perfect matching consists of a number of alternating cycles relative to S. Moreover, from the definition of $\delta_{i,j}$, every edge of S has zero weight. Therefore, $H_{(M,k)}$ contains a positive weight perfect matching if and only if it contains a positive weight alternating cycle relative to S. Hence, the problem becomes equivalent to detecting a positive weight alternating cycle (relative to S) in $H_{(M,k)}$. Note that this property holds for every RP instance, not limited to only MP.

However, the special property of MP is that the graph $H_{(M,k)}$ is bipartite, so we can divide the vertices of $H_{(M,k)}$ into two parts H_1 and H_2 with no edge between vertices in the same part. We then orient the edges of S toward H_1 and all other edges toward H_2, hence the problem of detecting a positive weight alternating cycle becomes equivalent to detecting a positive weight directed cycle (see Example 4), which can be done in $O(m\sqrt{n})$ time using the shortest path algorithm of Goldberg [9]. Therefore, by performing a binary search on $u(M)$, the total running time for RP is $O(m\sqrt{n}\log n)$.

Example 4. Consider the following matching M' in an MP instance.

Preference Lists
$m_1 : w_1, w_2$
$m_2 : w_1, w_2$
$w_1 : m_2, m_1$
$w_2 : m_1, m_2$

$M' = \{\{m_1, w_1\}, \{m_2, w_2\}\}$

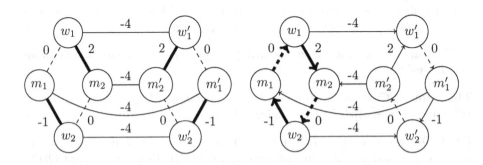

On the left, $H_{(M',2)}$ has a positive weight perfect matching consisting of the bold-faced edges, while S consists of the dotted edges. On the right, since $H_{(M',2)}$ is a bipartite graph with parts $H_1 = \{m_1, m_2, w_1', w_2'\}$ and

$H_2 = \{w_1, w_2, m'_1, m'_2\}$, we orient the edges of S (dotted arrows) toward H_2, and the rest toward H_1. This directed graph has a positive weight directed cycle consisting of the bold-faced arrows. Both figures imply $u(M') > 2$.   □

In a way similar to RP, we have the following lemma and theorem for MP.

Lemma 3. *Given an* MP *instance I, a matching M of I, and a number $k = x/y$, where $x \in [0, n-1]$ and $y \in [1, n]$ are integers, there is an algorithm to determine whether $u(M) > k$ in $O(m\sqrt{n})$ time.*

Theorem 2. *Given an* MP *instance I and a matching M of I, there is an algorithm to compute $u(M)$ in $O(m\sqrt{n}\log n)$ time.*

4   Weighted Unpopularity Factor

The previous section shows the algorithm to compute an unpopularity factor of a given matching in an unweighted RP or MP instance where every person has equal voting weight. However, in many real-world situations, people might have different voting weights based on seniority, position, etc. Our algorithm can also be slightly modified to support a weighted instance with integer weights bounded by $N = \text{poly}(n)$.

In the weighted setting, each person $a_i \in A$ has a weight $w(a_i)$. We analogously define $\phi(M, M')$ to be the sum of weights of people who strictly prefer a matching M to a matching M', i.e. $\phi(M, M') = \sum_{a \in A_{(M,M')}} w(a)$, where $A_{(M,M')} = \{a \in A | r_a(M(a)) < r_a(M'(a))\}$. $\Delta(M, M')$ and $u(M)$ are defined the same way as in the unweighted setting. For each $a_i \in A$, we assume that $w(a_i)$ is a non-negative integer not exceeding $N = \text{poly}(n)$. Note that an unweighted instance can be viewed as a special case of a weighted instance where $w(a_i) = 1$ for all $a_i \in A$.

To support the weighted setting, we construct an auxiliary graph $H_{(M,k)}$ with the same set of vertices and edges as in the unweighted setting, but with slightly different weights of the edges. For each pair of i and j with an edge $\{a_i, a_j\}$, define

$$\delta_{i,j} = \begin{cases} w(a_i), & \text{if } a_i \text{ is unmatched in } M \text{ or } a_i \text{ prefers } a_j \text{ to } M(a_i); \\ -kw(a_i), & \text{if } a_i \text{ prefers } M(a_i) \text{ to } a_j; \\ 0, & \text{if } \{a_i, a_j\} \in M \text{ or } a_i \text{ likes } a_j \text{ and } M(a_i) \text{ equally}. \end{cases}$$

For each pair of i and j, the weights of $\{a_i, a_j\}$ and $\{a'_i, a'_j\}$ is $\delta_{i,j} + \delta_{j,i}$. Finally, for each edge $\{a_i, a'_i\}$, we set its weight to be $-2kw(a_i)$ if a_i is matched in M, and 0 otherwise.

The auxiliary graph $H_{(M,k)}$ still has the same relation with $u(M)$, as shown in the following lemma.

Lemma 4. *In the weighted* RP *instance, $u(M) > k$ if and only if $H_{(M,k)}$ contains a positive weight perfect matching.*

Proof. The proof of this lemma is almost identical to that of Lemma 1. We define the sets $A_1(M')$, $A_2(M')$, $A_1^+(M')$, $A_1^-(M')$, and $A_2^-(M')$ the same way as in the proof of Lemma 1, but for each such set B we will count the sum of weights of elements in B instead of the number of elements.

In both directions of the proof, we can use exactly the same argument as in the proof of Lemma 1, but with $|B|$ replaced by $w(B) = \sum_{a \in B} w(a)$ in each line of the equations. □

Since the weights of people are bounded by $N = \text{poly}(n)$, the unpopularity factor $u(M)$ must be in the form $k = x/y$, where x and y are integers not exceeding Nn. For a given value of k, if we multiply the weights of all edges of $H_{(M,k)}$ by y, they will be integers with magnitude $O(Nn) = \text{poly}(n)$. Therefore, we can still use the algorithm of Duan et al. [5] to find a maximum weight perfect matching of $H_{(M,k)}$ with the same runtime.

Moreover, there are at most $O(N^2 n^2)$ possible values of $u(M)$. By performing a binary search on the value of k, we have to run the above algorithm for $O(\log N^2 n^2) = O(\log n)$ times as in the unweighted setting, hence the total runtime is still $O(m\sqrt{n}\log^2 n)$.

The argument for MP instances still works for the weighted setting as well since $H_{(M,k)}$ is still bipartite, hence we have the following theorems for the weighted setting RP and MP.

Table 1. Currently best known algorithms

	Two-sided lists		One-sided lists
	Roommates Problem	Marriage Problem	House Allocation Problem
Find a popular matching	NP-hard [6,10]	$O(m)$ [8,11]	$O(m+n)$ [1]
Find a matching that minimizes unpopularity margin			NP-hard [18]
Find a matching that minimizes unpopularity factor	NP-hard [12]		
Test popularity of a given matching	$O(m\sqrt{n}\log n)$ [3,5]	$O(m\sqrt{n})$ [3]	$O(m+n)$ [1]
Compute unpopularity margin of a given matching			$O((g+1)m\sqrt{n})$ [18]
Compute unpopularity factor of a given matching	$O(m\sqrt{n}\log^2 n)$ [§3]	$O(m\sqrt{n}\log n)$ [§3]	$O(m\sqrt{n_2})$ [18]

Theorem 3. *Given a weighted* RP *instance I with integer weights bounded by* $N = poly(n)$ *and a matching M of I, there is an algorithm to compute* $u(M)$ *in* $O(m\sqrt{n}\log^2 n)$ *time.*

Theorem 4. *Given a weighted* MP *instance I with integer weights bounded by* $N = poly(n)$ *and a matching M of I, there is an algorithm to compute* $u(M)$ *in* $O(m\sqrt{n}\log n)$ *time.*

5   Discussion

We develop an algorithm to compute the unpopularity factor of a given matching in $O(m\sqrt{n}\log^2 n)$ time for RP and $O(m\sqrt{n}\log n)$ time for MP, which runs only slightly slower than McCutchen's algorithm to solve the same problem in HAP as well as the algorithm of Biró et al. to compute the unpopularity margin of a given matching in RP and MP. Our results also complete Table 1, which shows the running time of the currently best known algorithms related to popularity in RP, MP, and HAP in the case with strict preference lists, where m is the total length of preference lists, n is the total number of people and items, n_2 is the number of items (for HAP), and g is the unpopularity margin of a given matching.

References

1. Abraham, D.J., Irving, R.W., Kavitha, T., Mehlhorn, K.: Popular matchings. In: Proceedings of the 16th Annual ACM-SIAM Symposium on Discrete Algorithms (SODA), pp. 424–432 (2005)
2. Abraham, D.J., Kavitha, T.: Dynamic matching markets and voting paths. In: Arge, L., Freivalds, R. (eds.) SWAT 2006. LNCS, vol. 4059, pp. 65–76. Springer, Heidelberg (2006). https://doi.org/10.1007/11785293_9
3. Biró, P., Irving, R.W., Manlove, D.F.: Popular matchings in the marriage and roommates problems. In: Calamoneri, T., Diaz, J. (eds.) CIAC 2010. LNCS, vol. 6078, pp. 97–108. Springer, Heidelberg (2010). https://doi.org/10.1007/978-3-642-13073-1_10
4. Cseh, Á., Kavitha, T.: Popular matchings in complete graphs. arXiv preprint https://arxiv.org/abs/1807.01112 (2018)
5. Duan, R., Pettie, S., Su, H.-H.: Scaling algorithms for weighted matching in general graphs. ACM Trans. Algorithms **14**(1), 8:1–8:35 (2018)
6. Faenza, Y., Kavitha, T., Powers, V., Zhang, X.: Popular matchings and limits to tractability. In: Proceedings of the 30th Annual ACM-SIAM Symposium on Discrete Algorithms (SODA), pp. 2790–2809 (2019)
7. Gale, D., Shapley, L.S.: College admissions and the stability of marriage. Am. Math. Mon. **69**, 9–15 (1962)
8. Gärdenfors, P.: Match making: assignments based on bilateral preferences. Behav. Sci. **20**, 166–173 (1975)
9. Goldberg, A.V.: Scaling algorithms for the shortest paths problem. SIAM J. Comput. **24**(3), 494–504 (1995)
10. Gupta, S., Misra, P., Saurabh, S., Zehavi, M.: Popular matching in roommates setting is NP-hard. In: Proceedings of the 30th Annual ACM-SIAM Symposium on Discrete Algorithms (SODA), pp. 2810–2822 (2019)

11. Gusfield, D., Irving, R.W.: The Stable Marriage Problem: Structure and Algorithms. MIT Press, Cambridge (1989)
12. Huang, C.-C., Kavitha, T.: Near-popular matchings in the roommates problem. In: Demetrescu, C., Halldórsson, M.M. (eds.) ESA 2011. LNCS, vol. 6942, pp. 167–179. Springer, Heidelberg (2011). https://doi.org/10.1007/978-3-642-23719-5_15
13. Huang, C.-C., Kavitha, T., Michail, D., Nasre, M.: Bounded unpopularity matchings. In: Gudmundsson, J. (ed.) SWAT 2008. LNCS, vol. 5124, pp. 127–137. Springer, Heidelberg (2008). https://doi.org/10.1007/978-3-540-69903-3_13
14. Irving, R.W.: An efficient algorithm for the "stable roommates" problem. J. Algorithms **6**, 577–595 (1985)
15. Kavitha, T., Mestre, J., Nasre, M.: Popular mixed matchings. In: Albers, S., Marchetti-Spaccamela, A., Matias, Y., Nikoletseas, S., Thomas, W. (eds.) ICALP 2009. LNCS, vol. 5555, pp. 574–584. Springer, Heidelberg (2009). https://doi.org/10.1007/978-3-642-02927-1_48
16. Mahdian, M.: Random popular matchings. In: Proceedings of the 7th ACM Conference on Electronic Commerce (EC), pp. 238–242 (2006)
17. Manlove, D., Sng, C.T.S.: Popular matchings in the weighted capacitated house allocation problem. J. Discrete Algorithms **8**(2), 102–116 (2010)
18. McCutchen, R.M.: The least-unpopularity-factor and least-unpopularity-margin criteria for matching problems with one-sided preferences. In: Laber, E.S., Bornstein, C., Nogueira, L.T., Faria, L. (eds.) LATIN 2008. LNCS, vol. 4957, pp. 593–604. Springer, Heidelberg (2008). https://doi.org/10.1007/978-3-540-78773-0_51
19. Mestre, J.: Weighted popular matchings. In: Bugliesi, M., Preneel, B., Sassone, V., Wegener, I. (eds.) ICALP 2006. LNCS, vol. 4051, pp. 715–726. Springer, Heidelberg (2006). https://doi.org/10.1007/11786986_62
20. Roth, A.E., Sotomayor, M.A.O.: Two-Sided Matching: A Study in Game-Theoretic Modeling and Analysis, volume 18 of Econometric Society Monographs. Cambridge University Press, Cambridge (1990)
21. Ruangwises, S., Itoh, T.: Random popular matchings with incomplete preference lists. In: Rahman, M.S., Sung, W.-K., Uehara, R. (eds.) WALCOM 2018. LNCS, vol. 10755, pp. 106–118. Springer, Cham (2018). https://doi.org/10.1007/978-3-319-75172-6_10

AND Protocols Using only Uniform Shuffles

Suthee Ruangwises$^{(\boxtimes)}$ (iD) and Toshiya Itoh

Department of Mathematical and Computing Science,
Tokyo Institute of Technology, Tokyo, Japan
ruangwises.s.aa@m.titech.ac.jp, titoh@c.titech.ac.jp

Abstract. Secure multi-party computation using a deck of playing cards has been a subject of research since the "five-card trick" introduced by den Boer in 1989. One of the main problems in card-based cryptography is to design *committed-format* protocols to compute a Boolean AND operation subject to different runtime and shuffle restrictions by using as few cards as possible. In this paper, we introduce two AND protocols that use only *uniform* shuffles. The first one requires four cards and is a *restart-free* Las Vegas protocol with finite expected runtime. The second one requires five cards and always terminates in finite time.

Keywords: Card-based cryptography ·
Secure multi-party computation · Uniform shuffle · AND protocol

1 Introduction

1.1 The Five-Card Trick

The concept of card-based cryptography started in 1989 with the "five-card trick" introduced by den Boer [3]. In the original problem, Alice and Bob want to know whether they both like each other. However, no one wants to confess first because of fear of embarrassment if he/she gets rejected. Therefore, they need a protocol that only distinguishes the two cases where they both like each other and otherwise, without leaking any other information.

This situation is equivalent to Alice having a bit a and Bob having a bit b of either 0 or 1. Such protocol outputs the result of a Boolean operation $\text{AND}(a, b) = a \wedge b$ without leaking unnecessary information, i.e. if a player's bit is 1, he/she inevitably knows the other player's bit after knowing $a \wedge b$; if a player's bit is 0, he/she should know nothing about the other player's bit.

Following is the description of the five-card trick protocol, using three identical ♣ cards and two identical ♡ cards. Throughout this paper, we encode the bit 0 by the commitment ♣♡ and 1 by the commitment ♡♣. Initially, we give each player two cards, one ♣ and one ♡. We also have another ♣ card faced down on the middle of a table. Alice places her two (face-down) cards encoding a to the left of the middle card, while Bob places his two (face-down) cards encoding b

© Springer Nature Switzerland AG 2019
R. van Bevern and G. Kucherov (Eds.): CSR 2019, LNCS 11532, pp. 349–358, 2019.
https://doi.org/10.1007/978-3-030-19955-5_30

to the right of the middle card. There are following four possible sequences of the cards.

$$\clubsuit\heartsuit\clubsuit\clubsuit\heartsuit \qquad \clubsuit\heartsuit\clubsuit\heartsuit\clubsuit \qquad \heartsuit\clubsuit\clubsuit\clubsuit\heartsuit \qquad \heartsuit\clubsuit\clubsuit\heartsuit\clubsuit$$
$$a = 0,\, b = 0 \qquad a = 0,\, b = 1 \qquad a = 1,\, b = 0 \qquad a = 1,\, b = 1$$

Then, we swap the fourth and the fifth card, resulting in the following four possible sequences.

$$\clubsuit\heartsuit\clubsuit\clubsuit\heartsuit \qquad \clubsuit\heartsuit\clubsuit\heartsuit\clubsuit \qquad \heartsuit\clubsuit\clubsuit\clubsuit\heartsuit \qquad \heartsuit\clubsuit\clubsuit\heartsuit\clubsuit$$
$$\Downarrow \qquad\qquad \Downarrow \qquad\qquad \Downarrow \qquad\qquad \Downarrow$$
$$\clubsuit\heartsuit\clubsuit\heartsuit\clubsuit \qquad \clubsuit\heartsuit\clubsuit\clubsuit\heartsuit \qquad \heartsuit\clubsuit\clubsuit\heartsuit\clubsuit \qquad \heartsuit\clubsuit\clubsuit\clubsuit\heartsuit$$
$$a = 0,\, b = 0 \qquad a = 0,\, b = 1 \qquad a = 1,\, b = 0 \qquad a = 1,\, b = 1$$

Observe that there are only two possible sequences in a cyclic rotation of the deck, and the two $\heartsuit$ cards are adjacent to each other in the cycle only in the case that $a = 1$ and $b = 1$ (while all other three cases result in another same sequence), hence we can determine whether $a \wedge b = 1$ by looking at the cycle. We can obscure the initial position of the cards by shuffling the deck into a uniformly random cyclic permutation, i.e. a permutation uniformly chosen from $\{\mathrm{id}, (12345), (12345)^2, (12345)^3, (12345)^4\}$ at random.

Mizuki et al. [7] later improved the five-card trick protocol so that it requires only four cards instead of five. While both protocols are useful, the format of the output value $a \wedge b$ is different from the format of the inputs a and b ($\clubsuit\heartsuit$ for 0 and $\heartsuit\clubsuit$ for 1). Both protocols have drawback in the case that we want to compute an AND operation over three or more inputs. If a protocol is *committed-format*, i.e. the output is encoded in the same format as the input, we can perform that protocol on an AND operation over the first two inputs, and use the output as an input of another AND operation with the third input, then with the fourth input, and so on. Therefore, most studies so far have been focused only on committed-format protocols.

1.2 Properties of Protocols

In the formal computation model of card-based protocols developed by Mizuki and Shizuya [8], a shuffle of the deck is mathematically defined by a pair $(\Pi, \mathscr{F})$, where Π is a set of permutations and $\mathscr{F}$ is a probability distribution on Π. We call the shuffle *uniform* if $\mathscr{F}$ is a uniform distribution, and *closed* if Π is a subgroup (of the symmetric group) [1]. Uniformness and closedness have practical benefits. A closed shuffle can be securely performed by letting the first player rearrange the deck into his selected permutation from Π without the second player observing, then the second player do the same without the first player observing. Closedness guarantees that performing the shuffle twice still results in a permutation in Π, while uniformness makes it easier and more natural for a player to randomly select a permutation from Π.

In term of runtime, a protocol is called *finite* if it is guaranteed to terminate after a finite number of steps. Apart from finite protocols, many studies have been focused on other protocols that are Las Vegas with finite expected runtime and *restart-free*, i.e. players are required to put their commitments to the deck only once, not having to restart the whole process again.

1.3 Previous Protocols

In 1993, Crépeau and Kilian [2] developed the first committed-format AND protocol using ten cards with four colors. Niemi and Renvall [10] also developed another protocol using 12 cards but with only two colors. Stiglic [11] later reduced the number of required cards to eight. More recently in 2009, Mizuki and Sone [9] developed an AND protocol using only six cards. This was an important milestone since their protocol was the first one that has finite runtime.

Koch et al. [5] investigated a novel way of shuffles that are not uniform or closed. That reduced the number of cards to five for finite protocol, and four for Las Vegas protocol with finite expected runtime. Most recently in 2018, Abe et al. [1] developed the first Las Vegas five-card AND protocol using only uniform closed shuffles by modifying the original five-card trick protocol. The important protocols developed so far are shown in Table 1.

Table 1. Previous development of committed-format AND protocols

	Card		Properties		
	#colors	#cards	finite	uniform	closed
Crépeau-Kilian [2], 1993	4	10	no	yes	yes
Niemi-Renvall [10], 1998	2	12	no	yes	yes
Stiglic [11], 2001	2	8	no	yes	yes
Mizuki-Sone [9], 2009	2	6	yes	yes	yes
Koch et al. [5, §4], 2015	2	4	no	no	yes
Koch et al. [5, §5], 2015	2	5	yes	no	no
Abe et al. [1], 2018	2	5	no	yes	yes
Ours (§2)	2	4	no	yes	no
Ours (§3)	2	5	yes	yes	no

1.4 Lower Bound

On the other hand, several lower bounds of the minimum required number of two-color cards for an AND protocol subject to different restrictions have been proved. Koch et al. [5, §6] showed that there is no four-card AND protocol with finite runtime. Kastner et al. [4] later proved that there is no finite five-card AND protocol using only closed shuffles, and no restart-free Las Vegas four-card AND protocol using only uniform closed shuffles.

Regarding the runtime, finiteness of shuffles, and closedness of shuffles, there are eight possible combinations of restrictions. The best lower bound and upper bound of the minimum required number of cards subject to each possible combination are shown in Table 2.

1.5 Our Contribution

Previously, the bounds in Table 2 were all tight except in the third row (restart-free Las Vegas, uniform) where the trivial lower bound was four (since we need at

Table 2. Minimum required number of two-color cards for a committed-format AND protocol, subject to each combination of runtime and shuffle restrictions

Runtime	Shuffle	Min. #Cards	Lower Bound	Upper Bound
restart-free Las Vegas	–	4	trivial	Koch et al. [5, §4], 2015
	closed	4		
	uniform	4		**Ours (§2)**
	uniform closed	5	Kastner et al. [4, §7], 2017	Abe et al. [1], 2018
finite	–	5	Koch et al. [5, §6], 2015	Koch et al. [5, §5], 2015
	uniform	5		**Ours (§3)**
	closed	6	Kastner et al. [4, §6], 2017	Mizuki-Sone [9], 2009
	uniform closed	6		

least two cards for a commitment of each player's bit) but the upper bound was five (protocol of Abe et al. [1]), and the sixth row (finite, uniform) where the lower bound was five [5] but the upper bound was six (protocol of Mizuki and Sone [9]).

In this paper, by modifying the protocols of Koch et al. [5, §4–5], we introduce the first restart-free Las Vegas four-card AND protocol that uses only uniform shuffles, as well as the first finite five-card AND protocol that uses only uniform shuffles. This result also means that the lower bounds in the third and sixth rows of Table 2 now become tight, thus completely answering the problem about the minimum required number of two-colored cards for a committed-format AND protocol subject to each combination of runtime and shuffle restrictions.

Shortly after this paper was first made public, Koch [6, §6] also developed two protocols with the same properties as ours. This constitutes concurrent and independent work.

2   Four-Card AND Protocol

Starting at the four-card protocol of Koch et al. [5, §4], we replace the closed but non-uniform shuffles by uniform but non-closed shuffles that have similar effects to the sequence. In this protocol, Alice's commitment and Bob's commitment are placed on the table in this order from left to right.

2.1   Pseudocode

A card is represented by a number based on its position on the table, with 1 being the leftmost card, 2 being the second card from the left, and so on. The following notions are also used in the pseudocode.

- **(turn, A)** denotes flipping all cards in the set A.
- **visible** denotes a visible sequence of the cards from left to right, with ? being a face-down card.
- **(shuffle, Π)** denotes a uniform shuffle of the deck on the set Π of permutations.

- (**perm, σ**) denotes rearranging the deck into a permutation σ.
- (**result, x, y**) denotes outputting a commitment of card x and card y, in this order.

(shuffle, {id, (1 3)(2 4)})
(shuffle, {id, (2 3)})
(turn, {2})
if visible $= (?,\heartsuit,?,?)$ **then**
   (turn, {2})
   (shuffle, {id, (3 4)})
 † (shuffle, {id, (3 4), (1 4 2 3)})
   (turn, {4})
   if visible $= (?,?,?,\heartsuit)$ **then**
     (result, 3, 2)
  else
     (turn, {4})
     (shuffle, {id, (1 2)})
     (perm, (2 3 4))
     goto $\star$
else
   (turn, {2})
   (shuffle, {id, (1 3)})
 $\star$ (shuffle, {id, (1 3), (1 2 3 4)})
   (turn, {1})
   if visible $= (\clubsuit,?,?,?)$ **then**
     (result, 2, 3)
  else
     (turn, {1})
     (shuffle, {id, (2 4)})
     (perm, (1 2 3))
     goto †

2.2 Proof of Correctness and Security

We can easily verify the correctness of the protocol by keeping track of every possible sequence of the cards throughout the protocol. For the security, note that the shuffle and perm actions never reveal new information about the inputs; the only action that may reveal new information is the turn action. When we turn a set of cards face-up, we have to be sure that the probability to observe a visible sequence of cards is independent of the inputs a and b.

In this paper we use a KWH-tree, a tool developed by Koch et al. [5], to help verify the correctness and security of the protocol. X_{00}, X_{01}, X_{10}, and X_{11} denote the probabilities of (a, b) being $(0, 0)$, $(0, 1)$, $(1, 0)$, and $(1, 1)$, respectively, with shorthands $X_0 = X_{00} + X_{01} + X_{10}$ and $X_1 = X_{11}$ being used. Also, a polynomial denotes the conditional probability that the sequence of the cards is the one next to the polynomial, given the current view of the deck.

The KWH-tree of our four-card AND protocol is given in Fig. 1. From the KWH-tree, We can verify that a correct commitment to $a \wedge b$ is obtained as a result, and that the sum of polynomials in every box equals to $X_0 + X_1$, implying that no information about a or b leaks. This protocol is clearly restart-free. Also, at the final separating points (the boxes marked with an asterisk), the protocol terminates with probability $\frac{1}{3}$ and re-enter a branch on the other side with probability $\frac{2}{3}$. Therefore, the expected number of times it goes through the branches is

$$\frac{1}{3}\left(1 + 2\left(\frac{2}{3}\right) + 3\left(\frac{2}{3}\right)^2 + \ldots\right) = 3,$$

thus having a finite expected runtime.

3   Five-Card and Protocol

Starting at the five-card protocol of Koch et al. [5, §5], we replace the closed but non-uniform shuffles by uniform but non-closed shuffles that have similar effects to the sequence. In this protocol, Alice's commitment, Bob's commitment, and an additional $\heartsuit$ card are placed on the table in this order from left to right.

3.1   Pseudocode

```
(shuffle, {id, (1 3)(2 4)})
(shuffle, {id, (2 3)})
(turn, {2})
if visible = (?,♡,?,?,?) then
        (turn, {2})
        (shuffle, {id, (3 4)})
        (shuffle, {id, (3 4), (1 4 2 3)})
        (turn, {4})
        if visible = (?,?,?,♡,?) then
                (result, 3, 2)
```

```
        else
                (turn, {4})
                (shuffle, {id, (1 2)})
                (perm, (2 3 4))
                goto ⋆
    else
            (turn, {2})
            (shuffle, {id, (1 3)})
    ⋆   (shuffle, {id, (1 3), (1 2)(3 5 4)})
            (turn, {3})
            if visible = (?,?♣,?,?) then
                    (result, 2, 1)
            else
                    (result, 1, 4)
```

3.2 Proof of Correctness and Security

The KWH-tree of our five-card AND protocol is given in Fig. 2. From the KWH-tree, We can verify that a correct commitment to $a \wedge b$ is obtained as a result, and that the sum of polynomials in every box equals to $X_0 + X_1$, implying that no information about a or b leaks. This protocol clearly terminates in finite time since there is no cycle in the KWH-tree.

4   Conclusion and Future Work

In this paper, we introduce a restart-free Las Vegas four-card AND protocol and a finite five-card AND protocol, both using only uniform shuffles. This result also completely answers the problem about the minimum required number of two-colored cards for a committed-format AND protocol subject to each combination of runtime and shuffle restrictions.

The existing lower bounds, however, cover only the case with two-color cards. An interesting question is that whether the minimum required number of cards can be lowered if we allow more than two colors. For example, is there a finite five-card AND protocol using only closed shuffles if three-color cards are allowed?

KWH-Tree: Four-Card Protocol

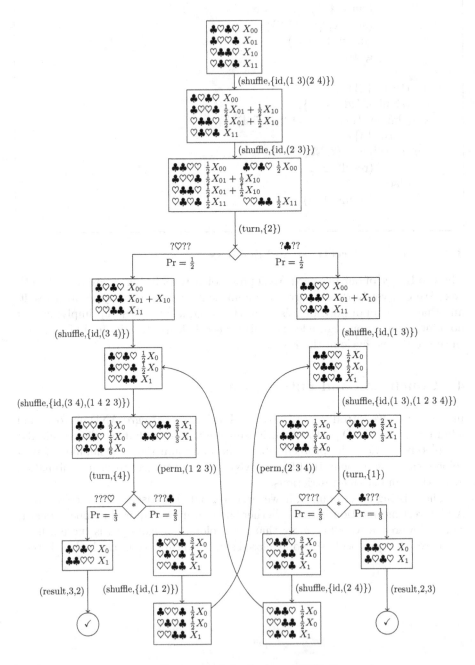

Fig. 1. A KWH-tree of the four-card AND Protocol

KWH-Tree: Five-Card Protocol

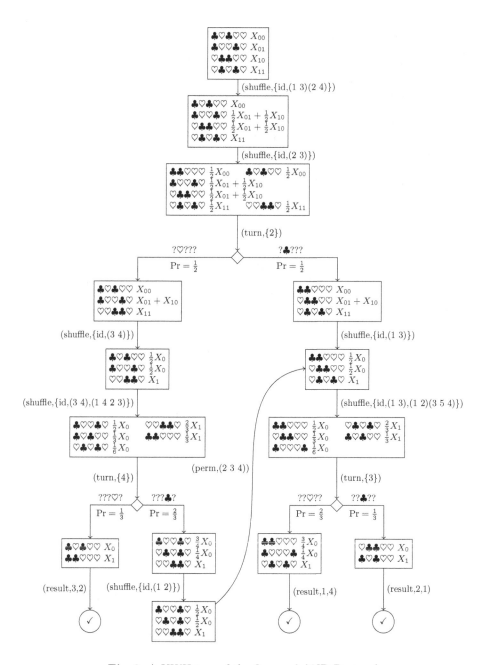

Fig. 2. A KWH-tree of the five-card AND Protocol

References

1. Abe, Y., Hayashi, Y., Mizuki, T., Sone, H.: Five-card AND protocol in committed format using only practical shuffles. In: Proceedings of the 5th ACM on ASIA Public-Key Cryptography Workshop (APKC 2018), pp. 3–8 (2018)
2. Crépeau, C., Kilian, J.: Discreet solitary games. In: Stinson, D.R. (ed.) CRYPTO 1993. LNCS, vol. 773, pp. 319–330. Springer, Heidelberg (1994). https://doi.org/10.1007/3-540-48329-2_27
3. Boer, B.: More efficient match-making and satisfiability *the five card trick*. In: Quisquater, J.-J., Vandewalle, J. (eds.) EUROCRYPT 1989. LNCS, vol. 434, pp. 208–217. Springer, Heidelberg (1990). https://doi.org/10.1007/3-540-46885-4_23
4. Kastner, J., et al.: The minimum number of cards in practical card-based protocols. In: Takagi, T., Peyrin, T. (eds.) ASIACRYPT 2017. LNCS, vol. 10626, pp. 126–155. Springer, Cham (2017). https://doi.org/10.1007/978-3-319-70700-6_5
5. Koch, A., Walzer, S., Härtel, K.: Card-based cryptographic protocols using a minimal number of cards. In: Iwata, T., Cheon, J.H. (eds.) ASIACRYPT 2015. LNCS, vol. 9452, pp. 783–807. Springer, Heidelberg (2015). https://doi.org/10.1007/978-3-662-48797-6_32
6. Koch, A.: The Landscape of Optimal Card-based Protocols. Cryptology ePrint Archive (2018). https://eprint.iacr.org/2018/951/20181009:160322
7. Mizuki, T., Kumamoto, M., Sone, H.: The five-card trick can be done with four cards. In: Wang, X., Sako, K. (eds.) ASIACRYPT 2012. LNCS, vol. 7658, pp. 598–606. Springer, Heidelberg (2012). https://doi.org/10.1007/978-3-642-34961-4_36
8. Mizuki, T., Shizuya, H.: A formalization of card-based crypto-graphic protocols via abstract machine. Int. J. Inf. Secur. **13**, 15–23 (2014)
9. Mizuki, T., Sone, H.: Six-card secure AND and four-card secure XOR. In: Deng, X., Hopcroft, J.E., Xue, J. (eds.) FAW 2009. LNCS, vol. 5598, pp. 358–369. Springer, Heidelberg (2009). https://doi.org/10.1007/978-3-642-02270-8_36
10. Niemi, V., Renvall, A.: Secure multiparty computations without computers. Theor. Comput. Sci. **191**, 173–183 (1998)
11. Stiglic, A.: Computations with a deck of cards. Theor. Comput. Sci. **259**, 671–678 (2001)

Sybil-Resilient Conductance-Based Community Growth

Ouri Poupko[1], Gal Shahaf[1(✉)], Ehud Shapiro[1], and Nimrod Talmon[2]

[1] Weizmann Institute of Science, Rehovot, Israel
{ouri.poupko,gal.shahaf,ehud.shapiro}@weizmann.ac.il
[2] Ben-Gurion University, Beersheba, Israel
talmonn@bgu.ac.il

Abstract. Preventing fake or duplicate e-identities (aka *sybils*) from joining an e-community may be crucial to its survival, especially if it utilizes a consensus protocol among its members or employs democratic governance, where sybils can undermine consensus, tilt decisions, or even take over. Here, we explore the use of a trust graph of identities, with trust edges representing trust among identity owners, to allow a community to grow without increasing its sybil penetration. Since identities are admitted to the e-community based on their trust by existing e-community members, *corrupt* identities, which may trust sybils, also pose a threat to the e-community. Sybils and their corrupt perpetrators are together referred to as *Byzantines*, and our overarching aim is to limit their penetration into an e-community. Our key tool in achieving this is graph conductance, and our key assumption is that honest people are averse to corrupt ones and tend to distrust them. Of particular interest is keeping the fraction of Byzantines below one third, as it would allow the use of Byzantine Agreement (see Lamport et al. *The Byzantine generals problem*, ACM Transactions on Programming Languages and Systems, 4(3):382–401, 1982) for consensus as well as for sybil-resilient social choice (see Shahaf et al., *Sybil-resilient reality-aware social choice*, arXiv preprint arXiv:1807.11105, 2019). We consider sequences of incrementally growing trust graphs and show that, under our key assumption and additional requirements, including keeping the conductance of the community trust graph sufficiently high, a community may grow safely.

Keywords: Graph theory · Conductance · Sybil resilience

1 Introduction

We wish to identify conditions under which an e-community of predominantly-*genuine* (truthful and unique) e-identities, may grow without increasing the penetration of *sybil* (fake or duplicate) e-identities. Our particular context of interest is e-democracy [16,17], where a sovereign e-community conducts its affairs via egalitarian decision processes, although another motivation is the task of growing a permissioned distributed system. We consider an initial community with

© Springer Nature Switzerland AG 2019
R. van Bevern and G. Kucherov (Eds.): CSR 2019, LNCS 11532, pp. 359–371, 2019.
https://doi.org/10.1007/978-3-030-19955-5_31

low sybil penetration that wishes to admit new members without admitting too many sybils. As it is not realistic to expect that no sybils will be admitted, our goal is to keep the fraction of sybils below a certain threshold. In a separate paper [15], we show that an e-democracy can tolerate up to one-third sybil penetration and still function democratically. Still, the fewer the sybils, the smaller the supermajority needed to keep the decision-making safe against them.

We model an e-community via a trust graph with a vertex for each identity and with edges representing trust relations between the owners of the corresponding identities (formal definitions in Sect. 2). We consider *genuine* and *sybil* identities, and refer to the genuine identities that do not trust sybils as *honest* and those that do as *corrupt*. Furthermore, as we are interested in an incremental admission process that grows the initial community, we consider sequences of trust graphs that capture such incremental changes.

We are interested in identifying sufficient conditions on such graphs, e.g. the type of identities in the graph, their relative fractions, and their trust relations, under which a community may grow while keeping the fraction of sybils in it low. To achieve this, our key assumption is that honest identities tend to trust honest identities rather than corrupt ones, and our key tool is graph conductance.

Related Work. We review existing work which helps in clarifying the differences to our model: A large portion of the literature on sybil attacks (see, e.g., [6,13, 14] and their citations) is focused on *sybil detection*, where the task is to tell the sybil agents from the honest ones. Of particular interest is the approach initiated by Yu et al. [23], which relies on structural properties of the underlying social network. Yu et al. show how to separate the honest and sybil regions by leveraging the relatively few number of edges between them. This framework was studied further [4,5,18–21]. As pointed out by Alvisi et al. [2], however, such attempts to recover the entire sybil region may potentially occur only in instances where the honest region is sufficiently connected, which is rarely the case in actual social networks. Consequently, Alvisi et al. suggest a more modest goal of producing a whitelist of honest vertices in the graph with respect to a given agent; that is, a *local* sybil detection scheme, in contrast to the *global* ones proposed before.

A problem of a similar flavor is that of *corruption detection* in networks, posed by Alon et al. [1] and later refined by Jin et al. [12]. This setting, inspired by auditing networks, consists of a graph with each of its vertices being either truthful or corrupt, where the overall goal is to detect the corrupt region. In contrast to the sybil detection problem, the corrupt agents are assumed to be immersed throughout the network, and the main assumption here is that each agent may accurately determine the true label of its neighbors and report it to a central authority. The authors show how good connectivity properties of the graph allows an approximate recovery of the truthful and corrupt regions.

We note that social networks have some special structure, e.g. having low diameters (a.k.a., the *small world phenomena* [7]) or fragmented to highly-connected clusters with low connectivity between different clusters. Moreover, as observed by some researchers [2,5,22], the attacker's inability to maintain

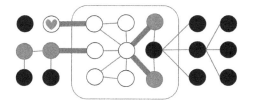

Fig. 1. Illustration of the general setting: The white vertices (honest identities) **and** grey vertices (corrupt identities) form the set of genuine identities, while grey vertices (corrupt identities) **and** black vertices (sybil identities) form the set of Byzantines. Bold edges represent attack edges. The white vertex with a grey heart in it represents an identity that is "corrupt at heart", as currently it does not trust any sybil, but in the future it will; thus, the edge connecting it to the honest identity to its right is an attack edge as well. The circled area contains the current community that wishes to grow. Notice that the nine identities in the community contain one sybil and two corrupt identities, thus in particular the community's Byzantine penetration is $\beta = 1/3$ and the sybil penetration is $\sigma = 1/9$. The fraction of internal attack edges to the volume of the honest part of the community graph, defined below, is $\gamma = 1/8$.

sufficiently many attack edges typically results in certain "bottlenecks", which can be utilized to pin-point the sybil regions.

Informal Model. While the problem we address is related to sybil detection, and indeed we incorporate some of the insights of the works discussed above, our main goal is different: Safe community growth. We aim to find conditions under which a community may grow without increasing the fraction of hostile members within it; but without necessarily identifying explicitly who is hostile and who is not. An additional difference from existing literature is our notions of identity and trust. Specifically, existing works consider identities or agents of only two types, "good" and "bad", with various names for the two categories. Our notion of identities is more refined and, we believe, may be closer to reality.

In particular, we consider genuine and sybil identities, with the intention that in a real-world scenario these would be characterized by the nature of their *representation*: genuine identities are truthful and unique, and sybil identities are fake or duplicate. We further distinguish between two types of genuine identities, based on their *behavior*: honest, which do not form trust relations with sybils, and corrupt, which do. This behavioral distinction is captured formally in our model. We naturally assume that the corrupt identities are the creators and operators of the sybils and that, in the worst case, all sybils and their corrupt perpetrators may cooperate, hence we label them together as *Byzantines*, and aim to limit their fraction within the community.

We thus begin with a unified formal model of such identities and their *trust graph*, consisting of vertices that represent identities and edges that represent trust relations among the owners of such identities. As we consider the task of sybil-resilient community growth, we define the *community history* that aims to capture the incremental changes a community trust graph undergoes over discrete time steps. In order to properly characterize identities, we first employ the basic distinction between genuine and sybil identities. Then, using the community history, we make a further delicate distinction within genuine identities between *honest* identities, which never trust sybils, and *corrupt* identities, which may trust sybils and, furthermore, may cooperate with other corrupt or sybil community members to introduce sybils into the community.

Some assumptions on the power of the sybils and their perpetrators is needed; otherwise there is no hope in achieving our goal. Intuitively, our *key assumption* is that honest identities are averse to corrupt identities, and hence are not likely to trust them. We call trust edges that connect honest and corrupt identities *attack edges*. So, loosely speaking, we assume that there are not too many attack edges. We view this assumption as a realistic relaxation of the assumption made in some related works [1,12], which assume that truthful agents are able to identify *precisely* whether a neighbor is corrupt or not. Figure 1 illustrates our general setting.

High-level Approach. Our approach for admitting a new identity to the community relies upon the *conductance* of the subgraph that includes the current community together with the potential candidate to be admitted. Our ability to protect the graph from Byzantine penetration is based on our key assumption that, while there could be arbitrarily many Byzantines wanting to enter the growing community, they will have limited connectivity to the current community. Indeed, this observation was applied in the context of sybil detection [2, 21–23]. In general, while the conductance of the whole network is typically fairly low, the conductance of the subgraphs restricted to each cluster may be high. In that sense, following Alvisi et al. [2], we adopt a local perspective and focus on the conductance of our community, regardless of the conductance of the entire network. In contrast to Alvisi et al. [2], however, we are interested in growing the community and not in whitelisting. Unlike the situation treated by Alvisi et al. [2], which can be viewed as whitelisting, initiated at a singleton community (i.e. from a single non-sybil vertex), here we consider arbitrarily-large communities and aim to bound, but not detect or eliminate, the sybils in them.

Specifically, our framework makes use of a "target conductance" parameter Φ, and aims to grow, i.e., admit new members, while retaining a conductance of at least Φ at the larger community. Assuming that the community harbors a limited number of attack edges and the relative fraction of Byzantines in the initial community is bounded, we show how to safely grow an initial community.

2 Preliminaries

We provide some needed definitions regarding graphs and conductance. Let $G = (V, E)$ be an undirected graph. The *degree* of a vertex $x \in V$ is $\deg(x) := |\{y \in V | (x, y) \in E\}|$. G is *d-regular* if $\deg(x) = d$ holds for each $x \in V$. The *volume* of a given subset $A \subseteq V$ is the sum of degrees of its vertices, $vol(A) := \sum_{x \in A} \deg(x)$. Additionally, we denote the subgraph induced on the set of vertices A as $G|_A$, by $\deg_A(x)$ the degree of vertex $x \in A$ in $G|_A$, and by $vol_A(B) := \sum_{x \in B} \deg_A(x)$

the volume of a set $B \subseteq A$ in $G|_A$. Given two subsets $A, B \subseteq V$, the size of the cut between A and B is denoted by $e(A, B) = |\{(x, y) \in E \mid x \in A, y \in B\}|$.

Definition 1 (Conductance). *Let $G = (V, E)$ be a graph. The conductance of G is defined by: $\Phi(G) = \min_{\emptyset \neq A \subset V} \frac{e(A, A^c)}{\min\{vol(A), vol(A^c)\}}$. (where $A^c := V \setminus A$ is the complement of A.)*

Remark 1. Generally speaking, graph conductance aims to measure the connectivity of the graph by quantifying the minimal cut normalized by the volume of its smaller subset. Conductance should be thought of as the weighted and irregular analogue of edge expansion [9], where both notions are essentially equivalent for regular graphs. To get a quantitative grip of this measure, notice that for all graphs, $\Phi \in [0, \frac{1}{2}]$. Intuitively, the conductance of a highly connected graph approaches $\frac{1}{2}$. E.g., cliques and complete bipartite graphs satisfy $\Phi = \frac{1}{2}$, while in a poorly connected graph this measure may be arbitrarily small; e.g., a disconnected graph satisfies $\Phi = 0$.

While determining the exact conductance of a given graph is known to be coNP-hard [3], the Cheeger inequality provides a direct relation between conductance of a graph and the second eigenvalue of its random walk matrix.

Lemma 1. *(Cheeger inequality [10]) Let G be a graph with conductance Φ, and let $\lambda_n \leq ... \leq \lambda_2 \leq \lambda_1$ denote the spectrum of its random walk matrix. Then, $\frac{\Phi^2}{2} \leq 1 - \sqrt{1 - \Phi^2} \leq 1 - \lambda_2 \leq 2\Phi$.*

Consequently, spectral methods, together with the Cheeger inequality, provide efficient approximation algorithms for measuring conductance. We refer the reader to [9,11] for comprehensive surveys and discussions.

3   Formal Model

Community Trust Graphs. The relation between people and their identities is rich and multifaceted. For the purpose of this paper, we assume that some identities are *genuine* and others are not, in which case they are called *sybils*. We represent trust relations among identities via a trust graph, in which vertices represent identities and edges represent trust among identities.

Definition 2. *A* trust graph *$G = (V, E)$ is an undirected graph with vertices that represent identities and edges that represent trust among them.*

As we are interested in a community that grows within such a trust graph, next we introduce the concept of a community trust graph.

Definition 3. *A* community trust graph *$G = (A, V, E)$ is a trust graph with vertices V, edges E, and a community $A \subseteq V$.*

Community Histories and Transitions. Our aim is to find conditions under which a community may grow safely. Hence we consider sequences of community graphs, obtained by applying elementary transitions of adding a member to the community or adding/removing an edge:

Definition 4 (Elementary Community Transition). *Let $G = (A, V, E)$ and $G' = (A', V, E')$ be two community graphs with the same set of vertices V. We say that G' is obtained from G by an elementary community transition, and we denote it by $G \to G'$, if:*

- *$A' = A \cup \{x\}$, $x \in (V \setminus A)$, $E' = E$, or*
- *$E' = E \cup \{e\}$, $e \in ((A \times V) \setminus E)$, $A = A'$, or*
- *$E' = E \setminus \{e\}$, $e \in E$, $A = A'$.*

Definition 5 (Community History). *A community history $\mathcal{G}_V$ over a set of vertices V is a (possibly infinite) sequence of community trust graphs $\mathcal{G}_V = G_1, G_2, \ldots$ with vertices V such that $G_i \to G_{i+1}$ holds for every $i \in 1, 2, \ldots$.[1]*

Types of Identities. We assume identities of two types, genuine and sybil. Next, we use community histories to distinguish between two types of genuine identities – honest and corrupt: we say that an identity is *honest* in a community history if it never partakes in an attack edge in this history, and *corrupt* if it does. We then lump together sybils and corrupt identities and call them *Byzantines*.

The rationale is that we are interested in bounding the number of sybils in the graph, not only at the present but also in the future. Hence, we need to bound also all potential sybil perpetrators, who may establish trust edges with sybils in the future, in an attempt to introduce them into the community. Hence, at any point in time (community graph in a community history), a corrupt identity may be only "corrupt at heart", with no action as-of-yet to demonstrate its corruption; and our key assumption is that honest identities are averse to corrupt identities even if they are only corrupt at heart.

Below and in the rest of the paper we use disjoint union $A = B \uplus C$ as a shorthand for $A = B \cup C$, $B \cap C = \emptyset$.

Definition 6 (Types of Identities, Attack Edges, Sybil Penetration). *Let V be a set of vertices that consist of two disjoint subsets $V = T \uplus S$ of genuine T and sybil S vertices, and let $\mathcal{G}_V$ be a community history over V. Then, a genuine vertex $t \in T$ is corrupt in $\mathcal{G}_V$ if it trusts a sybil at anytime in $\mathcal{G}_V$, namely, there is some $(t, s) \in E$, with $t \in T$, $s \in S$, for some $G = (A, V, E) \in \mathcal{G}_V$. A genuine vertex that is not corrupt is said to be honest. Thus, $\mathcal{G}_V$ partitions the genuine identities $T = H \uplus C$ into honest H and corrupt C identities. An edge $(h, c) \in E$ is an attack edge if $h \in H$ and $c \in C$. The sybil penetration $\sigma(G)$ of a community trust graph $G = (A, V, E) \in \mathcal{G}_V$ is*

$$\sigma(G) = \frac{|A \cap S|}{|A|} .$$

---

[1] As the set of vertices V is fixed in a community history, it does not explicitly model the birth and death of people; modeling this aspect is the subject of future work.

Remark 2. An important observation is that an attack edge (h, c) may be introduced into a community trust graph in a community history, and be defined as such, even if the corruption of c is still latent in this community trust graph, namely before a trust edge (c, s) between c and a sybil s is introduced.

Bounded Attacks and Resilience. Our key assumption is that honest people tend to trust honest people and distrust corrupt people; we capture the degree in which this is the case by the following parameter γ:

Definition 7 (γ-bounded attack). *Let $\mathcal{G}_V$ be a community history over $V = T \uplus S$ that partitions $T = H \uplus C$ into honest H and corrupt C identities. A community trust graph $G = (A, V, E) \in \mathcal{G}_V$ has a γ-bounded attack if*

$$\frac{e(A \cap H, A \cap C)}{vol_A(A \cap H)} \leq \gamma .$$

That is, γ is an upper bound on the fraction of attack edges within A to the volume of the honest subset of A. In the worst case, sybils and their corrupt perpetrators would cooperate; thus, to allow for incremental community growth we must bound their combined presence in the community, as defined next:

Definition 8 (Byzantines and their Penetration). *Let $\mathcal{G}_V$ be a community history over $V = T \uplus S$ that partitions $T = H \uplus C$ into honest H and corrupt C identities. Then, a vertex $v \in V$ is Byzantine if it is a sybil or corrupt and the Byzantines $B = S \cup C$ are the union of the sybil and corrupt vertices. The Byzantine penetration $\beta(G)$ of a community trust graph $G = (A, V, E) \in \mathcal{G}_V$ is*

$$\beta(G) = \frac{|A \cap B|}{|A|} .$$

As $A = (A \cap H) \uplus (A \cap B)$, it would occasionally be convenient to use the equivalence between Byzantine penetration to the community A and the fraction of Byzantines w.r.t. genuine identities in A. Formally,

$$\frac{|A \cap B|}{|A|} \leq \beta \quad \text{iff} \quad \frac{|A \cap B|}{|A \cap H|} \leq \frac{\beta}{1 - \beta} . \tag{1}$$

As mentioned above, it is necessary to have some assumptions on the quality of our trust graphs. The next definition, of α-solidarity, measures a certain aspect of the connectivity within the community; roughly speaking, it requires that each vertex in the community would have many neighbors inside the community:

Definition 9 (α-solidarity). *Given a community trust graph $G = (A, V, E)$, a vertex $a \in A$ satisfies α-solidarity if:*

$$\frac{|\{x \in A \mid (a, x) \in E\}|}{d} \geq \alpha.$$

where d is the maximal degree of G. The graph G satisfies α-solidarity if every $a \in A$ satisfies α-solidarity.

The following definition combines some of the definitions above. Note that in this definition β is a free parameter.

Definition 10 (Community Graph Resilience). *Let $G \in \mathcal{G}_V$ be a community trust graph in a community history $\mathcal{G}_V$ over V, $\alpha \in [0,1]$, and $\beta, \gamma \in [0, \frac{1}{2}]$. We say that G is (α, β, γ)-resilient if the following hold: 1) G satisfies α-solidarity; 2) G has a γ-bounded attack; and 3) $\Phi(G|_A) > \frac{\gamma}{\alpha} \cdot \left(\frac{1-\beta}{\beta} \right)$.*

Remark 3. To internalize the above definition, consider the options available to a community that wishes to achieve resilience to a given Byzantine penetration β: (*i*) to increase solidarity, α, up to 1; (*ii*) to increase conductance, Φ, up to $1/2$; (*iii*) or to curb the attack, γ, down to 0. However: (*i*) increasing solidarity may prevent the community from admitting additional members, defeating the purpose of community growth; (*ii*) increasing conductance may require community members who do not know each other to trust each other, defeating the notion of trust; and (*iii*) curbing the attack will require the community to expose sybils and their corrupt perpetrators, which is an effort. In summary, achieving (α, β, γ)-resilience for a given β is a challenge, and addressing it will require a community to craft a balance between its desire to grow, the measures it is willing to undertake to support such growth, and the risks it is willing to undertake in order to grow.

Lemma 2. *Let $G \in \mathcal{G}_V$ be a community trust graph in a community history $\mathcal{G}_V$ over V, $\alpha \in [0,1]$, $\beta, \beta(G) \in [0, \frac{1}{2}]$ and $\gamma \in [0, \frac{1}{2}]$. If G is (α, β, γ)-resilient, then G has sybil penetration $\sigma(G) \leq \beta - \gamma(1 - \beta)$.*

Proof. From the conductance of G we have that

$$\frac{\gamma}{\alpha} \cdot \left(\frac{1-\beta}{\beta} \right) < \Phi(G|_A) \leq \frac{e(H \cap A, B \cap A)}{vol_A(B \cap A)}$$

$$\leq \frac{d \cdot |C \cap A|}{\alpha d \cdot |B \cap A|} = \frac{1}{\alpha} \cdot \left(1 - \frac{|S \cap A|}{|B \cap A|} \right),$$

where the second inequality applies the definition of conductance and the fact that $vol_A(B \cap A) \leq vol_A(H \cap A)$, due to Byzantine penetration not greater than $\frac{1}{2}$; and the third stems from the requirement that each vertex is of degree at least αd, and the fact that only corrupt identities obtain attack edges, and each of them is of degree at most d. The last inequality applies the partition $B = C \uplus S$. Algebraic manipulations yield $|S \cap A| \leq |B \cap A| \cdot \left(1 - \gamma \cdot \left(\frac{1-\beta}{\beta} \right) \right)$. Thus,

$$\sigma(G) = \frac{|S \cap A|}{|A|} \leq \frac{|B \cap A|}{|A|} \cdot \left(1 - \gamma \cdot \left(\frac{1-\beta}{\beta} \right) \right)$$

$$\leq \beta \cdot \left(1 - \gamma \cdot \left(\frac{1-\beta}{\beta} \right) \right) = \beta - \gamma(1 - \beta) . \qquad \square$$

4   Safe Community Growth

The following main theorem provides a sufficient condition for safe community growth: It guarantees low Byzantine penetration in a community trust graph G' if it has certain properties and is obtained via an elementary transition from a community trust graph G that has low Byzantine penetration.

Remark 4. This theorem can be understood in at least two ways: (i) as offering formal guidelines to a community wishing to grow without increasing its sybil penetration; and (ii) as suggesting an algorithm that incrementally admits new identities to the community whenever the conditions of Definition 10 hold, thus guaranteeing low Byzantine penetration rate in the evolving community.

Theorem 1. *Let $\mathcal{G}_V$ be a community history over V and let $G, G' \in \mathcal{G}_V$ be two consecutive elements of $\mathcal{G}_V$, where G' is obtained from G by an elementary operation of vertex addition. Let $\alpha \in [0, 1]$, $\beta(G) \leq \frac{1}{2} - \frac{1}{|A|}$ and $\gamma \in [0, \frac{1}{2}]$.*

If G' is (α, β, γ)-resilient, for some free parameter $\beta \leq \frac{1}{2}$, then $\beta(G') \leq \beta'$.

Before proving the above main theorem we present a corollary that considers community growth over time.

Corollary 1. *Let $\mathcal{G}_V = G_1, G_2, \ldots$ be a community history over V. Let $\alpha \in [0, 1]$, $\gamma \in [0, \frac{1}{2}]$, $\beta \leq \frac{1}{2} - \frac{1}{|A_1|}$, and assume that $\beta(G_1) \leq \beta$.*

If Every graph $G \in \mathcal{G}_V$ obtained from its predecessor in $\mathcal{G}_V$ via vertex addition elementary operation is (α, β, γ)-resilient, then every member $G \in \mathcal{G}_V$ has Byzantine penetration $\beta(G) \leq \beta$ and sybil penetration $\sigma(G) \leq \beta - \gamma(1 - \beta)$.

Proof. Towards contradiction, let G_k be the first community graph for which $\beta(G_k) > \beta$. An elementary operation $G \rightarrow G'$ involving either edge addition or edge removal does not alter the proportion of Byzantines within the community, so G_k must be obtained via vertex addition from G_{k-1}. As a proper community history $\mathcal{G}_V = G_1, G_2, \ldots$ satisfies $A_1 \subseteq A_2 \subseteq \ldots$, we have $\beta \leq \frac{1}{2} - \frac{1}{|A_1|} \leq \frac{1}{2} - \frac{1}{|A_{k-1}|}$. As k is the minimal index for which $\beta(G_k) > \beta$, we have $\beta(G_{k-1}) \leq \beta$, contradicting Theorem 1. Sybil penetration follows Lemma 2. $\square$

Proof of Theorem 1. Let $d > 0$ be the maximal degree of the vertices in A'. Notice first that $|A' \cap B| \leq |A \cap B| + 1 \leq \left(\frac{1}{2} - \frac{1}{|A|}\right)|A| + 1 = \frac{|A|}{2} < \frac{|A'|}{2}$, where the second inequality follows from the Byzantine penetration in G. It follows therefore that $|A' \cap B| < |A' \cap H|$. Also, from α-solidarity we get that $vol_{A'}(A' \cap B) \geq \alpha d|A' \cap B|$ and $vol_{A'}(A' \cap H) \geq \alpha d|A' \cap H| > \alpha d|A' \cap B|$. Using the conductance of G' it follows specifically that:

$$\frac{e(A' \cap H, A' \cap B)}{\alpha d|A' \cap B|} \geq \frac{e(A' \cap H, A' \cap B)}{\min\{vol(A' \cap H), vol(A' \cap B)\}} > \frac{\gamma}{\alpha} \cdot \left(\frac{1 - \beta}{\beta}\right). \quad (2)$$

Since G' has a γ-bounded attack, we can write:

$$\frac{e(A' \cap H, A' \cap B)}{d|A' \cap H|} \leq \frac{e(A' \cap H, A' \cap C)}{vol_{A'}(A' \cap H)} \leq \gamma. \quad (3)$$

Combining the last two equations together we get:

$$\frac{|A'|}{|A' \cap B|} = \frac{|A' \cap H| + |A' \cap B|}{|A' \cap B|} \geq \frac{e(A' \cap H, A' \cap B)}{d\gamma |A' \cap B|} + 1 > \left(\frac{1-\beta}{\beta}\right) + 1 = \frac{1}{\beta},$$

where the first equality holds as $A = (A \cap H) \uplus (A \cap B)$, the second inequality stems from Eq. 3 and the third inequality stems from Eq. 2. Flipping the nominator and the denominator then gives $\beta(A') := \frac{|A' \cap B|}{|A'|} < \beta$. $\qquad\square$

5   Discussion

We discuss implications of the theorem above to sparse graphs, relations between the parameters α, β, and γ, and methods to estimate their values.

Sparse Graphs. Recall that the safety of the community growth relies upon the parameters α, β, and γ. While a given community may evolve wrt. any choice of parameters, some choices will inevitably yield degenerate outcomes; one case is as we require $\Phi(G|_{A'}) > \frac{\gamma}{\alpha} \cdot \left(\frac{1-\beta}{\beta}\right)$, while the conductance of any graph is upper bounded by $\frac{1}{2}$. Specifically, whenever $\gamma\left(\frac{1-\beta}{\beta}\right) > \frac{1}{2}$, the community cannot possibly grow, regardless of the choice of α. While complete graphs and complete bipartite graphs are the classic examples of graphs which satisfy $\Phi(G|_{A'}) = \frac{1}{2}$, the fact that their degree is of order $d = \Theta(n)$ makes them unrealistic in our setting, where agents may potentially trust only a uniformly-bounded number of identities. In this context, the main question seems to be the following: *Could a given community safely grow while retaining a given maximal degree d?* Surprisingly, not only that the answer is affirmative, it also holds for a plethora of trust graphs. We utilize Friedman's classical result:

Theorem 2. *(Friedman [8], rephrased) Let G be a random d-regular graph on n vertices. Then, for any $0 < \epsilon$, $\lambda(G) \leq \frac{2\sqrt{d-1}}{d} + \epsilon$ holds with probability $1 - o_n(1)$.*

Thus, almost all d-regular graphs on n vertices satisfy $\lambda_2 \leq \frac{2}{\sqrt{d}}$. Applying this term in Cheeger's inequality (see Lemma 1) yields that such graphs satisfy

$$\frac{1}{2} - \frac{1}{\sqrt{d}} \leq \Phi, \tag{4}$$

meaning that the choice of d affects the level of conductance one hopes to achieve.

Parameter Interplay. We consider numerical examples to better appreciate the analysis above. First, consider the realistic assumption where each identity is assumed to trust up to $d = 100$ identities (notice that this can be enforced by the system). Equation 4 now suggests that a random graph of degree d on n vertices (where d may be constant wrt. n) satisfies $\Phi > \frac{2}{5}$. For simplicity, we take this quantity as a benchmark. It follows that whenever $\frac{\gamma}{\alpha} \cdot \left(\frac{1-\beta}{\beta}\right) < \frac{2}{5}$, there exist a plethora of potential community histories for which a given community may potentially grow to be arbitrarily large. Some further examples: (1) if $\gamma = 0$, then any community history that begins with a connected Byzantine-free community would retain 0-Byzantine penetration; (2) the choice $\beta = 0$ is not attainable, corresponding to the intuition that we can never guarantee a completely Byzantine-free community growth.

Figure 2 illustrates the parameter interplay further. Notice that our key assumption, stating that honest people tend to trust honest people more than they tend to trust corrupt people, implies that $\gamma < \beta$ (as $\gamma > \beta$ implies that honest people trust corrupt people more than their relative share in the community).[2]

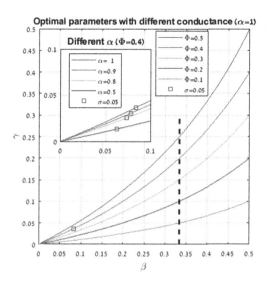

Fig. 2. Parameter Interplay. The large plot shows γ as a function of β, for $\alpha = 1$, where each line represents a different conductance Φ value. It shows, e.g., that if the community fixes $\alpha = 1$ and sets $\Phi = 0.4$, then to achieve $\beta = 0.2$ it can tolerate $\gamma = 0.1$. The small plot shows the effect of α, for $\Phi = 0.4$. In both plots, the red rectangles show respective β and γ values ensuring $\sigma = 0.05$.

Parameter Estimation. While α and Φ can be decided by the community, $\beta(G)$ and γ rely on the dynamics of the community history. To incrementally grow the community at a given time, one may settle for estimating the current state of affairs, as follows. Specifically, assuming that a thorough examination of a given identity could determine whether it is genuine or sybil, one may apply random checks to empirically estimate $\beta(G)$ and γ. This could be carried out in the following manner: (1) examination of an identity $x \in V$ determines whether it is genuine or sybil; (2) examination of the neighbors of a genuine identity

---

[2] In a separate line of research (in preparation) we consider processes and mechanisms that help lowering γ even further.

$x \in V$ (the ball of radius 1 around it) determines whether it is explicitly (but not latently) corrupt; and (3) examination of the ball of radius 2 around an honest identity x determines whether its neighbors are explicitly Byzantine.

6   Outlook

We proposed a method which allows an e-community to grow in a sybil-safe way. Future research includes mechanisms to penalizing the creation of attack edges, modeling the possibility of honest identities abandoning the community, and using simulations to better understand the dynamics of safe growth.

Acknowledgements. We thank the Braginsky Center for the Interface between Science and the Humanities for their generous support.

References

1. Alon, N., Mossel, E., Pemantle, R.: Corruption detection on networks. arXiv preprint arXiv:1505.05637 (2015)
2. Alvisi, L., Clement, A., Epasto, A., Lattanzi, S., Panconesi, A.: Sok: the evolution of sybil defense via social networks. In: Proceedings of SP 2013, pp. 382–396 (2013)
3. Blum, M., Karp, R.M., Vornberger, O., Papadimitriu, C.H., Yannakakis, M.: The complexity of testing whether a graph is a superconcentrator. Inf. Process. Lett. **13**(4–5), 164–167 (1981)
4. Cao, Q., Sirivianos, M., Yang, X., Pregueiro, T.: Aiding the detection of fake accounts in large scale social online services. In: Proceedings of NSDI 2012, pp. 15–15 (2012)
5. Danezis, G., Mittal, P.: SybilInfer: detecting sybil nodes using social networks. In: Proceedings of NDSS 2009, pp. 1–15 (2009)
6. Douceur, J.R.: The sybil attack. In: Druschel, P., Kaashoek, F., Rowstron, A. (eds.) IPTPS 2002. LNCS, vol. 2429, pp. 251–260. Springer, Heidelberg (2002). https://doi.org/10.1007/3-540-45748-8_24
7. Easley, D., Kleinberg, J.: Networks, Crowds, and Markets: Reasoning About a Highly Connected World (2010)
8. Friedman, J.: A proof of Alon's second eigenvalue conjecture. In: Proceedings of STOC 2003, pp. 720–724 (2003)
9. Hoory, S., Linial, N., Wigderson, A.: Expander graphs and their applications. Bull. Am. Math. Soc. **43**(4), 439–561 (2006)
10. Jerrum, M., Sinclair, A.: Conductance and the rapid mixing property for Markov chains: the approximation of permanent resolved. In: Proceedings of STOC 1988, pp. 235–244 (1988)
11. Jerrum, M., Sinclair, A.: The Markov chain Monte Carlo method: an approach to approximate counting and integration. In: Approximation Algorithms for NP-hard Problems, pp. 482–520 (1996)
12. Jin, Y., Mossel, E., Ramnarayan, G.: Being corrupt requires being clever, but detecting corruption doesn't. arXiv preprint arXiv:1809.10325 (2018)
13. Levine, B.N., Shields, C., Margolin, N.B.: A Survey of Solutions to the Sybil Attack, vol. 7, p. 224. University of Massachusetts Amherst (2006)

14. Newsome, J., Shi, E., Song, D., Perrig, A.: The sybil attack in sensor networks: analysis & defenses. In: Proceedings of IPSN 2004, pp. 259–268 (2004)
15. Shahaf, G., Shapiro, E., Talmon, N.: Sybil-resilient reality-aware social choice. arXiv preprint arXiv:1807.11105 (2019)
16. Shapiro, E.: Global cryptodemocracy is possible and desirable. arXiv preprint arXiv:1804.02049 (2018)
17. Shapiro, E.: Point: foundations of e-democracy. Commun. ACM **61**(8), 31–34 (2018)
18. Tran, D.N., Min, B., Li, J., Subramanian, L.: Sybil-resilient online content voting. In: Proceedings of NSDI 2009, pp. 15–28 (2009)
19. Tran, N., Li, J., Subramanian, L., Chow, S.S.M.: Optimal sybil-resilient node admission control. In: Proceedings of IEEE INFOCOM 2011, pp. 3218–3226 (2011)
20. Wei, W., Xu, F., Tan, C.C., Li, Q.: Sybildefender: defend against sybil attacks in large social networks. In: Proceedings of INFOCOM 2012, pp. 1951–1959 (2012)
21. Yu, H.: Sybil defenses via social networks: a tutorial and survey. ACM SIGACT News **42**(3), 80–101 (2011)
22. Yu, H., Gibbons, P.B., Kaminsky, M., Xiao, F.: Sybillimit: a near-optimal social network defense against sybil attacks. IEEE/ACM Trans. Netw. (ToN) **18**(3), 885–898 (2010)
23. Yu, H., Kaminsky, M., Gibbons, P.B., Flaxman, A.: Sybilguard: defending against sybil attacks via social networks. ACM SIGCOMM Comput. Commun. Rev. **36**, 267–278 (2006)

Author Index

Alhamdan, Yousef M. 1
Allender, Eric 13
Arseneva, Elena 25

Belbasi, Mahdi 38
Belovs, Aleksandrs 50
Blažej, Václav 60
Bose, Prosenjit 25
Braverman, Vladimir 70

Das, Bireswar 80
De Carufel, Jean-Lou 25
Demaine, Erik D. 93
Dolce, Francesco 106
Dudakov, Sergey 119
Dvořák, Pavel 60

Fürer, Martin 38

Gajjar, Kshitij 131
Ganardi, Moses 237
Glinskih, Ludmila 143
Goerdt, Andreas 156
Gryaznov, Svyatoslav 168
Guillon, Pierre 180

Hucke, Danny 237

Iacono, John 93
Ilango, Rahul 13
Itoh, Toshiya 337, 349
Itsykson, Dmitry 143

Jacob, Ashwin 191
Jeandel, Emmanuel 180

Kari, Jarkko 180
Karlov, Boris 119
Khadiev, Kamil 228
Klein, Shmuel Tomi 203
Kononov, Alexander 1
Koswara, Ivan 215

Koumoutsos, Grigorios 93
Kravchenko, Dmitry 228

Langerman, Stefan 93
Lohrey, Markus 237
Lorenz, Jan-Hendrik 250

Madathil, Jayakrishnan 262
Majdoddin, Ruhollah 275
Majumdar, Diptapriyo 191
Misra, Neeldhara 286, 299

Panolan, Fahad 262, 286
Perrin, Dominique 106
Petrova, Elena A. 311
Ponomaryov, Denis 325
Poupko, Ouri 359

Radhakrishnan, Jaikumar 131
Raman, Venkatesh 191
Rathi, Piyush 299
Ruangwises, Suthee 337, 349

Saadia, Shoham 203
Sahu, Abhishek 262
Saurabh, Saket 262, 286
Selivanova, Svetlana 215
Serov, Danil 228
Shahaf, Gal 359
Shapira, Dana 203
Shapiro, Ehud 359
Sharma, Shivdutt 80
Shur, Arseny M. 311

Talmon, Nimrod 359

Vafa, Neekon 13
Valla, Tomáš 60
Vanier, Pascal 180
Verdonschot, Sander 25

Ziegler, Martin 215

Printed in the United States
By Bookmasters